47 The value of the resale option for distributed resources is further enhanced by their divisibility into modules, of which as many as desired may be resold and the rest retained to a degree closely matched to new needs.   **48** Distributed resources typically do little or no damage to their sites, and hence minimize or avoid site remediation costs if red[...]el prices set by fluctuating market conditions represent a financial risk. Many distributed resources do not use fuels and thus avoid that costly risk. [...] se them more efficiently or dilute their cost impact by a higher ratio of fixed to variable costs, can reduce the financial risk of volatile fuel prices. [...]ts, such as renewables and end-use efficiency, incur less cost volatility and hence merit more favorable discount rates.   **52** Fewer staff may be ne[...]ts: contrary to the widespread assumption of higher per-capita overheads, the small organizations required can actually be leaner than large ones.   **53** Meter reading and other operational overheads may be quite different for renewable and distributed resources than for classical power plants.   **54** Distributed resources tend to have lower administrative overheads than centralized ones because they do not require the same large organizations with broad capabilities nor, perhaps, more complex legally mandated administrative and reporting requirements.   **55** Compared with central power stations, mass-produced modular resources should have lower maintenance equipment and training costs, lower carrying charges on spare-parts inventories, and much lower unit costs for spare parts made in higher production runs.   **56** Unlike different fossil fuels, whose prices are highly correlated with each other, non-fueled resources (efficiency and renewables) have constant, uncorrelated prices that reduce the financial risk of an energy supply portfolio.   **57** Efficiency and cogeneration can provide insurance against uncertainties in load growth because their output increases with electricity demand, providing extra capacity in exactly the conditions in which it is most valuable, both to the customer and to the electric service provider.   **58** Distributed resources are typically sited at the downstream (customer) end of the traditional distribution system, where they can most directly improve the system's lowest load factors, worst losses, and highest marginal grid capital costs—thus creating the greatest value.

59 The more fine-grained the distributed resource—the closer it is in location and scale to customer load—the more exactly it can match the temporal and spatial pattern of the load, thus maximizing the avoidance of costs, losses, and idle capacity.   **60** Distributed resources matched to customer loads can displace the least utilized grid assets.   **61** Distributed resource matched to customer loads can displace the part of the grid that has the highest losses.   **62** Distributed resources matched to customer loads can displace the part of the grid that typically has the biggest and costliest requirements for reactive power control.   **63** Distributed resources matched to customer loads can displace the part of the grid that has the highest capital costs.   **64** Many renewable resources closely fit traditional utility seasonal and daily loadshapes, maximizing their "capacity credit"—the extent to which each kW of renewable resource can reliably displace dispatchable generating resources and their associated grid capacity.   **65** The same loadshape-matching enables certain renewable sources (such as photovoltaics in hot, sunny climates) to produce the most energy at the times when it is most valuable—an attribute that can be enhanced by design.   **66** Reversible-fuel-cell storage of photovoltaic electricity can not only make the PVs a dispatchable electrical resource, but can also yield useful fuel-cell byproduct heat at night when it is most useful and when solar heat is least available.   **67** Combinations of various renewable resources can complement each other under various weather conditions, increasing their collective reliability.   **68** Distributed resources such as photovoltaics that are well matched to substation peak load can precool the transformer—even if peak load lasts longer than peak PV output—thus boosting substation capacity, reducing losses, and extending equipment life.   **69** In general, interruptions of renewable energy flows due to weather can be predicted earlier and with higher confidence than interruptions of fossil-fueled or nuclear energy flows due to malfunction or other mishap.   **70** Such weather-related interruptions of renewable sources also generally last for a much shorter time than major failures of central thermal stations.   **71** Some distributed resources are the most reliable known sources of electricity, and in general, their technical availability is improving more and faster than that of centralized resources. (End-use efficiency resources are by definition 100% available—effectively, even more.)   **72** Certain distributed generators' high technical availability is an inherent per-unit attribute—not achieved through the extra system costs of reserve margin, interconnection, dispersion, and unit and technological diversity required for less reliable central units to achieve the equivalent supply reliability.   **73** In general, given reasonably reliable units, a large number of small units will have greater collective reliability than a small number of large units, thus favoring distributed resources.   **74** Modular distributed generators have not only a higher collective availability but also a narrower potential range of availability than large, non-modular units, so there is less uncertainty in relying on their availability for planning purposes.   **75** Most distributed resources, especially renewables, tend not only to fail less than centralized plants, but also to be easier and faster to fix when they do fail.   **76** Repairs of distributed resources tend to require less exotic skills, unique parts, special equipment, difficult access, and awkward delivery logistics than repairs of centralized resources.   **77** Repairs of distributed resources do not require costly, hard-to-find large blocks of replacement power, nor require them for long periods.   **78** When a failed individual module, tracker, inverter, or turbine is being fixed, all the rest in the array continue to operate.   **79** Distributed generation resources are quick and safe to work with: no post-shutdown thermal cooling of a huge thermal mass, let alone radioactive decay, need be waited out before repairs can begin.   **80** Many distributed resources operate at low or ambient temperatures, fundamentally increasing safety and simplicity of repair.   **81** A small amount of energy storage, or simple changes in design, can disproportionately increase the capacity credit due to intermittent renewable resources.

82 Distributed resources have an exceptionally high grid reliability value if they can be sited at or near the customer's premises, thus risking less "electron haul length" where supply could be interrupted.   **83** Distributed resources tend to avoid the high voltages and currents and the complex delivery systems that are conducive to grid failures.   **84** Deliberate disruptions of supply can be made local, brief, and unlikely if electric systems are carefully designed to be more efficient, diverse, dispersed, and renewable.   **85** By blunting the effect of deliberate disruptions, distributed resources reduce the motivation to cause such disruptions in the first place.   **86** Distributed generation in a large, far-flung grid may change its fundamental transient-response dynamics from unstable to stable—especially as the distributed resources become smaller, more widespread, faster-responding, and more intelligently controlled.   **87** Modular, short-lead-time technologies valuably temporize: they buy time, in a self-reinforcing fashion, to develop and deploy better technologies, learn more, avoid more decisions, and make better decisions. The faster the technological and institutional change, and the greater the turbulence, the more valuable this time-buying ability becomes. The more the bought time is used to do things that buy still more time, the greater the leverage in avoided regret.   **88** Smaller units, which are often distributed, tend to have a lower forced outage rate and a higher equivalent availability factor than larger units, thus decreasing reserve margin and spinning reserve requirements.

89 Multiple small units are far less likely to fail simultaneously than a single large unit.   **90** The consequences of failure are far smaller for a small than for a large unit.   **91** Smaller generating units have fewer and generally briefer scheduled or forced maintenance intervals, further reducing reserve requirements.   **92** Distributed generators tend to have less extreme technical conditions (temperature, pressure, chemistry, etc.) than giant plants, so they tend not to incur the inherent reliability problems of more exotic materials pushed closer to their limits—thus increasing availability.

93 Smaller units tend to require less stringent technical reliability performance (*e.g.*, failures per meter of boiler tubing per year) than very large units in order to achieve the same reliability (in this instance, because each small unit has fewer meters of boiler tubing)—thus again increasing unit availability and reducing reserves.   **94** "Virtual spinning reserve" provided by distributed resources can replace traditional central-station spinning reserve at far lower cost.   (Continued on rear endpapers.)

OTHER PUBLICATIONS
BY RMI:

Natural Capitalism:
Creating the Next Industrial Revolution

Factor Four:
Doubling Wealth, Halving Resource Use

Cleaner Energy, Greener Profits:
Fuel Cells as Cost-Effective Distributed Energy Resources

The New Business Climate
A Guide to Lower Carbon Emissions and Better Business Performance

The Economic Renewal Guide

The Community Energy Workbook

Green Development: Integrating Ecology and Real Estate

Green Developments CD-Rom

A Primer on Sustainable Building

Greening the Building and the Bottom Line:
Increasing Productivity Through Energy-Efficient Design

Homemade Money:
How to Save Energy and Dollars in Your Home

Re-Evaluating Stormwater:
The Nine Mile Run Model for Restorative Redevelopment

Daylighting:
New Life for Buried Streams

Small
IS PROFITABLE

"Amory Lovins has done it again—

by thinking 'out of the box' he has greatly expanded our understanding

of the benefits and uses of distributed energy resources.

Everyone who cares about the electricity system of the 21st century should read this book."

T.J. Glauthier
President & CEO, Electricity Innovation Institute

London • Sterling, VA

RMI Published by Rocky Mountain Institute

1739 Snowmass Creek Road
Snowmass, CO 81654-9199, USA

phone: 1.970.927.3851
fax 1.970.927.4178

www.rmi.org

Cover art: Thomas Banchoff, Brown University, and Davide Cervone, Union College
(See http://mam2000.mathforum.com/master/dimension/cone/index.html for more information about the cornucopia image)

Production Credits:
Editor: *Beatrice Aranow*
Graphic Designers, Production Editors: *Ben Emerson, Robin L. Strelow*
Production Assistants: *Chris Berry, Wendy Bertolet, Cameron Burns, Doreen Clavell, Katherine Grimberg, Vinay Gupta, Betsy Hands, Ginny Hedrich, Jeremy Heiman, Joanie Henderson, Tim Olson, Jennifer Sweeting, Jeremy Sweeting, Josh Terry*
Webmaster (www.smallisprofitable.org): *William Simon*

Type: Palatino (body text) and Univers (supporting text elements)
Paper: Printed on recycled paper (75% minimum de-inked post consumer waste content) using vegetable based inks.

Printed in the United Kingdom by Cambrian Printers.

Second Edition

Reprinted 2007

ISBN: 978-1-881071-07-5

Grants from the Shell Foundation, The Energy Foundation, and The Pew Charitable Trusts partially supported the research, editing, production, and marketing of this publication, and are gratefully acknowledged. The authors, not the sponsors, are solely responsible for the content.

SHELL
FOUNDATION

The Energy Foundation
Toward a sustainable energy future

THE PEW CHARITABLE TRUSTS

Dedicated to

the memory of
the British National Coal Board's
chief economic advisor 1950–1970,
Ernst Friedrich Schumacher
(1911–1977),
who taught us,
far ahead of his time,
to ask the right questions about scale.

Part 3   A CALL TO ACTION:
POLICY RECOMMENDATIONS AND MARKET IMPLICATIONS
FOR DISTRIBUTED GENERATION

3.5 WHY DISTRIBUTED GENERATION MATTERS TO EVERY CITIZEN   381

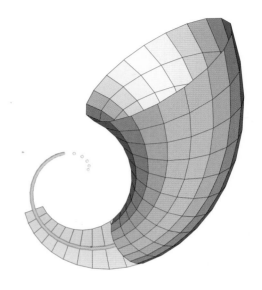

Executive Summary

This book describes 207 ways in which the size of "electrical resources"—devices that make, save, or store electricity—affects their economic value. It finds that properly considering the economic benefits of "distributed" (decentralized) electrical resources typically raises their value by a large factor, often approximately tenfold, by improving system planning, utility construction and operation (especially of the grid), and service quality, and by avoiding societal costs.

The *actual* increase in value, of course, depends strongly on the case-by-case technology, site, and timing. These factors are so complex that the distribution of value increases across the universe of potential applications is unknown. However, in many if not most cases, the increase in value should change investment decisions. For example, it should normally far exceed the cost differences between, say, modern natural-gas-fired power plants and wind-farms. In many applications it could even make grid-interactive photovoltaics (solar cells) cost-effective today. It should therefore change how distributed resources are marketed and used, and it reveals policy and business opportunities to make these huge benefits explicit in the marketplace.

The electricity industry is in the midst of profound and comprehensive change, including a return to the local and neighborhood scale in which the industry's early history is rooted. Through the twentieth century, thermal (steam-raising) power stations evolved from local combined-heat-and-power plants serving neighborhoods to huge, remote, electricity-only generators serving whole regions. Elaborate technical and social systems commanded the flow of electrons from central stations to dispersed users and the reverse flow of money to pay for power stations, fuel, and grid. This architecture made sense in the early twentieth century when power stations were more expensive and less reliable than the grid, so they had to be combined via the grid to ensure reliable and economical supply. The grid also melded the diverse loads of many customers, shared the costly generating capacity, and made big and urban customers subsidize extension of electric service to rural customers.

By the start of the twenty-first century, however, virtually everyone in industrialized countries had electric service, and the basic assumptions underpinning the big-station logic had reversed. Central thermal power plants could no longer deliver competitively cheap and reliable electricity through the grid, because the plants had come to cost *less* than the grid and had become so reliable that nearly all power failures originated *in* the grid. Thus the grid linking central stations to remote customers had become the main driver of those customers' power costs and power-quality problems—which became more acute as digital equipment required extremely reliable electricity. The cheapest, most reliable power, therefore, was that which was produced at or near the customers.

Utilities' traditional focus on a few genuine economies of scale (the bigger, the less investment per kW) overlooked larger *dis*economies of scale in the power stations, the grid, the way both are run, and the architecture of the entire system. The narrow vision that bigger is better ended up raising the costs and financial risks that it was meant to reduce. The resulting disadvantages are rooted in an enormous difference of scale between most needs and most supplies. Three-fourths of U.S. residential and commercial customers use electricity at an average rate that does not exceed 1.5 and 12 kilowatts respectively, whereas a single conventional central power plant produces about a million kilowatts. Resources better matched to the kilowatt scale of most customers' needs, or to the tens-of-thousands-of-kilowatts scale of typical distribution substations, or to an intermediate "microgrid" scale, thus became able to offer important but little-known economic advantages over the giant plants.

The capital markets have gradually come to realize this. Central thermal power plants stopped getting more efficient in the 1960s, bigger in the '70s, cheaper in the '80s, and bought in the '90s. Smaller units offered greater economies from mass-production than big ones could gain through unit size. In the '90s, the cost differences between giant nuclear plants—the last gasp of '70s and '80s gigantism—and railcar-deliverable combined-cycle gas-fired plants, derived from mass-produced aircraft engines, created political stresses that drove the restructuring of the industry. At the same time, new kinds of "micropower" generators thousands or tens of thousands of times smaller—microturbines, solar cells, fuel cells, wind turbines— started to become serious competitors, often enabled by information and telecommunications technologies. The restructured industry exposed the previously sheltered power-plant builders to brutal market discipline. Competition from micropower, uncertain demand, and the inflexibility of big, slow-to-build plants created financial risk well beyond the capital markets' appetite. Then in 2001, longstanding concerns about the inherent vulnerability of giant plants and the far-flung grid were reinforced by the 9/11 terrorist attacks.

The disappointing cost, efficiency, financial risk, and reliability of large thermal stations (and their associated grid investments) were leading their orders to collapse even before the cost difference between nuclear and combined-cycle costs stimulated restructuring that began to delaminate utilities. That restructuring created new market entrants, unbundled prices, and increased opportunities for competition at all scales—and thus launched the revolution in which swarms of microgenerators began to displace the behemoths. Already, distributed resources and the markets that let them compete have shifted most new generating units in competitive market economies from the million-kilowatt scale of the 1980s to the hundredfold-smaller scale that prevailed in the 1940s. Even more radical decentralization, all the way to customers' kilowatt scale (prevalent in and before the 1920s), is rapidly emerging and may prove even more beneficial, especially if it comes to rely on widely distributed microelectronic intelligence. Distributed generators do not require restructured electricity markets, and do not imply any particular scale for electricity business enterprises, but they are starting to drive the evolution of both.

Some distributed technologies like solar cells and fuel cells are still made in low volume and can therefore cost more than competing sources. But such distributed sources' increased *value*—due to improvements in financial risk, engineering flexibility, security, environmental quality, and other important attributes—can often more than offset their apparent cost disadvantage. This book introduces engineering and financial practitioners, business managers and strategists, public policymakers, designers, and interested citizens to those new value opportunities. It also provides a basic introduction to key concepts from such disciplines as electrical engineering, power system planning, and financial economics. Its examples are mainly U.S.-based, but its scope is global.

A handful of pioneering utilities and industries confirmed in the 1990s that distributed benefits are commercially valuable—so valuable that since the mid-'90s, most of the best conceptual analyses and field data have become proprietary, and government efforts to publish methods and examples of distributed-benefit valuation have been largely disbanded. Most published analyses and models, too, cover only small subsets of the issues. This study therefore seeks to provide the first full and systematic, if preliminary, public synthesis of how making electrical resources the right size can minimize their costs and risks. Its main findings are:

- The most valuable distributed benefits typically flow from financial economics—the lower risk of smaller modules with shorter lead times, portability, and low or no fuel-price volatility. These benefits often raise value by most of an order of magnitude (factor of ten) for renewables, and by about 3–5-fold for nonrenewables.

- Electrical-engineering benefits—lower grid costs and losses, better fault management, reactive support, etc.—usually provide another ~2–3-fold value gain, but more if the distribution grid is congested or if premium power quality or reliability are required.

- Many miscellaneous benefits may together increase value by another ~2-fold—more where waste heat can be reused.

- Externalities, though hard to quantify, may be politically decisive, and some are monetized.

- Capturing distributed benefits requires astute business strategy and reformed public policy.

Emerging electricity market structures can now provide the incentives, the measurement and validation, and the disciplinary perspectives needed to give distributed benefits a market voice. Successful competitors will reflect those benefits in investment decisions and prices. Nearly a dozen other technological, conceptual, and institutional forces are also driving a rapid shift toward the "distributed utility," where power generation migrates from remote plants to customers' back yards, basements, rooftops, and driveways. This transformation promises a vibrantly competitive, resilient, and lucrative electricity sector, at less cost to customers and to the earth—thus fulfilling Thomas Edison's original decentralized vision, just a century late.

Preface

The story told here is fascinating, unusual, and challenging. Like many good novels, it is necessarily extensive, both in its breadth of historical sweep and in its depth of detail. We must therefore ask the reader's patience as we seek to develop, piece by interlinked piece, some big ideas that have important consequences for the world's largest industry—providing electricity.

This discussion of a complex and wide-ranging field is presented in the way we feel will best serve diverse readers, ranging from interested laypeople and citizen activists to technical specialists and market participants. Those with a background in the essential concepts of the relevant disciplines are invited to skip familiar introductory material. Those lacking it may be glad of the introduction provided to help with basic terminology and navigation in fields perhaps unfamiliar to them. And those expert in these fields we ask to forgive the occasional simplifications made to increase clarity for nonexperts.

To simplify navigation and to avoid breaking up the narrative flow, certain details appear in color-coded boxes and sidebars: yellow for tutorials, gray for definitions, pale yellow for examples, and white for summaries. Technical notes appear in small italics. The 207 distributed benefits are numbered consecutively throughout Part Two, highlighted in green, and summarized on the front and rear endpapers. A detailed Table of Contents serves in place of an Index. Information about the authors and publishers is at the end of the book. Parenthetical reference numbers appear in blue throughout the text. These numbers correspond to an alphabetical Reference List at the end of the book.

The book is organized in three main parts:

- **Part One** introduces the history of the extraordinary transition now underway from very large to mainly small power plants, reviews the origins and course of our research, clarifies semantic issues, describes the existing U.S. electricity system and the main kinds of distributed resources, and concludes with brief discussions of some important background issues.

- **Part Two** introduces and launches a systematic and detailed survey of scale effects (how size affects value) and the corresponding 207 distributed benefits, explaining technical and economic concepts as needed. Although many distributed benefits could be classified in a variety of ways, we use some license to describe them under three main headings: system planning, construction and operation, and other sources of value. The system planning benefits, though they have important engineering content, are expressed mainly from the perspective of financial economics to make the narrative more coherent. In contrast, the construction and operating benefits use mainly the concepts and language of electrical engineering, as do most of the "other sources of value" (except such externalities as avoiding social and environmental costs).

- All the fine-grained analysis in Parts One and Two must ultimately be applied in a real business and policy context. Those seeking to harvest distributed benefits must understand market evolution, and those making the rules within which markets function

must understand distributed benefits. **Part Three** therefore describes public-policy
initiatives that could help distributed benefits to realize their value in the marketplace;
explores threats and opportunities for the private sector; and recommends attention
and action by citizens.

We offer these findings with the humility of discovering, after diligent effort but with
resources inadequate to the size of the task, that we have only mapped, not fully delved into,
a very rich lode of ideas. But we have persevered in the hope of encouraging a far wider,
deeper, more public, and more widely applied base of understanding of this perennial yet
badly neglected question of what's the right size for the job. And although for specificity we
have focused here on electric power systems, analogous scale issues clearly apply throughout many other technical and economic systems; indeed, Rocky Mountain Institute has
already begun to apply them fruitfully to water and wastewater systems.

As with any survey of a vast and tangled web of ideas, we have drawn freely—though, we
hope, with due and grateful attribution—on the work of hundreds of other researchers and
practitioners. Our many intellectual debts will be evident from the hundreds of references
cited throughout the text and consolidated at the end. But we want here to express special
appreciation to those who have particularly lightened our task by providing obscure information, patiently correcting our errors, or kindly reviewing drafts and offering helpful
suggestions for improvement. Though any remaining errors and omissions are solely our
responsibility, any value of this book springs from the courtesy and insight of these and
scores of other valued colleagues:

Nancy Mohn (ABB and Alstom Power), Bernard Chabot (ADEME), Daniel Shugar (Advanced
Photovoltaic Systems and PowerLight), Michael Margolick (ARA Consulting Group), Bob Shaw
(Arete Ventures), Peter Fox-Penner (Brattle Group), Pat McAuliffe, Sanford Miller, Commis-
sioner Arthur H. Rosenfeld, John Wilson, and Eric Wong (California Energy Commission),
Chris Robertson (Chris Robertson & Associates), Joe Iannucci (Distributed Utility Associates),
Joe Galdo, Dick Holt, and Philip Overholt (DOE), Greg Kats (DOE and Capital E Group), Greg
Motter (Dow Chemical), Roger Pupp (Econix), Vijay Vaitheeswaran (*The Economist*), Michael
McGrath and Chuck Linderman (Edison Electric Institute), Nancy Bacon (Energy Conversion
Devices), Elliot E. Mainzer (Enron), Howard Learner (Environmental Law and Policy Center),
Michael Shelby and Jim Turner (EPA), Clark W. Gellings and Vito Longo (EPRI), Gary Cler, Bill
Howe, Nicholas Lenssen, and Michael Shepard (E SOURCE), Caes Daey Ouwens (Government of
Haarlem), Elizabeth Teisberg (Harvard Business School), D. Gordon Howell (Howell-Mayhew
Engineering), Scott Gates (Idaho Power), Doug Koplow (Industrial Economics), Shimon
Awerbuch (International Energy Agency), S. Chauham (Joe Wheeler Electric), Charlie Komanoff
(Komanoff Energy Associates), Jon Koomey (Lawrence Berkeley National Laboratory), Tom
Stanton (Michigan PUC), David Schoengold (MSB Associates), Ralph Cavanagh (Natural
Resources Defense Council), Jim Welch (The Nature Conservancy), Mike Curley (NERC), Jeff
Petter (Northern Power Systems), Lynn Coles, Dick DeBlasio, and Yih-huei Wan (NREL),
Roland Schoettle (Optimal Technologies), Richard Ottinger (Pace Law School), H. J. Wenger and

Tom Hoff (Pacific Energy Group and Clean Power Research), Paul Gipe (Paul Gipe & Associates), John Carruthers, Bob Lambert, Bob Stewart, and David Turner (PG&E), John Fox (PG&E, Ontario Hydro, Perseus LLC, and RMI Board), Carl Weinberg (PG&E and Weinberg Associates), Landis Kannberg (PNL), Demetrio Borja (Polydyne), Gary Wayne and Tom Dinwoodie (PowerLight), John Mungenast (*Power Quality*), Tom Casten (Private Power LLC), Brian Farmer (PVUSA), Doug Pratt and John Schaeffer (Real Goods), Kevin Best, Dan Cashdan, and Paul Slye (RealEnergy—they generously contributed the first draft of the real-estate discussion in Section 3.4.6), Peter A. Bradford and David Moskovitz (Regulatory Assistance Project), Michael Vickerman (Renew Wisconsin), Jerrold Oppenheim (Renewable Technology Analysis), Paul Chernick, Adam Auster, and Rachel Brailove (Resource Insight), Chris Lotspeich (RMI and Second Hill Group), Brett Williams (RMI and University of California at Davis), Michael Edesess (RMI Board), Bent Sørensen (Roskilde University), Walter C. Patterson (Royal Institute of International Affairs), Don Wood (SDG&E), Jim Harding (Seattle City Light), Eric Daniels (Siemens Solar), Donald Osborn and Ed Smeloff (SMUD), Steven J. Strong (Solar Design Associates), Karl E. Knapp and William F. Sharpe (Stanford University), Georg Furger (Sustainable Asset Management), Christopher Freitas (Trace Engineering), Michael Tennis (Union of Concerned Scientists), Daniel Kammen (University of California at Berkeley), Ewald Fuchs (University of Colorado), John Michael Byrne (University of Delaware), Mike Russo (University of Oregon), Richard F. Hirsh (University of Vermont), Jim Hewlett (USEIA), Michael Mulcahy (Utility Free), Jason Edworthy (Vision Quest Wind Electric, Inc.), Andy Ford (Washington State University), Michael Totten (World Resources Institute and Conservation International), Chris Flavin (Worldwatch Institute), Dick Baugh, Janet Ginsburg, Paul Maycock, Neal McIlveen, Kelso Starrs, and Jeff Williams. A major debt is also owed to the numerous peer reviewers of several drafts from 1997 onward. Most importantly, the senior coauthors built on a great deal of hard work by three dedicated research assistants—Dr. André Lehmann, 1995–97; Ken Wicker, 2000–01; and Daniel Yoon, 1993–94. We are deeply in their debt.

This book was produced through the extraordinary effort and meticulous professionalism of graphic designer Ben Emerson and editor Beatrice Aranow. They were ably supported by numerous production assistants (listed in the colophon of this book) and by RMI's information-systems wizard Marty Hagen. All were led by RMI's Communications Director Norm Clasen, Executive Director Marty Pickett, and co-CEO (until June 2002) L. Hunter Lovins. I am grateful to them all for their faith, hope, and clarity.

Rocky Mountain Institute, as an independent nonprofit applied research center, is also grateful to the sponsors of this research and publication. The roots of this research go back more than two decades: scale issues were the subject of a chapter in *Soft Energy Paths* (1977) and an appendix in *Brittle Power* (1981/82). An RMI project to update and assemble a systematic survey of distributed benefits was launched in 1993 with partial funding from The Pew Charitable Trusts, which patiently awaited its long gestation. The research then made sporadic progress through the 1990s, sustained by the Institute's general-support donors, notably The William and Flora Hewlett Foundation and The Surdna Foundation, as well as by numerous private donors. Parts One and Two were drafted and peer-reviewed in 1997, but got stuck in the production queue behind *Natural Capitalism* (1999). In 2000–01, new grants from The Energy Foundation for editorial completion and the Shell Foundation for production and dissemination permitted RMI to resume and complete the project. Special thanks to The Energy Foundation's Hal Harvey and Eric Heitz and to Shell's Kurt Hoffman for their vision and persistence. RMI senior energy consultants Karl R. Rábago and Tom Feiler drafted portions of Part Three around the turn of 2001–02, when we also incorporated new insights from RMI energy team leader Dr. Joel N. Swisher PE's 2002 monograph *Cleaner Energy, Greener Profits: Fuel Cells as Cost-Effective Distributed Energy Resources*, funded chiefly by the W. Alton Jones Foundation. E. Kyle Datta generously contributed most of Part Three in the spring of 2002. With the help of all our generous and tolerant donors, editing and lay-out were finally completed in summer 2002 for August publication. Without the loyal support of all these friends, none of this work would have been possible. As primary author and final editor, I am responsible for all the deficiencies that doubtless remain.

Finally, a request to the reader: we need and solicit your help to improve this work. Please send your criticisms, comments, suggestions, references, contacts, examples, and any additional concepts or evidence on distributed benefits to sipcomments@rmi.org. Only by enlisting the distributed knowledge of the many emerging expert practitioners in this new field can we hope to advance the state of the art as quickly as its importance deserves. Corrections, updates, and related papers will be posted periodically in the Library/Energy section of www.rmi.org.

—ABL

Old Snowmass, Colorado
15 July 2002

Part One

NEEDS AND RESOURCES

1.1 THE INFLECTION POINT

The electricity industry is widely considered the highest-investment sector of the economy, and among the most important and mature. Electricity now enables a vast range of societal functions, from the most mundane to the most sophisticated. Yet as often happens in the history of technology, just as this industry seems to be at the pinnacle of its achievement, its own structure, design assumptions, and technological content are also becoming fundamentally obsolete.

Providing electricity is an almost unimaginably vast enterprise. In the United States alone, its half-trillion dollars' worth of net assets generates more than $220 billion of sales per year, or nearly 3% of GDP. It also consumes 38% of the nation's primary energy. By burning fossil fuels, which produce about 70% of U.S. electricity, the industry also releases more than one-third of the total oxides of carbon and nitrogen and two-thirds of the sulfur oxides emitted in the U.S. For many years until the late 1980s, the electricity industry's investments, plus roughly equal Federal subsidies (291–2), were about as large as those of the nation's durable-goods manufacturing industries, and today on a global scale it consumes for its expansion approximately one-fourth of all development capital.

By many measures, these prodigious commitments of resources have been successful. Although electricity is only 16% of all energy *delivered* to final users in the United States, it is such a high-quality, versatile, convenient, controllable, clean-to-use, and generally reliable form of energy that it has become a disproportionately pervasive and essential force in modern life. Though electricity has

so far been beyond the reach of the two billion people who still lack it (except for costly batteries), widespread aspirations to get it symbolize the path to modernity. Its use in the United States has grown each year but three (1974, 1982, and 2001) for the past half-century. During the second half of the twentieth century, the U.S. population grew 86% while electricity usage grew by nearly tenfold, so average per-capita use of electricity more than quintupled (191, 200). (Remarkably, there are no government statistics for total U.S. generation or consumption of electricity before 1989, because previous records were not consistently kept on production or disposition by non-utility entities, and electricity industry statistics don't exactly match government data.)

Producing and delivering electricity is extremely capital-intensive—several times as capital-intensive as the average manufacturing industry. Per unit of delivered energy, the electricity system is about 10–100 times as capital-intensive as the traditional oil and gas systems on which modern economies were largely built (414). Generating electricity by traditional means is also very fuel-intensive. Classical power stations that raise steam to turn turbines that run generators that ultimately deliver electricity through the grid necessarily consume 3–4 units of fuel per unit of electricity delivered, and even the most efficient combined-cycle plants decrease this ratio to only about 1.8. Electricity is therefore a far costlier form of energy than direct fuels: in 2000, for example, the average kilowatt-hour (kWh) of U.S. electricity was delivered at a price of $0.0666—the same price per unit of heat content as oil at $114 per barrel, about 3–6 times the recent world price of crude oil (not yet refined and delivered).

Electricity is only one-sixth of the quantity, but two-fifths of the cost, of all energy delivered to final users in the United States. This high price makes electricity an unjustifiably costly way of doing low-grade tasks like heating space or water. Yet the higher-quality services that electricity best provides, such as running motors and electronics, are a bargain. For example,[1] the lifecycle cost of an electric motor per horsepower-hour is on the order of 5% that of equivalently powerful horses. It is thus not surprising that a modern American household, or even a car, may easily contain several dozen motors. Modern life without electric light, shaftpower, and electronic equipment would be very different—for most people, much worse. Ultimately, electricity's value depends entirely on how it is supplied and used. New approaches to both the supply and the use of electricity therefore offer enormous and rapidly expanding opportunities for innovation and improvement.

Despite this vast global industry's remarkable success, and because of its recent history, its competitive and regulatory structures are rapidly shifting in many countries. Meanwhile, an even more fundamental change is emerging largely unnoticed: *a shift in the scale of electricity supply from doctrinaire gigantism to the right size for the job.* As one industry team stated in 1992, "From the beginning of [the twentieth] century until the early 1970s, demand grew, plants grew, and the vertically integrated utilities' costs declined. There is evidence that this trend may be fundamentally reversing in the 1990s." (629) Looking back on the 1990s, it is now obvious that this reversal has actually

occurred. In 1976, the concept of largely "distributed" or decentralized electricity production (412) was heretical; in the 1990s, it became important; by 2000, it was the subject of cover stories in such leading publications as the *Wall Street Journal*, the *Economist*, and the *New York Times* (229, 234); and by 2002, it was emerging as the marketplace winner.

This change is exactly the sort of "inflection point" described by Andrew Grove of Intel in his 1996 book *Only the Paranoid Survive: How to Exploit the Crisis Points That Challenge Every Company and Career* (278). Grove describes an inflection point as a pivotal, wrenching transformation that sorts businesses between the quick and the dead. If properly understood and exploited, an inflection point is the key to making businesses survive and prosper. In the technical system that invisibly powers the modern world, the shift of scale now underway has profound implications, both in its own right and as a harbinger of similar shifts toward appropriate scale in many other technical and commercial systems.

The change of scale dissolves the old pattern of the electricity industry; yet a clear vision of the new pattern is still struggling to be born. The shift has so far been motivated less by an understanding of appropriate scale's opportunities than by unpleasant experience of inappropriate scale's dangers. But with a more balanced appreciation of the opportunities that spring from making electrical resources the right size, the transition could be far faster, smoother, and more profitable. This book explores the issues that will define the new pattern as they emerge from radical changes of technology, analytic methods, and institutional attitudes already well underway. Properly understood, these

[1] A horse is about as powerful as seven strenuously exercising or twenty ordinarily laboring people. But a 50-horsepower motor might cost only ~$50/hp to buy and around $2/h to run, while 50 good draft horses with equivalent nominal total power and operating life might cost on the order of $1,500/hp to buy and $38 per working hour to feed (426). How one values the relative functionality, intelligence, feeding and waste characteristics, reliability, conviviality, self-reproducing and -repairing abilities, etc. of these options is a far more complex question.

issues could greatly accelerate and intensify the shift of scale by revealing many unexpected forms of value waiting to be captured by alert practitioners.

1.2 CONTEXT: THE PATTERN THAT CONNECTS

The electricity industry, to a degree still only dimly realized by many of its participants, has ended a long and illustrious chapter and is beginning the next. As we shall see, its history has created powerful forces that now compel this shift from highly centralized toward highly distributed—decentralized—physical and organizational patterns.

The shift of scale in electricity systems is accompanied by a shift toward renewable energy sources, and toward those that might not be renewable (such as fuel cells using hydrogen derived from fossil fuels) but can still be environmentally benign, either at the point of use or throughout the fuel cycle.[2] Not all renewables are either distributed or benign, but since all three shifts are occurring simultaneously, and many renewables are both distributed and benign, this discussion inevitably blends elements of all three. Its main focus, however, is on the size and interconnection of generating units.

1.2.1 A dozen drivers of distributed utilities

The electricity industry is starting to experience what might be ironically called the "market-driven withering away of the state." The vast arenas being prepared for the gladiatorial combat of wholesale power competitors may soon become echoing, windswept shells populated by the ghosts

of long-dead economic theorists—blind-sided yet again by technology.

As often happens, the generals are re-fighting the previous war, and the planners are too distracted by one recent change in technology to notice the even greater next change bearing down on them. Just as we are getting used to the idea that cheap, fast-to-build, factory-produced, and extremely efficient combined-cycle gas turbines (§ 1.2.4) are already finishing off classical central steam power stations, an even greater threat to both old and new generating technologies is creeping up unseen. Far smaller-scale ways to save, store, and make electricity are becoming spectacularly cheaper and more valuable.

These "distributed resources" could displace new bulk power generation, bulk power trade, and even much transmission[3] as new technologies, market forces, institutional structures, analytic methods, and societal preferences propel a rapid shift to "distributed utilities," operating on a scale more comparable to that of individual customers and their end-use needs. At least a dozen such forces are now massing to create an expanding and cavernous discontinuity:

1. **Efficient end-use.** Big savings of electricity can now often cost less than small savings, thanks to whole-system engineering that milks multiple benefits from single expenditures and hence "tunnels through the cost barrier." (288, 429, 433)

[2] For example, a fuel cell using hydrogen derived from natural gas can be climate-safe throughout its fuel cycle if the carbon dioxide produced when a reformer separates hydrogen from the natural gas is stored underground or in some other "sink."

[3] Ultimately some distribution could be displaced too, although in most plausible futures this would take decades. Most distribution capacity would simply last longer and become omnidirectional. The main potential exceptions arise if distributed electricity storage becomes really cheap. In that case, interconnection may be less advantageous than it now appears. Even if that never happens, microgrids (§ 2.3.2.12) could probably displace many of the larger parts of the distribution network.

2. **Small-scale fueled (co)generation**. Commercial gas-turbine co- and tri-generation can deliver electricity at an effective price of ~$0.005–0.02/kWh net of waste-heat credit. These benefits can be captured by microturbine, engine-driven, and, imminently, increasingly affordable packaged fuel-cell technology systems (88, 132–4).

3. **Cheap kilowatt-scale fuel cells**. Exploding volume and plummeting cost both seem inevitable for proton-exchange-membrane (PEM) fuel cells, driven by the interaction between two huge markets—buildings, where the waste heat can provide building services often about big enough to pay for natural gas and a reformer, and vehicles, at first standalone and later easily connected to the grid as portable generators when parked (440, 758).

4. **New fuels**. The traditional fuel slate is about to be transformed by adding more biofuels, and soon natural gas converted at the wellhead to pipeline hydrogen (with the added benefit of cheaply sequestered CO_2) (759); renewable hydrogen; and hydrogen made at old hydro-electric dams ("hydro-gen")—for which it will be a far more lucrative product than electricity (440). Indeed, not just natural gas but other hydrocarbons, even coal (92), may be able to produce competitive hydrogen and sequestered carbon— a combination that may be worth more than the hydrocarbons themselves. There are increasing signs that the transition to hydrogen as a major energy carrier, already being welcomed by major oil and car companies, could be unexpectedly rapid (590).

5. **Cheap, easy-to-use renewable sources**. Building from a trickle to a flood in vast global markets are "vernacular" renewables such as "AC-out, plug-into-the-wall-socket" photovoltaics, building-integrated photovoltaics that displace buildings' normal roof or wall structure or that are coated onto ordinary windows, and all kinds of renewables that will continue to become steadily cheaper as they are built in larger volumes.

6. **Distributed electric storage**. The move toward distributed energy systems has encouraged the development of small-scale, mass-producible, potentially quite affordable electricity-storing devices, notably ultracapacitors and superflywheels (341). Capable of efficiently storing and releasing electricity on demand and more efficiently than chemical battery storage, lightweight high-speed flywheels and other innovative storage devices will be used:

   • in conjunction with such intermittent renewable sources as photovoltaics and wind;

   • as electrical storage for peak-shaving and load-leveling;

   • for power quality and ride-through in uninterruptible power supplies and similar applications; and

   • in hybrid systems with fuel cells and microturbines, and for hybrid-electric traction in transportation (388, 584).

   Ultracapacitors—like but larger than the ones that can keep a portable computer operating for a few seconds while its battery is being changed—are also rapidly emerging as potent competitors to both superflywheels and chemical batteries (81), and so are reversible fuel cells.

7. **Grid improvements**. Much better thyristors and other solid-state switching devices guided by better control theory and incorporated into distribution automation are starting to change the electricity grid into a smarter, faster, cheaper way to convey and control electricity flows in all directions—facilitating the easier, more efficient, and more cost-effective integration of distributed resources into the grid.

8. **Distributed information**. Pervasive real-time price information, other forms of information such as stability signals, and bidirectional customer communications (303, 480, 515) could create a potential for distributing grid intelligence and control functions not only to the substation level but perhaps all the way to the customer level.

9. **Distributed benefits**. The ~207 kinds of hidden economic benefits surveyed in this book make all the distributed resources (#1–3, 5, 6) manyfold more valuable.

10. **Competition**. Market structures and forces for the first time are starting to attach economic value to many distributed benefits.

11. **Shifts in electricity providers' mission, structure, and culture**. In particular, planning and resource acquisition processes like Local Integrated Resource Planning (§ 1.4.1) are systematically prospecting for distributed benefits.

12. **Unbundled service attributes**. Customers' increasing desire for reliability, power quality, control and predictability of cost, and other aspects of electrical services can often be best met by distributed resources.

These developments form not simply a list of separate items but a web of developments that *all reinforce each other.* Their effect is thus both individually important and collectively profound. Together, they will not only continue the trend toward increasingly distributed energy resources, but also can greatly accelerate the shift to distributed utilities. This transition will probably continue regardless of the outcomes of the restructuring debate, which will affect some details but not the general pattern of change.

The distributed utility concept itself is also rapidly evolving. Its traditional embodiment was the deployment of distributed generators at the substation or in some other distribution system location that served many customers. This would be viewed as simply a substitute for expanding general supply capacity. Such use of distributed generation involved a significant utility investment, but was made at somewhat smaller scale and in a location targeted to optimize system benefits. Schematically (324), rather than building more power stations and grid capacity (Figure 1-1), utilities would add, for example, photovoltaics or a fuel cell at a heavily loaded substation.

Increasingly, however, this model, though valid and important, is starting to shift toward still a third one, in which distributed resources—both supply- and demand-side—may increasingly be located all the way downstream *at or near the customers' premises* (Figure 1-2), *e.g.,* on the roofs or in the basements of houses or in the form of insulation, superwindows, and other design features aimed at reducing peak space-conditioning loads and improving overall end-use efficiency. Under this approach, the traditional model of utility dispatch—a skilled

---

Figure 1-1: Traditional supply expansion
Utilities traditionally forecast growing demand and build more of all kinds of facilities to meet it.

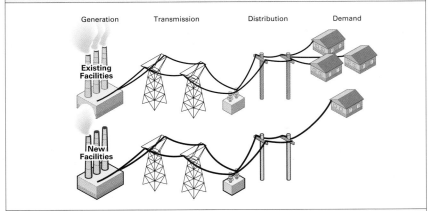

Source: Hoff, T. E., "Distributed Generation: An Alternative to Electric Utility Investments in System Capacity" (*Energy Policy* 24, no. 2, 1996), p. 2, fig. 1

Figure 1-2: Distributed generation
Adding grid-supporting distributed resources instead saves capital, operating, and external costs systemwide.

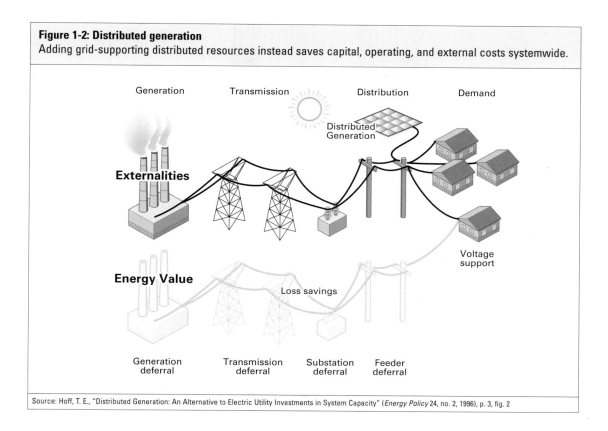

Source: Hoff, T. E., "Distributed Generation: An Alternative to Electric Utility Investments in System Capacity" (*Energy Policy* 24, no. 2, 1996), p. 3, fig. 2

operator sitting in a control room and telling all the electrons where to flow—may gradually morph into a highly distributed set of intelligent microelectronic devices, many at the customer level, whose collective interaction provides some, most, or even all of the system's control functions. Rather than incrementally shifting scale, then, this model re-creates the distribution grid in the fashion of a self-regulating ecosystem or a self-equilibrating market economy. Large and medium-sized generators may be not merely supplemented but supplanted, much as the tasks that formerly required mainframe computers are now often done by networked microcomputers. The upstream/downstream distinction could ultimately blur or dissolve as many dispersed generators and savings displace central stations, causing electricity to flow in not just one but all directions through the web of the distribution system.

1.2.2 The menu: three kinds of distributed resources

What are the "distributed resources" that can shift the predominant scale of the electricity system partway perhaps, and ultimately all the way, toward the right unit size for each task? They are not only generating technologies. Rather, they include all three main ways to meet additional demand for electrical services: *demand-side resources* (how services are derived from electricity), *grid resources* (how electricity is delivered from generator to user), and *supply-side resources* (how electricity is generated). The neutral term "resources" is used for all three classes of options to emphasize that they are comparable and fungible for most planning purposes, with no automatic preference merited for one over the rest.

Each of these kinds of resources contains a taxonomy of sub-options:

Demand-side resources include:

- end-use efficiency (reducing the number of kWh used to deliver a unit of service, such as hot showers or cold beer, illumination or comfort)

- load management (affecting *when* and in what temporal patterns those kWh are used)

- onsite storage (ultracapacitors, superflywheels, batteries, flow batteries [which store chemical energy in a reservoir], electrolyzers/fuel cells, reversible fuel cells, and others)

- fuel-switching (*e.g.,* from electricity to natural gas, liquefied petroleum gas, wood, passive or active solar heat, or wind-powered mechanical work such as water-pumping)

- power-factor management (see box)

- service substitution (*e.g.,* opening the curtains and turning off the lights, or storing rainwater in an uphill reservoir for later delivery rather than pumping it from a well)

- service redefinition (*e.g.,* e-mailing information instead of photocopying it, or, more fundamentally, sending the information only to people who actually want it)

Grid resources include:

- increased transmission or distribution capacity

- decreased transmission or distribution losses (*e.g.,* transmission reconductoring or amorphous-iron transformers)

- improved reactive power control, voltage and frequency regulation, phase control, etc.

- improved controls, sensors, algorithms, switchgear, etc.

- improved management of distribution circuits

- improved maintenance, such as infrared detection of loose or corroded connections

- reductions in "nontechnical losses" (theft of service) and unaccounted-for losses

- grid-sited storage, typically at the substation

Supply-side resources comprise three main categories:

- extending the lifetime, efficiency, or availability of existing generating capacity

- importing power from elsewhere (via transmission whose capacity may be increased through expansion or extension of stability limits)

- building new generating capacity

Power factor measures the extent to which two attributes of an alternating current of electricity—current (flow of electrons) and voltage ("pressure" of electrons)—are in or out of step with each other. Mathematically, power factor is the cosine of the phase angle between current and voltage. A lagging power factor, caused by inductive loads (those which store energy in a magnetic field), such as ordinary induction motors, means that current lags behind voltage. A leading power factor, caused by capacitance (which stores electric charge), means that current runs ahead of voltage. Since only current that is in phase with voltage can turn electric meters and do work, power factors other than unity measure how much the utility must generate, and provide capacity to deliver, out-of-phase current for which it incurs costs but receives no revenues. This topic is further explained in Section 2.3.2.3, which notes that two decades ago, one-fifth of all U.S. grid losses (which would scale today to about $3 billion worth per year) were believed to be caused by poor power factor. Yet only half of U.S. utilities today have any power-factor incentives or penalties in their tariffs, and probably none have economically optimal ones.

New generating capacity, in turn, embraces a spectrum of resources organizable by size:

Highly centralized, slow, costly, and monolithic *units* (GW-range electrical output):

- fossil-fueled steam plants

- nuclear fission steam plants

- large hydroelectric or geothermal plants

- large-scale electricity storage (typically hydroelectric pumped storage) when generating

- large cogeneration stations (*e.g.,* at oil refineries or petrochemical plants)

Midsized units (around a few hundred MW):

- packaged combined-cycle gas-fired plants

- classical combined-heat-and-power plants in the European style

- upgrades of old big hydropower plants with modern turbines and generators

Smaller units (under 100 MW):

- traditional combustion turbines (typically tens to 100+ MW/unit, usually clustered) and their steam-injected variants

- most biomass cogeneration (typically MW to tens of MW in pulp/paper mills, furniture factories, sawmills, etc., but can be larger or smaller)

- traditional internal-combustion engines (~5-MW diesels)

- repowered minor hydropower plants (often in the low MW range)

- wavepower arrays in suitable sites (tens of kW per lineal meter are often available)

- solar-thermal-electric modular plants using tracking or nontracking[4] optical concentrators, including cogeneration versions[5] (typically MW-range and upwards depending on aggregation)

- industrial bottoming cycles using Rankine turbines, Stirling engines, thermoelectric converters, or other devices to recover electricity from fairly low-temperature waste heat

Truly decentralized units—not simply scaled-down big ones, but a basic reoptimization nearly or fully to the scale of most customers' needs:

- wind machines (from roughly 1 MW or, more commonly, hundreds of kW downwards)

- fuel cells (200-kW packaged phosphoric-acid units down to kW-range and smaller proton-exchange-membrane or other

---

Cogeneration is the simultaneous co-production of electricity and useful heat. The heat may be at relatively high temperatures, typically for industrial use, or at relatively low temperatures, typically for space- or water-heating, or both. (The low-temperature arrangement is commonly called Combined Heat and Power, or CHP, in Europe, where it is widely used.) Additionally co-producing other services, typically cooling or dehumidification or both, is called **trigeneration**.

Combined-cycle power plants typically burn fuel to run a **gas turbine** (also called a **combustion turbine**)—essentially a converted aviation jet engine—whose shaft spins an electric generator; then the hot gas emerging from the gas turbine boils water to run an additional steam turbine and produce even more electricity. Combining these two thermodynamic cycles, as described in Section 1.4.1, nearly doubles the efficiency of converting fuel into electricity, as compared to a classical **simple cycle** power station, which uses a fueled steam boiler and steam turbine but without the gas turbine in front of it.

Megawatts (millions of watts) of electric power are abbreviated MW; if it is necessary to distinguish between megawatts of electric and of thermal energy, they are respectively written MWe and MWth. It's similar for kW (kilowatts or thousands of watts) and GW (gigawatts or millions of kW or thousands of MW). One watt is a rate of flow of energy equivalent to one joule per second. There are 1054.8 joules in a BTU and 3.6 MJ in a kWh.

---

[4] Examples of nontracking concentrators include saline-gradient solar ponds and Winston collectors. The latter are split parabolic troughs whose sides are not parts of the same parabolic section. Without tracking, they can provide several or even many suns' concentration onto a cylindrical focal zone. Or tracking can be in only one axis: for example, an ingenious echelon-lens array invented by Dr. Johannes Laing (Pyron Energy Products, La Jolla CA) can be assembled into a raft, floating on a pond and rotating slowly to face the sun's azimuth. The lenses then automatically focus the solar rays—regardless of the sun's elevation angle—down onto small, water-cooled photovoltaic cells.

[5] For example, providing electricity from photovoltaics and recovering waste heat into domestic hot water.

types of units expected to enter the mass market within a few years) operating on reformed natural gas, electrolytic hydrogen, biogas, or liquid fuels

- small engine-generators (typically converted car engines of up to ~100 kW), fueled by biogas, wood gasification, etc. and often incorporating cogeneration

- small hydropower and geothermal-heat Stirling or -Rankine engines (usually kW to hundreds-of-kW range)

- photovoltaics (from roadside-phone or -sign or single-household scale upwards)

Behind each of these options, and others not listed, lies a rich tapestry of technological, economic, environmental, and social characteristics. Many of these are described in standard reference works (62) or in special technological "snapshots" and surveys (416, 444), and will not be repeated here. Specific engineering texts on distributed generating technologies are also available (356, 761). We assume either a basic acquaintance with the relevant generating, storage, and end-use-efficiency technologies or an interim willingness to overlook those details and focus on distributed benefits that apply generically to most or all of the decentralized technologies.

More important than these technical details is a basic point about decision-making. The menu of generating options, like the complete menu of all options of every kind, is rather like the menus shown in some restaurants that list a great many items but no prices. Finding out more about each item is helpful, indeed essential, but not sufficient. No matter how many enticing offerings there are, diners seldom have an unlimited appetite or purse, and will therefore seek the choices that will together be most tasty,

attractive, nourishing, and affordable. Moreover, some options go especially well with others, others badly, so each choice influences the optimal mix of choices. It is therefore vital to integrate choices from the vast menu. How well that integration is done will determine whether the whole meal adds up to more or less than the sum of its dishes.

To understand the nature and importance of distributed resources, especially for generating electricity, we must start with history, about which George Santayana warned that those who don't remember it are condemned to repeat it. For specificity, this discussion uses the example of the United States. Many U.S. conditions are unusual; some are unique. Yet similar stories could be told worldwide—all different in details and in timing, but with analogous casts of characters and the same basic five-act plot:

1. The disappointing cost, efficiency, risks, and reliability of large thermal stations led to a collapse in orders for these plants…

2. even before the embarrassing price gap between nuclear and combined-cycle electricity stimulated restructuring and began the delamination of utilities (because powerful customers wanted to get the cheap new power and let others pay for the costly old power),…

3. creating new market entrants, unbundled prices, and increasing opportunities for competition at all scales…

4. and thereby launching the scale revolution, introducing new technologies, modes of thinking, and institutional arrangements for distributed resources,…

5. which made distributed generation important, and ultimately dominant, in new orders for generating capacity.

Distributed resources, in short, have emerged not simply as a spontaneous technological development but as an evolutionary reaction to the shortcomings and costs of overly centralized resources. The trends have accelerated in recent decades, but have a history spanning more than a century. Over that period, large generating units first achieved and then forfeited economic advantage.

1.2.3 Outrunning the headlights: the pursuit of illusory scale economies

Thomas Edison opened the world's first central thermal power station in London in January 1882, and the first American one in New York nine months later (the first U.S. hydro station opened 26 days after that at Appleton, Wisconsin). The New York station, at Pearl Street, was powering 1,300 light bulbs within a month, 11,000 within a year—"each a hundred times brighter than a candle. Edison's reported goal was to 'make electric light so cheap that only the rich will be able to burn candles.'" (191) For the next century, generating units got ever bigger. By 1903, Samuel Insull had commissioned the largest steam-driven generator yet—five megawatts. A quarter-century later, the largest generator was rated at 200 MW. During the 1960s, the size of the largest new generators went from about 500 to over 1,200 MW. Ever larger unit size seemed justified and beyond question, and trend was assumed to be inexhaustible destiny. Skilled engineers using better designs and alloys to handle hotter and higher-pressure steam enabled the unit size of the largest turbo-alternators to double every six and a half years through a size range of five orders of magnitude (479). The Federal Power Commission's *1970 National Power Survey*

(230) envisaged an extrapolation of then-recent trends, with 1.8-GW units dominating and 3-GW units entering the market by 1990:

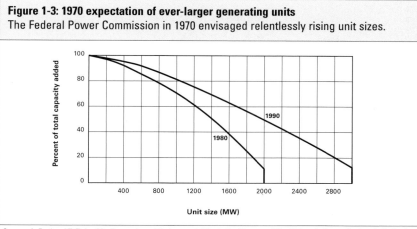

Figure 1-3: 1970 expectation of ever-larger generating units
The Federal Power Commission in 1970 envisaged relentlessly rising unit sizes.

Source: A. Ford and T. Flaim, "An Economic and Environmental Analysis of Large and Small Electric Power Stations in the Rocky Mountain West" (Los Alamos National Laboratory, October 1979), p. 9, fig. 2-2

Had this adventure continued, around the year 2064 the largest single turbo-alternator would have had an electric capacity of eight billion kilowatts, or approximately the total global rate at which human beings converted energy in all forms in the mid-1970s. But any such process is subject to limits long before such an *expansio ad absurdum*, and in fact, the power engineers' heroic efforts ran out of steam, so to speak, around 1970, at a mere 1,400 MW of electric capacity:

Figure 1-4: Ever-larger generating units (all types, all U.S. utilities)
Generating units' size saturated around 1970 after growing 7-fold in ~15 years.

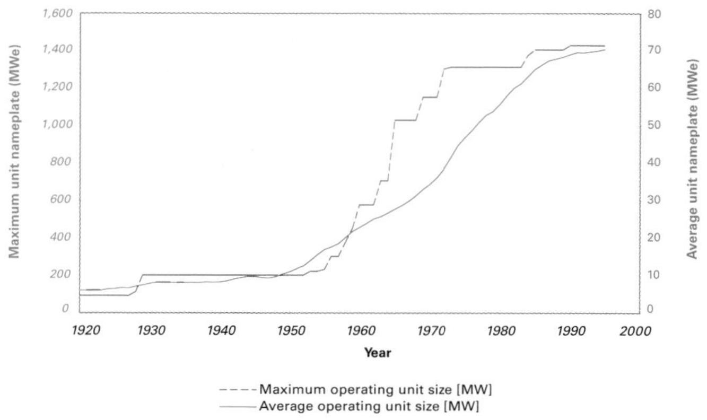

Maximum unit nameplate (MWe)

Average unit nameplate (MWe)

Year

- - - - Maximum operating unit size [MW]
——— Average operating unit size [MW]

Source: EIA (Energy Information Administration), *Inventory of Power Plants in the United States as of January 1, 1996* (December 1998)

The size of the largest operating units, having leaped up to gargantuan levels in the 1960s, coasted to a halt at scarcely larger sizes during the 1970s and 1980s as appetites for both engineering and financial risk were sated. The average unit size (held down by many small non-steam units such as combustion turbines and engine-generators) followed the classic s-curve of saturation. So did the average size of power stations, which often clustered multiple generating units at one site. During 1938–57, for example, the average total capacity of power stations had risen two-thirds faster than the average capacity of the generating units they contained. But the number of units per station later saturated too.[6]

Meanwhile, the total capacity of the electric grid was doubling slightly more slowly than power stations were getting bigger (electrical demand and generating capacity doubled about every seven years until the 1970s), so generating capacity became steadily more centralized. At the pinnacle of this process, the standard generic new North American thermal power station was rated at somewhat more than one million kilowatts (kW) of electricity—one gigawatt (GW), about enough to power San Francisco in 2000.

Maximum transmission voltages also rose from a few kilovolts (kV or thousands of volts) around 1890 to 765 kV in the late 1960s, not only because of giant thermal plants but also, and more often, to exploit highly centralized hydropower sites. The highest voltages have risen little since then, except short 800-kVDC lines and small experimental lines in the megavolt range. Even today, only 0.4% of America's trans-mission circuit-miles of at least 22-kV line carry at least 765 kV, strategically placed as the spines of regional grids. Increasing transmission voltages reduces losses as the square of voltage, allowing more centralized plants to be sited upwards of 500 km from their loads. (For example, 765-kV transmission for 500 km is as effective as 138-kV transmission for only 16 km.) The larger generating units and longer distances also promoted concentration of utility ownership, creating giant companies that could meet the huge thermal power plants' financial, technical, and management demands. Those large and capable organizations in turn were not only able but also inclined to build ever larger generating units. Because average costs for power tended (for a time) to decline with each new unit built, they also gained the economic and political power needed to establish and enforce the franchise-monopoly system. This power allowed giant power stations to be financed with assurance of cost and profit recovery from captive customers.

However, the economic logic of these ever-larger power stations rested on a perilously narrow foundation created by the way utility executives thought and the way policy-makers approached the industry (297, 299). Traditional utilities' resource allocation decisions were largely driven by:

- the discipline of power engineering;

- the objective of expanding generating capacity to meet projected aggregated demand with a safe reserve margin in case of unusual weather, demand patterns, forced outages (unscheduled plant or transmission-line failures), or other exigencies;

---

[6] This was mainly because it didn't make sense to put too much capacity at a single place—it excessively concentrated risks of failure in both generation and transmission, and often raised problems with siting and cooling.

- the focus on capital cost—almost exclusively for the generating plant;

- surprisingly limited sophistication in financial and economic risk concepts, so that, for example, the same discount rate was applied to the financial analysis of every kind of power station and fuel type, as if they all had exactly the same amounts and kinds of financial risk (§ 2.2.3); and

- electricity *cost* measured at the *busbar* (generator output) rather than electricity *price* measured at the retail customer's *meter*—reflecting a tacit assumption that the grid and everything else downstream of the plant would be identical no matter what sort of central station were selected. Since central stations were presumed to be the only reasonable way to make electricity, the major costs of the grid were virtually ignored as a sort of unchangeable overhead at the crucial time when approval to construct the plants was granted.

Within this cultural context, the power engineers understandably strove to minimize capital cost per kilowatt ($/kW). Ever larger generating units and power stations, they thought, could keep on doing this by taking advantage of two well-known factors:

- in large projects, the fixed costs—those incurred by the project regardless of its size—would become smaller relative to the variable costs that were proportional to size, so total costs per unit of capacity should shrink as the fixed costs were diluted (spread over more units of capacity and output); and

- the costs of the materials and labor needed to build objects would depend partly on geometrical relationships. For example, the cost of building a vessel depends mainly on its surface area, while its capacity depends mainly on its volume, which rises more rapidly with size than does surface area. This logic leads to the classical rule-of-thumb that cost per unit of capacity for boilers,

---

Reserve margin is the difference between total installed **generating capacity** and expected **peak load**, expressed as a percentage of expected peak load. Peak load is the maximum rate at which electricity is being demanded from the provider at any one time, typically measured over an interval of 15 or 30 minutes, although it is usually managed over even shorter intervals. It can be either actual or "weather-normalized"—mathematically adjusted to what it would be in an average-weather year; part of the purpose of reserve margin's extra capacity is to cope with exceptionally high loads due to unusual weather. Generating capacity is conventionally based on the generating unit's **"nameplate"** rating for safe and continuous output, and is usually expressed in **net** terms, after subtracting several percent (or more if elaborate emissions-reducing equipment is used) for the electricity used within the power plant itself. A power plant's output capacity is often rated under the conditions in which the peak demand normally occurs, since, for example, thermal power plants can generally produce less output on the hottest days when their condenser water is warmer. In that case, capacity is usually called the maximum summer **capability**.

The **thermal efficiency** of a power station is how much electricity it produces (usually net) from each unit of fuel it consumes. It is normally evaluated on the assumption that the plant is operating under certain conditions specified in its design, but actual results may differ, and depend on many variables including fuel quality, air and cooling-water temperature, and age. Thermal efficiency is usually expressed as a percentage or as a decimal fraction of one, where 100% efficiency (unachievable in principle) would mean converting fuel into electricity with zero losses, so that each 3.6 kilojoules or 3,413 BTU of fuel would yield one kilowatt-hour (kWh). The reciprocal of thermal efficiency is called **heat rate**. A common sort of heat rate for a fossil-fueled steam-driven power station would be about 10,000 BTU/kWh, corresponding to a thermal efficiency of 34%.

Discount rate expresses the time value of money. A dollar placed in a bank account that yields 5% annual interest is worth $1.05 next year, $1.1025 the year after that, and so forth (assuming no inflation). Therefore, going in the other direction—discounting **future value** back to **present value** at a discount rate of 5% per year—$1.1025 two years from now has the same present value as $1.00 today, *i.e.*, $1.1025 / (1.05)^2 = 1.00. Discount rates are used to calculate the present value of long-term investment decisions, such as power plants and securities. For example, a U.S. Savings Bond denominated at $1,000 (its future value when it matures) and earning 5% annual interest over a period of ten years would have a discounted present value of $1,000 / (1.05)^{10} = 614. A stream of annual $100 payments sustained for 20 y, discounted at a real (inflation-adjusted) discount rate of 5%/y, has a present value of $100(1.05^{20} - 1) / 0.05(1.05^{20}) = $1,246$; the $2,000 value is thus discounted by x 0.623.

chemical plants, etc. rises only as roughly the two-thirds power of size, so that doubling capacity increases total cost by just over half.

The orthodoxy of the day simply extrapolated along dashed lines from experience into a heady mixture of expectation and hope. As the Federal Power Commission showed the extrapolations in 1964:

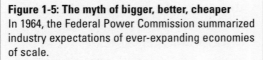

Figure 1-5: The myth of bigger, better, cheaper
In 1964, the Federal Power Commission summarized industry expectations of ever-expanding economies of scale.

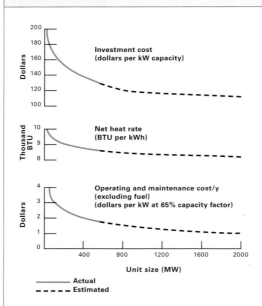

Source: R. F. Hirsh, *Technology and Transformation in the American Electric Industry* (Cambridge University Press, 1989), p. 59, fig. 16

But reality soon proved different, not only in future projections but also in interpreting past experience. In practice, the economies of scale in $/kW turned out to be mostly exhausted by the time a power plant got as big as about 100 MW of electric output, not 500 MW as claimed, and they often became trivial or even negative above a few hundred megawatts. Unfortunately, that was only starting to be understood in the 1970s

(346, 484), after most of the giant plants had already been ordered. For the industry as a whole, construction economies of scale—the first of the three sanguine 1964 graphs just shown—declined a few years later, vanished in the 1970s, and radically reversed in the 1980s, when real construction cost, deflated using indices specific to escalation in power-plant ingredients, simply stood up on end:

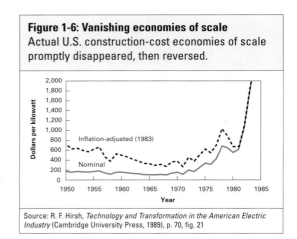

Figure 1-6: Vanishing economies of scale
Actual U.S. construction-cost economies of scale promptly disappeared, then reversed.

Source: R. F. Hirsh, *Technology and Transformation in the American Electric Industry* (Cambridge University Press, 1989), p. 70, fig. 21

So great were the funds and prestige committed particularly to nuclear expansion in the 1960s and '70s that the empirical data, showing economies of scale 2–3 times smaller than the vendors, utilities, and government had assumed, were ignored or rejected until too late (384). Much the same was true for coal-fired power plants, which showed *no* statistically significant correlation between size and cost—at best, the weak possibility that very large units might yield a 3% gross cost saving, reduced to 2% net by the longer construction time's increase in financing and escalation costs (384).

Moreover, classical steam plants' thermal efficiency topped out after units' electrical capacity reached about 400 MW. As plants continued to grow far beyond that size, thermal efficiency collided with practical limits around 1960 (Figure 1-7)—partly because new pollution-reducing equipment used more energy, but mostly for fundamental reasons of metallurgy and other engineering factors described below.

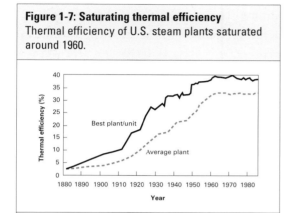

Figure 1-7: Saturating thermal efficiency
Thermal efficiency of U.S. steam plants saturated around 1960.

Source: R. F. Hirsh, *Technology and Transformation in the American Electric Industry* (Cambridge University Press, 1989) , p. 4, fig. 1

Indeed, as projects based on supposedly inexhaustible economies of scale moved from paper to construction sites to operational experience, it started to become clear that the divergence between assumed and actual costs went far deeper than $/kW. The engineers had assumed that *other* key variables, such as reliability, operational flexibility, construction time, and ease of siting, *would not vary significantly with unit size.* This unfounded extrapolation from limited experience was odd behavior for an otherwise conservative industry. It also proved a colossal error, for reasons that became obvious in hindsight, for example:

• Such gigantic plants were not easy to build or site. Since they needed to reject about twice as much low-temperature heat as they produced in electricity, they were typically sited on the shore of the ocean or of large lakes or rivers—sites often preferred by other sectors of society for other purposes. Giant plants also required ever larger transmission lines

that became harder to site and, carrying more power through particular corridors over ever longer distances, became more prone to fail with graver consequences when they did inevitably fail. (As an extreme example, on 21 January 2002, failed transmission from the 12.6-GW Itaipu hydroelectric dam, the world's largest, cut off 18% of Brazil's electricity for several hours, blacking out six major cities in five states.)

• Big plants often yielded less operational flexibility—ability to vary their output widely and quickly—than smaller plants. This reduced the big plants' ability to respond to loads that were meanwhile tending to become more variable with time (especially with the spread of air-conditioning in inefficient buildings) and increased many kinds of system costs. It also led to poor utilization of grid capacity, with less than 40% of capacity in use at least half the time (Figures 1-35–1-37).

• Because of their prodigious scale, each such plant entailed either confining billions of curies of radioactivity and hundreds of kilograms of plutonium (which one hoped could not escape through accident or malice) or a fossil-fuel massflow equivalent to nearly 130 kg of coal per *second* or hundreds of railcars per dy (which would then turn into climate change and acid rain). These sorts of numbers made more prominent the plants themselves and public perceptions of their potential risks of accident or pollution.[7] Those perceived risks then became internalized through the political and regulatory processes, forcing ever larger investments that won diminishing returns in safety or cleanliness, and hence a geometric rise of real construction cost per kW as more plants were built or planned (79, 384, 493). This process resulted in precisely what was observed (Figure 1-8)—the very opposite of the "learning curves" and "scale economies" that were supposed to make plants cheaper as more were built. Rather, build-

[7] For example, the more large plants are built, the more likely one is to be near you, the more likely something is to go wrong among the larger population of plants, and the more likely you are to notice it and make a fuss about any accident or emission hazards that you perceive to flow from nearby and other plants. This natural effect can be abated only by making each plant at least proportionately cleaner and safer (both in reality and in public perception) as more plants are built. (Small plants using inherently benign technologies tend to avoid this problem.)

ing or proposing more plants made each *more* expensive because of increasing intensity of inputs (labor and materials). This escalation was less severe for coal-fired plants, perhaps because the perceived risks that must be abated as more plants are built are more tangible and understandable; but it was contrary to the cost decreases that had been theoretically assumed for both kinds of plants.

The analysis explains 92% of the observed cost variation among 46 nuclear plants totaling 39 GW, and 68% among 116 coal-fired plants totaling 70 GW. The data set includes all U.S. commercial units >100 MW entering service between 31 December 1971 and 31 December 1977 (nuclear) or 1978 (coal).

The solid curve showing empirical data (384) was striking enough, but its projection proved conservative: as nuclear projects later started to be canceled, the nuclear "supply curve" actually bent backwards toward the upper left, with real $/kW construction costs rising as the order books shrank. That seems to be partly because some specialized workers saw no further prospect of selling skills like nuclear welding, and hence were in no hurry to finish the job.

Yet the myths of learning curves and economics of scale died hard. To the very end of the nearly industry-busting debacle from the mid-1970s to the mid-1980s, official government and industry assessments continued to deny the reality of the field data, preferring to rely instead on far more optimistic and unfounded projections. For example, the U.S. Department of Energy's 1983 Electricity Policy Project assumed nuclear completion costs about one-third below the average of those actually estimated at the same time by all the utilities then building such plants (422). The nuclear industry, in particular, often claimed that U.S. capital-cost escalation was due to peculiar regulatory conditions that could be fixed by "reforming" the siting and licensing processes; yet comparable escalation was also occurring throughout the world's market economies and even in centrally planned ones (422). The industry blamed everything except the obvious culprit: boldly scaling up to 800–1,200-MW plants based on technical, organizational, and societal experience that was typically in the 100–200-MW range.

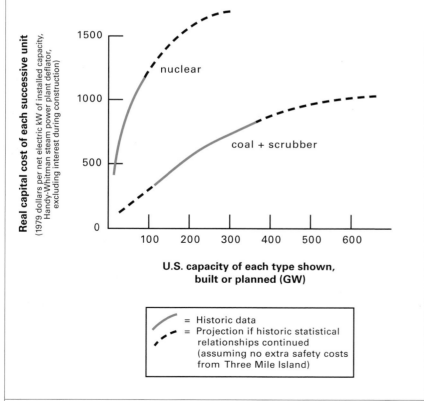

Figure 1-8: Building more coal and nuclear plants made them costlier
The empirical "forgetting curve" that made nuclear and coal plants costlier, not cheaper, based on Charles Komanoff's 1981 multiple-regression analyses.

Source: A. B. and L. H. Lovins, *Brittle Power* (Brick House, 1982), p. 378, fig. A3; C. Komanoff, "Power Plant Cost Escalation" (Komanoff Energy Associates, 1981)

Reliability tended to decline with size (Figure 1-9): for example, Edison Electric Institute found that coal and steam plants' forced outage rates (what fraction of the time they failed unexpectedly and had to be shut down for repairs) were directly related to unit size.

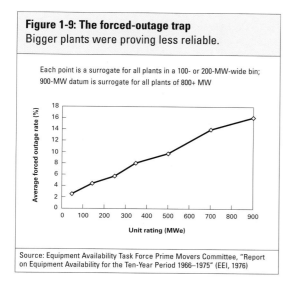

Figure 1-9: The forced-outage trap
Bigger plants were proving less reliable.

Each point is a surrogate for all plants in a 100- or 200-MW-wide bin; 900-MW datum is surrogate for all plants of 800+ MW

Source: Equipment Availability Task Force Prime Movers Committee, "Report on Equipment Availability for the Ten-Year Period 1966–1975" (EEI, 1976)

This was happening for fundamental reasons. Highly efficient boilers relying on "supercritical" steam conditions, and bigger turbines with hotter steam, meant hotter and more highly stressed blades. More stress on blades required more exotic alloys that turned out to have unexpected metallurgical properties, such as new ways to crack and corrode. Bigger boilers meant more kilometers of tubing that did not become proportionately less failure-prone per kilometer. Bigger nuclear reactors meant new and worse kinds of potential accidents requiring more complex safety and control equipment. As the more and newer things that could go wrong started to go wrong more often and in newer ways, small hoped-for gains in thermal efficiency (electricity wrung from each unit of fuel) often turned into not-so-small *losses* in efficiency, partly because ener-

gy was lost in reheating the vast boiler after it cooled down during operational glitches. Furthermore, the failure of a big plant became a more serious event than the failure of a small plant, and carried the potential to trigger wider failures that could cascade across a whole region. A big plant therefore required more backup instantly ready to step into the breach in case that big block of capacity should suddenly fail.

Since many of the same causes of decreased reliability with size also depended on age, such as metal fatigue and corrosion, a striking double correlation emerged: "broken-in" mature plants tended to become less reliable, and to have a greater scatter in plant-to-plant reliability, as they aged. This was especially pronounced, as suggested by Figure 1-10, for the larger plants that do most of the generating. Detailed multiple-regression analysis would doubtless reveal more of the causal factors (multiple units, seawater cooling, supercritical steam conditions, etc.).

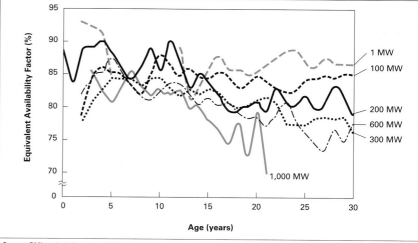

Figure 1-10: Big steam plants age ungracefully
As steam plants age, their availability deteriorates more, and becomes more volatile, in larger than in smaller units (RMI analysis from 1,347–1,527 units, smoothed as Markovian series to reduce noise, based on National Electric Reliability Council raw data).

Source: RMI analysis based on NERC, "Fuel, EAF and Dependable Capacity for 1982–1993 Power Plants" (29 July 1994; kindly provided by Resource Insight, Inc.)

Still, this gross correlation between smaller units, higher median availability, and less volatile unavailability seems strongly suggestive. Later evidence for small units is consistent with Figures 1-9 and 1-10: for example, Trigen Corporation experienced availability over 96% with 4-kW to 160-MW units—over 99% with small backpressure turbines in the 40-kW to 6-MW range, the same as Turbosteam's 150 units up to 17 years old (94). Section 2.2.9.2 will further discuss the exceptionally high availability of many small-scale resources as a potential source of reliability benefits.

Equivalent availability factor is the fraction of its full-time, full-power output that a power plant is actually available to generate if desired. For example, if a plant is available to run at its full rated power half the time, or at half its full rated power all the time, it would have a 50% equivalent availability factor.

Technical Note 1-1: Fossil-fueled steam power plants' reliability deteriorates more for big units

The accompanying graph (Figure 1-10) shows the median equivalent availability factor for fossil-fueled steam power station generating units in the United States, analyzed by André Lehmann at RMI from 1982–93 data of the National Electric Reliability Council, kindly provided by Resource Insight, Inc., a Boston consultancy. The number of units in each size range varies somewhat during that period, but is shown in the following table, as is the mean availability of all plants in that size range over their various lives, unweighted for capacity or lifetime. The number of years in the data set for each size range varies because smaller plants were introduced earlier than large ones; some units also retired earlier than others. The units are each analyzed by age and size, so because they were brought into service in various years, the median EAF plotted by age (for all plants of that age and in that size range) does not correspond directly to the calendar years for which the data were measured. Data for 900-MW units have been omitted because there were only two plants in that range (they also happened to perform poorly and erratically). The graphed data have been smoothed by a Markovian process that weights data inversely by variance and by how long ago they occurred, so as to reveal underlying trends without requiring a curve-fitting exercise. This method assumes that with so many plants, median EAF will not jump abruptly between one year and the next. Data for the 400-, 500-, 700-, and 800-MW units have been removed from the plot because they make it hard to read but add little to its message. (The 800-MW units do appear to show slightly higher median availabilities than the smaller unit classes—reversing the otherwise durable correlation—but the sample is substantially smaller than for all but the 900- and 1,000-MW units, and a high-availability datum for the oldest 800-MW units contains only two units.) Astonishingly, the 1-MW-range units over a half-century old continue in many years to exhibit availabilities in the high 90s of percent; but of course they are the successful products of the commercial version of natural selection, which retains only those units that give the least trouble. To correct for such survival bias, one would need hard-to-get statistics for the history of every fossil steam unit ever commissioned.

Table 1-1			
maximum nameplate capacity (MWe) of unit size range	*number of years in data series*	*number of plants (varies during period analyzed)*	*unweighted mean equivalent availability factor (%)*
1	60 *(years 6–8 of 63 years are missing)*	326–379	87.33
100	46	388–406	84.22
200	41	173–192	83.29
300	36	137–142	78.14
400	34	90–104	78.59
500	32	106–135	78.58
600	31	51–71	80.08
700	27	45–53	79.46
800	27	23–33	83.59
1,000	25	8–12	79.36

Complexity, especially in construction management, turned out to rise steeply with size. Size directly created new kinds of engineering and logistical problems. Building such a huge, intricate, and technically demanding artifact was more like building a cathedral than like manufacturing a car. Cost was dominated by costly craft labor at the site (making parts too big to make elsewhere and move to the site) rather than by mass-production labor prefabricating parts in a factory and hence benefiting from economies of production volume. Each GW-range plant typically took about a decade to build, and both its cost and its completion date became far less manageable and predictable. And because many large plants were under construction at once, there were not enough truly gifted managers in the industry to handle so many projects of such complexity. In such big projects, field labor and overhead, both time-consuming and both completely dependent on custom design and custom-planned building methods, came to total about four-fifths of total construction cost—enough by itself to wipe out the hoped-for economies of scale, according to the former Chairman of Consolidated Edison Company (485).

The longer construction intervals meant higher interest payments during construction—a rapidly fatal condition for many projects when interest rates and real construction costs unexpectedly soared in the 1970s. Longer borrowing periods were not offset by cheaper money, because big loans carried about the same interest rate as smaller loans: *i.e.*, the cost of money showed almost no economies of scale (140); and the bigger power stations were made increasingly of money.

Each GW-scale plant ended up costing around $1–2 billion. That huge lump of investment, strung out over the decade of construction in an increasingly turbulent and unpredictable business environment, often represented a bet-your-company decision. Some companies lost the bet; many others had near-death experiences. In the 1970s, the average U.S. investor-owned utility increased its construction expenditures eight times as fast as its cash earnings, and borrowed about two-fifth of its dividend payments: some hard-pressed nuclear utilities even borrowed to make their interest payments. These symptoms, however, were partly masked by direct Federal subsidies to electrical expansion.[8]

For these and other reasons, capital cost per kilowatt of generating plant—the main driver of scale decisions for decades—turned out to be a profoundly misleading metric. Most utilities tried to optimize in isolation this one variable or component within a complex interactive system, but thereby ended up pessimizing the whole system—not minimizing but maximizing cost and risk. This unhappy result might have been avoided if utilities' strategic choices had been informed less by engineers, accountants, and lawyers—important though their skills and insights were—but more by financiers, economists, and social scientists. For deeply rooted historical reasons, including an "edifice complex" tradition of building monuments to senior utility executives who became personally identified with projects they had launched (and which often got named for them), this seldom occurred.

[8] In FY1984 alone, these totaled some $30 billion, $16 billion of it just for nuclear fission (almost as large as retail revenues from nuclear electricity). This subvention nearly equaled electric utilities' annual investments; made electricity look about one-fifth cheaper than it really was; and per unit of delivered heat content, was over 11 times the subsidy to directly used fossil fuels and at least 48 times the subsidy to more efficient energy use (291–2). Current subsidies are smaller but scarcely less lopsided: the Renewable Energy Policy Project's somewhat less detailed 2000 analysis (274) found that 50 years of Federal subsidies to wind, solar, and nuclear power totaled $150 billion but went 95% to nuclear power.

The combined effects of some of these factors on project costs, as seen during the painful hangover from the nuclear ordering binge, were described by six Los Alamos National Laboratory researchers (676). Their quantitative examples, although hypothetical, approximate actual (especially U.S.) experience.

> An electric utility orders a nuclear plant with a $1 billion overnight cost, a 6% cost of capital, a 6-year construction duration, and no anticipated cost escalation. The expected cost of this plant is $1.29 billion. Throughout construction, [well-publicized mishaps at other plants]...result in extensive safety regulations that require retrofitting. The construction expands from 6 years to 10 years and costs escalate at the rate of about 12% per year. The anticipated cost of the plant has now risen to $2.46 billion....With additional capital expenses the utility must now return to the bond market. Additional financing is obtained at a much higher rate, perhaps 16%, either because all interest rates have risen or because the bond rating of the hypothetical utility has deteriorated. With the higher interest rate, the anticipated cost of completing the plant becomes $3.62 billion, which is almost three times the initial estimate.
>
> A cost overrun by a factor of 3, not atypical for recently completed nuclear plants, has serious repercussions for the electric industry. With a large outstanding debt, the interest coverage ratio falls, indicating that the firm is in serious financial trouble. When the plant enters the rate base at $3.37 billion instead of $1.29 billion, a very substantial rate shock [electricity price increase] is required. For instance, [LILCO] is requesting [as of 1985] a 60% rate increase to help pay for the Shoreham nuclear plant [which was subsequently abandoned after its completion]. If the demand for electricity has a [long-run own-] price elasticity of –1...[i.e., a 1% decrease in long-run demand for each 1% increase in the electricity's price—a rough number well supported by econometric literature cited in (248)], total revenues to the utility will remain unchanged [despite the higher tariff], and the realized rate of return [on capital employed] will decline. Even if the price elasticity of demand is less [in absolute value] than –1, it may be impossible for an electric utility to recover an adequate rate of return on a plant that has experienced large cost overruns.

The above illustration, which generally parallels the recent history of investing in large nuclear plants, also represents a worst-case scenario for a risk-averse electric utility. Electric utilities will be motivated to avoid repeating such an investment experience. The most desirable properties of a baseload investment are low and predictable capital cost and short and predictable lead-times. Short lead-time plants do not necessarily have lower overnight capital costs, but they offer substantially less risk of cost escalation. To the extent that smaller plants have shorter lead-times, they will be a preferred investment for future baseload generating capacity.

Thus engineering and logistical flaws led inexorably to managerial, financial, and political consequences that made most utility executives as wary of major projects as Mark Twain's cat "that sits on a hot stove lid [and] will not do so again; neither, however, will it sit on a cold one."

As these overlooked drawbacks of larger power stations became painfully evident and consequential, the triumphal progress of ever more centralized power stations—one of the greatest achievements in millennia of engineering—ground to a halt around 1970. Planning, building, and operating new power stations suddenly became a less happy affair. As marginal costs gradually worked their way through into average-cost prices, the real price of delivered electricity, after falling for nearly a century, leveled out for a few years, then began nearly a decade of steady rise after the macroeconomic shock of the 1973 oil embargo—dismaying utility regulators, whose task turned from allocating the pleasure of ever lower prices to allocating the pain of ever higher ones.

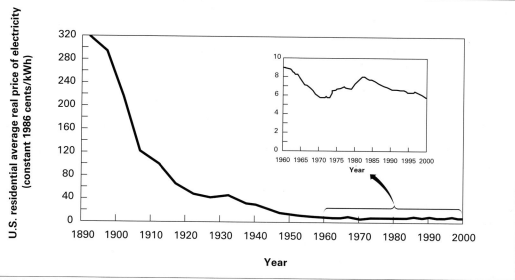

Figure 1-11: U.S. residential average real price of electricity
The controversial "electric rate shock" of the 1980s, caused largely by the nuclear building binge, was a tiny blip in long-term historical context, yet helped trigger the utility restructuring of the 1990s. To convert from 1986 to 2000 dollars, multiply by 1.42.

Source: R. F. Hirsh, *Technology and Transformation in the American Electric Industry* (Cambridge University Press, 1989), p. 9, fig. 7, with data extended and magnification added based on EIA, *Annual Energy Review 2000* (EIA, 2001)

Baseload power plants are those that have the lowest operating cost (regardless of how much it cost to build them) and are thus **dispatched** (meaning that their output is sent out to the grid) whenever available. Traditionally, "baseload" was synonymous with large steam plants.[9] But since the definition is actually economic, not technological or size-based, this association is incorrect. For example, a renewable resource such as a windfarm, or even a small solar generator, that had an even lower operating cost would be dispatched in preference to a big coal or nuclear plant, even if the intermittence of wind or sun made it available for fewer hours. In a system with enough renewable resources, previously base-loaded big steam plants could even be displaced in **merit order**—the sequence of increasing operating cost in which (subject to other constraints) plants are brought into service as load rises—and could thus end up not running at some times when they are available, because enough renewable output is available at even lower operating cost. Power-system managers or competitive markets are supposed to do **economic dispatch**—operate plants in their merit order, best buys first—so as to minimize total system operating cost. Actual operating sequence may be influenced by many other factors, and must take account of the interactions between operation, maintenance needs, and plant lifetime. A given unit's position in the merit order can also change on many timescales for many reasons.

Peaking or **"peaker"** generating units have the highest operating costs in the system and are therefore run as little as possible—typically <20% of the time, and ideally just to meet rare peak loads that would otherwise exceed the system's generating capacity. The commonest peakers are simple-cycle gas- or oil-fired combustion turbines. Many steam plants operate at an **intermediate load factor**, running more than peakers but less than baseload plants, because of their intermediate operating costs. Many of these units are oil- and gas-fired steam plants with lower efficiencies or higher pollution than newer or larger plants. In 2000, total oil- and gas-fired plants—both steam plants and simple or combined-cycle turbines—totaled 35% of U.S. generating capacity but provided only 19% of net electricity generation (206), consistent with the generally higher cost of these fuels.

---

[9] The common engineering concept that "Baseload plants are those that have a very high load factor" is found even in such otherwise excellent treatments as Wan & Parsons (699). It is true only insofar as high load factor is a *result* of low operating cost: a windfarm, for example, should normally be baseloaded (dispatched whenever available) because of its nearly zero operating cost, even though it may have a capacity factor of only about 0.3 because the wind is intermittent. The economic definition of baseload is the plants with lowest operating costs, regardless of their availability or how often they operate. Thus windpower is more a baseload plant than nuclear because windpower has a lower operating cost, even though its load factor is also lower. Some system operators' choice to dispatch hydropower as an intermediate-load-factor or even as a peaking resource is for convenience, and represents economic dispatch only in a much more convoluted sense (taking account of ramp rates, maintenance schedules, etc.) than traditional straight-operating-cost merit order. Incidentally, the "baseload" concept should in princple include all resources, not just generating resources, but it is not traditionally applied to demand-side or grid resources.)

The rise of U.S. utilities' average retail electricity prices from the mid-1970s to the 1980s was similar for all classes of customers in most parts of the country, and so was the subsequent fall back to real prices below even those of 1960, as summarized in the following national-average data:

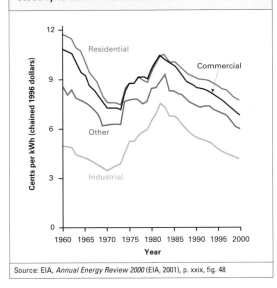

Figure 1-12: Electricity prices have retreated
Substantially higher average retail real prices for electricity hit all U.S. customer sectors in the 1970s, but since the early 1980s, have fallen back steadily to below the 1960 levels.

Source: EIA, *Annual Energy Review 2000* (EIA, 2001), p. xxix, fig. 48

The causes of the mid-1970s reversal in the previously downward price trend were many and interactive. They included cost-lier fuel, exhausted improvements in thermal efficiency, flagging reliability, siting disputes, environmental cleanup, excess capacity, costlier capital, real factor-cost escalation, and (especially) serious misjudgments in nuclear ordering and construction management. These problems plagued utility executives until the early 1980s, laying the foundations for the profound changes of scale discussed in Section 1.2.4. Those technical changes became visible starting around the early 1980s, when many of the unfavorable cost trends moderated or were digested through cost recovery from customers. Both forms of relief allowed electricity prices to start falling again, as they continued to do through the 1990s. But meanwhile, with a further lag, the price rises had triggered political forces that later emerged as the underpinning of the competitive restructuring movement of the 1990s.[10] This in turn led ultimately to the California fiasco of 2000–01 with its supply disruptions and sky-high prices. And the restructuring movement then began, as we shall see, to create new market conditions in which fundamentally new technological options, market entrants, and ways of creating value could start to express themselves. Despite some reflexive returns to ordering big

[10] The price volatility and spikes experienced in California and some other parts of the United States in 2000–01 similarly created short-term political pressures to slow or reverse competitive restructuring. Many who found competition and price deregulation attractive in theory found it less enticing in practice when short-term supply deficits were translated by poorly structured markets into dramatically higher market-clearing prices.

The **marginal cost** is the cost of the next unit of electricity. **Short-run** marginal costs are incurred by operating existing capacity more; **long-run** marginal costs represent an amount of additional electricity that exceeds available capacity and thus entails building new capacity (whose marginal cost may be less or greater than that of older "embedded" capacity).

Real cost means cost corrected for monetary inflation—that is, expressed in dollars (or other currency) of constant purchasing power. **Nominal cost** is measured in the currency of whatever year it happens to be. **Deflators** measure inflation and are used to convert between real and nominal costs. Unless otherwise noted, throughout this book, dollars of different years are converted using the GDP Implicit Price Deflator series published by the U.S. Department of Commerce and available on its website. **Factor costs** are the costs of specific inputs to building or making something; for example, building a power plant incurs costs for such "factors of production" as concrete, steel, and craft labor. The "steam-plant deflator" used to create Figure 1-8 corrects not just for general monetary inflation but also for specific changes in the cost of each factor, so if a cost expressed using that deflator rises over time, it means that a larger quantity of one or more factors is being used, or that the mix is shifting from less costly to more costly factors.

plants because they are familiar, highly visible, and politically comforting, restructuring seems, on balance, likely to seal the fate of the giant plants whose costs and risks had originally nourished such responses.

1.2.4 Discontinuity: a century of size trends reverses

Four causes of the steady fall in real electricity prices since 1982 are obvious. They include the influence of energy efficiency that damped fuel prices; low monetary inflation; gradual depreciation or write-off of surplus and unusually costly (especially nuclear) stations; and cheaper, more efficient combined-cycle power plants. But concealed among these and other causes is a discontinuity that shakes the electricity industry to its foundations. The era of the giant thermal power plant has quietly ended.

Historians may well come to view this as an event as momentous as when dinosaurs—highly evolved, superbly designed creatures that utterly dominated their landscape through superior size, strength, and skill—suddenly gave way some 65 million years ago to little scurrying mammals. To be sure, the mammals probably had a lot of help from a giant asteroid, but the outcome was inevitable because the mammals were more adaptive to the resulting rapid changes in the environment. In the utilities' case, the asteroid's role was played by the confluence of internal and external forces, including emergent new species of technologies, that together created a new business environment requiring adaptive, flexible, agile technologies.

The collapse of orders for gigawatt-range power stations in the United States, being echoed with some delay in other market economies, is a clean break with a century of tradition based on devout belief in economies of unit scale. Early signs of that break started to be explicitly recognized in the business press as early as 1978, when *Fortune* featured "The Little Engine that Scared ConEd." In 1980, a *Business Week* story (83) headlined "The Utilities Are Building Small" summarized an early warning signal, and concluded that, "The giant plant is fading. Small units spread risk and avoid excess capacity." In 1978, the article noted,

> ...almost half of the boilers ordered were larger than 650 MW. But [in 1979]...not one of the 12 fossil-fuel boilers ordered by utilities was larger than 650 MW, and half were under 400 MW....One year, of course, does not prove a significant statistical trend. Still, many utility analysts believe that the recent numbers reflect the start of a transition by utilities from reliance on large centralized units to systems based on large numbers of smaller generating units. "Utilities are starting to think smaller instead of larger," says Richard E. Rowberg...of Congress' Office of Technology Assessment....

> "If you're wrong with a big one, you're really wrong," says Jerry Peterson...of General Electric Co. "If you're wrong with a small one, you can just put up another."..."We have avoided the large units because they would mean too much capacity coming on stream at one time," says Vice-President Frank N. Davis [of Utah P&L]...."When you go to very large units, you put too many eggs in one basket," says Harvey H. Nelken, vice-president of Foster Wheeler Energy Corp., a utility engineering firm."..."The problem is that we just haven't built enough of them," insists John W. Landis, senior vice-president of Stone & Webster Inc. [a major builder of power plants]....Landis' contention may never be tested. "Uncertainty over demand is the main reason for the appeal of small plants," says GE's Peterson, "and I don't see any improvement ahead."

Now the speculation is over. Energy Information Administration data through 2000 on every U.S. power station in service reveal the astonishing dimensions of the giant-plant collapse. *Business Week* had simply been observing the next-to-last gasp of utilities' big-plant ordering and, as it happened, the all-time peak of the average unit size entering utility service. The following graph (Figure 1-13) analyzed from this huge EIA database shows that for all kinds of U.S. generating units being commissioned by utilities in a given year, the largest turbo-alternator units, previously hovering around 1.2–1.4 GW, suddenly fell to ~400 MW in 1994 and have not exceeded 600 MW since then. Meanwhile, the number of units utilities commissioned each year, which had twice peaked at around 400, plunged to levels reminiscent of the late Victorian period. Moreover, the average size of newly added utility units, having peaked at around 200 MW in the late 1970s, fell back to as little as about 6% of that level.

Was this discontinuity due to smaller steam units, or only to a change in the *mix* between large steam plants and smaller gas-turbine and other non-steam plants? This can be easily determined by looking only at the steam plants. For further clarity, the data can be recharted without the nuclear plants (which averaged over 1 GW through the 1980s but stopped being added to the grid in 1993). This leaves only the fossil-fueled stations. And since the number of generating units being added each year—having trended downward from more than 160 just after World War II to very low values in 1999–2000—became so small in the 1990s that average unit size started oscillating wildly according to the size of individual plants, the data can be smoothed using a five-year rolling average. (These data are just for utility units, reflecting competition under PURPA [1984-]— and the 1992 Energy Policy Act. But Figures 1-17 to 1-20 will show below, nonutilities didn't find GW-scale units attractive either.)

Figure 1-13: Maximum and average size of operating units (all types, all U.S. utilities) by year of entry into service
The era of adding giant new utility generating units—ordered upwards of a decade earlier—ended in 1990.

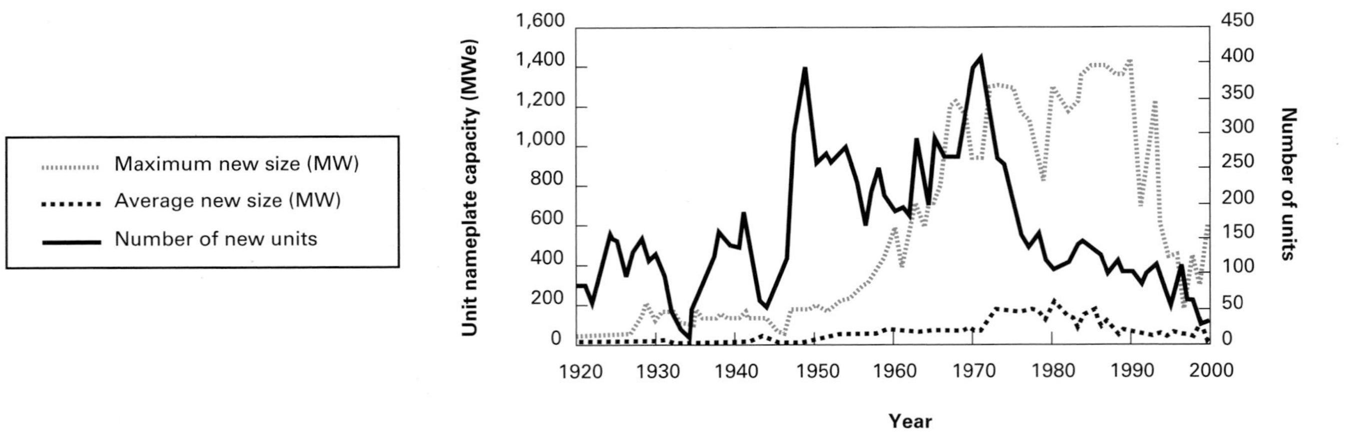

.... Maximum new size (MW)
• • • • • • Average new size (MW)
———— Number of new units

Source: RMI analysis from EIA, *Annual Energy Review 2000* (EIA, 2001), www.eia.doe.gov

Figure 1-14: Maximum and average size of new generating units (fossil-fueled steam, all U.S. utilities, five-year rolling average) by year of entry into service
On a rolling-average basis, big power plants have been fading since about 1970.

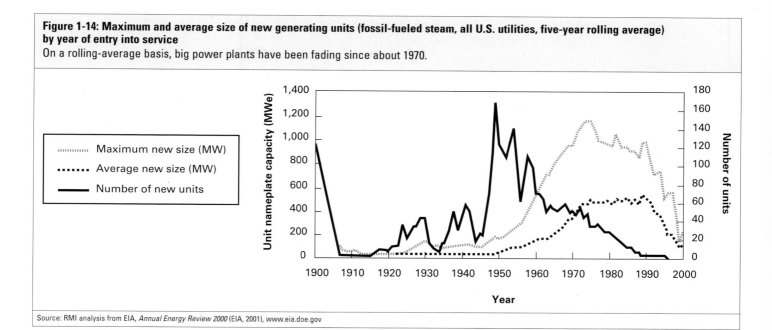

Source: RMI analysis from EIA, *Annual Energy Review 2000* (EIA, 2001), www.eia.doe.gov

The result (Figure 1-14) tells a dramatic story. In the 1990s, both the largest and the average size of new utility steam-plant generating units added fell by four-fifths—before such additions ceased altogether.[11] In hindsight, the central condensing-steam-plant business has been dying since the early 1970s—just as the dinosaurs, because of the narrowness of their outwardly successful environmental adaptation, were doomed even before the ecological shocks from the asteroid impact administered the *coup de grâce*.

If that's what happened to units being brought into service, ordered generally in or before the mid-1980s, then what is expected for the next decade of installations based on orders placed since then? A combined look at steam plants brought online through 1995 *and* ordered for 1996–2005 commissioning by all U.S. utilities shows a strong recent trend from very large to medium-sized and smaller steam plants. At first glance, this

seems to be only a retreat from units in the 1.01–2.15-GW range in favor of the 0.46–1.0-GW range (Figure 1-15):

Figure 1-15: Capacity distribution by date in service (all U.S. utility-owned steam units)
At first it appears that the most recently ordered plants have only retreated from the largest size range...

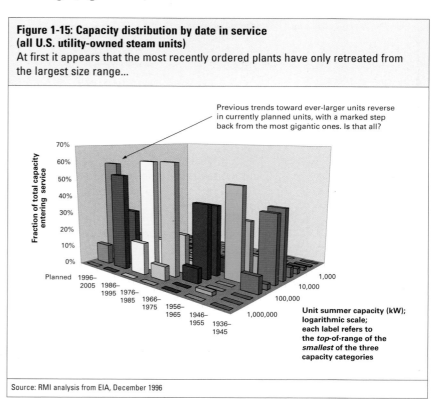

Source: RMI analysis from EIA, December 1996

[11] Unfortunately, the data set is too small and the date information available from EIA too vague to disclose a significant decrease in lead time accompanying the smaller unit sizes in the past few years.

But if the tall bars near the front are rendered transparent, a more striking pattern is revealed (Figure 1-16):

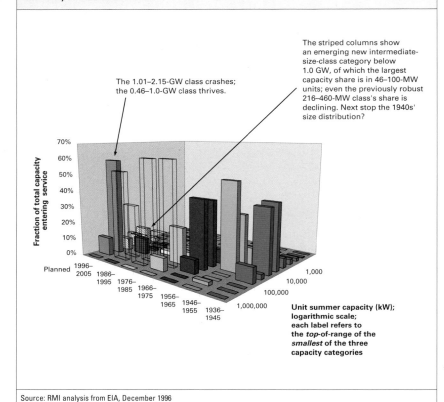

Figure 1-16: Capacity distribution by date in service (all U.S. utility-owned steam units)
...But look what's coming up in the garden! Behind the scenes, new plant size shifts down by tenfold and more.

The 1.01–2.15-GW class crashes; the 0.46–1.0-GW class thrives.

The striped columns show an emerging new intermediate-size-class category below 1.0 GW, of which the largest capacity share is in 46–100-MW units; even the previously robust 216–460-MW class's share is declining. Next stop the 1940s' size distribution?

Fraction of total capacity entering service

70%
60%
50%
40%
30%
20%
10%
0%

Planned
1996–2005
1986–1995
1976–1985
1966–1975
1956–1965
1946–1955
1936–1945

1,000
10,000
100,000
1,000,000

Unit summer capacity (kW); logarithmic scale; each label refers to the *top*-of-range of the *smallest* of the three capacity categories

Source: RMI analysis from EIA, December 1996

Clearly, steam units in the one-to-two-gigawatt class are becoming less attractive and less common. But one could fairly infer that even the two size classes below (215–460 and 460–1,000 MW) may also be heading for trouble. That is, most of the 0.46–1.0-GW plants shown were ordered as much as a decade before their planned 1996–2000 in-service date, and hence no longer reflect the market trends of the late 1990s, let alone the early 2000s. Instead, steam plants an order of magnitude smaller than were recently dominant are suddenly burgeoning, even in the utility sector.

A closer look at the unit-by-unit data posted on the U.S. Energy Information Administration's website reveals that during the 1990s, the number of large U.S. utility-owned units commissioned dropped off significantly. Figure 1-17 shows that the addition of coal-fired and nuclear power plants stalled in the late 1980s. Even utility companies, long the main proponents of building large coal-fired power plants, cut back drastically on orders for these plants, adding only 22 in the 1990s. This compares to an average of 268 plants ordered during each of the previous four decades (189). That this sudden decrease was due to more than tightening environmental controls can be inferred from Figures 1-15 and 1-16, which show that not only did utilities' orders for coal-fired plants plummet, but the few that were added after 1990 were smaller than 1,000 MW.

Meanwhile, the non-utility sector has been growing rapidly. During 1990–2001, its total net generation increased by 414%, even though the vast majority of the units it built were smaller than 100 MW. (However, as noted below, roughly half of non-utilities' 209 GW of total capability in 2000 had been built by utilities and then sold to non-utilities under restructuring [193].) Figures 1-18–1-19, reflecting this evolution, offer a window into the future, although unfortunately the federal government refused to release any data on non-utilities' pre-2002 construction plans. Nonetheless, natural gas is the fuel of choice for most of these smaller, more efficient and modular non-utility plants, and renewables are also important contributors.

Figure 1-20 reveals a startling development. The size range (up to 100 MW) in which U.S. utilities virtually stopped adding

Figure 1-17: U.S. utility generating capacity commissioned 1920–2007

All units and all sizes reported to U.S. Energy Information Administration; units 1920–97 actual, 1998–2007 projected at the end of 1997.

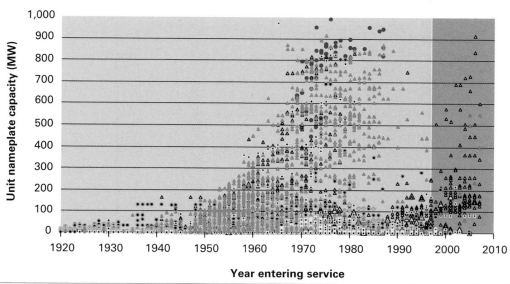

◆ Wood
■ Solar
△ Natural gas, LNG, LPG, Syngas
◆ Geothermal

■ Wind
✳ Refuse, Landfill methane, Refinery gas
✳ Multi-fuel
• Fuel oil, Jet fuel, Kerosene

△ Waste heat
● Nuclear
• Hydroelectricity
▲ Coal (coal, lignite, coke, culm)

Source: RMI analysis from EIA, *Annual Energy Review 1999* (EIA, July 2000), www.eia.doe.gov

Figure 1-18: U.S. utility generating capacity commissioned 1920–2007 (logarithmic scale)

All units and all sizes reported to the U.S. Energy Information Administration; units 1920–97 actual, 1998–2007 projected at the end of 1997.

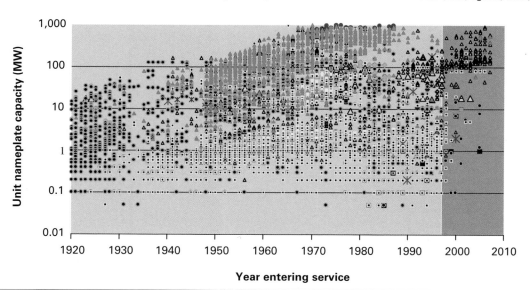

◆ Wood
■ Solar
△ Natural gas, LNG, LPG, Syngas
◆ Geothermal

■ Wind
✳ Refuse, Landfill methane, Refinery gas
✳ Multi-fuel
• Fuel oil, Jet fuel, Kerosene

△ Waste heat
● Nuclear
• Hydroelectricity
▲ Coal (coal, lignite, coke, culm)

Source: RMI analysis from EIA, *Annual Energy Review 1999* (EIA, July 2000), www.eia.doe.gov

Figure 1-19: U.S. non-utility generating capacity commissioned 1920–1997

All units and all sizes reported to U.S. Energy Information Administration; units 1920–97 actual, 1998–2007 projected at the end of 1997.

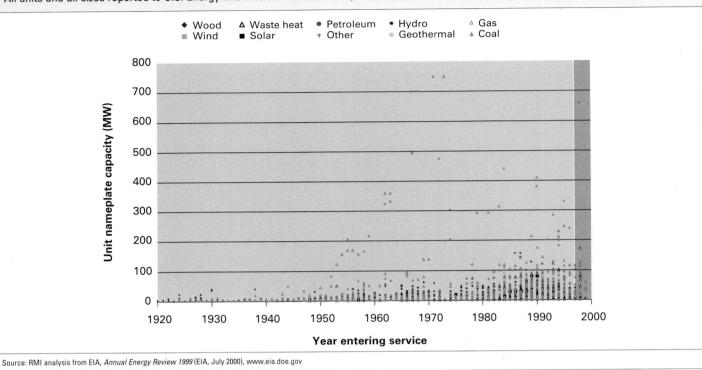

Source: RMI analysis from EIA, *Annual Energy Review 1999* (EIA, July 2000), www.eia.doe.gov

Figure 1-20: U.S. non-utility generating capacity commissioned 1920–1997 (logarithmic scale)

All units and all sizes reported to U.S. Energy Information Administration; units 1920–97 actual, 1998–2007 projected at the end of 1997.

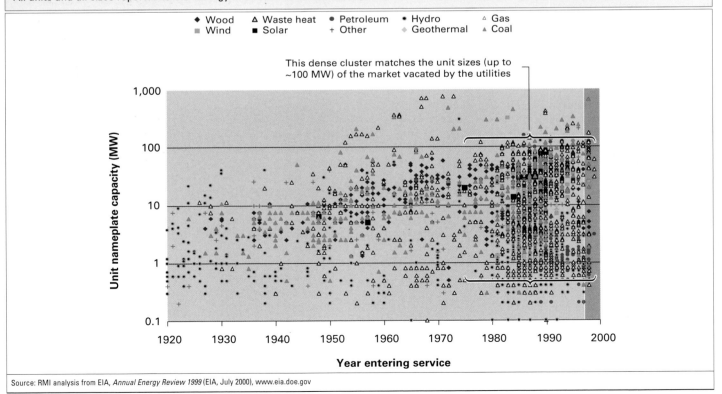

Source: RMI analysis from EIA, *Annual Energy Review 1999* (EIA, July 2000), www.eia.doe.gov

capacity starting in the late 1990s is precisely the market heavily occupied by non-utilities. The only difference is that they began earlier—a natural consequence of their shorter lead times. Of course, there are many plausible reasons for this shift in which operators were adding units under 100 MW. Their intensive installation by non-utilities began immediately after the Supreme Court, in 1984, upheld the Public Utility Regulatory Policies Act of 1978 (PURPA), which forced utilities to accept and buy back any power generated by non-utilities. One interpretation would be that after a suitable lag time, the utilities discovered that private entrepreneurs could execute such projects more cheaply than they could themselves, so it made more sense to buy power back from them and then use the utilities' political power to persuade regulators to set a lower buyback price. In fact, as the modest number of 1990s large plants came online, Qualifying Facilities contracts dried up by the late 1990s and regulator-approved avoided-cost buyback prices dropped dramatically. This may explain the thinning of 10–100-MW additions by non-utilities in the late 1990s. Utilities' interest in building such plants themselves may also have been decreased by a perception that as restructuring delaminated them, their distribution companies would be prohibited in many states from owning distributed generators, no matter how much sense their integration with the distribution system made.

Figure 1-21, from a major international vendor of power-supply equipment, shows a similar worldwide gain in orders for MW-scale units, generally windpower or diesel. The 1998 orders in the ≤1-MW unit size range reached 12 GW, while formerly dominant orders for steam turbines >200 MW fell about

one-fourth from their average 1995–97 level. These trends have since intensified. For example, in the year 2000, just a single major vendor of diesel generators—Caterpillar, Inc.—reported shipping more than 60,000 generator sets totaling nearly 20 GW, or nine times the capacity of Hoover Dam, increasing its global fleet to more than 300,000 units. Its sales grew by more than 20%/y during 1995–2000 (95). By 1997, Electricité de France was using 0.61 GW of distributed diesel generators as dispatchable reserve (360). Similarly, preliminary figures indicate (76) that global installed windpower capacity grew 5.5 GW in 2001 alone, from 17.8 to 23.3 GW (three-fourths of it in Europe), and the European Wind Energy Association increased its 2010 European projection from 40 to 60 GW.

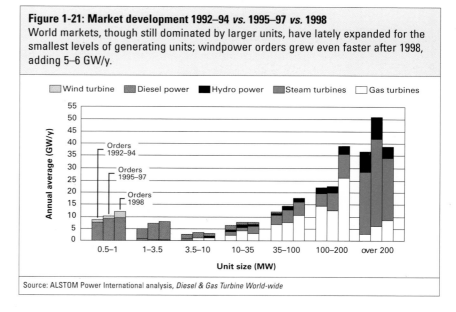

Figure 1-21: Market development 1992–94 vs. 1995–97 vs. 1998
World markets, though still dominated by larger units, have lately expanded for the smallest levels of generating units; windpower orders grew even faster after 1998, adding 5–6 GW/y.

Source: ALSTOM Power International analysis, *Diesel & Gas Turbine World-wide*

To see the main triggering event for the demise of the giant steam units, we must look beyond steam plants to a different technical innovation. The critical event was the emergence of combined-cycle natural-gas-fired power stations modified from mass-produced jet aircraft engines. (Figures 1-17–1-18 include their steam but not their gas-turbine portions.) These "aeroderivative" plants came from the likes of General Electric, Westinghouse, Asea Brown Boveri (ABB), and major Japanese vendors as packaged units with an astonishing capacity of a quarter or a third of a gigawatt per ship-deck frame, barge, or railcar. They can be ordered, installed, and commissioned on an established U.S. site in two years, only slightly slower than the generally accepted 1.5 years for a far less efficient simple-cycle gas turbine. Combined-cycle plants in early 2001 cost between $600/kW and $700/kW for a 700-MWe unit completely installed. (The lower figure was for a big unit with duct firing and inlet chilling. Smaller units or those in certain Northeastern states cost about $750/kW. Prices recently spiked upwards as panic buying tightened the market, but the shortage seems temporary.) Combined-cycle plants burn natural gas quite cleanly with an impressive efficiency of ~50% (on the same basis on which classical steam plants are ~30% efficient). At expected natural-gas prices over a 20-year planning horizon (conservatively, $4.0/GJ—vs. lowest late-1990s spot prices of ~$1.5/GJ), and if operated at high capacity factors, they can generate electricity at a total busbar cost around 2.9–3¢/kWh (levelized 1999 $). (478, 580) It is no accident, then, that 52% of the generation and 71% of the non-utility capacity additions in 1999, 95% of the total capacity additions in 2000, and 83% of the non-utility capacity additions planned for 2000–2004 commissioning were gas-fired (196, 207).

At the beginning of 1999, U.S. utilities' 195 installed combined-cycle gas plants represented only 14 GW of net summer capability or 15 GW of nameplate capacity. This represents only 2% of the total national nameplate

capacity. But utilities' planned natural gas capacity additions for 1996 through 2005 were slightly higher than total planned steam unit additions plus simple cycle turbine units (182, 198). The relatively new combined-cycle technology has rapidly grabbed half the entire utility market and is aiming at the other half too, by combining gas turbines' low capital cost and short lead time with steam plants' reliability and low fuel cost. The latest base-case Federal energy forecast envisages all U.S. combined-cycle plants' summer capability increasing from 31 GW in 2000 to 60 GW in 2005, 140 GW in 2010, 182 GW in 2015, and 214 GW in 2020—equivalent to 70% of all 2000 coal capacity, or 15% more than all nuclear, hydroelectric, and other renewable capacity in 2000 (199).

Such growth sounds superficially plausible (if adequate gas deliverability keeps pace) because the next generation of combined-cycle plants—even more powerful, efficient, and inexpensive—will beat the busbar cost of power from new central steam plants by about twofold. They'll also undercut just the operating cost[12] of most nuclear plants (436). But two other categories of resources make combined-cycle gas plants a bad buy (§ 3.4.2.2.1). The first of these are most end-use efficiency and grid improvements. The second are some distributed generators, renewable or non-renewable, that either are very well designed and mass-produced or installed in a way that yields substantial "distributed benefits" unavailable to any hundreds-of-MW plants. As we shall see, properly counting distributed benefits— previously uncounted economic values of right-sized ways to make (or sometimes also to store, move, and use) electricity—can make the relatively large combined-cycle gas plants vulnerable to such competition from

[12] Including maintenance, for which the biggest repair bills are misleadingly booked as capital costs rather than as operating costs; please see Section 1.3.3, note 64.

options roughly ten times to a hundred thousand times smaller, once the market starts to perceive and reflect those benefits.

Combined-cycle gas plants are an extraordinarily tough new competitor that brought a new player (aircraft-engine companies) into the highly mature power-plant business. Their key concept comes directly from an idea older than Henry Ford. Yet that old idea apparently struck General Electric's turbine makers with the force of a revelation only around 1979. That's when a few foresighted turbine makers realized that with smaller units "it becomes possible to standardize a design and replicate a large number of identical units," opening up "the possibility of a new dimension in scale economy" which "may be of considerable significance" and hence "an entirely new and profoundly different avenue for reducing the capital cost of generating capacity." (237) The combined-cycle plants also turned out to be far more efficient than the best classical steam plants, and their ideal fuel, natural gas, was unexpectedly found in the 1980s and '90s to be not scarce at all but rather ubiquitous, abundant, and cheap. The evolution of optimal unit size thus took an abrupt U-turn from large to small scale in the 1990s. This new trend has not yet run its course: as the question-mark bubble at the lower left corner of Figure 1-22 indicates, smaller gas turbines may in time be displaced by still smaller and cheaper fuel cells produced in even larger volumes and shorter manufacturing cycles. Compared with combined-cycle plants' two-year ordering and installation cycle, even smaller, more modular units like wind turbines, microturbines, solar arrays, and fuel cells can be cranked out in immense numbers and installed at a given site at rates of MW per *day*.

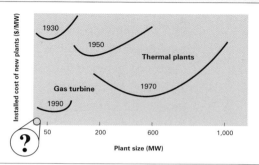

Figure 1-22: Cheaper meant bigger
Cheaper meant bigger—until advanced gas turbines suddenly made the curve buttonhook back to smaller units. Will fuel cells ultimately reach the lower left corner?

Source: C. Bayless, "Less is More: Why Gas Turbines Will Transform Electric Utilities" (*Public Utilities Fortnightly*, 1 Dec. 1994), pp. 21–25. Cited in G. L. Cler and M. Shepard, "Distributed Generation: Good Things are Coming in Small Packages" (E SOURCE, 1996)

Interestingly and counterintuitively, the largest gas turbines on the market are not even necessarily the most efficient:

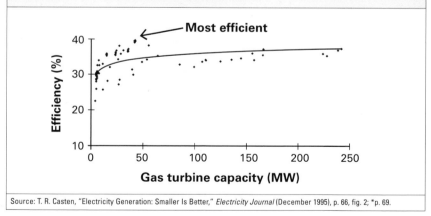

Figure 1-23: Bigger isn't always more efficient
Smaller gas turbines can be more efficient than big ones. This graph of all available gas turbines above 3 MW shows that the highest simple-cycle efficiency comes from an aeroderivative 40-MW unit (GE LM6000), not from the largest units at 250 MW. Taking account of both capital cost and efficiency, therefore, "economy of scale is largely missing with respect to single cycle efficiency. Many offerings below 50 MW compare well with 250 MW machines," especially counting potential cogeneration and avoided high-voltage step-up.*

Source: T. R. Casten, "Electricity Generation: Smaller Is Better," *Electricity Journal* (December 1995), p. 66, fig. 2; *p. 69.

Moreover, the trend in *who* generates electricity, and where, has also shifted (Figure 1-24). In 1900, about 60% of U.S. electricity was generated onsite by non-utilities, mainly industries; by 1920, only 30%; by 1980, a mere 3%. Since onsite generation often permits industries or building operators to capture and use valuable heat instead of wasting it, industrial and commercial sources have been generating an exponentially growing amount of electricity themselves since the early 1980s. From its 1983 nadir until 1999, non-utility generation increased by 797%, equivalent to compounded annual growth of 14%. By 1999, 16% of *all* U.S. net electricity generated came from non-utilities (161, 183, 190); by 2001, 29.5% (209). Even more astonishingly, the 21% of all U.S. generating capacity in 1999 owned by non-utilities was set to increase dramatically. According to data from the EIA, non-utilities plan to add 146 GW of capacity during the period 2000–04 while utilities plan a mere 2 GW (193). It is not clear whether all those increases are to be built or partly bought from utilities. However, most of the 1998–2001 jump in non-utilities' share of generation is due to their acquisition of capacity divested by utilities as part of restructuring. Such transfers totaled 23 GW in 1998, 51 GW in 1999, 48 GW in 2000, and 28 GW (79% of it involving the Exelon group) in the first eleven months of 2001 (206, 208)—a total of 150 GW. (Many such

transactions simply transferred ownership from a regulated utility to an unregulated subsidiary of the same holding company.) Although an exact comparison isn't possible because these figures are stated in terms of nameplate capacity, the capacity transferred to the non-utility sector during 1998–2000 was equivalent to 58% of that sector's net summer capability in 2000. The electricity generated by these transferred plants cannot be determined without a plant-by-plant calculation, but must account for a substantial share of the increased non-utility generation shown in the graph. There does appear to be significant net addition of capacity built by non-utilities too, but lately it has been dwarfed by their plant purchases.

Thus the seemingly inexorable trends ever since 1882 have quietly reversed in recent years. Before the 1998–2001 restructuring-induced plant transfers, that reversal was making non-utility generation increase about twice as fast as it had previously decreased. Since then, non-utility growth

Figure 1-24: The fall and rise of U.S. non-utility generation
Since 1983, non-utility generation has grown twice as fast as it previously fell, and since 1998, far faster as non-utilities bought plants sold by utilities under restructuring. The public/private utility split for 2001 is estimated.

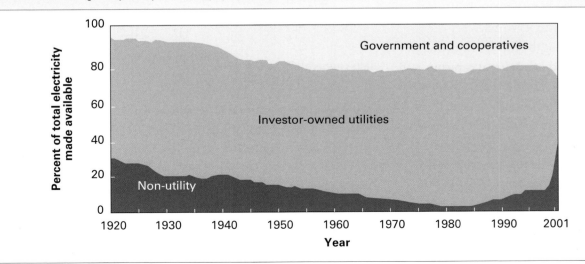

Source: EIA, December 1998; EIA, *El. Pwr. Annual 2000, Ann. En. Rev. 2000, Monthly En. Rev. March 2002; El. Pwr. Monthly Feb. 2002;* EEI, *Stat. Rev. El. Pwr. Ind. 2002*

other than by plant transfers has continued. The same increasingly competitive conditions that have led non-utilities to buy utilities' divested power plants and operate them outside utilities' regulatory framework will also presumably induce non-utilities to build more capacity of their own.

At its peak in the first half of the 1980s, just after the second oil shock, the speed of non-utilities' capacity additions illustrated the extraordinary capability of decentralized market actors, no matter what they're called, to install relatively small electricity-generating or electricity-saving technologies quickly. For example (420):

- between 1981 and 1984 inclusive, U.S. central-station orders were 65 GW smaller than their cancellations, yet new orders and firm letters of intent[13] totaled 25 GW for cogeneration (one-fourth of it renewable) and upwards of 20 GW for small hydro, wind power, and other noncogeneration renewables;

- in California, electricity sold per real dollar of Gross State Product fell by about 17% during 1975–83, and was officially projected to fall by another 30% during 1985–2004 just from existing market forces and such policies as building and appliance standards;

- when California had 37 GW of peak demand and 10 GW of utility-owned in-state hydro and geothermal capacity, and its utilities started offering an attractive price for privately generated power, through the first quarter of 1985 they were firmly offered 20.3 GW of independent small power production (mostly renewable, with a 12-MW average unit size), increasing by 9 GW, equivalent to one-fourth of total peak demand, per *year*—until the resulting power glut forced suspension of new contracting the following month;

- by autumn 1988, small power commitments covered more than 48% of Maine's and 15% of New Hampshire's total peak loads—and a decade later, non-utility producers' output was equivalent to 68% of all electricity sold by utilities in Maine (of which more than two-thirds was renewable), 19% in New Hampshire, and 41% in California (188);

- by 1994, more than a third of both Southern California Edison Company's and Niagara Mohawk Power Corporation's energy production came from independent producers (310);[14]

- by 1996, ~60% of all new generating capacity being built in the United States, and 100% in California, was non-utility (302); and

- by the end of 1998, California had installed 23.5 GW of non-utility generation with a summer capability of 21.7 GW, equivalent to over half the state's peak load, and a further 1.1 GW was slated for addition by 2003—a figure that has almost certainly risen dramatically since then.

To be sure, the stunning success of that "Wild West" period in the early 1980s—stimulated by high offering prices and perceptions that energy supply was a serious national problem—led to reactions that soon slowed the pace of development. Yet the experience proved that exposing monopolists' generating assets to increased competition can elicit remarkably vigorous expansion of relatively decentralized generating options. Indeed, more than half of the United States have run auctions to see if independent producers might like to undercut the utilities' offered price. All were promptly offered far more power than they wanted—by an average of fourfold, and for many states, eightfold—essentially all of it from relatively small plants, many of which were renewable.

[13] Mostly but not all fulfilled, largely because utilities later sought to block or unwind some of the contracts.

[14] Partly because (312) they paid respective average prices of $0.08 and $0.065/kWh for that output, or roughly twice the market-clearing price they estimated would prevail in a competitive environment.

Curiously, however, this progress remains invisible to many. In 2000, at the height of California's power shortages, many politicians were introducing proposals to short-cut licensing, relax environmental standards, or subsidize new large power plants. They evidently believed what seemingly knowledgeable people in senior energy policy positions, including the state's Independent System Operator, were telling the media—that California had built "no generation" for the past decade. This statement, trumpeted worldwide, is simply false. During 1990–99, California actually commissioned several hundred new generating units whose capacity probably exceeded that of its four operational nuclear units (189).[15] The new units were invisible only because they were non-utility-owned and mainly distributed. With a half-dozen exceptions, none over a quarter-gigawatt, the largest single unit was 80 MW. Most were much smaller; the average unit size was only about 30 MW (or about half that according to some databases). At least 30% of the new capacity was renewable. Far more was built in the 1980s when the utilities encouraged it more strongly, but the 1990s too were an unheralded success story for California's distributed resources.

1.2.5 Scale: what's the right size?

This historic discontinuity between highly centralized and relatively dispersed generating technologies is no accident. It is merely the latest chapter in an old story that has been unfolding for decades. It is the story of how the intricate balance between economies of scale and diseconomies of scale determines the right size for the job.

There are enormous differences between the scale of most energy (especially electricity) uses and the scale of most supply technologies. Using round numbers for illustration, the following table illustrates typical scales of using or producing electricity:

[15] The exact capacity California added during 1990–99 is uncertain; surprisingly, no database deals gracefully with changes in units' name or ownership, so disentangling the data is not easy. A March 2001 RMI analysis of the standard public- and private-sector databases found 1990–99 California additions of 4.532 GW (USDOE), 3.683 GW (California Energy Commission, which often omits smaller units), 4.965 GW (EGrid), or 4.710 GW (FTEnergy). The EIA's *Inventory of Power Plants* is not useful for this purpose, since its 1990 California capability of 43.681 GW is utility-only, while its 53.157-GW 1999 capability includes non-utilities. (All these figures exclude all business and household onsite standby generators and all other distributed generators smaller than ~100 kW.) Clearly, however, the correct number is not zero as was widely claimed. When told this, some changed their claim to "no major power plants were built"—as if megawatts from smaller units were somehow less effective than those from large units!

Table 1-2: Electricity supply and use scale over fifteen orders of magnitude	
average electricity used per m² of U.S. land [0.0797 W]	10^{-2} watt (W)
small portable radio	10^{-1} W
handheld cellular phone	10^0 W
portable computer; average electric use by 1 m² of very efficient U.S. office or of normally inefficient U.S. home	10^1 W
desktop computer or television; large household incandescent lamp; average electric use by 1 m² of inefficient U.S. office; one resting adult person's metabolic rate; average rate of solar energy falling on 1 m² of U.S. land (year-round, day or night: ~181 W)	10^2 W
average U.S. household's electricity use; 1-hp motor's input; bright noon sunlight falling on 1 m² of land	10^3 W
peak heating load of a normally inefficient U.S. house; peak demand of large electric stove or clothes-dryer	10^4 W
U.S. car engine's peak shaftpower; big supermarket's input	10^5 W
peak power used by a typical medium-sized office building	10^6 W
power typically used by a medium-to-large factory	10^6–10^7 W
peak power used by largest buildings	10^8 W
...or by the largest industries (smelters, uranium enrichers,...)	10^9–10^{10} W
compared, on the supply side, with:	
one *typical* central thermal power station's electricity output	10^9 W
output of a large hydroelectric dam or power-plant cluster	10^{10} W
energy output of all Alaskan oilfields in 2000 (0.97 million bbl/d)	10^{11} W
North America's total electrical generating capacity	10^{12} W
total world primary energy production	10^{13} W

It is especially interesting that the average density of electrical usage in the lower 48 United States is only about 1/5,000th of the average density of solar energy falling on the same land (averaged over all states of the Earth's rotation and orbit), and that the average electric power density of a nominal U.S. house, about 6 W/m², is only about 1/20th of its average solar input. This issue was first raised 20 years ago and still resonates (415):

> Most of the end-use devices important to our daily lives require 10^{-1} to 10^3 W and are clustered within living or working units requiring 10^3 to 10^5 W. Most production processes of practical interest can be, and long have been, carried out in units of roughly that scale.

Thus it is not obvious, *prima facie*, that energy must be converted in blocks of order 10^8–10^{10} W. The arguments usually given for such large scale include reduced unit capital cost (typically by a two-thirds-power scaling law), increased reliability through interconnection, sharing of capacity among nonsimultaneous users [*i.e.*, load diversity], centralized delivery of primary fuel, ease of substituting primary fuels without retrofitting many small conversion systems, localization and hence simplified management of residuals and other side effects, ability to use and finance the best high technologies available, ease of attracting and supporting the specialized maintenance cadre that such systems require, and convenience for the end user, who need merely pay for the delivered energy purchase as a service without necessarily becoming involved in the details of its conversion.

These contentions are not devoid of merit. Big systems do have some real advantages—though advantages are often subjective, and one person's benefit can be another's cost. But...many of the advantages claimed for large scale may be doubtful, illusory, tautological, or outweighed by less tangible and less quantifiable but perhaps more important disadvantages and diseconomies.

This possibility gains force from graphing two new data sets that show in detail the distribution of average electrical consumption per U.S. residential unit (not necessarily each meter), based on a statistical sample of 7,111 households in all 50 States and D.C. in July 1993 (170), 75% of which used no more than 1.5 average kilowatts.

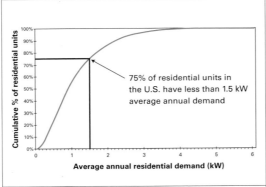

Figure 1-25: Average electricity demand of U.S. residential units, 1997

75% of residential units in the U.S. have less than 1.5 kW average annual demand

Source: RMI analysis based on EIA, *Commercial Buildings Energy Consumption Survey* (EIA, 1997)

Similarly, 75% of U.S. commercial buildings (based on a sample of 6,751 buildings in 1992) used no more than 12 average kilowatts.

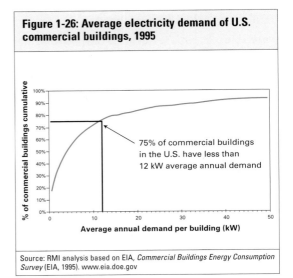

Figure 1-26: Average electricity demand of U.S. commercial buildings, 1995

75% of commercial buildings in the U.S. have less than 12 kW average annual demand

Source: RMI analysis based on EIA, *Commercial Buildings Energy Consumption Survey* (EIA, 1995). www.eia.doe.gov

Of course, generating capacity must suffice to meet coincident peak loads, not just average loads; still, we are sipping power from a firehose, and spilling a great deal in the process. Thus a single one of the largest individual steam turbo-alternators (1.4 GW) could serve nearly a million typical households among the lower three-fourths in usage, or more than 100,000 typical commercial buildings in the lower three-fourths. Such enormous discrepancies between most uses and most supply technologies invite the obvious question whether such a big mismatch of scale really makes economic sense. It "seems to require more justification than a mere appeal to custom." (451) Yet astonishingly few analysts have sought to address this question in any quantitative detail, and very few have even examined the scale spectrum of actual customer uses, on which measured data (like the two graphs just shown) remain rare. This dearth of data and analysis is especially startling for an industry as big and sophisticated as the electricity industry, whose capitalization probably approaches $2 trillion worldwide. It is almost certainly due to the absence of incentives in the old system of monopolistic, monopsonistic fiefdoms where regulators were often compliant, error bore modest penalty, and customers could be taken pretty much for granted.

1.2.6 The origins of this study

Among the first to raise the scale question persuasively, albeit mostly philosophically, was E.F. Schumacher, originally chief economist of Britain's National Coal Board. His *Small Is Beautiful* (1973) aroused a storm of controversy among those who thought he was calling for everything to be small rather than, as he made quite clear, to be *the right*

size for the particular task at hand, most but not all of which happened to be small. As he remarked (personal communications, ~1972), it would be just as silly to run a huge metal-smelter with lots of little wind machines as to heat houses with a giant power station. Both would be a mismatch of scale that would naturally incur economic penalties.

The seemingly simple but richly complex scale question gained prominence in 1976 when an influential article (410), expanded technically in an accompanying Oak Ridge National Laboratory symposium paper (413–4), criticized the historic approach of providing energy in excessive amounts, inefficiently used, at the highest quality without regard to whether the task required or justified that quality and expense, and at a scale typically around one million to one

Figure 1-27: U.S. energy intensity has fallen by 40% since 1975, to once-heretical levels
Two energy futures envisaged in 1976 *vs.* actual evolution (the heavy black line).

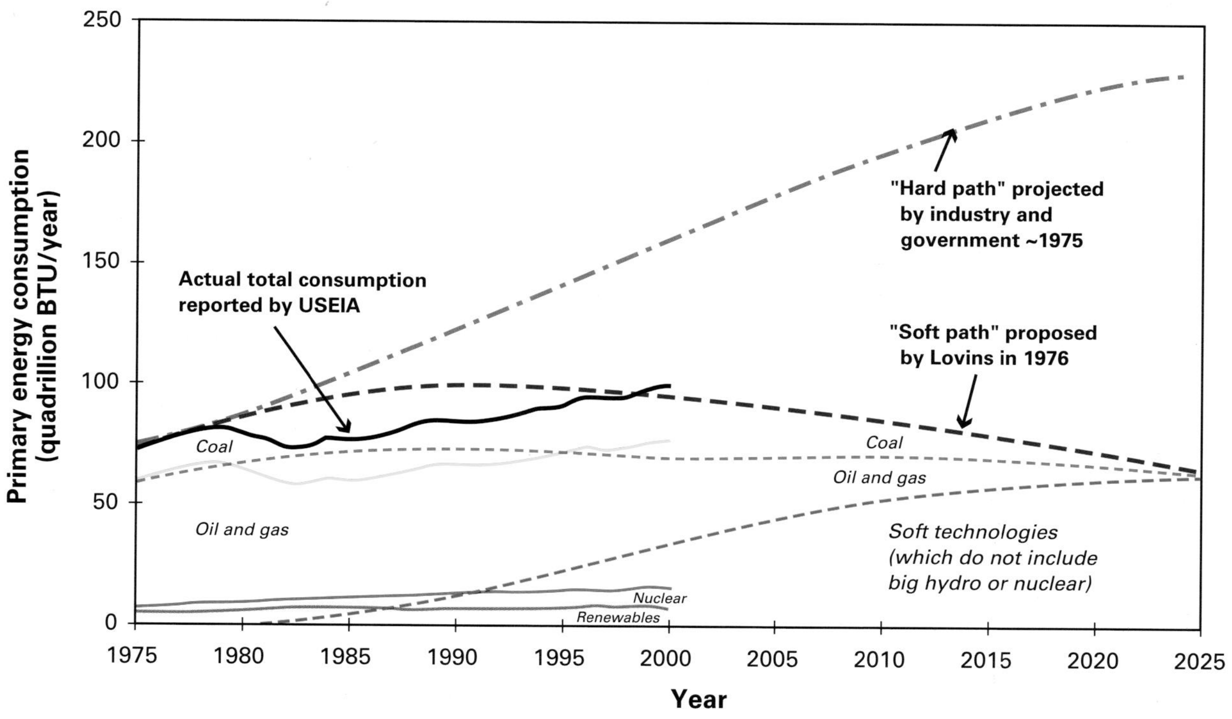

Source: A. B. Lovins, "Energy Strategy: The Road Not Taken?" (*Foreign Affairs*, October 1976); EIA data from *Annual Energy Review 2000* (EIA, 2001)

hundred million times the size of end-use needs. The article suggested that energy systems based instead on efficiency, appropriate renewables, and providing energy at the right quality and scale for each end-use task would work better and cost less. This least-cost approach turned out to provide gratifyingly accurate foresight into long-term U.S. energy demand (Figure 1-27).

Yet at first this seemingly common-sense approach was intensely controversial. Prominent critics denounced it as "the onset of a New Dark Age," "naked nonsense," "silly," "flaccid and flatulent," "chilling," "appalling," "fantasy," "Shangri-La," "never-never land," a return to medieval peasantry, and worse. Suggestions that contemporary energy systems, particularly power stations, might already be too big—might have overshot their optimal size—elicited an especially intense emotional response. These vivid reactions were conveniently assembled by Edison Electric Institute in a June 1977 special issue of its magazine *Electric Perspectives* that its contributors could now re-read with surprised amusement, and in a book of 17 critiques and responses (502) digesting roughly 30 of each from a voluminous Congressional hearing record.

That 1976 *Foreign Affairs* paper started to make rigorous the hypothesis that appropriate scale "can achieve important types of economies not available to larger, more centralized systems." Its 1977 expansion in a more technical book (414) offered a whole chapter quantifying scale effects in energy systems, including some of the first published analyses suggesting that power plants had already exceeded their most cost-effective size.

In 1981, an unclassified Pentagon study of domestic energy vulnerability (446) compactly assembled persuasive evidence on how the scale of electric power systems affected their total net cost. Several economies of scale (the bigger, the cheaper) were found, but so were nearly fifty *dis*economies of scale in about ten main categories. Taking all these effects together, the study found that:

> ...very large unit scale can typically reduce the direct construction costs (per unit of capacity) by tens of percent—at extreme sizes, even by sixty or seventy percent. But most of the *diseconomies* which inevitably accompany that increase in unit size are *each* of that magnitude....[There are] nearly fifty such diseconomies....*Almost any combination of a few of these documented effects could tilt the economic balance toward small scale* for all but the most highly concentrated applications. Thus there is a *prima facie* case that big energy technologies are not inherently cheaper, and may well be costlier, than those scaled to match their end uses, most of which are in fact relatively small and dispersed (452).***

> [T]he evidence of compensatory diseconomies of large scale which favor smaller technologies is so overwhelming that no rational decision maker can ignore it. However these many competing effects are balanced, it is difficult to imagine a way—save in the most centralized applications, such as operating a giant smelter—that they can yield lower net costs of delivered energy services at very large scale than at moderate, and often quite small, scale. Thus the relatively small, dispersed modules of energy supply required for a genuinely resilient energy system do not appear to be incompatible with reasonable cost, and may indeed be one of the cheapest ways of achieving it (457).

Despite the previous decade's painful lessons of runaway project budgets for giant plants, the power industry in 1981 was not yet quite ready for such heresy. That had to await further, equally painful, learning experiences in the marketplace. But by 1994, those experiences started to become widespread, creating a "teachable moment," so Rocky Mountain Institute returned to this theme in a project sponsored by The Pew Charitable Trusts. Reporting the early stages of the research that this book completes, Amory Lovins and Daniel Yoon refreshed the 1981 analysis in a December 1994 video-conference presentation to the Australia/ New Zealand Solar Energy Societies' joint congress in Perth, Western Australia (441).

Yet between 1981 and 1994 there was little new conceptual work to report, for two reasons. In public-sector and public-interest organizations, almost nobody was asked to do such work or even left to do it: the teams that had done pioneering topical studies in the 1970s (*e.g.*, Dr. Andrew Ford's group at Los Alamos National Laboratory) had been long disbanded under the twin influences of negative to erratic Federal attitudes (1980–92) and the 1986 oil-price crash. Congress, in the 1978 PURPA law, had ordered the Economic Regulatory Administration to assess "the cost effectiveness of small versus large [electrical] generation, centralized versus decentralized generation, and intermittent generation, to achieve desired levels of reliability," but for all practical purposes, that work was never done. A 1980–85 exploration by Congress's Office of Technology Assessment (which was disbanded by Congress in 1995) nicely consolidated, but scarcely extended, previous knowledge (537). A handful of government and independent researchers continued to indulge their personal curiosity about scale effects, but through the 1980s they found few sponsors, audiences, or market opportunities.

But the second reason there was little new conceptual work to report during 1981–94

was more encouraging: some utilities, at first a pioneering handful and then a swelling herd, started to realize that distributed benefits could have major business value. The genesis of this discovery was not so much curiosity about the economics of scale as it was the approach to "technology-push" (102):

> In 1988 several PG&E researchers expressed the idea that placing small photovoltaic generators at weak points of the utility system might somehow be helpful. No one knew how to quantify that value, or how to find these weak spots, but there was hope that this "grid-support" application might be an early niche for cost-effective applications of solar technology.
>
> We have come a long way from those early concepts. While photovoltaic technology is still a prime candidate for grid-support applications, it has now been joined by solar thermal electric, small generator sets, fuel cells, battery and [superconductive]... storage units, and even targeted demand management programs. Grid-support applications, perhaps augmented by an intelligent distribution management system, have coalesced into the "Distributed Utility" (DU) concept....
>
> The concept's appeal is clear. Utilities today are investing substantial amounts of money in transmission and distribution. Yet these assets are poorly utilized, since they are built for infrequent, but large[,] peak loading. To date there has not been any other alternative to line reconductoring, larger transformers, and line extensions for reinforcing or expanding distribution service. The DU concept suggests that perhaps small amounts of generation, storage, and/or specially tailored customer efficiency programs can be used to handle these infrequent peaks, while simultaneously being dispatchable for system-wide needs as well. It is almost too good to be true: a way to get double-duty from new generation, storage, and/or customer efficiency programs....

Among the early-1990s projects related to this idea and managed by Pacific Gas and Electric Company's visionary research director Dr. Carl Weinberg (734), the Delta

Project explored how targeted demand-side resources could defer transmission expenditures (523), while photovoltaic studies at the Kerman substation and elsewhere found that distributed benefits could double (576) or treble (595) the value conventionally calculated from energy and capacity savings alone. The leader of PG&E's demand-side activities, John C. Fox, then moved to senior positions at Ontario Hydro, where similar efforts soon demonstrated up to 90% capital savings from targeted distributed resources.

Such startling results help to explain why the most useful official effort at a synthesis—the Distributed Utility Valuation Project of the National Renewable Energy Laboratory, the Electric Power Research Institute, Pacific Gas and Electric Company, and Pacific Northwest Laboratory—was halted in summer 1995 after several years' effort, despite its puny budget of under $1 million per year. In summer 1993, its leaders had forthrightly written (596):

> The problem is both exciting and frustrating—exciting because, after considerable scrutiny, the concept still appears feasible and economically attractive; frustrating because every answer leads to three more questions. Perhaps this is to be expected from an attempt to turn the utility system in-side-out in search of a more efficient way of operating.
>
> We have only uncovered the tip of the iceberg at this point; the more we explore the DU concept, the more interesting it becomes, and the more encompassing our discussions and explanations need to be.

In the following two years, those researchers became so successful that what they found was far too interesting to publish. In essence, EPRI, PG&E, and other utilities became so engaged in distributed utility concepts that they felt it would be commercially imprudent to share their data and insights with competitors—an innovative concept that was itself starting to emerge in discussions of the future shape of the electricity industry. After all, if the project's hypothesis[16] were confirmed, it would represent "a fundamental shift in electricity production economics" and "could restructure [the] power industry." (6)

This was an excellent sign for market recognition of distributed benefits' value, but bad for public understanding: it meant that major public goods could actually reach the public only indirectly, inferentially, and rather slowly, as competitive market conditions were created and exploited. (In the wake of the failed California experiment with restructuring, and the less-than-truly-competitive restructuring underway in some other states, the process of revelation may be slowed even more.) By the mid-1990s, however, the increasingly intriguing little secret of valuable distributed benefits was starting to leak out. A few utilities' and many industries' brave and largely successful experiments with decentralized power sources were gaining notice and emulation. Starting in 1994, Detroit Edison's new CEO, Anthony Earley, started to make distributed resources the core of its growth strategy, executed through a separate business unit, DTE Energy Technologies, that seeks to "become the dominant player in the distributed energy market." (354)

The good news about the mid-1990s utility experiments was that they demonstrated that the basic principles of "distributed benefits" were sound and could make small be profitable. The bad news was that convincing field *measurements* of those benefits were few, sketchy, and seldom well reported, for four reasons:

[16] Namely, that the distributed utility is cost-effective, is technically feasible, and offers economic benefits "sufficiently large to warrant changing the way utilities plan and operate." (98)

- obviously cost-effective applications, like remote photovoltaics, weren't worth measuring because everyone could see why they made sense without needing any analysis;

- the more arguable applications that were really interesting would require such a complex, sophisticated, and fine-grained analysis of local conditions to propose and test their detailed cost-justification that this would cost too large a fraction of the small project's budget;

- internal skeptics of distributed resources could and did use budgetary pressures to de-fund such efforts and hence reduce competition with their own favorite options; and, most importantly,

- internal advocates of distributed options often felt that in a more competitive market environment, the proprietary value of such information to their own companies outweighed the public interest in disclosure, so any detailed measurements should remain secret.

Nonetheless, today one overriding factor forces the electricity industry to pay very careful attention to "distributed resources" (DR), making them suddenly relevant and fashionable. That is the movement begun in the mid-1990s to restructure the electricity industry and to foster wholesale and even retail competition among providers. This effort to scrap a century of industry structure and regulated-monopoly principles is *de facto* creating an utterly new business psychology and logic. As explained in Section 1.2.12.3, it could enable some previously suppressed DR benefits to express their market value. It could also encourage technical trends, such as bidirectional distribution automation (DA), that would make DR deployment much easier and cheaper: indeed, DA and DR are as intimately related as the two sides of the same coin. And it

forces all utilities to consider whether DR assets might better meet the diverse customer needs that will increasingly determine competitive success.

For these reasons, interest is rising rapidly. The Electric Power Research Institute—think-tank of the North American utility industry—sponsors regular proprietary conferences on DR, publishes a topical newslet-

> **Distribution automation** "matches load and supply and attempts to improve system performance over the entire load cycle. To achieve the benefits of automatic control, it is necessary to develop the appropriate system state model and determine the optimal feedback law used to modify the natural system inputs." This requires research on, among other topics, "modeling approaches, control formulations and hierarchies, control algorithms, communication requirements, and data requirements," plus a detailed understanding of how best to handle automatically most or all faults in or entering the distribution system. The aim of the automated controls is thus to "achieve peak-shaving and valley-filling through generation, storage and load control; ensure proper voltages and minimize losses through reactive power controllers; and respond to either generation [or transmission] outages or overloading by issuing the appropriate load transfer commands." (113)

ter, issues proprietary reports for its members (www.disgen.com), and has spun off a consulting house (www.primen.com). The U.S. Department of Energy, under new leadership, similarly conducts modest but excellent research, mainly at the National Renewable Energy Laboratory, where it established a Center for Distributed Power in early 2001. Skilled consultants, many formerly with pioneering DR utilities like Pacific Gas and Electric Company (PG&E), have developed and now sell proprietary models for analyzing distributed benefits. (Such consultants can offer important insights, but sell them for a living and hence regard public interest, open literature efforts like ours as direct competitors, regrettably limiting our cooperation from this knowledgeable and otherwise friendly group.) In 1997, E SOURCE,[17] the premier source of technical and strategic information on electric efficiency, launched a special service focus-

[17] At 3333 Walnut Street, Boulder CO 80301, 720/548-5000, fax -5001, esource@esource.com, http://www.esource.com. E SOURCE was incubated within Rocky Mountain Institute 1986–92, spun off in 1992, and sold in 1999 to Pearson LLC, parent of the *Financial Times* and many technical information services. In 2001, its owner FTEnergy was resold to the Platt's division of McGraw-Hill.

ing exclusively on distributed resources. By 2000, distributed utilities were among the most popular topics of international energy conferences, and industrial inquiries and alliances focused on this business opportunity had emerged in places from China to Brazil, Australia to Switzerland.

By the late 1990s, distributed generation was already important to some countries' electricity supply. An international body, CIRED, (350, 555) reported examples of actual distributed generation (in nameplate MW) already installed around the world by 1997. While the data below appear incomplete and sometimes inconsistent with IEA statistics on non-utility generators, it is striking that at least five industrial countries were found to get 12–28%, and a further three got 7–9%, of their system capacity from distributed resources—even before the recent rapid expansion of European windpower.

In 2000, power-equipment giant ABB began switching its strategy from big to small plants, though it then hit a downdraft for other reasons. By 2001, Standard & Poor's *Creditweek* was commenting that distributed

generation would probably start to put downward pressure on pure grid companies' credit ratings over the long term (756). *Business Week* was commenting that "many energy analysts and market watchers predict that distributed power could account for as much as one-fifth of all electric generation in the U.S. by 2010." (271) By 2001, distributed generation had become a booming commercial reality and a potential savior of power-short California. In short, the market verdict that huge, centralized power stations are no longer the most cost-effective choice is belatedly starting to spawn an infrastructure of understanding what *is* the right size.

The interest in scale issues is at long last catching up with the subject's potential importance. Besides widely available and applicable field data, there's only one thing missing: *synthesis*. Nearly every paper at nearly every DR conference focuses on only one kind or a few kinds of distributed benefits without putting them in the context of all the rest. The proprietary quantitative models appear to be similarly short on context and comprehensiveness. Confusions of concepts and terms are rife. Commercial

Table 1-3: Illustrative international distributed generation capacity installed by 1997										
	Dispatched diesel & gas turbine	Non-dispatched cogeneration	Wind	Steam	Hydropower	Photovoltaic	Other, incl. waste	TOTAL distributed capacity	System capacity	System peak demand	TOTAL distributed capacity
Denmark		2,000	1,450					3,450	12,150	6,400	28
Netherlands		4,736	427		37		80	5,280	18,981	12,000	28
Poland		3,000			2,008			5,008	33,400	23,500	15
Belgium	214	1,174	5		97		448	1,938	14,693	11,972	13.2
Australia	718	1,747	5	2,754				5,224	42,437	29,841	12.3
UK		3,732	300		1,494		421	5,977	68,340	56,965	8.7
Spain		2,500			1,500 (all renewables)			4,000	50,311	27,251	8
Germany		2,800	1,545		3,333	17	904	8,599	114,100	106,290	7.5

secrecy casts its cloak over all work in this field—even the most basic codification and synthesis of concepts.

This pervasive lack of public synthesis leads naturally to understatements of the importance of the full range of distributed benefits. For example, a fairly typical 1994 analysis (54) found that

> Integrating renewable energy systems into electric power distribution systems [in seven utilities studied] increased the value of the benefits [of those distributed options] by about 20 to 55% above central station benefits in the national regional assessments...[with some values ranging up to] near[ly] 80% for a case where costly investments were deferred. In general, additional savings of at least 10 to 20% can be expected by integrating at the distribution level.

However, only careful readers would note that "the distributed utility benefits considered in this study are not necessarily a complete set" (55), consisting of only about seven of the roughly 207 benefits considered in this study. The rest aren't mentioned. Similarly, virtually all industry studies of distributed benefits pick one or another small subset of the full range of distributed benefits. None seeks to identify, let alone quantify, all those benefits so readers will understand what is missing from the subset and how much it matters. Like the fable of the blind men who each touch part of the elephant, they fail to give a proper picture of its nature and size as an integrated whole.

These conditions made it important and urgent for an independent party to try to organize in an orderly, *public* framework *all* the relevant links between scale and value. This book was undertaken with precisely that ambitious goal. It extended what is by now a 27-year line of inquiry launched by

the *Foreign Affairs* paper (410), *Soft Energy Paths* (414), *Brittle Power* (442), and the summary of first steps in the present research (441). It is meant to stimulate many other students of this subject, in many institutions, to improve on RMI's initial work. We hope readers with specialized knowledge will find this book limited not by our imagination so much as by our restricted access to proprietary data—and perhaps helpful in organizing those data into a more inclusive and systematic form whose market value will make it gradually permeate industry practice.

Before becoming immersed in the specific details of distributed benefits, we address some broad semantic and philosophical issues that if left unsaid might cause confusion. Then we shall survey the existing U.S. power system and the menu of distributed resources, and conclude Part One with further background issues and an introduction to the fine-grained perspective that reveals the value of distributed benefits.

1.2.7 Proximity: how close to home?

A century ago, the first electric power systems powered a building, a neighborhood, or a town. But by around 1980, the average power station delivered its output over an average distance of roughly 343 km (213 miles). This difference of proximity creates many important technical, economic, and political-economy differences that are often confusingly lumped together into abstract debates about the alleged virtues of "decentralization." But that term, as science historian Langdon Winner has remarked, is a "linguistic trainwreck," defining itself by what it is *not*. "Worse," notes the Pentagon study, *Brittle Power* (447), "it is ambiguous. In the literature of appropriate technology, alternative development concepts, and 'post-industrial' patterns of settlement, production, and politics, the term 'decentralized' has been used to mean everything from 'small' to 'agrarian/utopian' to 'individually controlled.'"

The study continues: "Even confining the discussion to energy and specifically to electrical systems—not to industrial, urban, or governmental patterns—still leaves at least eight dimensions of 'decentralization' to be distinguished." Each of these dimensions is linked to the rest; each is a spectrum, not a pair of polar values; and each depends on a particular context of use (448):

> An energy system [that] is small in the context of running smelters, for example, may be large if the use is running a television. A system [that] is distributed across the country may nonetheless be clustered in localized clumps, not spread evenly. A device [that] is comprehensible to farmers may be mysterious to physicists and vice versa. A source [that] is local in the city may be remote in the countryside (and possibly vice versa). Accordingly, it is important to remember, even in a specific context, that all the dimensions of "decentralization" are relative, not absolute.

The first four dimensions of "decentralization" are essentially technical and geographic (448)

- *unit scale*—the output capacity or output rate of a single unit of supply;

- *dispersion*—whether individual units of supply are clustered or scattered, concentrated or distributed, *relative to each other* (but this property of spatial density does not specify the scale of each unit, nor how they may be interconnected);

- *interconnectedness* (which likewise says nothing about unit scale, dispersion, or distance from the user); and

- *texture*—ranging from monolithic (comprising inseparable parts, like a central thermal station) to granular (combining separate multiple modules with analogous functions, like turbines in a windfarm), regardless of unit scale.

It is seldom necessary to apply these distinctions in practice, but they may help to reduce the confusion arising from new and inconsistent usages still common in this new field. For example, a standard U.S. text unhelpfully defines "distributed generation" to include "all use of small electric power generators, whether located on the utility system, at the site of a utility customer, or at an isolated site not connected to the power grid"—but distinguishes "dispersed generation, a subset of distributed generation," as referring to "generation [typically 10–250 kW/unit] that is located at customer facilities or off the utility system." (761) Apparently if it's on a utility system, it's not considered dispersed even if it's small or far-flung. Similarly, the standard British text for power engineers is entitled *Embedded Generation*—a term that "comes from the concept of generation embedded in the distribution network while 'dispersed generation' is used to distinguish it from central generation. The two terms can be considered to be synonymous and interchangeable.... There is, at present, no universally agreed definition of what constitutes embedded or dispersed generation and how it differs from conventional or central generation." (359)

Adding to the confusion, the attributes commonly used in Europe to distinguish distributed from central resources include whether they're centrally planned by a utility, whether they're centrally dispatched, how big they are, and whether they're normally connected to the distribution rather than the transmission system (359). These are pragmatic features important to power engineers, but they do not capture the four dimensions noted above, nor those described next.

1.2.8 **Control: the center and the periphery**

Four additional dimensions of "decentralization" are sociological and psychological:

* *locality*—used here to mean not a technical property of a unit in isolation, but rather expressing its users' perception of its physical and social relationship to them (whether they feel remote from it or close to it physically, geographically, or both)—again, regardless of unit scale;

* *user-controllability*—how closely and readily users can autonomously control their use of the device, and whether decisions about it are participatory and pluralistic or more dominated by a remote or unaccountable technical elite;

* *comprehensibility*—Whether a unit or system is a tool or a machine—whether it's vernacular and understandable enough for ordinary people to make an informed choice about whether they want it (even if they couldn't build one themselves); and

* *dependency*—how far users feel a humiliating inability to repair, adjust, or modify the device, to control its presence or price, to obtain it from diverse and competitive sources, and to serve and suit primarily their own interests rather than the possibly different interests of its providers.

These issues of accountability—of the tension between the sovereign citizen or consumer and what Jefferson called "remote tyranny"—are hard to measure in engineering and economic terms, but that makes them no less important. They determine acceptance, social order,[18] even the longevity of political careers and of governments. While energy technologies with modest unit scale and dispersed geography do not automatically lead to a more democratic, accountable, or pleasant society, there is good reason to suspect that energy technologies with the opposite attributes may have the opposite tendency (414) Investments that take proper account of the political economy of siting, customers, corporate reputation, and brand equity will therefore pay as careful attention to these "fuzzy" social-science issues as to engineering and cost attributes. We return briefly to this topic in Section 2.4.10.

While on this subject, it is important to address the caricature, common in the 1970s (502) and occasionally still encountered, that small-scale and dispersed energy sources, like the photovoltaic arrays that are popping up everywhere from military hardware to highway signs and roadside telephones, are somehow a covert plot to "decentralize society" in order to cause fundamental changes in our way of life and the dissolution of national power. This resembles the canard that by suggesting an appropriate choice and mix of scale for the range of tasks society needs done, one is seeking, romantically but unrealistically, to power an advanced industrial society with billions of backyard windmills, analogous to the micro-steel-mills of China's ill-fated "Great Leap Forward."

These bizarre fun-house-mirror versions of the appropriate-scale thesis can be quickly dealt with. There is no evidence that smaller-scale or decentralized energy systems would require people to live or to manage their affairs in a different fashion; rather, such

[18] As at Gorleben, where nuclear waste shipments in the 1990s—before the Federal government finally adopted a schedule for phasing out nuclear power altogether—entailed the largest police operation in Germany's postwar history.

technologies preserve a complete range of choice in social and political structure and scale. "The confusion between the choice of technologies and the choice of patterns of social organization arises in part from sloppy terminology...and in part from some advocates' failure [chiefly in the 1960s and 1970s] to distinguish their technical conclusions from their ulterior political preferences." (450) This report, like all RMI's and the senior author's previous works, considers the optimal scale of units of electrical energy supply only in the context of "how to construct an energy system with maximal economic (and national-security) benefits to meet the needs of a heavy-industrial, urbanized society—a society, moreover, that is assumed to wish to continue rapid economic and population growth" (450) and to sustain all the historic goals and structures of a democratic market economy. Exploring other social goals or other forms of social organization is far beyond our scope. Appropriate scaling of electric power systems certainly does not restrict, and should expand, the range of such separate social choices.

Lest any offense be inadvertently given to those responsible for past choices of large scale, it is also worth clarifying that our suggestion here that optimal scale may differ widely from gigawatt scale (460)

> ...does not mean that decisions to build large plants in the past were always irrational. Rather it means that, taking all relevant economic factors into account, such decisions would no longer be cost-effective in today's altered circumstances. Nor does it deny that big projects may have real economies of scale in *construction cost per kilowatt of installed capacity.* But where this economy of scale exists, it is a gross, not a net, effect. It must be tempered by other effects[,] which may, for example, make each *installed* kilowatt of capacity *send out* or *deliver* less energy than at smaller scale. Other tempering effects may increase the

costs of other parts of the energy system, or they may increase indirect costs or inefficiencies. The object, after all, is to *deliver* energy—or, more precisely, to enable particular services to be performed by using energy—rather than merely to install the *capacity* to put the energy into a distribution system. The goal should therefore be to build the energy system [that] will perform the desired energy services at the lowest possible economic cost. If bigger technologies decrease construction costs by less than they increase other costs, then the[y]...are too big....Of course, there are still tasks for which big systems are appropriate and cost-effective....Mismatching scale in *either* direction incurs unnecessary costs...[but it] appears that a more sophisticated and comprehensive view of the economics of whole energy systems would lead to a very different balance of sizes between demand and supply.

1.2.9 Vulnerability: brittle power

Brittle Power, an extensive, 1,200-reference 1981 analysis based on these considerations— still probably the definitive unclassified discussion of domestic energy vulnerability— ascribes the ease of disrupting and difficulty of repairing centralized systems to their architectural qualities (419, 460, 467). Today's predominantly centralized energy systems (448):

- consist of relatively few but large units of supply and distribution;

- compose those units of large, monolithic components rather than of redundant smaller modules that can back each other up;

- cluster units geographically, for example near oilfields, coal mines, sources of cooling water, or demand centers;

- interconnect the units rather sparsely, with heavy dependence on a few critical links and nodes;

- knit the interconnected units into a synchronous system in such a way that it is difficult for a section to continue to operate if it becomes isolated—that is, since each unit's operation depends significantly on the synchronous operation of other units, failures tend to be system-wide;

- provide relatively little storage to buffer successive stages of energy conversion and distribution from each other, so that failures tend to be abrupt rather than gradual;

- locate supply units remotely from users, so that links must be long...;

- tend to lack the qualities of user-controllability, comprehensibility, and user-independence. These qualities are important to social compatibility, rapid reproducibility, maintainability, and other social properties...important...to resilience.

These attributes contradict the fundamental requirements for resilient design (445), to such a degree that as "a recipe for disaster, its design could hardly be more expert and comprehensive." (449) Around 1980 it was true, and it remains true in 2002, that a handful of people, for example, could shut off three-fourths of the oil and gas supply to the eastern U.S. in one evening without even leaving Louisiana. Electric grids were and remain more vulnerable still, as accidentally illustrated by a number of regional and national blackouts continuing to the present. And the study documented smaller attacks that, by the early 1980s, were already occurring somewhere every few *days*, and had been reported in more than 40 countries and at least 26 of the United States. By the 1990s, attacks on key nodes of energy systems ranked high on most military planners' target lists, as illustrated in the Persian Gulf and Kosovo conflicts. Such built-in brittleness, however, was not necessary. An extensively documented synthesis of design principles drawn from biology and from military, nuclear, aerospace,

and other engineering disciplines revealed the practical potential for a very different architecture in which major failures would become impossible (442). The same qualities that can create such a highly secure energy supply system (§ 2.4.10.1) also happen to be compatible with the economic thinking behind optimal scale: that is, essentially the same distributed architecture that creates resilience can also reduce system cost. In an age where causes of serious disruption do not seem likely to decrease, and may on the contrary become endemic and acute, reexamining optimal scale can also offer important opportunities to make society safer from devastating disruptions. That it can also save money is especially good news, because then national security can be improved not at a cost but at a profit, and therefore can gradually be done in the marketplace just by choosing the best buys first.

The benefit to national security is not what sells micropower. Yet as Assistant Secretary of Energy David Garman says (267), "Aside from its obvious environmental benefits, solar and other distributed energy resources can enhance our energy security." Garman adds:

> Distributed generation at many locations around the grid increases power reliability and quality while reducing the strain on the electricity transmission system. It also makes our electricity infrastructure less vulnerable to terrorist attack, both by distributing the generation and diversifying the generation fuels. So if you're engaged in this effort, it is my view that you are also engaged in our national effort to fight terrorism.

1.2.10 Diversity: monocultures *vs.* ecosystems

Evolutionary biology and ecology—which

the unfortunate dinosaurs experienced but had no opportunity to study—teach the transcendent value of diversity for sustaining resilience in the face of surprises. The history of energy systems teaches the same lesson. Over-dependence on any particular fuel, source, route, or technology has typically led to exploitation or embarrassment. Weather, climate change, wars, terrorism, epidemics, technical failures, strikes, market instabilities—whatever the cause, the disruption should be limited *by design* to affecting only an acceptable fraction of one's total supply capability. This requires, however, the sense to avoid simply substituting one set of risks for another when other options can avoid them all. It also risks indiscriminately trying one of everything whether it fits or not—as one might choose one item from each section of the vast restaurant menu of energy options, not because it will improve the meal but through mere indecision.

Appreciation of these lessons leads prudent planners to pay a premium for diversification: to prefer a slightly costlier system that is virtually guaranteed to work, come what may, to a cheaper one whose monocultural choices make it prone to major shutdowns. That much has been known and widely practiced for decades if not centuries. But a riper examination of optimal scale now opens for serious consideration a large range of technologies, chiefly renewables, that are inherently far more diverse—and, incidentally, far less prone to external disruptions—than traditional centralized resources. Many, such as onsite solar heat and photovoltaic power, are so relatively simple and close to the user that otherwise dominant failure modes, such as grid failure, can become unimportant or irrelevant. Thus counting distributed benefits can

greatly expand the policy maker's palette of affordable choices, indirectly increasing the diversity and hence resilience of energy supply, and permitting the capture of the widest range of advantages with the most limited range of flaws. Moreover, as Section 2.2.6 will describe, this engineering goal of diversification can also gain important diversifications of *financial* risk.

1.2.11 Governance: concentrated *vs.* dispersed

Another important lesson of the biological metaphor is that ecosystems disperse their control into a myriad local and systemic feedback loops rather than a rigidly centralized hierarchical control. The human body does much the same: breathing, heartbeats, digestion, etc. are routinely controlled by local physiological and endocrine feedback mechanisms rather than requiring constant control by the higher functions of the brain, which usually has better things to do.

Dispersed control is one of many important design lessons that the world of the made is increasingly borrowing from the world of the born. Kevin Kelly's provocative book *Out of Control* (378) surveys the expanding range of technical systems in which the biological control strategy is advantageously displacing the mechanistic, hierarchical one, simply because it works better: several billions of years' design experience has created an extremely effective and resilient bio-logic.

Though few economists know much about biology, economic markets are supposed to work (though traditional utility systems were never markets in this sense) on the same decentralized, "out-of-control" principle. Ever-varying prices reflect the instantaneous balance between supply and demand and hence instruct everyone how much of each item is worth making or using, so those decisions reequilibrate supply and demand. This approach has been successfully mimicked within technical systems, notably by "agoric" (marketplace) programming. This technique uses shadow prices to allocate the resources of a computer in real time to achieve the user's computing priorities. Conversely, "genetic algorithms" simulate biological evolution to refine the design of computer programs or technological designs by calculating numerous "generations" of successively improved outcomes whose "reproductive" success is aligned with their functional fitness. Both these programming techniques underscore the conceptual convergence between many market-economics and biological concepts.

Considerable dispersion of control has already occurred in the U.S. natural-gas and airline industries: the latter used flexible pricing to increase asset utilization by 30% in a decade (515). Now important advances on these lines are being made in electric load management too. The electric system typically disperses its control through permeation by real-time price information, which the BBC, for example, has long broadcast every half-hour, which some utilities e-mail via AT&T's EasyLink, and which Georgia Power offers by EnerLink electronic transmission directly from its real-time dispatch data (515). (Some electric systems also add, and any could add, specific information about voltage, frequency, phase, and site-specific information on weather, occupancy, etc.) Numerous electric utilities are already implementing dispatched or locally intelligent controls for highly distributed

demand-side resources, using diverse communications methods, control protocols and interfaces, pricing schemes, and end-use devices (268, 340, 575, 696). Interconnection LLC, which operates PJM (the first fully functional regional transmission organization in the U.S.), joined Converge Technologies in 2001 to use cellphone technology to connect small power producers (up to 10 MW) to the dispatcher, eliminating costly communications equipment (776). But an important and underappreciated point (242) is that *there is little if any basic difference between applying such controls to demand- and to supply-side resources*, whether dispatched according to price or to direct command.

Despite this encouraging analogy, power engineers are naturally nervous about any scheme that loosens or abrogates central control over hundreds of extremely large, delicate machines, each taking about a billion dollars and a decade to build, that must continuously rotate in exact synchrony throughout an intricate net of aerial arteries spanning half a continent. Their concerns are legitimate, and come from people who are directly responsible for those machines and for the vital national functions supported by their output. As the Energy Information Administration summarizes (174), dispersing control through telecommunication of price or technical variables or both would redefine what an "electric power system" traditionally is (emphasis added):

> An electric power system is a group of generation, transmission, distribution, communication, and other facilities that are physically connected and operated as a single unit *under one control*. Transmission and distribution lines and associated facilities are used to transmit electricity from its point of origin (the generator) to the ultimate consumer. Although, due to its phys-

ical characteristics, electricity flows along all available paths, it follows the path of least resistance. The flow of electricity must be closely monitored to ensure that sufficient generating capacity is available and on-call to satisfy all demand (load) for electricity placed on the power system. In addition, for system standardization and reliability purposes, the flow is maintained at a frequency of 60 cycles per second.

The flow of electricity within the system is maintained and monitored *by dispatch centers* having control and security responsibilities. Historically, the dispatch center inventoried and prioritized all generating capacity available to it, tracked transactions involving the buying or selling of either electric power or capacity, monitored current load, and anticipated future load on the system. In the future, this responsibility may be handled differently. How, is now being determined by participants in the new electric power industry.

Advances in power electronics, microelectronics, telecommunications, control systems, control theory, and institutional arrangements now make it feasible and often profitable to distribute the control intelligence of the grid—determining the flow of power into and through the grid, and regulating associated matters such as voltage, frequency, and phase—from large regional dispatch centers to a much more decentralized pattern focused on individual substations. Further decentralization, moving control from the substation to or at least toward neighborhoods and even individual

customers, may also be possible and worthwhile. Regardless of the degree to which the execution of control functions is dispersed, the information used to guide those functions is not inferior to the information available to a traditional central dispatcher; indeed, it may be far more locally relevant and fine-grained. The difference is largely in the psychological perception (and ultimately perhaps also the physical reality) of whether that information flows from the top down or from the bottom up.

Any degree of decentralization of control would be a technically and psychologically major step in the decentralization of the electricity system, because it implies a transition from a centrally controlled system to an "out-of-control" one. Actually, in *neither* world are outcomes fully predictable; if they were, regional blackouts wouldn't happen, and problems like Pacific Northwest loop flow wouldn't exist either (see box). Relatively recent severe disturbances in which grid voltage has suddenly collapsed, or in which equipment faults have interacted over huge distances (Arizona to British Columbia and Colorado to San Francisco), show that the idealized, linearized world of conventional grid control theory is quite different from the nonlinear and even chaotic reality (281). But at least the direct-control, centralized-dispatch world provides the *illusion* of control, and is hence a more comforting world for power engineers than its laissez-faire, localized, automated, bottom-up, "self-organizing" alternative. The technical issues of decentralizing grid control are all fascinating, and many of them are unresolved. This is a frontier topic with strong emotional as well as technical dimensions.

Happily, the complex issues of control

Loop flow is the flow of electricity through two or more transmission paths when it was intended to flow along only one path. It commonly occurs in the Pacific Northwest. Since electricity follows the line of least resistance, some power meant to flow south through the Pacific intertie from the Bonneville/BC Hydro system may instead flow through parallel lines through other systems, so it can end up meandering east through Idaho, south through Colorado, and thence to the Southwest by a circuitous and unintended route. Under suitable loadings, it can then even flow back north again. Power sloshing around in this loop can cause instability and contribute to operational failures; and of course it may limit the amount of power that the system through which the inadvertent loop flow occurs can transfer for its own purposes. Power flows use the same capacity whether they occur intentionally or not, so by following a circuitous rather than direct route, they tend to use grid capacity unintentionally and often inconveniently.

decentralization can be deferred or finessed for as long as desired until empirically resolved (unless restructuring accelerates the issue by bringing in new competitors that dispatch themselves). Using the means described in Section 2.2.10, and assuming the continuation of traditional monopolistic institutional arrangements, *distributed resources can be efficiently and reliably operated indefinitely under direct central-dispatch control*, just as generators, switchgear, substations, and other key elements of the grid are operated today. The only difference is that the dispatcher would use modern telecommunications (wire, fiber, or wireless) to control a much larger number of points spread around the network, rather than using similar telecommunications to control only a modest number of points at higher hierarchical levels, such as power stations, transmission switchgear, and substations.

Direct central-dispatch control of many dispersed, customer-level devices *is already a routine reality* for dispatchable load management on water heaters, air conditioners, etc., and for dispatchable onsite backup generators. (Many utilities already use telecommunications-linked hardware to start, run, and dispatch backup engine-generators and fuel-cell generators at customers' sites.) Dispatching distributed resources can stabilize transmission grids at significantly lower cost—in one case examined, with 28% less device capacity—than a centralized option (399). These distributed technologies are well-established and highly cost-effective: indeed, once "smart" retail meters are installed (which quickly pay for themselves through a variety of benefits), using their wireless spread-spectrum communications capabilities to add load control to end-use device can cost as little as ~$10 per point.

The communications technologies, having been originally developed for military data and voice traffic, are highly reliable, resilient, and secure.

Dispatchable load-management and stand-by-generator resources typically have availability in the high 90s of percent—comparable to or better than the availability of conventional generating resources. The difference is that the one-to-several-percent random failure of distributed resources to respond to dispatch commands would matter and cost far less than a corresponding unavailability of large-scale resources. More of the dispersed units could fail, but if properly integrated with more modern distribution and control equipment, those failed units would comprise far less capacity, more widely dispersed, affecting fewer customers, and with more options for alternative supply from either local or remote resources. This logic suggests that we can examine the appropriate scale of electric resources, and evaluate most of the benefits of distributed resources, *without* having to resolve longer-term issues of how those resources' technical operation is to be controlled, in which direction the information predominantly flows, and who feels or needs to feel "in control."

1.2.12 Transition: the forces of renewal

The idea that right-sized energy technologies may make sense and make money at the same time is hardly new. In 1978, Congress's independent Office of Technology Assessment found (534–5) that

> If energy can be produced from on-site solar energy systems at competitive prices, the increasing centralization which has characterized the equipment and institu-

tions associated with energy industries for the past thirty years could be drastically altered; basic patterns of energy consumption and production could be changed; energy-producing equipment could be owned by many types of organizations and even individual homeowners.

But this renewable-energy context was only part of a larger picture. A year later, a diverse panel of government, industry, and academic experts found (18):

> [D]ecentralized [electricity] generation systems are likely to confer major consumer benefits. These may include shorter lead times in planning and construction, easier siting, reduced capital requirements, greater efficiency in fuel use, and reduced vulnerability to fuel shortages....We find a number of such options are at, or are approaching, a state of technical and economic competitiveness with larger centralized systems.

The panel also found that "on balance,...the climate for the development of small, diversified, and dispersed supply and [efficiency]...options is likely to improve." Just a year later, those improvements had sufficiently impressed the nation's third-largest investor-owned utility, coal- and nuclear-oriented Southern California Edison Company, that it announced it would henceforth aggressively pursue an efficiency/renewables strategy as the cheapest option for future expansion. That imprimatur helped to spark a 5–6-year run of commercial success and rapid growth for dispersed technologies: during 1979–86, the United States got five times as much new energy from savings[19] as from all net changes in supply, and renewable output increased by roughly one-fourth, or 7% as big an absolute contribution as savings. Those very successes, and their ability to outpace the expansion of central supply technologies, contributed to the 1986 oil-price crash (469–70). That in turn drove

down the deregulated U.S. prices of natural gas; slowed efficiency investments; and speeded commercialization of combined-cycle gas turbines, deployed largely by the new and fast-growing independent generating industry (§ 1.2.4). In 1982–84, only 2.5% of the U.S. generating capacity in service was non-utility. By the end of 2000, thanks in part to PURPA (and to favorable California contract terms based on assumptions of high future prices for oil and gas), that figure had soared to 21%, and 209 GW of non-utility capability was in operation—some of it generating useful heat for onsite use as well as power. Non-utilities were then planning to build about as much new capacity (9 GW over three years) as utilities were adding (3 GW in 1998). Both independent (non-utility) generators and distributed generation had become powerful market realities.

By 1999, with 16% of all U.S. electricity coming from non-utility generators, most utilities were rapidly exiting from the business of ordering and building power stations, and many utilities, concerned about the competitiveness of older plants in a restructuring marketplace, were trying to sell the ones they'd previously built.[20] The results were revealing. Free-market sales of U.S. nuclear plants realized about a tenth of book value (essentially their fresh-fuel and uncontaminated-scrap value), central plants were discounted well below replacement cost, and only combined-cycle gas plants sold at or above replacement cost. (Non-nuclear plant sales averaged about twice book value.) With its legal monopoly and monopsony under assault, its expertise in building and running giant plants decreasingly relevant, its obligation to serve starting to erode, and its balance sheet less compelling, the tradi-

[19] In the crude sense of reductions in aggregate primary energy consumption per dollar of real GDP.

[20] This is gradually causing the historic plant stock to be marked to market value, which can be higher or lower than book value for both generating and transmission capacity—at least until the market fully internalizes the implications of distributed resources and their benefits.

tional utility represented a concept whose meaning and future were becoming steadily less clear.

1.2.12.1 New technologies

Meanwhile, more technologies than just combined-cycle gas turbines were getting better and cheaper. One-fifth of non-utilities' 98 GW of capacity at the end of 1998 was powered by renewable or waste sources. Windpower, the world's fastest-growing energy supply source (averaging 27% annual growth in the 1990s), was officially recognized at both state and Federal level as the cheapest new generating resource in appropriate sites; in 2001 it added 1.6 GW, more than twice its biggest previous annual increment. Both wind turbines and photovoltaics (the world's second-fastest-growing source)[21] continued relentlessly down the standard "experience curve" of higher volumes yielding lower costs. So did many other distributed renewable sources. Royal Dutch/Shell Group, in widely noted 1995–96 planning scenarios, felt it was therefore plausible that renewables could supply half the world's primary energy by 2050, just through direct cost competition as niche markets expanded production volumes. Within a few years, this was considered highly plausible if not conservative.

Other supporting technologies emerged too. Photovoltaic shingles, standing-seam metal roofs, windows, and other integrated roof and wall structures added important benefits from saved construction materials and labor. Power electronics made many renewables more convenient to integrate stably and reliably into the grid. By late 1996, advanced inverters the size of a cigarette-

Table 1-4: Sales of U.S. generation assets, 1997–98*		
Type of plant	**Amount sold**	**Selling price ($/kW)**
Gas-fired combined-cycle and cogeneration	1 GW	$900–$1,400
Coal plants	16 GW	$500–$1,000
Gas-fired simple-cycle or condensing	16 GW	$200–$370
Wind energy plants	0.16 GW	$240
Nuclear plants	2 GW	$30–$100

*All sales January 1997 to early September 1998. A total of more than 35 GW was sold for $16.5 billion. Analysis by Prof. John Byrne (Center for Energy & Environmental Policy, U. of Delaware, personal communication, 24 August 2000), based on EIA, The Changing Structure of the Electric Power Industry 1999, fig. 11, ch. 6, and separating out combined-cycle/cogeneration plants using data from www.energycentral.com. There were only two nuclear sales in the data set. Some subsequent nuclear sale prices were higher as short-term regional power shortages seemed to raise the value of old generators. In 2002, increasing corrosion and security concerns may again be depressing the market.

pack were making possible "vernacular" photovoltaic arrays that could simply plug into a wall socket just like an appliance, only backwards—they'd put electricity back *into* your house.[22] By 2001, photovoltaic manufacturer AstroPower had joined merchant homebuilder Shea to offer a solar electricity option in a 250-house tract development of efficient, solar-water-heated homes near San Diego. The PV homes sold better and made up two-fifths of the total despite their $6,000 (~1.5%) higher price. Some buyers want to add more PV capacity than the original 2 kW, and some non-PV buyers want to retrofit it. With a state credit covering about half the cost, net metering, and inclusion in mortgage financing, such systems cost less than the electricity they save (666). In Sacramento, five of the nine homebuilders offering PV roofs in 15 developments made them standard equipment. By the end of 2001, at least a half-dozen other major merchant homebuilders announced PV-powered on-grid housing developments across the country (530). And the technology continued to improve, with a 25-MW/y plant to make 13%-efficient tricolor amorphous PVs—

[21] Its global growth rate averaged about 19%/y in the 1990s, but about 26–42%/y as the 1990s were ending. PV sales grew 43% (to 288 MW) from 1999 to 2000 (481).

[22] In early 2000, such 100–250 W models as the Trace MicroSine and AES MI-250, produced in small quantities at correspondingly high prices, were withdrawn because of slow sales and because they hadn't been designed to meet the UL-1741 specification, which came into effect in November 2000 (D. Pratt, Real Goods, personal communication, 26 December 2000). European versions remained available. Updated U.S. versions are likely to re-enter the market around 2002–03 to exploit net-metering and other PV-intertie opportunities.

glassless and flexible—nearing completion.

In many respects, the technological future accelerated at an accelerating rate. The first large-scale commercial phosphoric-acid fuel cells came on the market in 1992, and the first commercial power-conditioning superfly-wheel in 1995, with more sophisticated ones in hot pursuit: by 2000, Trinity Flywheel was shipping three models. Packaged gas-fired microturbines in the tens-of-kW range began shipping in the late 1990s; by 2002, 30- and 60-kW modules were in widespread use and modules up to 400 kW were nearing production, all about 29–30% efficient without counting cogeneration potential. And in mid-2001, Target, a giant discount retail chain, started selling a 1-kW Bergey wind machine for home use as part of a new Target Energy Savers campaign to market innovative ways to save or produce home energy.

In 2001, Honda began testing with Osaka Gas Co. a 1-kWe + 3-kWt home-scale gas-fired engine generator cogenerating at 85% system efficiency and said to be as quiet as a home air conditioner. Comparable Stirling systems entering field tests were said to use 105 units of natural gas to deliver 10 of electricity and 90 of useful heat—enough to be cost-effective in much of northwestern Europe (778). And coming up fast on the outside track were fuel cells. By 1997, three independent studies found that high-volume mass production could cut the cost of very efficient, clean, silent, reliable, and modularly scaleable polymer fuel cells to only ~$30–50/kW (7, 60, 408). By 2000, many other analyses had reached the same conclusion, and some of the ~84 firms in the field were shipping initial pilot-produced units. By 2002, engineering for volume production was well underway; portable and home-scale units were entering the market (albeit at high initial prices); Electrolux had announced a fuel-cell-powered vacuum cleaner; and several flavors of solid-oxide fuel cell were emerging from the laboratory.

Market expectations were changing too. Offered such options—and others, such as power-conditioning superconductive storage loops, brought to market a few years earlier—many customers started more stridently to demand premium-quality power (§ 2.3.3.8), more individual control, and other unbundled forms of "mass customization." These were all attributes that this diverse stable of new technologies could deliver better than could their homogeneous, single-flavor predecessors. In particular, premium power quality and reliability could clearly be best delivered from onsite generators, because most power glitches came from the grid itself, and hence could not be avoided in any central-station-based model no matter how abundant and reliable the central stations might become.

1.2.12.2 Competitive restructuring

While this technological revolution was flowering, a potent political stew, spiced by a deregulatory ideology that peaked around 1992–94, was simmering in the heat of the 1980s rise in electricity prices—a rise largely caused, as we have seen, by the previous three decades' central-station (especially nuclear) construction binge. Around 1994, the stewpot boiled over in an unprecedented flurry of proposals for restructuring the entire electricity industry.
This effort was partly led by ELCON, an organization of large industrial customers that used 4% of all U.S. electricity. To be sure, the average real price of industrial

electricity had already fallen sharply (in 1994 it was 35%, and in 2000 it was 45%, below its 1982 high),[23] and 1994 electricity costs averaged only 1.3% of total manufacturing value shipped.[24] Nonetheless, the emergence of still cheaper marginal power from combined-cycle gas turbines made some firms passionately eager to capture that cheap power for themselves and to shift to other customers the burden of the uncompetitive, chiefly nuclear, older plants. This "big dogs eat first"[25] principle did not commend itself to the other customers, so battles were joined in many state regulatory commissions and legislatures over whether to preserve or abandon a century's practice of fairly sharing all utility assets, costs, and benefits among all customers.

Wholesale competition—required by Federal law since 1992 and being implemented under a massive 1995 Federal Energy Regulatory Commission (FERC) order—suddenly became an idea whose time had finally come. Many vertically integrated utilities were invited or required to "delaminate"—to sell off or separate their generating plants from their transmission and distribution businesses—so their own capacity would have to compete fairly against all comers. In a far more radical step, *retail* choice of supplier, and market-determined retail prices, were also widely proposed to replace regulated franchise monopolies as the dominant structural form in the United States. Even though wholesale competition already captures virtually the same benefit of competitive generation, and it can be captured only once, this idea enjoyed in some quarters the attractive political resonance of "choice" and "competition,"[26] and appeared to be gaining partial success in some other countries' field experiments, such as in Britain, Norway, New Zealand, and Chile. Ultimately, the European Union began to require increased competition ("liberalization") among its entrenched utilities, mainly large state monopolies, with mixed success but major psychological and institutional shifts (487).

The resulting U.S. debate was salutary and educational, and healthily shook up many moldy old paradigms and managements. But of course there were practical complications: for example, what to do with the perhaps $100–300+ billion worth of "stranded assets" nationwide, paid for by investors who thought they had been compensated for the use of their capital but not for its confiscation.[27] In contrast, many competitors felt those investors had been amply compensated for accepting such business risks as the obsolescence of their assets, and needn't be paid twice if those risks materialized and rendered old plants uncompetitive.

---

[23] According to the same source (202), the 2000 real price of electricity for residential customers averaged 26%, and for commercial customers 35%, below its 1982 high.

[24] The range was from 0.2% (SIC 21, Tobacco Products) to 3.4% (SIC 26, Paper and Allied Products). (162) It was, of course, tens of percent for a handful of industries such as light-metal smelting, but these generally held long-term low-price hydropower contracts.

[25] This phrase is due to Jon Hockenyos and Brian O'Connor (730).

[26] As with U.S. telephone deregulation, these mantras may become less attractive as telemarketers interrupt family suppers with "customer-choice" electricity offers that customers can't understand and may not want.

[27] A fairly typical estimate as restructuring entered the height of fashion, as reported in the 14 February 1997 *Energy Daily*, was the Resource Data International study estimating U.S. stranded assets at $202 billion ($147 billion of it held by investor-owned utilities, $33 billion by municipal utilities, and $22 billion by rural electric cooperatives). This total included $86 billion for nuclear plants, $54 billion for power-purchase contracts that had become above-market due to declining spot prices, and $49 billion for "regulatory assets" booked but deferred for potential future entry into rate base and then rendered unlikely ever to get there for actual cost recovery. Such analytic estimates, typically reflecting the present values of unamortized or undepreciated assets, are obviously sensitive to the assumed prices with which those assets' output must compete, typically set by combined-cycle-gas-turbine proxies that are sensitive to fluctuating natural-gas prices.

Diverse actors with a wide range of motives proposed, and some states approved, various ways to share symbolic partial write-offs with continued (at least temporary) socialization of these costs, sometimes refinanced with tax-exempt bonds, and relying on the silently ticking depreciation clock to make most of the costs disappear automatically. Some other states, believing they could defend the position that the investors had already been compensated for the business risk of changes in technology and regulation, restructured their power sectors without allowing stranded-asset cost recovery—a concept nearly unknown outside the historically peculiar conditions of the United States. Most states that did restructure, however, bowed to the incumbent utilities' superior political power, and not only guaranteed their recovery of stranded assets' sunk costs on terms comparable to or better than they enjoyed under the previous regulated regime, but also often entrenched their monopoly or monopsony status, all the while calling it "competition." The practical effect was to avoid political pain while also discouraging real competition, since customers who switched to other suppliers could save little or no money. Few switched.

Some other issues proved more profound and less negotiable. For example, the uniquely complex institutional form and legal context of the American utility system could hardly have been better designed if its primary goal were to make basic structural changes impossible. Those eager for "deregulation" soon discovered that the FERC, to which the Federal Power Act grants fundamental jurisdiction over "wholesale prices and no others," viewed "retail wheeling" as simply a transfer of jurisdiction over prices, terms, and condi-

tions from the state commissions to itself.

For good reasons (428, 433), few if any states embraced the classical "retail wheeling" agenda originally proposed (and widely misreported to have swept the country before any state had actually adopted anything like it). Nonetheless, major restructuring did start to occur, often on the more thoughtful lines adopted by Pennsylvania, Massachusetts, and Rhode Island. It usually comprised vibrant wholesale competition through power pools and bilateral trading arrangements, delamination, prohibition of self-dealing (favoritism to self-owned capacity), and special arrangements to continue to serve the "public-goods" interests—equity to and among investors and customers, health, safety, reliability, farsighted R&D, etc.—that a private market of self-interested individual bidders would otherwise be prone to overlook.

In startling developments still unfolding at this writing in summer 2002, but sure to influence profoundly the politics of U.S. restructuring, its initial implementation in California proved disastrous in 2000–01 (439). In 1999, electricity prices and availability had been normal. Yet from December 1999 to December 2000, the price of wholesale electric energy dispatched from the new statewide pool soared by 13-fold, and that of spinning reserve (bought a day ahead) by 120-fold, even though the pool's load rose only 0.7% and its monthly peak load *fell* 1.9%. The increase in wholesale prices was enormously more than could be explained by higher natural-gas prices. Since the two largest distribution utilities, Pacific Gas and Electric Company and Southern California Edison Company, were paying sky-high wholesale prices that they couldn't pass through to their retail customers (thus preventing customers from responding to a price signal until mid-2001), they both teetered on the edge of bankruptcy, which PG&E ultimately declared. Californians paid $7¼ billion for electricity in 1999, $33½ billion in 2000, and $7½ billion just in the first six weeks of 2001—the greatest interstate wealth transfer in U.S. history.

Complicating state and federal efforts to address the crisis, punctuated by rolling blackouts, was systematic misreporting by uninformed or, often, deliberately disinformed media:

- California was reported—chiefly by those anxious to resume building the large power plants that the market had rejected, but also by top state officials—to have added no capacity in the 1990s. As noted earlier, the state had actually added at least 4.5 GW, more than its nuclear capacity, but because it was non-utility-owned and largely distributed, nobody noticed. (The utilities themselves had not been adequately encouraged to enter distributed generation, some restructuring rules discouraged it, and some historic utility obstacles to private distributed generation have lingered.[28]) In addition, those opposed to environmental regulations falsely blamed environmentalists for blocking power-plant construction that in fact the private market didn't propose because it was uneconomic.[29] On the contrary, the state's main environmental organizations had pressed for 1.4 GW of new renewable and gas-fired capacity in the early 1990s and gotten the state to obtain attractive bids for it—only to be frustrated when the Federal Energy Regulatory Commission voided the auction, saying the capacity wouldn't be needed.

- California was said to be experiencing soaring demand. In fact, during 1999–

2000, the state pool's wholesale energy sales rose only 0.5%, and peak demand *fell* 4.5%; in 2001, they fell 4.7% and 5.4% respectively. Average 1990–99 kWh sales growth was 1.15%/y—half the growth rate of the state economy. (In Silicon Valley, often cited as the focus of runaway electricity growth, kWh sales grew 1.31%/y.) Per-capita electricity demand was nearly flat for a quarter-century. The year 2000 was exceptional—sizzling weather and economy (Gross State Product grew ~8.7%), and a leap year, which adds 0.3% to the length of the year—but not that exceptional. Compared with 1999, the state's average hourly peak load fell 4.6%, the average daily peak load rose 4.8%, and kWh sales grew about 4.6%. In short, nothing very unusual happened to demand in 2000, let alone earlier in the 1990s.

- The nonexistent soaring demand was claimed to be due to the huge electricity needs of the Internet, said by advocates Mark Mills and Peter Huber to total 8–13% and to be heading for 50% of total U.S. electricity use. (The actual figure for all office and network equipment is 2%, or adding phone-company switches and all the equipment's manufacturing energy, 3%, and is rising slowly if at all [10, 385].) This fiction was propagated as a disinformation campaign sponsored by the Western Fuels Association, the leading anti-climate-protection coal lobby, which sought to persuade the public that a prosperous digital economy required more coal-fired power plants. Though authoritatively rebutted (385), the lie continued to spread, reinforced by deliberately sown confusion between the quantity and quality of digital power needs and by occasional (though generally much exaggerated) local distribution requirements of server-farm data centers, which actually use less than an eighth of one percent of U.S. electricity and less than 1.6% of Bay Area electricity. Also overlooked was the provocative but plausible finding that E-commerce was probably *de*creasing total

[28] For example, even in summer 2001, a wealthy PG&E customer who installed a 31-kW, $400,000 photovoltaic system was told by PG&E to pay about $600,000 to upgrade its distribution equipment before he could connect it. That was because the California Public Utilities Commission, combating statewide power shortages, had raised the ceiling for free interconnection and net metering from 10 kW to 1 MW, but hadn't also updated "Rule 21," which requires customers to pay for upgrades needed at the end of a distribution circuit to accommodate their injection of power into the grid (86).

[29] Siting was never easy in California, but was perfectly feasible, as was quickly proved when the California Energy Commission dutifully issued licenses through the 1990s, and, when prices rose and economic interest returned, >6 GW from April 1999 to mid-2001, with >7 GW more poised to follow.

U.S. energy intensity, and possibly electric intensity too (568).

- The President declared that "We're running out of energy," and the White House claimed California was in the grip of a "desperate fuel shortage," reinforcing the supposed need to drill for oil beneath the Coastal Plain of the Arctic National Wildlife Refuge. In fact, only 1% of California's and 3% of U.S. electricity is made from oil, and only 2% of U.S. oil, chiefly residual oil for which there's no other use, made electricity. Oil and electricity are almost completely unrelated.

- California was said to have run out of generating capacity because of a decade of rapid growth in electricity demand. Actual growth was lower than officially forecast and two-fifths slower than the national average. The grain of truth in the assertion—though it was rarely stated—is that the 16 other states and provinces connected to the same regional power pool generally had brisk population and economic growth in the 1990s but did little or nothing about demand-side management, so they did run down the region's reserves as only 16 GW of new capacity got added to the pool in the 1990s.[30] Since California imports about 15% of its electricity—the largest net importer in both absolute and percentage terms—it was whipsawed most by the resulting price volatility.

- The public debate was framed initially in terms only of supply, as if California had no demand-side options. In fact, until the mid-1990s, when the restructuring debate derailed them, California was the world leader in demand-side management, having saved 10 GW (a fifth of peak demand in 2000) and billions of net dollars. This is partly because for over a decade, the state decoupled private utilities' profits from their sales volumes, so they were not rewarded for selling more electricity nor penalized for selling less. The Legislature undid this, and returned to

rewards for greater sales, in an ill-advised rate freeze voted in 1996 with effect from 1998. (The Legislature reauthorized the sensible old incentive system in April 2001, and it's expected to be implemented in 2002.) In addition, in the late 1980s and early 1990s, those utilities were allowed to keep as extra profit part of any savings they achieved for their customers, so they were rewarded for cutting customers' bills rather than for selling more energy. This alignment of shareholders' with customers' interests emulated efficient market outcomes so everyone chose the best buys first. It gave California such enviable demand-side success that the state's two largest private and two largest public utilities envisaged by the early 1990s getting most or all of their future service needs without building *any* new generating capacity. Had the demand-side momentum been maintained, this would have been a realistic expectation. But when restructuring derailed the demand-side efforts in the mid-1990s and the Legislature returned to the old system of rewarding increased electricity sales, the state's robust economic growth did tighten the supply/demand balance. In the mid-1990s, utilities' demand-side budgets were slashed by 40% or more, losing the equivalent of at least 1.3 GW of supply. The demand-side programs were revitalized by new laws in September 2000 (too late to avoid a year or two of shortages), and remained under attack by some regulatory staff even in 2002—when the State Power Authority still was prohibited from buying end-use efficiency on the same terms as generated kWh.

- Most reports focused only on supposed physical shortages of electricity. In fact, the problem was not a shortage of installed generating capacity, as is easily proven: rolling blackouts occurred in mid-January 2001 at peak pool loads of only 29 GW, 24 GW below the previously met 1999 summer peak of 53 GW. To be

[30] California accounted for about 40% of the Western System Coordinating Council's load, but only ~15% of its rise in peak demand during 1995–99. Ten other Western states average over twice California's growth in kWh usage. A typical Las Vegas house—ten times less energy-efficient than could be built at the same cost—used two or three times the annual electricity of a typical Bay Area house. In effect, the booming areas around Las Vegas, Phoenix, Albuquerque, and Denver free-loaded on the pool's shared reserves.

sure, a bad hydro year had reduced Northwest exports by up to 5 GW. But half the remaining capacity didn't suddenly disappear; rather, about a third of the plants started calling in sick. Some units went down for deferred maintenance, which the new owners had every incentive to schedule when it would help raise the price. Some suppliers apparently contrived reasons to withhold supply to drive up the price. Some old units had been run hard and probably had legitimate maintenance needs, but when fourth-quarter average daily forced or scheduled outages rose from 2.44 GW in 1999 to 8.99 GW in 2000, exceeding 10 GW in November 2000, suspicion naturally arose that some of the plants calling in sick were malingering. (Litigation will ultimately sort out what happened and whether the behavior was wrongful or merely opportunistic.) In the circumstances, suppliers' profit would theoretically be maximized if they dispatched only half their capacity—a close match to actual behavior. Starting in late summer 2000, ~10–15 GW of previously healthy generating units out of a pool total of about 48 GW called in sick at critical moments, reporting forced outage rates averaging two or three times higher than the same units had exhibited when utility-owned a few years earlier. In addition, many suppliers hesitated to provide power for which the near-bankrupt utilities might never pay, some independent providers had been bankrupted by non-payment (cutting supply by upwards of a tenth) or prevented by anticompetitive practices from selling their power, and some suppliers apparently gamed their bids to create transmission bottlenecks.

- The chief underlying reason for the extraordinary prices appears to be that the California bidding system's rules rewarded gaming and price-gouging. The system's structure was seriously defective in both architecture and detail. So much

of the supposedly competitive volume was "pre-met" by "must-run" (chiefly nuclear) capacity that a relatively modest number of players and transactions would set the marginal price paid to all bidders—who naturally often bid zero or negative prices knowing that when the half-hourly market cleared, they would get the "uplift" to the highest accepted bid price and would be guaranteed to be dispatched. The bidding system was therefore gamed skillfully and extensively. Two-thirds of the bidding space was occupied by a mere seven suppliers who had concentrated their market power by buying fossil-fueled plants divested by the utilities under restructuring. *Any one* of those seven firms could move the market all by itself without any collusion (although policymakers' curious choices of which data were published or secret also permitted "virtual collusion" in undetectable forms). The suppliers soon figured out that they could make more profit by selling less electricity at a higher price rather than selling more at a lower price. In the resulting ticket-scalpers' paradise, the market performed brilliantly; suppliers followed the incentives they were given; and nobody looked after the public interest.

- In addition, the state put nearly the entire burden on the malstructured and heavily gamed spot market (actually two of them, gamed against each other) by discouraging utilities from entering major long-term power purchase contracts.[31] The utilities, perceiving no important upside price risk, chose not to hedge with unbundled financial contracts, which were still permissible. The traders proved smarter than the utility planners and pool operators.

- Little-noticed, another crisis subtly drove the electricity shortage: botched restructuring of natural gas had meanwhile destroyed all parties' incentive to store gas for the winter, when Southern California had too little deliverability to

[31] Long-term purchases up to 20% of needs were permitted, but only Southern California Edison Co. did so.

meet demand without storage. Stored gas at the end of November fell 89% from 1998–99 to 2000; then a cold winter and a pipeline explosion (which cut deliverability by another 5%) further tightened the market. There are also strong suggestions that a major pipeline contract may have been gamed. The result was a huge midwinter spike in the Southern California gas price, exceeding $50/million BTU at times in January 2001. Since nearly a third of the state's electricity is gas-fired, this passed straight through into electricity prices, enabling generators to blow past price caps. After that, the sky was the limit as frantic bidders sought to keep the lights on. And it is also possible that odd behavior meanwhile in the Southern California market for reduced emissions of NO_x may have been gamed too, further multiplying electricity prices.

- That the extreme price volatility was driven largely by California's restructuring rules is a logical inference from the relative price stability enjoyed (at least until wholesale-market distress began to spill over around the start of 2001) by the neighboring states—and the public utilities within California—that share the same regional power pool but didn't do California-style restructuring. Unlike other commodities, electricity can't be readily stockpiled, and has been provided by large regulated-monopoly utilities for nearly 100 years; yet these obviously unique features were inadequately reflected in California's restructuring policy, which was driven largely by a dangerous mix of economic ideology and political accommodation. The West Coast had enjoyed a vibrant wholesale market since about 1980, and California had ample power supplies with reasonable and stable prices, but Governor Wilson wanted to refinance nuclear debt with cheaper public debt, and economic rationality was an early casualty to legislative dealmaking.

- As the California authorities had been warned, their restructuring—commonly misnamed "deregulation"—simply transferred much of their authority to the Federal Energy Regulatory Commission in Washington, D.C. The FERC's free-market ideology, and its disinclination (especially under the incoming Republican administration) to help Democratic Governor Gray Davis out of his difficulties, soon clashed with the state's need for sympathetic intervention. FERC therefore simply ignored its 1935 core duty of ensuring "just and reasonable" wholesale prices, declaring in effect that whatever the market would bear was just and reasonable.

- The forced sales of most non-nuclear, non-hydro generating assets meant that even though the Legislature soon wanted to "put the toothpaste back into the tube," this was no longer possible. However, three other major forces intervened to damp down the crisis in spring 2001. First, the state bought over $40 billion worth of previously prohibited long-term power contracts, though inevitably the prices were high and the state is now trying to escape from its own contracts. Second, the FERC grudgingly began to do its job by imposing weak after-the-fact price caps, and national politics shifted in ways that probably worried price-gougers. Third, and perhaps most important, just in the first six months of 2001, customers undid the previous 5–10 years of demand growth, cutting the weather-adjusted peak load per dollar of Gross State Product by a remarkable 14% even before they received higher bills. Of course, most of this saving was temporary and behavioral, but as surcharges bite and several billion dollars' worth of state-funded demand-side programs gain traction, the share of permanent technological savings is expected to rise (Figure 1-28).

Figure 1-28: Peak demand reduction in California, voluntary and program-induced
Astounding reductions in peak load were achieved voluntarily, even before non-San Diego customers saw any increase in retail electricity prices (which first hit in early July 2001). Presumably this was motivated by feelings of civic solidarity, anger at suppliers, etc. The peak demand reductions shown are all adjusted to normal weather and constant economic activity. State efforts to install more efficient technologies and to encourage load management (partly through initial deployment of real-time meters that permit price responsiveness) should accelerate to over 1 GW in 2002, when voluntary curtailments in the presumed post-crisis atmosphere are expected to be less dramatic.

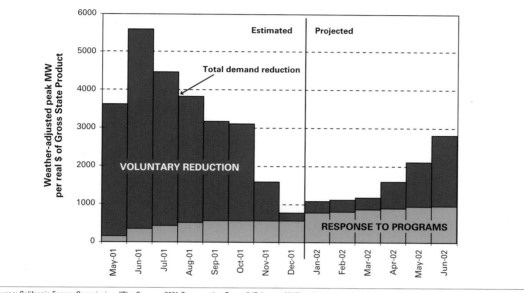

Source: California Energy Commission, "The Summer 2001 Conservation Report" (February 2002), p. 14

- California's electrical fire appears to have been put out; now the politicians are merely assessing the water damage and arguing about whether it was arson. But the recovery is far from over. Although California reversed much of its restructuring—thereby killing the green power industry it had sought to promote—many elements of the failed system persist, and the outcome of litigation and PG&E's bankruptcy is unpredictable. Even after supplies and prices are stabilized, the shock to the system will continue to reverberate for years as complex regulatory, legal, and financial issues play out. For example, the costs of purchased power contracts—perhaps a new form of stranded asset—increase the incentive to leave the grid as distributed resources become cheaper. The price spike also encouraged developers to propose by early 2001 to build new generat-

ing capacity equivalent to 83 percent of California's current total demand,[32] 96 percent of the western region's, and at least one-third of the nation's—consistent with Vice President Dick Cheney's call to build at least one power plant a week. But in August 2001, *Barron's* cover story noted the coming glut of electricity. By spring 2002, scores of plants had been canceled for lack of demand,[33] and their irrationally exuberant builders are reeling as Wall Street, stung by Enron's collapse, downgrades their bonds. Newly revitalized demand-side programs, macroeconomic uncertainties, and the uncounted engine-generator and other onsite backup capacity lately installed will all be happening at once. This would risk an overshoot into a power glut, plummeting prices, and more painfully stranded assets, much as occurred for U.S. energy supply as a

[32] Much of this new capacity was or is to be built by the same firms whose concentrated market power was a fundamental causes of the crisis. Having more capacity to withhold, and no less reason to do so, is not obviously a sound solution. If not very diversified in ownership and preferably in scale, more supply can actually exacerbate shortages created by uncompetitive market structures.

[33] R. Smith, in the *Wall Street Journal* (658), reports data from Energy Insight (Boulder, CO), showing that at least 18%, or 91 out of a total announced portfolio of 504 billion watts planned for construction, had been cancelled or tabled by the end of 2001. (The 504-billion-watt portfolio included longer-term projects than those just summarized.) Ms. Smith interprets the reductions as likely to create power shortages; we interpret them as likely to reduce financial losses when demand assumptions prove exaggerated—especially if saving electricity is allowed to compete fairly with producing it.

whole in 1985–86.[34]

In sum, then, California inflicted on itself a painful technical and economic disruption that may take quite a while to heal. While this was not a necessary result of restructuring, it has dampened many others' enthusiasm to experiment with something as vital and complex as the electricity system. And it certainly reinforces the necessity of seeking all three potential major benefits of restructuring—competitive generation, cheaper end-use efficiency and load management, and an optimal mix of generating scale—rather than sacrificing the latter two benefits in pursuit of the first.

By spring 2002, after nine years of the greatest turbulence in the electricity industry's history, about all that could be said with confidence was that the retail-wheeling *Blitzkrieg* and the demise of the traditional utility structure had been exaggerated. In the United States, massive lobbies and institutional, legal, and technical obstacles were still stalemating most major changes, and California's unhappy experience, plus problems emerging in such supposed success stories as Pennsylvania, added a powerful cautionary tone. By April 2002, 24 of the United States and the District of Columbia had enacted restructuring laws or regulatory orders to implement retail access, but one of those (California) had reversed and seven delayed it, and 26 had ignored or rejected it (774), leaving only about 17 states still implementing retail choice. This hodgepodge made coherent national policy still

more difficult to achieve. Simple questions are being belatedly asked, such as: If we have wholesale competition, why do we need retail competition? How can we prevent excessive market power and gaming of power auctions? And is restructuring really leading to greater overall economic efficiency than would a well-regulated monopoly rewarded for minimizing customers' bills?[35]

The deregulatory urge seems to have passed in the U.S., especially after the scandalous 2001 collapse of giant energy trader Enron. Nonetheless, in the U.S. as in many other countries, North and South, the paradigm of restructuring, competition, lighter regulation, and capturing the benefits of both least-cost generation and efficient end-use has begun to take hold. With it, unexpectedly and unintentionally, have come new elements of emerging market and regulatory practice that are starting to allow the economic benefits of distributed resources to express themselves in the marketplace. That may ultimately prove to be restructuring's most unambiguous and powerful benefit.

1.2.12.3 Distributed benefits start to emerge in the market

Some examples illustrate this potential for utility restructuring to help distributed benefits become commonplace elements of market pricing:

- Electricity sold to and bought from the wholesale market would be priced in

---

[34] The same risk became clearly visible in national policy in 2001–02 as a new Administration committed to stimulating energy supply, but apparently with little appreciation of the rapid pace of "invisible" energy savings meanwhile occurring in the marketplace, risked ruining the energy industries it was seeking to help. If customers bought even a tiny fraction of the "overhang" of unbought energy efficiency, they'd stick the suppliers with unsaleably costly surpluses, as happened in 1986 (469). Credit-rating agencies agreed in 2002 (§ 3.4.2.2)

[35] New York State regulators, for example, were told by a consultant (GE) that to guard against excessively concentrated market power by independent suppliers, the system's reserve margin should be raised from the traditional 15% to around 30%. It was far less clear that the system cost of doing so would be justified by the supposed cost reductions promised—and not consistently delivered—by competitive restructuring.

real time, reflecting the balance (typically every half-hour or so) between supply and demand. This is bound to make prices much higher around peak periods, such as hot summer afternoons, when certain renewable sources (photovoltaics and, in suitable topography, windpower) yield the highest output (§ 2.2.8.1).

- Tolls for using the transmission and perhaps the distribution grid as common carriers would depend on real-time congestion: the scarcer the grid resource at a given time, the more its users would be charged. But conversely, this implies (432) that distributed resources, which make, store, or save electricity *at or near the load center* without requiring grid capacity to deliver that service to customers, should get paid a symmetrical "decongestion rent" to reflect their "Dristan[36] value" (§ 3.3.3.1.4). This would start for the first time to internalize the fair market value of an important class of distributed benefits, reflecting the peak transmission capacity avoided by putting sources near customers (§ 2.3.2.6)—though it still wouldn't capture the often larger benefit of decongesting local distribution capacity.

- Scarcities of generation or deliverability would be immediately signaled in market-clearing prices.[37] This was dramatically illustrated in winter 2000–01, when bulk spot prices in California, for example, soared to \$1.50/kWh, up from a normal 2–3 cents per kilowatt-hour, eliciting strong interest in distributed generation because of its short lead times (99).[38]

- Many customers buying electricity at prices based on fluctuating and unpre-

dictable wholesale real-time market prices might choose to buy price-risk insurance.[39] But how would the writers and underwriters of such price-risk insurance protect their positions? They could be expected to buy financial instruments such as electricity or fuel futures and options. But they would also have a strong incentive to underwrite their contracts with physical assets that produce electricity at constant prices. Renewables, and in a sense also efficiency resources, have exactly this property of immunity to fuel-price fluctuations. Thus a kWh of renewable electricity is more valuable than a kWh of fossil-fueled electricity because it carries no price risk (§ 2.2.3). This is especially important if restructuring causes electricity prices to be based on value rather than cost of production, because value approximates GDP (little of which can be produced without electricity), so suppliers can raise prices by about 10–100-fold whenever there is an actual or artificial scarcity. Price is in fact limited in these conditions only by customer assets or FERC intervention (which, as California found, cannot be relied upon). Customers averse to such extreme volatility—or to the long-term rents into which power contracts convert it when market power is concentrated among a few suppliers—may well find renewables, especially those in their own neighborhoods and under their own control, especially attractive.

- Since the parties buying such constant-price assets for their portfolios would be not power engineers but market analysts versed in option theory and portfolio the-

---

[36] This registered trademark of Whitehall Laboratories was used for a popular over-the-counter nasal decongestant, before it was withdrawn from the market due to safety concerns about one of its ingredients, phenylpropanolamine.

[37] The obvious inconsistency between this prospect and the promise of lower prices (428) was not widely noted when restructuring was first proposed.

[38] Even where prices were not deregulated, other considerations—such as decrepit and unreliable distribution infrastructure in Chicago—could strongly motivate local generating capacity, such as the cleanup and revival of old standby engine-generators.

[39] In Britain, the old practice of buying "contracts for differences" might appear to have this function. Its real purpose, however, was often to circumvent the legal requirement that all generators above 50 MW sell to the pool. In effect, a bilateral CfD between generator and user established a fixed price between them via reciprocal compensation payments-based on whether the pool price was higher or lower than the agreed fixed price. This structure permitted the functional equivalent of a fixed-price bilateral power sale. It was transactionally simpler than a sale through the pool at variable prices, with price risk sold to a third-party insurer.

ory, they could be expected to apply different discount rates to different resources, according to their relative financial risk. This important benefit of renewable sources was never before counted by utilities, which drew no distinction between the financial risk of resources that did or didn't (for example) need price-volatile fuels (§ 2.2.3). Just properly counting this attribute could increase the economic value of renewables, compared with natural gas-fired generators, by as much as severalfold within an optimal mix of both resources. Unfortunately, the firms that supplied California with one-third renewable electricity going into its power crisis couldn't capture this value, which benefited all customers but not its providers.

- Similarly, market actors versed in financial economics rather than engineering and accountancy will understand that since the future is not deterministic, technologies that come in small modules with short lead times can greatly reduce investment risks, and that the value of that reduced risk can be quantified and internalized using option theory or decision theory—again increasing some distributed resources' value by up to severalfold (§ 2.2.2).

- Such actors will also understand that risk reduction through fuel diversification, in the sense understood by financial economists rather than by engineers, encourages and even requires that the portfolio include a significant share of riskless (renewable or efficiency) investments (§ 2.2.6).

- New categories of market actors will emerge. For example (286), public- or private-sector "renewable aggregators" can aggregate, firm, transmit, and resell renewable generation, so that a diversity of sources and sites can collectively provide firm power (§ 2.2.8.1) that is more valuable in the wholesale market. For instance, green power marketers, such as Green Mountain Power, aggregate customers with a particular preference and then deliver blended power certified to meet those customers' "green" requirements. Such aggregators could also greatly reduce transaction costs that inhibit marketing power from small generators, and could better negotiate long-term power sales contracts. And aggregators could even match up intermittent renewable generation with interruptible or dispatchable loads—thus increasing their option value—if firming up the generation through diversification or backup proved costlier than a demand-side solution (306).

- An increasing fraction of customers need and are willing to pay for premium reliability. Electricity providers can respond in at least three ways. First, they could help customers to use electricity more efficiently, install onsite storage, or install onsite or near-site generators. All these distributed resources would therefore acquire extra value expressing their reliability contribution (§ 2.2.8). Second, providers will unbundle their service package to offer customers wider choices between different levels of reliability, power quality, etc. at corresponding prices, thus making explicit certain distributed resources' advantages in these respects. And third, providers will probably find it highly advantageous to install and improve distribution automation (§ 2.3.2.10)—which in turn provides the ideal technical conditions for more easily and closely integrating distributed resources into the grid, partly by making

Interruptible loads are those whose users do not require or expect electricity to be always available on demand, and are willing to sell the right to have their electricity interrupted when the utility hasn't enough to serve all loads. **Dispatchable loads** are those controlled by the utility, such as water heaters or air conditioners that the utility can briefly cycle off, say for a quarter of each hour, by remote control when it wishes; this kind of interruptibility too is typically compensated by a periodic payment to the customer. **Firm loads** aren't expected to be interrupted.

the distribution system flexible enough to handle power flows in any direction. To the extent (which could in practice range from zero to unacceptably large in some regions) that restructuring of the electricity industry degraded the perceived or actual reliability of retail supply, providers' incentive to pursue all three of these avenues would expand. From the United States to Taiwan, dwindling system margins—previously an overhead borne by captive customers of monopolies—had by 2000 created important market perceptions that distributed generation's greater reliability could create important customer benefits meriting major customer investment. And of course in California in 2000–01, where power supplies became both very costly and unreliable, many customers scrambled for whatever kind of onsite generation they could find and afford. Long after the crisis, those assets will still be there and will probably still be used.

These are not the only ways in which more market-oriented and competitive utility structures could make resources that are dispersed, renewable, or often both look considerably more valuable than they did traditionally when these attributes were ignored. But they suggest that a judicious mix of wholesale competition, public-goods investment, and incentives that emulate socially efficient market outcomes—chiefly rewarding utilities for cutting customers' bills, not for selling more electricity—could bring distributed benefits rapidly up the list of attributes to which investors pay careful attention.

This is encouraging not only in its potential results but also in its cause: for the first time, new electricity market structures can provide the market incentives, the tools and systems of measurement and validation, and the more diverse, chiefly financial-eco-

nomics, disciplinary perspectives needed to give distributed benefits a market voice and reality. We hope that the analysis of distributed benefits' economic value in Part Two, and the strategic opportunities and policy options in Part Three, will further encourage power brokers and other new market actors to evaluate and internalize the full range of these benefits, and thus more closely to match true economic value with expressed prices.

It is too early to say how the electric utility restructuring debate—a clash of titanic forces, interests, and political lobbies—will play out. But whatever the outcome, the debate is for the first time focusing close business attention on the fine-grained structure of power flows, customer needs, and new technological options, and how the related economic values are constantly shifting in time and space. Now these new questions about distributed resources will inevitably yield new answers that will begin to bring distributed benefits into market consciousness and everyday practice.

To help understand what those benefits are, what they are worth, and how they could be expressed, we turn next to short primers on the existing electricity system, distributed electric resource options, and key issues faced by electricity planners, investors, and engineers. These primers will form the essential background for the specific examination of distributed benefits in Part Two. But first, a little musing on energy history is needed to round out our discussion of the forces of renewal.

1.2.12.4 What next?

In 1973–74 and in 1979, oil-price shocks

launched major improvements in energy efficiency. Just so, in the 1970s and 1980s, a power-plant-price shock unleashed commercial forces—and now political forces reinforcing the commercial ones—that are launching major improvements in the assessment and application of distributed resources. And in 2000–01, California learned the same lesson in a different form: resources that provide constant-price electricity under effective local control are the best way to keep the lights on in the face of lucrative but antisocial supplier behavior.

That much is not new: the economic system is routinely reequilibrating itself, just as a perturbed ecosystem exerts selective pressures on how its organisms behave and, ultimately, how they evolve. But reequilibration is a never-ending journey, not a destination. The oil-price shock had another chapter: it reversed itself in the 1984–87 oil-price collapse, partly because the remarkable success of energy efficiency created a supply glut:
Low fuel prices then triggered stunning technical advances on the supply side, echoed on the demand side, that many ana-

lysts believe will keep fuel prices generally low for at least decades to come (though always subject to shocks from supply interruptions). Energy efficiency still remains cheaper than fuel or electricity, and its margin of advantage is widening (it's becoming even cheaper even faster),[40] but there is less sense of urgency to adopt it than in times of scarcity. Efficient end-use will therefore be bought increasingly not because it saves energy *costs* but because it provides qualitatively *superior service*—a trend most evident today in green buildings and Hypercar® vehicles (434, 437, 474, 492, 775). Such side-benefits as a ~6–16% increase in labor productivity (58, 571, 769) can easily be worth an order of magnitude more than eliminating the entire energy bill, conferring strong competitive advantage. This may help to explain why during 1996–99, the United States nearly beat its own all-time record for the three-year speed of improving aggregate energy efficiency (reducing the ratio of primary energy consumption to real GDP)—3.2% a year—despite record-low and falling energy prices. (During the same three years, electric intensity sustained a similarly surprising decrease at about 1.6% a year.)

Similarly, we daresay, the distributed resources elicited by the "power-plant price shock" will increasingly be bought not so much to save energy costs as to achieve other, important, but previously unrecognized benefits—distributed benefits. Thus will the two-decade binge in electrical gigantism, like the pre-1973 binge in profligate gas-guzzling, ultimately be seen to have been a salutary cause of its own undoing. The *next* chapter in this never-ending evolutionary process can be only dimly

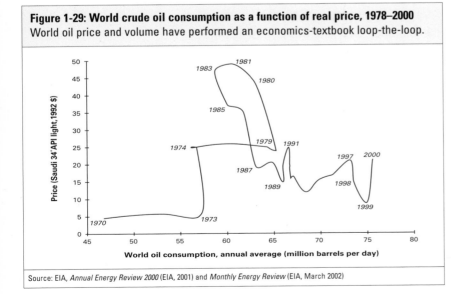

Figure 1-29: World crude oil consumption as a function of real price, 1978–2000
World oil price and volume have performed an economics-textbook loop-the-loop.

Source: EIA, *Annual Energy Review 2000* (EIA, 2001) and *Monthly Energy Review* (EIA, March 2002)

[40] This is now less because of new technologies than because of better whole-system design integration (288, 429, 433).

foreseen. But just as lower energy prices made energy efficiency and its delivery more sophisticated and integrated but did not lead people to take insulation out of their roofs nor to reinstall the previously removed inefficient motors, we suspect the trend will remain in one direction: a more

efficient, diverse, dispersed, and renewable energy system dominated by the breadth of benefits, and ultimately by the bio-logic, that this book seeks to synthesize.

1.3 WHERE WE START: THE EXISTING POWER SYSTEM

The U.S. electric power industry is a combination of private, public, cooperative, and federal utilities; when distinguished from those owned by private shareholders (formerly called "investor-owned utilities" or "IOUs"), those last three categories can all be considered public utilities. In 1999, the 242 investor-owned utilities, or 8% of the total number of electric utilities, accounted for more than three-quarters of sales to end-use consumers.[41] Historically they have served the large, consolidated markets.

The nearly 3,000 publicly owned electric utilities—including municipals, public utility districts, irrigation districts, federal agencies, state authorities, and other state organizations—accounted in 2000 for 22% of electricity generation and probably a comparable share of total retail sales. Most public utilities do not generate their own power but purchase and distribute it to end-use customers. Most are non-profit local government agencies established to serve their communities. Rural electric cooperatives, owned by their members, currently operate in 46 states and in 1999 represented some 29% of the total number of electric utilities.

The federal government is primarily a hydroelectric generator and wholesaler of electricity, rather than a distributor to retail customers. As required by law, most of the generated electricity is sold to public utilities, cooperatives, and other non-profit utilities. The primary producers of the hydro-electric power are two federal agencies: the Army Corps of Engineers and the Bureau of Reclamation of the Department of the Interior. Electricity generated by these producers is marketed by the four federal power-marketing administrations: Bonneville, Southeastern, Southwestern, and Western Area Power Administrations. Electricity is also generated by the Tennessee Valley Authority, the largest federal producer, and by other federal entities, such as the Bureau of Indian Affairs.

Except for the three large federal agencies—Tennessee Valley Authority, Bonneville Power Administration, the Western Area Power Administration—which own or market the output of over 40% of the total U.S. hydro capacity, very few other public utilities have sales comparable to any of the 50 largest privately owned companies. Only the New York State Power Authority, the Salt River Project, and the new California Consumer Power and Conservation Financing Authority (commonly referred to as the California Power Authority) are of comparable size.

[41] Surprisingly, these data are the most recent available, because as of 31 March 2002, the U.S. Energy Information Administration had not yet published the disposition of non-utility electricity generated in 2000 or 2001, nor the statistical second volume of the *Electricity Annual* 2000. However, net generation figures had been published through 2001 (209), showing that non-utilities' generation had reached 29.5% of the national total, as described in Section 1.2.4. The most recent public data on the public-power share of generation and sales, for 2000, were from the American Public Power Associa-tion's website www.appa.org, and showed 28% of utility, implying 22% of total, net generation.

Despite the large number of utilities that distribute electricity in the retail markets, generation is far more concentrated. The ten largest utilities in 1999 owned more than one-half of the total power plant capacity. The other half was supplied by 628 other private and public utilities. The remaining 2,600 utilities, mainly small cooperative utilities, purchased power in the wholesale markets and resold it to end-use customers in their communities.

It is important to note that no single entity generates more than 4.2% of the U.S. total. This fragmentation is one of the characteristics of an industry that has given rise to speculation about a potential wave of mergers and acquisitions.

The history of the U.S. electric power supply system—a story masterfully told elsewhere by industry historian Professor Richard Hirsch (297)—emphasizes how its evolution of scale long satisfied all the parties through coupled and seemingly endless growth in unit size, thermal efficiency, cost savings, reliability, demand, revenues, and profits. The main historic trends, as summarized by the U.S. Energy Information Administration, were (297):

> Early in the 20th century, more than half of all electricity produced in the United States came from industrial firms. However, during the first half of the 20th century, major changes occurred in the industry: economies of scale in generation, decreased [electricity prices]..., and greatly improved reliability made electricity inexpensive and demand soared. Most industrial plants shifted away from generating their own power and opted to purchase electricity from their local utilities [which had gained franchise monopolies around the 1920s]. By 1950, the electric utility industry was serving virtually all electricity demand, except for a few industries that generated small amounts for their own

use. Electricity was inexpensive, capacity growth appeared to be limitless, and electric utilities were strictly regulated to protect the consumers.

> By the late 1970s, changing economic conditions and legislation made non-utility generation attractive again for many industrial facilities and power project developers.During the 1970s [actually the 1960s (170)], however, the electric utility industry changed from one characterized by decreasing marginal costs to one of increasing costs. Inflation, the energy crises, environmental concerns, and the rising costs of nuclear power led to increased electricity [prices]...and reduced growth in capacity. The oil-price shocks in the 1970s led to a dramatic rise in energy prices, while high interest rates and stricter Federal air quality regulations increased the cost of building power plants. These factors led to a re-examination of alternatives such as non-utility electric power.

> Non-utility power producing facilities seeking to establish an interconnected operation with an electric utility faced three major obstacles. First, utilities were seldom willing either to purchase the electric power output of non-utility producers or [to] pay a fair [price]...for that output. Second, some utilities charged high [prices]...for backup services to non-utility power producers. Third, facilities that provided electricity to a utility connected to the grid risked being considered a public utility and subject to extensive State and Federal regulation.

> Congress acted to relieve a nationwide energy crisis by enacting [five laws in 1978]....Some of the provisions of [these laws]...were designed to encourage the development of cogeneration and small power production by loosening the economic, regulatory, and institutional barriers that discouraged cogeneration and the use of renewable energy resources.

Professor Hirsh adds an important further insight (171) that concisely summarizes our narrative so far:

> After improving steadily for decades, the technology that brought unequaled productivity growth to the industry appeared to stall [in the 1960s], making it impossible to mitigate the difficult economic and regulatory assaults of the 1970s. Unfortunately,

most managers did not recognize (or did not want to believe) the severity of technological problems, and they dealt instead with financial and public relations issues that appeared more controllable. Partly as a result, the industry found itself in the 1980s challenged by the prospects of deregulation and restructuring.

These trends have created a system still dominated by the traditional institutional and technical structure—mainly large utilities operating mainly large, centralized, fossil-fueled power stations—but with more diverse structures and technologies rapidly oozing up through the cracks, as noted in the earlier graph of the fall and rise of non-utility generation (Figure 1-22). A snapshot of the industry reveals the following major elements.

1.3.1 Basic characteristics

At the end of 1999, the U.S. power system consisted of (164, 192):

- Utility-owned power plants: 10,207 generating units of utility-owned generating capacity in active service, totaling 677 GW of nameplate-rated capacity: 43% coal-fired steam plants, 19% gas-fired steam and combustion turbine plants,

15% nuclear fission steam plants, 12% hydroelectric, geothermal, and other renewables (chiefly windfarms and biomass), 8% oil-fired steam, combustion turbine, and internal-combustion (mainly diesel) plants, and 3% hydroelectric pumped storage;

- Non-utility-owned power plants: an additional 167 GW[42] of non-utility generating capacity,[43] of which 13% was renewable and 32% gas-fired;[44]

- Central dispatch: dispatch coordinated by nine regional Reliability Councils[45] organized within three power grids— eastern, western, and Texas;[46]

- Peak load: a noncoincident utility peak load (in the lower 48 States) of 680 GW summer and 594 GW winter, implying a –6% reserve margin from the 642 GW of utility summer[47] capability alone or +15% including also the 140 GW of non-utility summer capability (a total of 782 GW);[48]

- Generation: annual utility generation of 3,182 TWh (billion kWh) (net of ~5% in-plant uses and losses)—derived 56% from coal, 23% from nuclear fission, 9% from natural gas, 9% from hydroelectricity, and 3% from oil—of which ~3% was lost in transmission, ~4% in distribution, and the remaining ~93% (supplemented by 29 TWh of net imports and 344 TWh of purchases from non-utilities) was sold

[42] Preliminary Edison Electric Institute data from May 2000 show 175 GW. Some of the discrepancy might be due to EIA's exclusion of units under 1 MW.

[43] Of this, 44% was transferred from utility ownership during 1998–99, essentially all the rest built by non-utilities originally. See Section 1.2.4 for further discussion of utility-to-non-utility transfers.

[44] Only four years earlier, 24% of non-utility capacity was renewable and 51% gas-fired, but these got heavily diluted by non-utilities' purchases of utility capacity, which were largely responsible for non-utility increases of 14 GW of coal-fired and 12 GW of oil-fired capacity in 1999 alone.

[45] Spanning the contiguous U.S., Canada, and Baja California Norte. Parts of Alaska are in effect a tenth Council, and Québec also has an independently controlled grid. However, industry restructuring is eroding traditional collaborative relationships within these Councils and destroying their traditional planning function, and the FERC is trying to consolidate planning and dispatch into a smaller number of Regional Transmission Organizations.

[46] A westward synchronous link from Texas is now being studied. The eastern and western grids now have an asynchronous DC link. The U.S.-wide statistics shown here include not only the U..S. portions of the three regional North American grids, but also the minor quantities supplied and used in Hawai'i, Puerto Rico, and U.S. territories and possessions overseas.

[47] The country as a whole is summer-peaking due to ~200 GW of peak air-conditioning loads, but some utilities and regions are winter-peaking due largely to cooler climates or the predominance of electric space heating or both.

[48] The associated demand-side statistics must be interpreted with caution, because slightly over half of the output of these non-utility, chiefly industrial, generators was devoted to their own use rather than being resold to utilities, and some of that own-facilities use may not be considered a normal utility load. Industrial generation data are weak too.

Capacity factor is the fraction of a generating unit's or plant's full-time, full-power output that it actually produces. Capacity factor can be less than one (or 100%) through any combination of being **unavailable** (out of service); **derated** to less than its full rated capacity due to deterioration, regulatory restriction, or unusual operating constraints such as very warm condenser water; or **not dispatched** because its power was not needed or not economically competitive at the time.

to retail customers, more than half of it via one or more intermediaries, the rest directly;

- Power plant utilization: for utilities' total capacity, a capacity factor[49] averaging 57%, ranging from only a few percent for some peaking plants to 85.5% for the average nuclear plant;[50]

- Fossil-fuel consumption: utility fossil-fuel consumption costing ~$48 billion, including 894 million short tons of coal, 144 million barrels of oil, and 3.1 trillion cubic feet of natural gas—a total of 33 quadrillion BTU, or 33% of total U.S. primary energy consumption, converted to delivered retail electricity at an average efficiency of 33%—thereby creating utility power-plant emissions of 12 million short tons of SO_2, 7 of NO_x, and 2,192 of CO_2;[51]

- Organization: 3,187 separate utilities, no two alike—242 investor-owned utilities with about three-fourths of utilities' total capacity (77%), sales (75%), revenues (79%), and ultimate consumers (75%), plus the other one-fourth held by publicly owned utilities comprising nine Federal utilities such as the very large Bonneville Power Administration and Tennessee Valley Authority, 900 cooperatives, and 2,012 public utilities ranging from the Los Angeles Department of Water and Power to the tiniest municipals and power districts, plus over 400 power marketers (over two-thirds of

them licensed but inactive) and 2,168 non-utility generating entities responsible for 16.7% of total net generation (perhaps an understatement given discrepancies with other data sets and regarding industrial generation);

- Regulation: a public utility regulatory commission elected or appointed in each state except Nebraska,[52] plus an intricate mix of public governance at the cooperative, municipal, regional, or large-public-utility Board scale, plus the interstate and wholesale jurisdiction of the Federal Energy Regulatory Commission, all amounting to an immensely intricate regulatory context differing between utilities, between the states, and over time;

- Local siting: an additional and at least equally complex set of relationships with environmental, financial, land-use, and other regulatory bodies at every level of government;

- Transmission: more than $50 billion worth, or 0.7 million circuit-miles, of 22-or-more-kV transmission lines, of which 37% of the circuit-mileage was 22–50 kV, 29% 51–131 kV, 13% 132–188 kV, 17% 189–400 kV, and 4% 401–800 kV (165);

- Distribution: an inventory of distribution facilities with a net book value probably around $140 billion,[53] with an astonishing ~43% of the total line length, much of it single-phase, owned by rural electric cooperatives (330) (§2.3.2.1.1);

---

[49] Capacity factor figures can be ambiguous because of changes in plant rating, condenser water temperature, in-plant usage, etc.

[50] The U.S. nuclear fleet's capacity factor rose from a miserable 47.8% in 1974—meaning that the plants collectively produced only 47.8% of their full-time, full-power potential output (with no stops for maintenance or refueling)—to the mid- to upper 50s of percent in the 1980s. Average capacity factor then improved dramatically, to 70+% from 1991 onward and to a remarkable 88.1% in 2000. This was due partly to better management and operational practices, perhaps partly (say critics) to less attentive and rigorous safety regulation, and certainly to the shutdown of 28 units through 1999. (These included many of the least reliable performers, but some of the units retired through mere age were relatively reliable. Age was influential for many, and dominant for some, of the ten retirements of U.S. operating units that had already occurred by August 1994 and the four more that were then expected before 2000 according to Resource Insight's nuclear-plant mortality model (121)—actually five units retired.) At the end of 2001, of the 259 nuclear generating units originally ordered in the U.S. 124 had been cancelled before completion, 28 were shut down after some operation, 104 remained operational, and 3 were pending but unlikely to be completed (210).

[51] Non-utility emissions (197) were respectively 1.4, 0.9, and 323 million short tons—respectively 11%, 12%, and 15% of the utility values, while total net generation was 17% as much. This suggests that as a whole, non-utility generators were consistently cleaner, through some combination of thermal efficiency and renewable content. This conclusion would be even stronger if the cogenerations' useful heat byproduct, displacing boilers and furnaces, were also credited.

[52] Because its utilities are all publicly owned. Interestingly for the debate over regulated *vs.* "deregulated" utilities and the alleged benefits of free-market competition, coal-dominated Nebraska's electricity prices are among the lowest in the nation except for a few states particularly rich in hydroelectricity or cheap coal.

[53] This, the transmission system's value discussed above, and the whole electric system's value discussed below, are all estimated at four-thirds of the respective values declared by major investor-owned utilities, which are about three-fourths of the total utility industry. Non-utility T&D facilities are relatively minor, poorly reported, and not counted here.

- Historical construction expenditures: for investor-owned utilities, 1998 construction expenditures of $22.6 billion—$6 billion for generation, $2.5 billion transmission, $10.2 billion distribution, and $2.3 billion general and miscellaneous (165);

- Planned construction expenditures: planned construction of 2 GW (nameplate ratings) by utilities (2001–04) and 22 GW by non-utilities for 2000 plus 146 GW planned for 2000–04 (193);

- Current additions: 1999 addition of 3.7 GW of utility capacity and 6.8 GW of non-utility capacity;

- Equity: equity or its public-sector equivalent, and total net book value, each total on the order of half a trillion dollars, with an asset turnover ratio[54] for the investor-owned utilities around 0.37 and for major public utilities, 0.25;

- Retail electricity prices: an average retail price of $0.066/kWh, continuing a six-year decline in real terms, and averaging $0.082/kWh residential, $0.073/kWh commercial, $0.044/kWh industrial—the lowest in real terms since 1973—and $0.064/kWh other. There are dramatic differences in price between regions, however, and even between companies within regions. For instance, during 1994–1999, electricity prices in New England increased by more than 20%, averaging 9.7¢/kWh, while in the Great Plains states, prices rose by only 2% during the same period, to a modest 6.0¢/kWh. Average regional prices can differ by more than 60%. Individual company prices, however, can vary by more than 1,000%. In a recent year, the least expensive residential price in the country was 1.5¢/kWh in Douglas

County, WA, while the most expensive was on Long Island, NY, at 16.1¢/kWh;

- Variable operating costs: for the fossil-fueled steam plants that dominated the system, operating costs averaging 77% for power plants' fuel and 23% for their operation and maintenance (but both are typically less important than capital costs, and both have labile definitions and accounting conventions);

- Non-utility capacity and generation: within the non-utility generating sector (counting only units with at least 1 MW of capacity), capacity equivalent to 24% of the total utility-owned capacity,[55] and generation totaling 555 TWh (17% as much as all utilities generated)—of which 370 TWh was sold to and 90 bought from utilities, 43 TWh sold to third-party end-users, and 250 TWh was used onsite for power-plant operation and industrial processes;

- Retail sales: $215 billion worth of electricity sold to 125 million ultimate customers, with the kWh sales divided 35% to households, 30% to the commercial sector, 31% to industry, and 3% to other sectors (street and road lights, railroads and subways, miscellaneous public authorities, and interdepartmental sales);

- Direct employment: 0.5 million people employed by investor-owned utilities, somewhat more by the entire electricity sector;

- Investor-owned utility net profits: $17 billion on operating revenue of $214 billion (representing 8.0% profit margin, 10.1% return on common equity, and 2.92% return on investment), public utility net surplus of $2.4 billion on operat-

---

[54] *I.e.*, $2.70 of assets was necessary to generate $1 of annual revenue—about three times the capital intensity typical in manufacturing industries. The public utility ratio was therefore ~$4 for each $1 of annual revenue, largely because public utilities typically have more scattered customers needing more grid investment. Undepreciated utility book value would approach $1 trillion, about a tenth of the underlying U.S. asset base.

[55] These are classified by a thicket of confusing administrative rules into cogenerators that may or may not be Qualifying Facilities under PURPA; small and chiefly renewable producers under PURPA (though the original 80-MW size limit was removed in 1990); Independent Power Producers under the 1992 Energy Policy Act; and other commercial and industrial establishments. These classifications are obscure and subject to change.

Demand-side management is a catch-all term for all efforts to alter how much electricity customers use or when they use it. **End-use efficiency** (less electricity yielding the same or more service) and **load management** (changing the time pattern of electricity usage) are types of DSM. So are promotional practices that aim to sell more electricity, even at onpeak times, if that suits the utility's financial objectives—not an infrequent practice where it is rewarded, as it is in all but a few of the United States.

ing revenue of $36 billion, plus unknown profits and revenues to non-utility generators; and

- Demand-side management: utility demand-side management (DSM) investments of $1.4 billion (about matched by customers' own investments) resulting in incremental savings of 3.1 TWh/y plus 7.3 potential (in average weather) or 2.3 actual (at the time of actual system peak load) peak GW.

1.3.2 **Scale of existing utility generating units**

Different types of generators dominate in different ranges of nameplate capacity. While wind, hydro and internal-combustion generators dominate the lower ranges, nuclear generators and fossil-fueled steam turbines dominate the 1 GW range.

The unit sizes of the generators in traditional utility service range from around 5 MW

for most engine-powered generators to around 75–100 MW for combustion turbines (both together constitute 8% of utility capacity) and up to ~1,400 MW for steam-turbine generating units (which in all sizes constitute 76% of total utility capacity). To give a snapshot of the historical system, we performed an analysis of all 8,922 generating units reported in utility service in the United States at the end of 1994, before reporting became incomplete and unduly complicated by non-utility expansion and intercompany transfers (169).[56] Those units' shares of the capacity of all kinds of units in each range ("bin") of unit size show the typical unit-size range of each technology (windpower falls into a smaller kW-range bin *and* a larger, ~50-kW, commercial wind-farm range).[57] The histogram of unit sizes plotted on a horizontal scale further shows the lowest-capacity peak dominated by small hydro and internal-combustion engines, then the larger combustion turbines, blending into the large steam plants.

Figure 1-30: Share of U.S. utilities' 1994 capacity by technology and unit size
Different generating technologies dominate at different unit sizes.

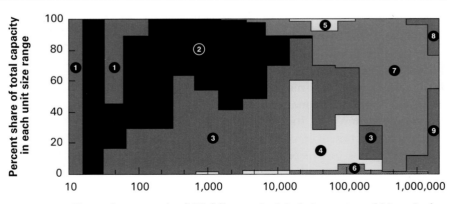

❶ Wind turbines
❷ Internal combustion
❸ Hydraulic turbine (conventional)
❹ Combustion turbine
❺ Jet engine
❻ Combined-cycle combustion turbine
❼ Fossil-fueled steam turbine
❽ Boiling-water nuclear reactor
❾ Pressurized-water nuclear reactor

Percent share of total capacity in each unit size range

Nameplate capacity (kWe) (log scale; label shows top-of-bin value)

Source: RMI analysis based on EIA (December 2000)

---

[56] The big difference between this number of units and the 10,207 reported by the same data source for the end of 1999 appears to be due to a change in EIA's reporting requirements, and does not represent a substantive difference relevant to our analysis. A difference of nearly 200 units is in steam plants, probably small ones.

[57] Using unit count rather than capacity share for the vertical axis yields a nearly identical graph because of the relatively narrow bins used.

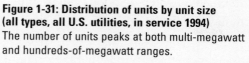

Figure 1-31: Distribution of units by unit size (all types, all U.S. utilities, in service 1994)
The number of units peaks at both multi-megawatt and hundreds-of-megawatt ranges.

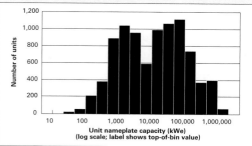

Source: EIA, *Inventory of Power Plants in the United States as of January 1, 1996* (December 1996)

However, even though many of the *units* are small, *total capacity* is dominated by a relatively small number of large units:

Figure 1-32: Distribution of capacity by unit size (all types, all U.S. utilities, in service 1994)
Units with capacity around 1 GW (one million kilowatts) are the workhorses of the fleet.

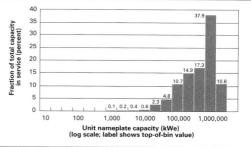

Source: EIA, *Inventory of Power Plants in the United States as of January 1, 1996* (December 1996)

The distribution of capacity is quite different for non-utility generators, whose data are not yet as well reported,[58] and could be expected to become more so for all capacity, regardless of ownership, if distributed benefits were properly taken into account.

1.3.3 Operating cost and dispatch of existing power stations

Each of the power stations whose constituent units are described above operates at different costs and for a different number of hours—variables that are closely related.[59] The following scatter-plot published by the president of Synapse Energy Economics (68) shows U.S. power stations' short-run marginal running costs, measured at the busbar (generator output terminals), and classified by plant type. This one-year "snapshot" includes 676 plants totaling 579 GW and generating 2,719 TWh in 1995, or 91% of the total national utility generation in that year. The graph omits all 114 plants of 100 MW or less (totaling 7 GW), all 54 renewable or geothermal plants (8 GW) which had operating costs close to zero, an unstated amount (perhaps around 112 GW) of peaking plants that cost more than $0.09/kWh to operate but were run for very few hours, and apparently all 70 GW of the non-utility capacity operational in 1995. Nonetheless, it usefully illustrates the wide range of operating costs in the main utility fleet of fossil-fueled and nuclear plants, excluding the cost of delivering the power to customers.

---

[58] The Energy Information Administration has declined to release even the most basic data on the size distribution of units on order, since apparently these data, reported to the Federal government, are treated as proprietary for plans through 2000. Plans starting in 2001, however, will not be so treated, and the data should soon become available. The average unit size of the 4.5–6 GW installed in California in the 1990s is approximately 20 MW (a range of about 14 to 35 MW depending on the database used), and even that average is raised by a half-dozen outliers in the hundreds-of-MW range.

[59] The less a unit costs to run, broadly, the more hours it will be run, as explained in a moment; but also, complex relationships between fixed and variable operating-and-maintenance (O&M) costs may make the number of hours run indirectly affect the operating cost. The analyst here describes his graph as showing running cost consisting of "fuel plus O&M" but does not distinguish between fixed and variable O&M—a somewhat slippery concept in any case, since whether an O&M cost is fixed or variable is an accounting convention that may depend on the timescale over which it is assessed.

Figure 1-33: 1995 busbar operating costs *vs.* capacity factor of U.S. utilities' nonrenewable, non-peaking units >100 MWe
Plants with the lowest operating cost are run the most, although location, transmission constraints, and other factors leave a considerable range of cost among the plants operating simultaneously at a given capacity factor.

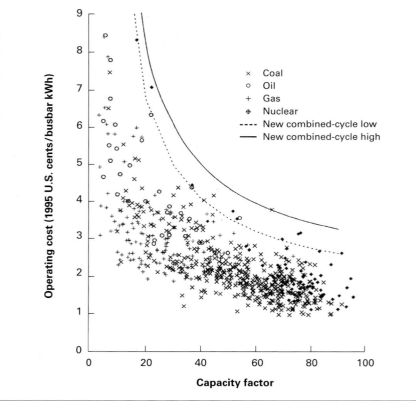

Source: B. Biewald, "Competition and Clean Air: The Operating Economics of Electricity Generation" (*Electricity Journal*, January 1997)

paid. Such dispatch decisions can become extremely complex.

Moreover, on the grounds that they cannot be turned on and off without neutronic instability and undue thermal fatigue, many nuclear power plants have received the right to dispatch their power whenever it is available, even if—as appeared to be the case for about 20–25% of U.S. nuclear plants in 1999 (436)—their operating cost was uncompetitive on the spot market.

Because of these operational constraints, and because transmission capacity is not unlimited or free, cheaper-to-run plants often cannot displace costlier ones as perfectly as market theorists might suppose. Even by 2000, for instance, only about half of U.S. electricity had become subject to genuine wholesale competition.

This dynamic is revealed by the vertical scatter among the points in Figure 1-33: under perfect competition, all the points would fall along a line, without plants of widely varying costs running simultaneously. However, the broad trend that the cheapest-to-run plants are operated for the most hours is consistent with the principle of economic dispatch (§ 1.2.3). To the extent that vertical (cost) scatter *within a particular region* represents an arbitrage opportunity, infusing distributed generation—by delivering power when and where it's needed—can help to displace out-of-merit-order plants, significantly reducing the system's total operating cost.

Current wholesale power trading prices in the coal-dominated Midwest closely approximate the short-run marginal costs shown: during a summer week in 1997, for example,[60] the simple average of the daily

This graph illustrates the basis on which utility dispatchers decide which plants to run in which order—in theory, lowest running costs first. Actually it is more complex than that, because plants may be run outside their strict "merit order" according to their maintenance or nuclear refueling schedules, technical characteristics such as ramp rate and turndown capability, location that might be critical for voltage support in weak portions of the grid, pollution restrictions or "environmental dispatch" requirements, or other considerations such as who owns and/or operates them and under what contractual arrangements they are

maxima, minima, and mean were respectively $0.0354, $0.0132, and $0.0226/kWh, while the absolute weekly maximum per kWh was $0.042 and the minimum $0.0085.[61]

Although bulk power transfers are expanding in geographic scope—a Washington State utility, for example, has contracted to wheel power across seven states to Wisconsin (141)—such trading costs remain regionally specific: in the glutted Pacific Northwest in 1997, $0.016/kWh was the official estimate of long-run marginal wholesale generating cost (98). Such prices were far below those experienced a few years later as the California restructuring fiasco (§ 1.2.12.2) bid up wholesale prices all the way to British Columbia.

The two curves overlaid onto the data points in Figure 1-33 above show a range of estimates of levelized cost for new combined-cycle plants under assumptions that seemed rather conservative (tending to overstate costs) around 1997.[62] It might at first appear that only about nine operating U.S. nuclear plants could be cost-effectively displaced by new combined-cycle plants. However, even though those nuclear running costs, as of 1995, averaged 72% O&M[63] and 28% fuel, they omit a kind of maintenance cost that for certain plants can be far larger than routine O&M: "net capital additions." Those are major repair costs, such as re-tubing steam generators, that are conventionally capitalized rather than expensed.[64] It would be more transparent to expense them just like other O&M.[65] On this basis, more careful regulatory scrutiny of most re-tubing projects would probably disclose, as market competition will ultimately reveal anyhow, that any nuclear plant needing such a major repair should simply be abandoned as not worth fixing. Around one-third or more of the 98-GW 2001 U.S. nuclear fleet is thus economically ripe for abandonment, and that fraction could rise rather quickly under increased competitive pressure and safety scrutiny. If waste-management and decommissioning costs, both of which increase more or less proportionally to kWh generated,[66] were fully internalized rather than partly (perhaps largely) socialized, this conclusion would strengthen.

---

[60] All CPEX (Continental Power Exchange) trades during hours ending 0700–2200, Wednesday 25 June through Wednesday 2 July 1997, excepting Sunday 29 June.

[61] Curiously, that was below the lowest 1995 value reported in the chart, even though no trading occurred at the lowest-volume hours, 2200–0600, nor on Sunday.

[62] Assuming $2.35/million BTU gas, $28/kWy fixed and $0.009/kWh variable O&M costs, a 13%/y fixed charge rate, and ranges of 45–60% for efficiency and (respectively) $635–500/kW for construction cost. However, around mid-1997, before temporary scarcity bid up prices, the actual installed costs were approaching $400/kW (580), partly through higher volumes and keener competition that briefly cut uninstalled plant cost nearly to $300/kW (405), but mainly through more streamlined installation processes. At the high-end capacity factor shown, the low-cost curve for combined-cycle plants in a truly competitive equipment market could therefore be up to $0.0016/kWh cheaper than shown.

[63] Operation-and-maintenance cost escalated steeply, to >$100/kWy for the average nuclear plant, through the 1980s, but has lately stabilized. It is part of the price of the gratifying increase in nuclear capacity factors reflecting better plant management. In the 1980s it might also have reflected more stringent regulatory oversight, though that has lately been hard to detect.

[64] This unusual accounting practice is rationalized on the grounds that such major projects should be added to the utility's rate base and amortized or depreciated over the plant's remaining estimated engineering or accounting lifetime—even though that often results in the plant's appreciating rather than depreciating, because new expenditures each year are, on average, exceeding its straight-line depreciation of the original investment. This is especially odd because those investments are often being made only in an effort to achieve the original depreciation life rather than to extend it.

[65] Indeed, on a long time-scale, all repair-and-maintenance costs could be considered variable, in the sense that if the plant weren't operated at all, even the biggest repairs needn't be made. To be sure, not doing certain mandated repairs or safety upgrades could endanger the operating license, but that wouldn't be needed if the plant weren't intended to be operated again. Presumably this subject is taboo not only because many existing operating licenses could not be obtained today de novo, but also because any plant no longer considered "used and useful" could be required to be removed from rate base.

[66] Decommissioning costs rise with increased neutron fluence (time-integrated neutron flux) because the reactor's materials become more activated and hence more intensely radioactive. This increases the cost and difficulty of decommissioning; to first order, the increase is probably about linear once the plant has become "hot" in the first place.

This is not a uniquely nuclear issue. Similar plant-specific considerations about upgrade *vs.* abandonment apply to fossil-fueled plants requiring definite or contingent retrofits for pollution abatement, especially for any plants emitting the superfine particulates now being more stringently restricted.[67] Merit order and operational competitiveness could also be radically changed by a carbon tax or trading price that disadvantages the carbon-intensive coal plants: a $20 tax or price per metric ton of carbon would allow combined-cycle gas plants (with long-term fixed-price gas contracts) to displace many, and $40 most, of the coal plants now operating—which may help explain why the coal lobby is leading the fight against carbon taxes. Another possible cause of considerable shifts would be changes in actual (or, for the combined-cycle-turbine projections shown, in the assumed) price of natural gas, or in its deliverability to certain constrained areas; gas-price increases in 2000–01 have raised this concern in some quarters. Still another possibility is repowering of inefficient old boilers, perhaps combined with fuel-switching to save cost or pollution.

Competition in the generation industry, although genuine, has been rather limited in scope. The Public Utility Regulatory Policies Act of 1978 and the Energy Policy Act of 1992 have introduced a measure of competition in the market for new generating resources, but have not affected the embedded costs of generation for utilities. Most of the generating capacity operated by utilities today was built in the previous era of power plant construction: over one-half of the operational power plants larger than 50 MW are over 25 years old. They were ordered and built at a time when increased

scale continued to lower the average unit cost of generation and therefore the price to customers.

This basic trend reversed in the early and mid-1970s as limits to the economies of scale were discovered, and strict new environmental and safety requirements raised the costs of building and operating large coal and nuclear power plants. In addition, during the building boom of the mid- and late-1970s, the electric utility industry, the world's most capital-intensive industry, was battered by historically unprecedented interest rates. These basic forces combined to increase vastly the cost of new power plants and reverse the declining cost curve for generation that had prevailed for most of the twentieth century.

What evolved from this volatile period was a fragmented electric industry with rapidly diverging cost structures among companies. The balance between increasing prices and/or satisfying shareholders led to a widening gap between prices across the country.

Today, the generation industry is increasingly competitive, made up of a combination of traditional utilities, utility affiliates, and independent developers, all competing for financing and market opportunities. Because the marginal costs of new generation technologies, particularly gas-fired turbine generators and distributed resources, continue to fall, new market entrants have the means to seriously undercut most utilities' average costs of generating electricity. This basic set of circumstances—the difference between the embedded costs of generating capacity owned by established utilities and the marginal costs of new

[67] Increasing epidemiological evidence suggests that very fine particulates are considerably more hazardous to public health than previously believed; *e.g.*, the half-million-adult, >100-city study announced in March 2002 that found ambient exposure had a long-term lung-cancer risk comparable to that of a non-smoker living with a smoker (552). Sooner or later, regulation will catch up.

resources—is creating a powerful new competitive environment.

But perhaps the most important "wild card"—one that, unlike these, is *not* mentioned in the analysis built around Figure 1-33 (68)—is competition from distributed resources that require little or no distribution, offer superior power quality and reliability, and provide the dozens of other advantages described in Part Two. For as we shall show next, just *delivering* the average kWh to the average customer costs considerably more than the generating costs shown in the graph for most U.S. generating plants. Distributed resources, being already at the load center, can avoid essentially all of that distribution cost. So how big *is* that cost of getting a kWh from the busbar to the customer's meter?

1.3.4 The invisible grid

While extensive data are publicly available on the generation sector, data are astonishingly sparse on the allocation of costs downstream of the generator.

The utility industry historically focused almost all its attention on how to *produce* electricity. It treated the grid as a necessary but relatively uninteresting accessory that transported large amounts of electricity from the power station to the customers in return for payments. This emphasis on the generator far more than on the grid spawned a curious bias, persistent to this day, against careful accounting for the costs of *delivering* electricity. The result is an industry whose economists and accountants know almost everything about plant-by-plant generation, but little or nothing

(especially in comparably facility-specific detail) about transmission and even less about distribution.

For example, one might suppose that for utilities in general, or for some class in particular such as large investor-owned utilities, one could readily look up:

- how much of a kWh's retail price goes to generation or power purchase, transmission, distribution, and other costs of customer service such as billing and sales; or

- for wholesale power, how much of the generator's total cost is for capital, for fuel, for other operating and maintenance costs of the generator, and for *delivery* to the point of sale; or

- for the grid, how much money goes to capital cost, losses, operation, and maintenance, and how all those costs differ between average "embedded" capacity and new or "marginal" capacity, or even better, between different locations.

Yet one searches in vain through the voluminous statistics of the Energy Information Administration, Edison Electric Institute, and other organizations for these fundamentals.[68] The absence of comprehensive data on transmission and distribution costs is clear evidence of an industry mindset that is largely inattentive to delivery costs, and a pervasive lack of feedback from those costs to influence investment choices. That lack of information is itself important information: it says what people aren't noticing. And under traditional regulation with its distribution monopoly, why should they?

In any other industry, such a "blind spot" would be extraordinary. In 1999, roughly half of all electricity sold in the U.S. was sold in the wholesale market before it was

[68] Capital accounts and operating-and-maintenance expenses are expressed separately, and both generally lump together generating, transmission, and distribution assets. By drilling deeply into the published accounts, one can more or less accurately calculate the quantities bulleted above, but they are not directly displayed.

sold to ultimate customers (195); the whole-sale price averaged about $0.034/kWh.[69] Yet its ultimate retail price averaged $0.0666/kWh (194)—outwardly equivalent to a more than 100% markup.[70] The composition of the costs that cause this markup is not normally reported for the industry as a whole, and is very hard to find, requiring calculation from intricate and obscure reports filed by each utility separately. This anomaly reveals a culture whose focus remains on production, not delivery, and whose generation, transmission, and distribution planning philosophies and practices are disjointed, as an EPRI/NREL/PG&E report describes (111):

> Generation planners identify the need for new generation or storage facilities and the size and type of such facilities based on projections of system-wide load increases and the cost of these facilities. Transmission planners identify transmission system needs to accommodate new generating and storage facilities, load growth, or wheeling requests based on the study of a few transmission system reinforcement alternatives, without much regard to the existence of the distribution system. Distribution planners concentrate on meeting local load growth without an in-depth examination of the capabilities and the constraints of the transmission system.***

> Most fully-integrated utilities have substantially separate generation, transmission, and distribution resource planning organizations....[M]anagement attention has been on generation resource planning involving large, single, expensive generation acquisitions. Bulk transmission planning and design has focused on issues such as power system stability, secure operation[,] and interface with neighboring utilities. Like generation, bulk transmission projects also represent large, single investments which are closely scrutinized by upper management and regulators. Distribution[, in contrast,] is at the tail end of this planning process. Distribution planners, typically located in distribution divisions physically removed from company headquarters, must respond to changing customer needs and coordinate these with their utility's marketing strategies and practices....The distribution system finds itself responding to outside needs rather than having its technology and budgets driven by strategic planning.

This dis-integrated approach creates blind spots—and hence, in a more competitive environment, huge new business opportunities to identify and wring out waste at an enticing profit. These include grid losses—as we shall show in Section 2.3.2.1, poorly known but probably worth, in the U.S., on the order of $15 billion per year at retail prices—and in the long run, much if not all of the capital and operating cost of the entire grid, which we shall calculate (§ 2.3.2.1.2) to average around $0.024/kWh in 1999, or one-third of the average retail price of electricity.

---

[69] Table 11 of EIA's 1999 *Electric Power Annual*, Vol. II, reports that large investor-owned utilities spent $43.26 billion for purchased power. The 1,636 TWh national total of sales for resale (p. 9), however, may well be for a broader category of utilities. If, as may well be the case, the wholesale transactions should also include power purchases by publicly owned utilities, which bought power for an average of $0.032/kWh (generators) or $0.041/kWh (non-generators) according to Tables 16 and 20, then the average wholesale price could rise to about $0.03/kWh, since public utilities had about a quarter of the retail market.

[70] This is not quite a correct interpretation for several reasons, including many utilities' preference to buy power from their own generating departments even if outside wholesalers are cheaper, and the cost and profit structure of the ~51% of retail electricity that was provided in 1999 through a vertically integrated company rather than undergoing at least one sale-for-resale along the way. The structure of sales for resale is also extremely complex.

1.4 FINE-GRAINED THINKING

The "invisibility" of the grid in traditional utility economics conceals not only the cost of delivering electricity, but also the enormous *variability* in that cost over time and space. Drilling down into that variability reveals startlingly large opportunities for distributed resources not merely to avoid the costs of distributing electricity, but also to avoid those costs specifically, and first, at the places and times where they are greatest.

1.4.1 Tapping the area- and time-specific bonanza

Traditional utilities project aggregated customer demands and build to meet them. Status and attention follow budgets, making generation the core activity and the grid a mere appendage planned and run by minor functionaries (§ 1.3.4). The glamor and drama of huge generating stations makes stringing wires, digging up cables, and maintaining substations seem dull by comparison. However, a handful of utilities have lately turned this traditional cultural assumption on its head, led first by PG&E and then by Ontario Hydro.[71]

These practitioners of "Local Integrated Resource Planning" (LIRP) reject the traditional planning approach from the generator downstream. Instead they start with what *customers* want, then work back upstream toward the generator to see what mix of resources can meet customers' needs at least cost. The "wires business" then becomes not ancillary but central—not a conduit for electrons but a way of solving customers' problems at least cost. Or with sufficient emphasis on end-use efficiency and onsite or local generation, the wires business, especially bulk transmission, could even become less important. In either case, remote central generation becomes not central but ancillary, an uninteresting and generally not very profitable commodity business. In contrast, demand-side management and distributed generation become crucial and highly profitable extensions of distribution planning.

Under LIRP (or, as RMI renamed it in 2001 to reflect more accurately the nature of the endeavor, ERIS—Energy Resource Investment Strategy), demand-side management is aimed *at the specific end-uses and neighborhoods that will best defer or avoid costly grid investments*. Aimed like a rifle instead of a shotgun, the resulting "precision-guided programs" use the utility's fine-grained knowledge of customers and of which are the costliest avoidable grid investments to increase the DSM efforts' returns manyfold. And since grid, especially distribution, investments are driven by *local*, not systemwide, peak demand (the two may or may not coincide), anomalies in the timing, intensity, and composition of area-specific loads, and their relationship to costly increments of capacity in substations, cables, feeders, etc., become a profit opportunity that can be picked off a piece at a time, juiciest first.

In practice, demand-side and grid resources are typically so rich in opportunities that additional generating resources are not required at all. That is certainly what Ontario Hydro found in its first three case-studies, chosen from its roughly 200 distribution planning areas, to explore the possibility of deferring or avoiding[72] major grid

[71] To which John C. Fox migrated from managing PG&E's demand-side programs, then the world's largest, to running Ontario Hydro's downstream and then upstream half—taking with him some of the lessons of the pioneering distributed-resources work. (Mr. Fox, a Canadian civil engineer, is also Chairman of Rocky Mountain Institute and was Chairman of the Board of RMI's E SOURCE subsidiary until its sale in 1999.)

[72] It is more common to defer distribution investments and to avoid transmission investments. This difference can affect the economics of the distributed resource being compared with those investments, since transmission investments are traditionally compared with central generating stations lasting for several decades, while deferrals of distribution investments are usually much briefer and hence less valuable.

Figure 1-34: Seasonal and time-of-use range of PG&E's cost to produce power and deliver it to feeders
At PG&E's system level in the early 1990s, averaging over all locations, the marginal cost of production and delivery to the feeder was naturally higher in the summer, especially onpeak, but not wildly different from retail tariffs.

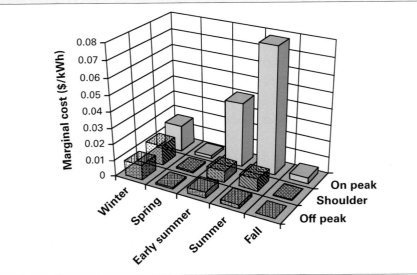

Source: J. N. Swisher and R. Orans, "The Use of Area-Specific Utility Costs to Target Intensive DSM Campaigns" (*Utilities Policy* 5, 1996), pp. 3–4

having deferred or canceled some C$1.7 billion in T&D [transmission and distribution] spending." (397) This idea has rapidly spread: by late 1995, more than 100 LIRP-type analyses were reported by North American utilities, plus other examples in Australia, Brazil, Ireland, New Zealand, and the United Kingdom (397). Some successes have been spectacular, as when New York State Electric & Gas Corporation avoided a $6.5-million grid upgrade by installing $0.045 million worth of communications and metering hardware to dispatch two customer backup generators at times of peak demand (397)—a 99.3% capital saving.

1.4.2 Basking in the "hot spots"

Part of the reason LIRP is so lucrative is that distribution assets typically have very low utilization, for an obvious but often overlooked reason: the smaller the area served, the less load diversity is available. Taking the argument to its extreme, a single household has a very low load factor because capacity to serve it must be sized for a peak load that is very seldom experienced, and

investments. In all three cases, the customers' needs could be most cheaply met by a mixture of demand-side and grid resources alone; marginal capital intensity decreased by up to 90%; and net savings totaled around C$0.6 billion. Through August 1995, Hydro credited LIRP "with

Load factor is the ratio of how much energy a load draws over a given time, such as a year, compared with how much it would draw if it drew at the rate of its maximum (peak) power continuously throughout that period. For example, a load that peaks at 2 kW but averages 1 kW has a load factor of 0.50 or 50%. Electricity providers must size their equipment to deliver peak loads, but collect revenue proportional to average loads—unless they use **time-of-use pricing** or **peak-load pricing** to charge more for electricity used when demand is highest and therefore costliest to meet (both because it is the peak loads that drive capacity requirements and because at times of peak loads, the costliest-to-run generators must be operated).

Load diversity is the ability of different customers to share a smaller amount of generating or grid capacity than they would require if their usage all peaked at the same time. Because different customers tend to use different devices in different ways at different times, one customer's peaks tend to offset another's valleys. Load diversity can be deliberately increased by education, load-management controls, tariff structures, or technical improvements. For example, more thermally efficient buildings change temperature more slowly and store heat or coolth better, so their space-conditioning peak loads will tend not to coincide with those of inefficient buildings that closely match the outdoor temperature.

A **load-duration curve** shows how much of the time a given asset in a utility system, such as a substation, transmission line, or power station, is being utilized to a given extent. Low load durations indicate constant ownership costs offset by scanty revenues.

the average load can easily be ten or tens of times smaller than that peak. The result: utility capacity that can easily be utilized to only 20–30% of its full year-round capacity. But as customers are aggregated at the level of a feeder, load diversity rises; at the level of a substation, it rises still more; and ultimately at the level of the entire utility or country it approaches its maximum. (On a continental scale, there is the additional diversity of different weather patterns, climatic zones, and time zones.) That maximum load diversity at the most aggregated level enables a utility to minimize generating capacity—which would provide a major economy of scale if transmission and distribution to reach all those customers were free to build and to operate. But *distribution* equipment *must reach every customer.* Being therefore inherently fine-grained, it suffers from the ever worsening load factors all the way out to the end of the system. Yet it is precisely at the end of the system that distributed resources are typically installed—just where they will serve the peakiest loads and hence save the biggest distribution costs and losses.

The resulting potential for improved utilization of distribution assets is illustrated by the following, increasingly detailed, graphs for PG&E in the early 1990s. These load-duration curves compare typical distribution feeders, and reveal much exploitable scatter between different segments of the 2,979-feeder "fleet":

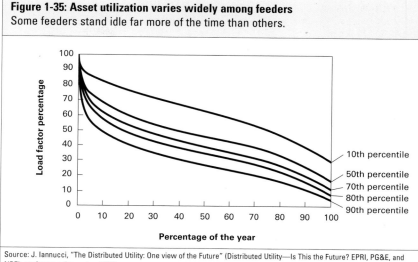

Figure 1-35: Asset utilization varies widely among feeders
Some feeders stand idle far more of the time than others.

Source: J. Iannucci, "The Distributed Utility: One view of the Future" (Distributed Utility—Is This the Future? EPRI, PG&E, and NREL conference; December 1992)

Such analysis is especially revealing for the feeders at the top (most peaky) 10% of the system load-duration curve:

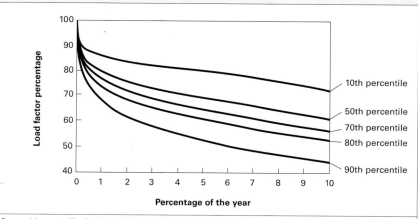

Figure 1-36: Differing feeder asset utilization is exacerbated near peak-load hours
The one-tenth of the year when feeders are most heavily loaded (magnifying the upper-left corner of Figure 1-35) reveals especially lucrative opportunities for distribution support or load displacement by distributed resources.

Source: J. Iannucci, "The Distributed Utility: One view of the Future" (Distributed Utility—Is This the Future? EPRI, PG&E, and NREL conference; December 1992)

Thus distributed supply- or demand-side (or grid-improvement) resources applied at the level where the load factor is worst can most improve distribution asset utilization and can best avoid costly distribution investments. Understanding which parts of the distribution system are least utilized can reveal where distributed resources are most lucrative to install.

Besides that broad principle is a specific circumstance often overlooked. The costs of electric generating capacity are often decreasing with time, due to changing technical and social conditions (combined-cycle gas instead of coal or nuclear steam plants). Much distribution expansion is on the contrary becoming costlier, mainly because it is necessarily installed in built-up areas that require undergrounding at many times the capital cost of overhead lines.[73] By one estimate, the real total cost of grid delivery in the U.S. probably increased by about 35% during 1955–2000 despite technological improvements (767). Grid installation becomes especially costly in areas where the grid capacity is fully utilized by rapid residential or business growth, raising both land prices and opposition to siting facilities in the very places where those facilities are required. Moreover, most such growth tends to increase grid-capacity requirements not with steady industrial baseloads but with the peaky, hard-to-predict loads driven largely by space-conditioning—implying more cost but less revenue. These factors generally increase the ratio of marginal grid costs to marginal generating costs, degrade grid asset utilization, and present an obvious opportunity for profitably rethinking the capacity problem.

When Australia's South East Queensland Electricity Board, for example, was contemplating an A$11 million grid upgrade to meet a load occurring only 50 hours a year (9% of the utility's local capacity was being utilized less than 0.6% of the time), it was strongly motivated to develop instead an A$1.5-million demand-side alternative to shave off that peak (767). Typical cases are less obvious—yet often highly profitable. PG&E, for example, found the disquieting pattern shown in Figure 1-37: a *typical* distribution circuit is used at under 50% capacity more than 60% of the time and reaches 70% utilization less than 10% of the time—whereas the company's average generating asset utilization *never* falls below 50%. The difference in asset utilization expresses the difference in load diversity between a huge utility and a particular, local, fine-grained service area that has fewer customers doing a smaller variety of things that are more likely to need electricity at similar times.

Moreover, PG&E found that very locally specific study often disclosed enormous disparities: marginal transmission and distribution capacity costs across the company's sprawling system (most of Northern California) were found to vary from zero to $1,173/kW, averaging $230/kW.[74] The maximum cost of new grid capacity was thus five times its average cost. Since marginal energy and power supplied to customers in these different areas would yield more or less identical revenues (even with more transparent pricing) but would incur such gigantic differences in delivery cost, demand-side interventions carefully targeted on avoiding the costliest capacity additions could disproportionately raise profits.

Not all utilities have similar opportunities: PSI Energy in Indiana found that 73% of its planning areas had zero marginal T&D capacity cost over a 20-year planning horizon,[75] bringing the system average down to $63/kW—only 27% of the PG&E average.

[73] The latter part of that hypothesis cannot be tested by examining aggregated time-series data on distribution investments *vs.* electric sales, because these two time series are clearly disconnected by the "inventory" of lumpy capacity, but is evident from specific utilities' field experience.

[74] However, this may be low, since another PG&E source (626) cites a system average cost cost of $282/kW for PG&E's transmission alone.

[75] Presumably because of excess capacity previously built, or slackening demand, or both.

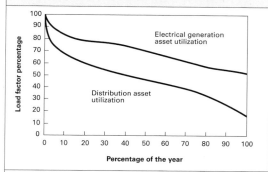

Figure 1-37: Distribution assets stand idle more than generation assets
A typical early-1990s PG&E distribution asset is less than half-used more than 60% of the time.

Source: J. Iannucci, "The Distributed Utility: One view of the Future" (Distributed Utility—Is This the Future? EPRI, PG&E, and NREL conference; December 1992)

However, Central Power & Light, a mostly rural Texas utility with hot spots on the Gulf Coast and in the rapidly growing Rio Grande Valley, had a maximum marginal T&D capacity cost of $1,801/kW and a mean of $550/kW, 2.4 times PG&E's (515, 681). Similar or even higher cost ranges can be found in rural areas of some developing countries (761).

Most strikingly, informal 1995 estimates from Southern California Edison company indicated that in some areas where old underground feeders need to be reconductored or deloaded—perhaps fancy neighborhoods or traffic-critical areas where excavation is costly and awkward—some grid-support applications are already val-

ued at $5,000–10,000/kW (525, 527). The *lower* end of that range is already a competitive opportunity for complete photovoltaic systems (§ 2.2.2.2); the higher end can support rather fancy ones. Practically any other kind of distributed generation would cost less than photovoltaics, and demand-side investments would cost even less. Similar opportunities leap out from area- and time-specific marginal costs reflecting both energy and capacity values: PG&E, for example, found that while *system-average* marginal *revenues* reached ~$0.08/kWh on hot summer afternoons (Figure 1-34), some "hot spots" in the system, while collecting no greater revenue, had actual *local* marginal *costs* each about $3.50/kWh (Figure 1-38) (515), nearly forty-fold higher than the revenues!

Figure 1-38: Peak power in a high-cost part of the distribution system can incur huge delivery costs
PG&E's average marginal cost (delivered to a feeder) in the early 1990s for a *specific and high-cost* distribution planning area can rise to as high as $3.50/kWh—a huge multiple of the price charged. Area- and time-specific analysis can identify distributed resources to lop off such costly peaks—in this case, over 40 times the systemwide average of such marginal costs (Figure 1-34)—exactly when and where they occur.

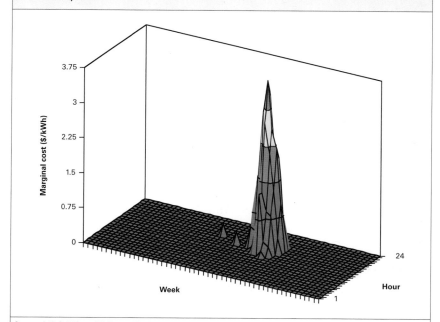

Source: J. N. Swisher and R. Orans, "The Use of Area-Specific Utility Costs to Target Intensive DSM Campaigns" (*Utilities Policy* 5, 1996), pp. 3–4

These area- and time-specific (ATS) costs can vary widely in time and space, creating important variations. They allow precise targeting of distributed resources in areas where the distribution utility costs are relatively high. This is further illustrated by data from a study of four U.S. utilities, in four different states, with a total of 378 utility planning areas (293). These utilities were quite diverse in customer mix, load profile, and size. Their differences in marginal distribution capacity cost (MDCC) were dramatic:

and by location. It is encouraging, however, that three of these four utilities, despite their wide variations, showed considerable opportunities worth at least $200–400/kW for deferred distribution capacity. Moreover, distributed resources need not meet an area's entire load to defer planned distribution capacity, because the needs are typically spotty. In fact, deferring distribution capacity in *all* high-cost areas shown in the previous graph would require distributed resources equivalent to less than one-tenth of the total existing load, yielding big benefits from modest investments.

It is also noteworthy that since local peak demand drives the MDCC value, that peak may occur at different times, and be caused by different customers or loads, than the system peak. Thus if the system peak occurs in the late afternoon, it may nonetheless be true that for a particular heavily loaded area, the local peak is actually at midday and thus suitable for (say) photovoltaics whose output does not coincide with the system peak.

The previous graph is a snapshot in time. But in fact, ATS costs change as power systems evolve:

Figure 1-39: Range of marginal distribution capacity cost for four U.S. utilities, 1994

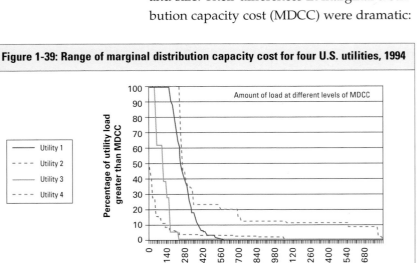

Source: J. N Swisher, "Cleaner Energy, Greener Profits: Fuel Cells as Cost-Effective Distributed Energy Resources" (RMI, 2002), www.rmi.org/sitepages/pid171.php

Utility 2 had built ample distribution capacity, so 72% of its planning areas had zero MDCC over the 20-year planning horizon, while Utility 4, with less spare distribution capacity, had MDCC above $320/kW in 75% of its planning areas. The MDCC for Utility 3 ranged from $50/kW to only $182/kW, while Utility 1 showed a range from zero to over $1,300/kW. The mean MDCC varied from $73/kW for Utility 2 to $556/kW for Utility 4. Sound planning to maximize the benefits of distributed resources thus requires utility-specific and fairly up-to-date information, differentiated by time of use

Figure 1-40: Area- and time-specific costs are important but not constant
Comparing conventional aggregated-cost siting of distributed resources with using area- and time-specific costs (ATS method).

Conventional approach:
Based on system-level costs, all areas look the same and stay the same.

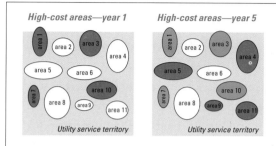

Based on area-specific costs:
Some high-cost (red) areas are attractive for DG now, but these areas become low-cost (green) areas later.

Source: J. N Swisher, "Cleaner Energy, Greener Profits: Fuel Cells as Cost-Effective Distributed Energy Resources" (RMI, 2002), www.rmi.org/sitepages/pid171.php

Siting distributed resources in the places where, and using them at the seasons and times of day when, they will yield the greatest value is clearly advantageous. But these optimal sites and times will gradually change as the distribution system and its loads evolve, making the optima into moving targets. Fortunately, many distributed resources can move too: they are portable, as described in Section 2.2.2.8, preserving their flexibility to remain in the right place at the right time as system needs change.

Such a fine-grained understanding of opportunities in specific utility systems is a rare but important business asset. Its value far outweighs the cost of collecting such time- and area-specific load data—data that can become almost automatically available to the distribution utility (and, one hopes, to its decentralized competitors) as a byproduct of distribution automation. Capitalizing on those local data could lead utilities to business strategies that successfully bypass the emerging wholesale bulk-power market with demand-side and grid resources "that aren't competitively bid because they don't flow through the grid at all: they are *already* at the load center." (427)

Because of its fine-grained geographic focus, LIRP is often called "Distributed Resources Planning." However, being focused on avoidable T&D investments, LIRP often neglects many other very important classes of distributed benefits discussed in Part Two. LIRP is thus an important driver of cultural change toward the customer focus, the attention to the grid, and the fine-grained thinking that assessment of distributed resources require. Yet LIRP offers only a modest part of the full range of distributed benefits. LIRP is therefore less a self-contained solution than it is a key to unlock the door into a new realm piled with a bewildering variety of riches. Part Two will explore this treasure-house, room by room. But first it is important to understand some of the major uncertainties that can further motivate electricity sellers and buyers to harness distributed resources.

1.5 Uncertainty Reigns

For nearly a century, the growth of demand for electricity was exponential. It could be rather accurately forecast by applying a straightedge to semi-logarithmic graph paper. A chimpanzee could do it. (The uncharitable were heard to mutter that load forecasting models simply semi-automated the chimps and disguised them as econometric equations.) But in the 1970s, previously durable trends came unstuck both on the demand side and, as described earlier, on the supply side.

By the late 1990s, essentially every rule of the comfortable pre-1970 world had been shredded by changes that seemed to be screaming into fast-forward. Today's electricity industry, still largely staffed by dedicated professional engineers with a deep commitment to reliable public service, faces a profoundly disquieting world. Old verities are vanishing into a vortex of pervasive turbulence. The turbulence is intensifying, and familiar rules and structures are vanishing.

The basic assumptions, methods, and actions needed for maintaining a prudent balance between supply and demand for electricity—a vital part of the challenge of keeping the lights on—are rapidly changing under at least ten main influences:

1. Tighter regulation and competition are gradually squeezing out the once-bloated reserve margins left over from the 1970s–1980s lag between pre-planned, long-lead-time construction projects and the slackening demand growth shown in the famous "NERC fan":[76]

Figure 1-41: Summer peak demand projections: comparison of annual 10-year forecasts
Successive industry forecasts of contiguous-U.S. summer-peak electric load ratcheted down until they bumped into reality around 1984. The actual non-coincident peak load in 2000 (a summer about as hot as the 1949–2000 average) was 686 GW, slightly above the "Actual" trend-line shown.

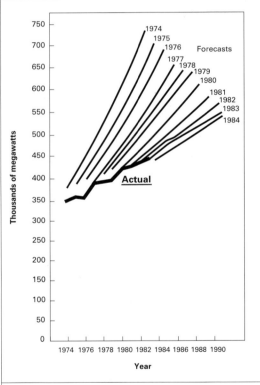

Source: OTA, "New Electric Power Technologies: Problems and Prospects for the 1990s" (OTA, July 1985), p. 45, fig 3.3

[76] This is the 1984 version of a durable classic (512). For comparison, the (non-weather-normalized, noncoincident) peak demand in 1990—a year with 10% more population-weighted cooling degree-days than the 1980–94 average—was 546 GW, slightly below NERC's 1984 forecast for 1990, so evidently NERC had learned its lesson by then. The approximate national reserve margin (a notional figure because what really matters is deliverable reserve margin in each region) fell from about 39% in 1986 to about 15% in 2000.

In response to the 1973 and 1979 oil-price shocks, the U.S. government strongly propelled a supply-side push, but it was soon overwhelmed by a successful market demand-side response that cut the nation's energy bills by roughly $200 billion a year and left much capacity standing idle. The subsequent collapse of new-capacity orders, and prolonged "coasting" on the surplus, has since eroded that cushion. With many local and regional variations, the national-average capacity reserve margin in 2000 was around 15%—approaching the traditional lower bound of prudence—including all the non-utility capacity, although about one-third of non-utilities' output serves onsite loads that may or may not be reported as part of the national total peak load.[77] The more important non-utility generation becomes, the more net supply available to meet general customer demand becomes unknowable until this ambiguity in the statistics is resolved. The whole concept of system reserve, who determines it, who pays for it, and who's responsible if it proves inadequate—all these are up for grabs. Typically the politicians who put them up for grabs were assuming that free-market pricing will be as attractive in practice as it seemed in prospect. So far that faith seems misplaced, and constituents seem disinclined to pay electricity prices that are as volatile as other commodity prices.

2. Traditional generation-portfolio planning methods extrapolated long-term demand trends, added a more or less fixed reserve margin to cope with the unexpected, and selected least-cost steam plants or combustion turbines to fit the load-duration curve,[78] perhaps subject to some diversification of the fuel portfolio. State commissions reviewed or oversaw this process, often requiring an overlay of integrated resource planning (mandated in principle by the 1978 PURPA law) to ensure, at least theoretically, that supply-side resources are properly compared with demand-side resources. But this whole elaborate structure is currently fashionable to attack as "central planning," and is being gradually displaced by market-driven transactions that rely on someone else—anyone—to figure out what's needed and to build it in time. That choice may well be duplicative, early, late, or absent. If it is absent, nobody is clearly accountable for the market's non-delivery of timely capacity.

This approach tends to emphasize short-term capacity additions in response to wholesale electricity prices (starting to be augmented by electricity futures and options prices) that are supposed to signal current or imminent scarcity. On a spot basis they certainly do this, as survivors of the enormous spike in British wholesale electricity prices in late 1995 found out and survivors of its even bigger California counterpart in 2000–01 will presumably discover in due course. Typically, however, suppliers stampede to earn high prices, and their overreaction is often exacerbated by governments' further overreaction. The

[77] The Energy Information Administration reports significantly different non-utility generation and disposition than Edison Electric Institute. In both systems, substantial ambiguities remain about how both supply and demand are reported for non-utilities and their transactions with utilities. However, the reporting is improving now that non-utilities' 29.5% of 2001 U.S. generation has become too large to relegate to a mere footnote. For example, EIA reports (193) that in 1999, U.S. non-utilities generated 569 TWh, received a further 90 from utilities and other non-utilities, delivered 413 to utilities and other end-users, and used about 246.0 themselves. Unfortunately, no updates had been published by the end of March 2002.

[78] This curve shows what fraction of capacity is dispatched what fraction of the time, typically with baseloaded plants toward the right and peakers toward the left. It is economically correct to base dispatch decisions only on operating cost, because the capital costs of available capacity have already been sunk; the only remaining decision is how much to run the plants. However, dispatch decisions are sometimes distorted, *e.g.*, when PG&E used to dispatch its own Diablo Canyon nuclear plant (which earned a high profit whenever operated) in preference to lower-operating-cost windfarms (whose power the company instead had to pay for), or when nearly all nuclear utilities capitalize major repair costs and exclude those costs from dispatch decisions, even though if the plant were not run, it wouldn't need to be repaired (note 65 above). (The correct economic decisions in most such cases would be to write off the plant, but accountants who think in terms of unamortized assets rather than the economic principle of sunk costs are reluctant to do this.)

result of bad market rules can even be higher prices and risks than under the old "central planning" method.

Whether electricity prices can anticipate scarcity early enough to elicit timely increases in supply (or decreases in demand other than by belt-tightening and curtailment)[79] remains to be seen. San Diego's experience in summer 2000 was not encouraging: politicians sought to limit or roll back market prices to suppress the unwelcome price signal, but that signal led few actors to appropriate remedies. As then soon became evident statewide, the shift from long-term planning to short-term incremental ad-hocracy makes nobody directly responsible or accountable for the portfolio, which instead emerges *de facto* from market decisions. That may turn out to increase the risk both of overbuilding and of underbuilding, and may change those risks asymmetrically: nobody knows. But many industry experts are unsettled by the uncertainty of market outcomes and the prospect of shrinking reserve margins.[80] They vigorously express their probably prescient concern that utilities will ultimately be held politically responsible even for other parties' omissions, as in this November 1996 editorial in *Power Engineering* (659):

> We're going to have start constructing new generating capacity soon, or we risk widespread and prolonged electrical blackouts in the United States....
>
> ...The only way to solve the capacity problem is to build new capacity. Starting now. Peaking capacity. Intermediate-duty capacity. Baseload capacity. Simple-cycle gas turbines. Combined-cycles. Coal-fired steam plants. And we're going to have to upgrade our existing capacity because we can't afford to retire any of it. And we're going to have to build new electrical [transmission] systems.
>
> We can't hide behind restructuring and deregulation. Even with unbundled generation, the obligation to serve the load remains. Those who neglect to prepare

now to meet that obligation will pay a terrible price if they fail to meet it. Tomorrow's politicians, regulators and consumer advocates will not listen to excuses about uncertainty and restructuring. We're asking for trouble and we're running out of time.

This seems to be saying that even if utilities are relieved of the legal obligation to serve, they will still, in public perception, bear that burden politically, so in a world where nobody is actually responsible for keeping the lights on, utilities had better do it anyway, and need to get busy. That this same refrain has been sung in many keys for the past few decades does not necessarily make it incorrect: one can certainly imagine a set of conditions, however uneconomic or unwise, that could make it come true. Indeed, in 2000–01, California demonstrated this as described above (§ 1.1.2). (One reliable method is, like the editorial writer, to ignore the demand side: even a few years' slackened momentum in California's demand-side efforts in the late 1990s rapidly eroded reserves, but only specialists noticed this ominous trend.) Some of those conditions closely match the vision of economic theorists and political ideologues whose zeal to "reform" the "centrally planned" electricity system is matched only by their ignorance of and indifference to its engineering, risk profile, financing, political economy, and regulation. If the lights do go out, they will probably be the last ones to receive due blame. When California invoked rolling blackouts in 2000–01, some politicians blamed restructurers...but for not implementing their theories thoroughly enough, while others sought to suppress the price signals that were meant to elicit more supply and more efficiency. Veering from one extreme to another, and risking overreaction and overshoot into gross overbuilding, the politicians mainly confirmed [Ken] Boulding's Law of Political Irony—"Whatever you do to try to help people hurts them, and vice versa."

[79] However, these options are far from trivial, as New Zealand showed in the severe 1991–92 drought mentioned in Section 2.2.6.1.

[80] At least if peak demand grows relatively rapidly, implying continued or worsened inattention to the vast and still largely untapped demand-side potential.

3. Predictions of available capacity—long the best-known side of the business—are becoming fuzzier as non-utility generators provide a larger share of new and total capacity (about half of all new capacity in the mid-1990s and an extraordinary 99% for the 2000–04 planned installations reported by EIA in 1999). Non-utility generators may not coordinate their planning with utilities or their regional and national planning organizations—they're not yet even members of the North American Electric Reliability Council (NERC)—and they are seldom under any obligation to help meet such organizations' planning targets. Indeed, they may seek to derive market advantage from *not* sharing their capacity plans with the utilities with which they increasingly compete in wholesale markets. Many non-utility generators' capacity plans are now commercial secrets, absent from government statistics. Even previously collegial utilities are becoming cagey about sharing market plans with their neighbors. The planning and coordinating role of the regional power pools, long the mainstay of prudent system forecasting, is in disarray. This dearth of basic information makes intelligent market behavior more difficult.

4. Barring a major terrorist attack (419, 442, 467) or a similarly cataclysmic event, the major fuel-supply disruptions that dominated energy planning in the 1970s and 1980s now seem unlikely—after all, a full-scale war in the Persian Gulf caused no gasoline lines—and increasingly deep forward markets[81] in oil and gas seem to be stabilizing spot prices. But new uncertainties have emerged to take their place:

- Both competitive and regulatory forces may force the premature retirement of tens of GW of nuclear capacity during the next 5–10 years—by one recent estimate, 40% of the nation's nuclear capacity (221). Supportive rhetoric from the Administration that entered office in 2001 may not overturn the market verdict. Nuclear power in 2000 provided 12% of the nation's total summer generating capability and 20% of its generation (including non-utility generators in both). In principle, therefore, a 40% nuclear retirement could be equivalent to removing nearly one-fourth of the utility-plus-non-utility reserve margin.

- Even more significant could be climatic concerns leading to taxation or restriction of carbon emissions—no small matter to a power system that burns coal, the most carbon-intensive fuel, to make 52% of its total electricity. However, a complete switch from old coal to new gas plants would cost at most about 8% of what America now pays for electricity, and in some cases could actually save money.[82]

- Hydroelectric capacity could drop steeply due to siltation from overlogging and poor watershed management, or conflicts over salmon and other anadromous fisheries (now a critical issue in the Pacific Northwest), or conversion to more lucrative production of

[81] One can now buy these fuels 10–20 years ahead at predetermined prices, with the seller, broker, or third-party underwriter bearing and being compensated for the price risk.

[82] A modern combined-cycle gas plant has only one-fourth the carbon emissions per kWh of a classical coal plant, but replacing all the 2000 coal-fired generation (1,965 TWh) would require, assuming 90% capacity factor, 249 GW of combined-cycle plants—about ten times the combined-cycle capacity existing in 2000, and 5% more than the Federal forecast for 2020. On the other hand, building and running the combined-cycle plants—especially if building that many made them cheaper by more than it made their gas costlier—would cost little more than just running most coal plants (and less than running some), as Figure 1-33 suggests. Part of the cost premium would also be offset by avoided sulfur and nitrogen oxide emissions, which can be traded at market value. As a first-order approximation, such a switch, with due attention to gas deliverability, should not be unduly expensive even in private internal cost—in round numbers, $20 billion a year if building and running an average combined-cycle gas plant cost $0.01/kWh more than just running an average coal plant (a reasonable estimate from Figure 1-33, assuming $400/kW combined-cycle capital cost and 60% efficiency). Of that, about two-fifths could be recovered by avoiding a carbon emission penalty at the $20/T tax rate commonly discussed, to say nothing of other avoided emissions that trade for even more. A more sophisticated analysis would naturally have to take account of many other factors, including the potential price and other effects of a ~48% increase in the national rate of consuming natural gas—a controversial subject, especially in view of the gas-price run-up in 2000–01, and even more importantly, the potential locational and cogeneration value of distributed generators.

hydrogen for a fuel-cell-powered Hypercar® fleet.[83] A rapid hydrogen transition is indeed envisaged in the October 2001 Royal Dutch/Shell Group Planning scenarios.

Surprises may also be indirect. For example, climate change can significantly change electricity demand: one major Sunbelt utility alone found that each Fahrenheit degree of increase in peak-day ambient temperature would raise its peak load by 300 MW. Conversely, changes in temperature, humidity, and ecological conditions (such as those favoring growth of certain clams that clog power-plant condensers) can affect electricity supply. Severe storms are already disrupting energy supply more frequently through such phenomena as Midwestern coal-barge freezeups, coal-rail-stopping Western blizzards, Gulf of Mexico hurricanes (which can shut down natural-gas production platforms), and hurricane, ice-storm, and lightning interference with transmission lines. Electrical delivery is already regularly upset by natural phenomena ranging from solar storms to earthquakes, but climate change could make weather-related disruptions systematically more frequent and intense.

5. The effects of utility restructuring and regulatory changes on various parties' incentives and performance is completely speculative. For example, if restructuring turns out to make inter-utility coordination less effective, degrading the reliability of supply, it is not clear whether the political and policy response would be to favor less or more regulation, more or less emphasis on new generating capacity (*vs.* institutional, grid, or demand-side solutions), more or less internalization of reliability costs, etc. But especially in an increasingly electronics-dominated society, public unhappiness is virtually certain, analogously to that observed in the wake of recent air

[83] Selling hydrogen as a vehicular fuel for fuel-cell cars to compete with $1.25/gallon U.S. retail taxed gasoline is roughly equivalent in heat terms to selling the hydroelectricity for ~9–12¢/kWh, delivered in a different form that attaches a proton to each electron. Even after paying for the electrolyzer, compression (which can be partly done free by the hydrostatic head at the foot of the dam), and delivery, that is still far more profitable than selling the electricity into an increasingly crowded and price-competitive market. The hydrogen's advantage as a vehicular fuel, ignoring its cleanliness, arises because the fuel-cell car converts hydrogen energy into traction several times as efficiently as current cars convert gasoline energy into traction (440).

disasters blamed on careless deregulation, competitive pressures, and resulting corner-cutting. Results like the following, presumably simulated, satellite images 35 seconds apart (259) during the Western blackout of 10 August 1996, clearly imply the potential for political explosions: an Oregon power line sagged onto a tree-limb and launched a cascading series of events that ultimately blacked out four million customers in part or all of nine states spanning one-third of the continental United States and parts of Canada (328).

Figure 1-42: The brittle grid
The world's most complex machine can fail in unexpected ways, and very quickly.

Source: EPRI

Americans reluctantly put up with deteriorating telephone service, but not with noticeably flawed electricity service. A public stampede toward onsite and renewable generation, perceived as more reliable than remote grid supply, is one plausible consequence if such events become more frequent and wide-spread—a trend that emerged vigorously during and after California's 2000–01 power crisis.

Conversely, some observers expect real-time pricing—a relatively early and widespread consequence of wholesale competition—to level peaky loads, mitigating traditional concerns about inadequate generating and grid capacity. Indeed, on 4 March 2002, the FERC released a consultant's study showing that real-time pricing's demand response is likely to be at least *twice* as effective at decreasing electricity costs through 2006, and half again as effective through 2021, as the regional dispatch and transmission initiatives on which the FERC has been focusing (348). And a consistent undercurrent is likely to be greater attention to diverse customers' needs and expectations. Efforts to unbundle power quality and reliability, for example, may please customers desiring premium performance, but may leave others facing degraded performance and feeling like second-class citizens who are no longer receiving the sort of universal service they have long grown to expect. Most industrial energy managers polled in 1998 by E SOURCE pragmatically recognize that commodity electricity is not reliable enough for such critical uses as computers (343)—it can't be, because it's delivered via the glitch-prone grid—but less sophisticated customers might assume that that's what their present electric bills are paying for.

6. Even larger indirect effects of restructuring could result from aggressive market actors' higher-risk business strategies—remember the Savings & Loan industry and Enron disasters—or from unanticipated consequences of changes in regulation, taxation, subsidies, and other foundations of business decisions. Anyone who enjoyed the deregulation of airlines, cable TV, and telephones will love

the unfolding deconstruction of the electricity industry. As regulatory economist Professor Alfred Kahn has remarked, there would be little point in restructuring utilities if all the consequences were predictable. But some economists' fondness for making intriguing and surprising discoveries may be less congenial for the rest of us—especially for utility executives, who, as one advocate remarked, were weaned on a predictable, closely regulated environment and "...have little experience working in dynamic, unsettled environments. This, combined with the large investments required, makes the potential for financial loss as large [as], if not greater...than[,] that which existed in the first diversification wave." (656)

This risk in the nation's largest economic sector—to investors, operators, and customers—emphasizes the importance of exceedingly thoughtful and well-informed changes in an electricity system that has long been the envy of the world. But that is hardly how any dispassionate observer would describe the legislators, regulators, consultants, advocates, ideologues, and others now meddling in, or trying to blow up, the underpinnings of that system. On the contrary, some of those now most fervently pressing to transform its supposedly obsolete structure appear to know very little about its technology, history, institutional arrangements, or other fundamentals. Although the political pendulum seems to be starting to swing away from doctrinaire deregulation of everything, and the cautionary tale of California will certainly dampen global enthusiasm for experimentation, there is still enough momentum for legislative tinkering or worse, at both federal and state levels, that major mistakes cannot be excluded, nor strong political reactions to them.

7. In particular, forms of restructuring that inhibit demand-side management—*e.g.,*

by returning to the old system of rewards for selling more electricity—could markedly increase planning risks by restoring lately weakened links between electricity demand and the vagaries of weather and of economic growth. This occurred in 1996 (effective 1998) even in a state with the long and sophisticated regulatory tradition of California: restoration of perverse incentives to distributors, plus a few years' faltering in previously exemplary demand-side efforts, quickly eroded reserve margins, contributing to huge price and political volatility. Conversely, strengthening rewards for "best buys first" could align utilities' and customers' incentives enough to release a pent-up flood of investments in end-use efficiency, load management, and distributed generation, largely—perhaps even more than—offsetting both plant retirements and the demands of a growing economy. Of course, there are ways to keep demand-side investments vibrant even without regulated utility monopolies, but semi-reforms that continue to reward distributors for selling more electricity certainly complicate efficiency vendors' task.

8. Globalization is rapidly expanding transnational utility takeovers: aggressive and cash-rich U.S. utilities trapped in mature markets have been seeking faster growth everywhere else, often in societies they know little about. This can often bring useful modernization in attitudes and skills. However, it can also have unexpected consequences, including a diversion of management talent at headquarters, from "sticking to the knitting" to trying to remake challenging overseas utilities under alien and sometimes unstable conditions.[84] Meanwhile, back home, many nontraditional executives are entering the once-stodgy utility industry, bringing both fresh thinking and a limited grasp of technical fundamentals. Before restructuring and global-

ization, there were not enough first-rate utility executives to go around; but these trends mean there will probably be even fewer as more demands, including training electric novices, chase a similar number of top-class leaders. And meanwhile, the gradual attrition and retirement of traditional utility engineers—especially in a country (the United States) whose top electrical engineering schools have swung so far toward a computer focus that most no longer teach fundamentals of power engineering, rotating machines, etc.—will deplete the knowledge pool needed to sustain the existing system. While this turnover of human capital may help to modernize certain attitudes, it also risks losing the intellectual underpinnings that keep the lights on; and once the cultural continuity of teaching those skills is lost, it cannot be easily restored. In short, although older power engineers may not have caught up with the latest thinking in industry restructuring the distributed generation, newer entrants enthusiastic about those novelties may overlook at their peril the half-century-old engineering knowledge of how big power grids work. The trick nobody has yet mastered is how to keep a solid base of engineering understanding while overcoming the inertia of old ways of doing business so we can move freely on to more modern arrangements in both engineering *and* business.

9. RD&D on electricity generation, storage, delivery, and end-use is creating an ever larger portfolio of potential technological surprises that could dramatically shift the traditional slate of options. Cheap fuel cells and photovoltaics are among the prominent examples of technologies—some perhaps still unknown—that could render thermal power plants fundamentally uncompetitive. However, innovation is increasingly likely to come from proprietary developers and from overseas, because in the U.S., both public- and private-sector R&D budgets relevant

[84] Such as the surprise of a U.S. utility executive, newly transplanted to his firm's South American subsidiary, who was told—by some big men with bulges under their arms who surrounded him in a bar—that he would be extremely unwise to try to bill their cousin for the electricity his firm was using.

to the electricity industry have been slashed to ribbons.[85] These budget cuts and the resulting diaspora of experience and talent have already severely damaged the electricity-related RD&D infrastructure, made its focus ever shorter-term, and privatized what was previously a largely shared public enterprise. The hollowing-out of the public good of utility RD&D, and the high discount rates that leave many worthy technologies stranded just short of commercialization, could have serious long-term consequences, not least for national competitiveness: America could end up—is *already* ending up in many cases—importing from Japan, Europe, and elsewhere many of the renewable technologies that it originally developed. By 2000, for example, the windpower industry—with more than 13 GW installed worldwide, doubling every few years—had no significant U.S. firm left; both of the two large firms had been bankrupted, largely by tax-law instability, though others picked up some of the pieces. (Congress never learned this lesson: at the end of 2001, after a record-breaking year with 66% capacity growth, over twice the world growth rate of 31%, Congress once more crashed the industry by allowing its production tax credit to lapse amidst unrelated political squabbles, though it was retroactively renewed in March 2002.) As American leadership became largely a memory, the new leaders came to include China, India, Germany, and Spain. Three-fourths of all wind machines in world trade came from the mighty manufacturing nation of… Denmark, providing more jobs than it had had from fishing and shipbuilding. Denmark fully deserved its success. It is less obvious that the American wind-power industry deserved its failure.[86]

10. The fleet of "workhorse" power plants is inexorably aging. Like a demographic age-structure, the dramatic peaking of power-plant construction in the 1970s,

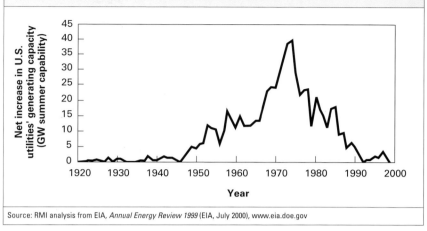

Figure 1-43: A past construction boom may have future echoes
Most plants now in U.S. service were built in the 1960s and 1970s.

Source: RMI analysis from EIA, *Annual Energy Review 1999* (EIA, July 2000), www.eia.doe.gov

graphed above, would normally be expected to be echoed by a peaking of plant retirements a few decades later, *i.e.*, early in the twenty-first century. This appears to be starting off with a bang: as of 2000, the rate of U.S. additions planned for 2000–04 averaged 30 GW/y, nearly equaling the early-1970s peak.

This new construction boom—though of very different types and sizes of plants than before, and 99% built by non-utilities—is motivated not only by power plants' retirement but also by their potential deterioration in reliability. Figure 1-10 showed from historic behavior that "broken-in" mature plants tend to become less reliable, and to have a greater scatter in plant-to-plant reliability, as they age—especially for the larger plants that do most of the generating. The degree to which the lives of these plants (particularly the non-nuclear ones) can be stretched and their reliability made higher and more consistent in their older age through more sophisticated maintenance and life-extension measures is hotly disputed, especially by those who prefer selling new plants to maintaining or refurbishing old ones.

In sum, the electricity future, more than at any time in history, is now dominated by

[85] Investor-owned utilities' R&D expenditures rose at an average rate of 4.5%/y during 1990–93, then fell 9.9% in 1993–94 alone, with far worse to come (683). Tired of spending most of their energy defending their budgets, the best people doing the most innovative work often left first. Some went where they could be more effective, others less. The net effect so far appears profoundly unfavorable: R&D capabilities that took decades to build were rapidly dismantled and scattered in the 1990s (376, 556).

[86] Ironically, at the start of 2002, a new Danish government shut down nearly all of its world-leading efficiency and renewables institutions after 15–25 years of consistent successes. The U.S. is no position to take advantage of this inexplicable blunder because of its Administration's generally indifferent-to-hostile attitude toward those same options: one of its first proposals on entering office in 2001 was to halve the already inadequate RD&D budgets for efficiency and renewables.

wild cards—all ten just listed, plus some more, plus others nobody has yet thought of. These wild cards could send peak loads up or down, could prolong or accelerate the retirement or derating of old power stations, and could limit or expand the supply-side, demand-side, grid, and institutional options available—all at the same time. Walter C. Patterson, the prescient electricity thinker at London's Royal Institution of International Affairs, even thinks that large-scale, integrated, synchronized AC networks may evolve into a far more diverse mix of relatively localized AC and DC networks requiring far less systemwide control.

Amidst such fundamental shifts, the sense of "flying blind," of trying to retain long-term prudence and wisdom amid the pressures of short-term improvisation and "precision guesswork," may turn out to be a psychological context in which distributed resources' cost-reducing and risk-managing opportunities may become attractive for both utility executives and public policy-makers. That is, the swirling uncertainty of the old system is creating a "teachable moment" in which the seeds of coherent new ideas could find fertile ground.

Distributed benefits may well be among those seeds. We suspect that the rapid shift away from lumpy, long-lead-time, highly centralized power plants has been driven more by aversion to their costs and risks than by a thorough understanding of the positive *benefits* of their decentralized alternatives—more by fear of the known than by attraction to the innovative. How much more might that shift accelerate and consolidate if most of the economic benefits of distributed resources, still unperceived or only vaguely sketched, became well-known? That is why we hope this work may be especially timely in defining and supporting the fundamental transition already underway from gigantism toward the right size for the job.

1.6 CAUTIONS AND HERESIES

Just before we explore each kind of distributed benefit in turn, five generic issues of assessment and technology merit mention to guard against confusion: cost, cost allocation, value, risk, and interactions (usually favorable) between different technologies. In addition, it is important to understand generically why small-scale resources can be quicker to bring into service than large-scale ones, and how the contributions of many small resources can add up to very large totals.

1.6.1 Cost and its allocation

Without counting any distributed benefits, just the bare capital cost of many technologies can depend very strongly on the simplicity of design and on other technical properties surveyed elsewhere (443). It is not unusual for different ways of providing the same type of energy service by the same general kind of technology to differ in cost by severalfold. Cost differences can be as large as one order of magnitude between different nuclear plants, and of even two or three orders of magnitude between different renewable resources of the same general

Utilities sometimes use a **Total Resource Cost** test to assess how much society as a whole pays for a given service. This is the correct method for examining societal economic efficiency, which does not take account of distributional effects. Those effects, which can be politically important (but are easily handled where problematic), can be assessed using **Rate Impact Measure** (RIM) tests. These, however, are sometimes misused to determine whether to make an investment, thereby guaranteeing economic inefficiency. The **All-Ratepayers** test examines cost-effectiveness from the perspective of the collectivity of utility customers (hence utility profits) without counting possible supplementary payments by others, whereas the RIM test asks whether a utility investment in a resource (usually end-use efficiency) that benefits those who choose to participate in a given program will raise the price of electricity, however slightly, to those who do not. If it did, this could of course be remedied by equal-opportunity participation in a portfolio of investments and programs, but the RIM test is typically applied to each micro-choice in isolation rather than to the portfolio. The result is to sacrifice large gains in economic efficiency for all customers in pursuit of tiny distributional effects.

Regulators are often urged to use these tests to support ideological perspectives, thereby pleasing some constituencies by distorting outcomes. For example, a utility that can generate its next unit of electricity in existing plants more cheaply than its average retail price (*i.e.*, almost any utility today) would fail the RIM test if it bought any end-use efficiency, even if that investment cost less than operating the existing power stations, or even if it cost zero, and thus would reduce all customers' bills (421, 424). Some states permit, and Florida has long required, this serious distortion: in essence, such states deliberately sacrifice large savings for all customers in pursuit of much (typically 10–100-fold) smaller distributional equity effects between certain customers, even though more thoughtful policies can achieve both benefits at once. This and similar issues are most commonly raised when utilities consider investing in more efficient use of electricity, but can also be important for supply-side investments—which uniformly fail this so-called "no-losers test" (more accurately called a "virtually-no-winners mandate") in these common circumstances.

Market pricing might at first appear to eliminate the issues underlying these differences of perspective by clearly defining their values and who is responsible for which costs. However, gaming could do the opposite, decreasing transparency and accountability. For example, deregulated prices could tempt electricity providers to look more competitive by lowering the variable part and raising the fixed part of their price structure. To some degree this may properly reflect costs and reduce the providers' perceived market risk, but it could also seriously, and deliberately, disadvantage competing end-use efficiency options, especially for the most efficient customers.[87]

type (466). This has nothing to do with distributed benefits, but everything to do with ingenuity, execution, and market structure (such as the number of markups between manufacturing and the retail customer).

Assessments of costs also depend on who pays for what. In the traditional and still largely prevalent U.S. regulatory framework, several different mechanisms are commonly used, and they can give very different answers.

1.6.2 Value

Value to customers has many dimensions other than the commodity cost and price of electricity. (Strictly speaking, electricity cannot be called a commodity unless it can be stored at minor cost until the price is right. This may become possible soon, but not

with currently commercial technologies.) In a more competitive and market-driven environment, customers' wishes become more differentiated and important. Branding, end-use services, satisfaction, responsiveness to customer wishes, psychological dependence, and many other dimensions become important, even decisive—and the less relatively important are the commodity aspects of the electricity.

Those aspects, too, become strongly differentiated. Reliability value is clearly very different depending on, say, whether the missing electricity was to run a water heater or a heart-lung machine, an airline reservations computer or an engine-block heater. Even as simple a value as locational rent can rise far above the ~$0.20–0.40/kWh common to Alaskan villages where fuel and mechanics arrive by Cessna: electricity at the South Pole (from engine-generators fueled with oil

[87] In mid-2000, a major electricity distributor around Christchurch, New Zealand, did exactly such an "anti-efficiency" change in its tariff structure, spurring public outrage that could prompt many customers to switch suppliers.

flown in by cargo plane), if properly costed, would run perhaps $2–4/kWh (though that doesn't seem to cause it to be used any more efficiently), and the cost of fuel-cell electricity aboard the Space Shuttle is...well, astronomical.

To pick an obvious if extreme example of the unexpected dimensions of value, Wayne Gould of Southern California Edison Company has estimated that the price people gladly pay for an ordinary AA alkaline battery is equivalent to ~$212/kWh; for a watch battery, ~$14,000/kWh. This portability premium is not confined to highly miniaturized batteries, and remains significant even for rechargeable versions: *e.g.*, a standard $139, 19.4-VAh lithium-ion battery for a subnotebook computer is equivalent at unity power factor (assuming an operating life equivalent to 200 full cycles) to $36/kWh, and comparable batteries for advanced cellular telephones can be priced at upwards of $100/kWh—interestingly, less than the throwaway AA alkaline battery's energy cost, despite the fancy lithium-ion battery's caviar-like cost per gram.

Conceptually, such examples of convenience and portability suggest that in at least some significant subset of markets, cases, and customers, the market value of distributed resources may be much higher than any commodity market could bear for the price of the electricity alone. People pay five orders of magnitude more per kWh for electricity from a hearing-aid battery than from the grid because they want to be able to hear, wherever they are, without having to plug their hearing aid into a wall socket. (That's why many developers consider hearing-aids, cellphones, and portable computers to be among the most promising

markets for miniature fuel cells that replace batteries.) Perhaps more surprising is the 3–4-orders-of-magnitude premium for the disposable AA batteries that almost every American routinely uses in all manner of gadgets, some of which provide only modestly more convenience than plug-in versions. Evidently what those batteries' buyers have in mind is the portable *function* or service provided, not the electricity. That is why photovoltaics, widely touted as uneconomic, now power the majority of the world's handheld calculators and are moving into the watch market. Oddly, nobody questions their economics because their functionality is considered their key attribute. Yet for many distributed resources, the opposite is historically true.

1.6.3 Risk

Section 2.2.3 will show that different resources that produce electricity at the same average cost may have radically different risks of price fluctuation. As a general rule, prices being equal, customers will prefer lower risk, and risk being equal, customers will prefer a lower price. Each customer is presumed by theorists to have some implicit mental function relating price to risk—how much lower the price must go to justify accepting a larger risk, or how much higher the risk can be without making the lower price unattractive. These risk/price relationships can in principle be discovered in the market, but will be different for each customer and may vary over time and according to many complex physical and psychological circumstances.

The value of risk for a given customer at a given time can be estimated using methods developed by stock-market and commodity analysts. Traditional engineering economics, however, has no comparable methods, and thus tends to neglect to value price-volatility risk, as well as most other kinds of planning risk (§ 2.2.2).

Over the next few years, electricity products are likely to emerge that let customers choose intermediate options between a fairly high fixed price and a lower but volatile real-time price, according to their individual risk aversion. Early experiments suggest that the revealed risk aversion can be about tenfold different if response to price is under the customer's control (via voluntary measures) than if it is compulsory and occurs unpredictably (via interruptible tariffs). (515)

It is also worth noting, though beyond the scope of this analysis, that different customers have widely divergent views of what a risk is, what it is worth, how much of it they want, and what they must be paid in order to tolerate it (411). This thorny issue—part of the wider problem that there is no valid way to compare different people's "utility functions" (367)—tempts many economists into improper attempts to evade it by asking people not how much compensation they would require in order to accept a risk, but how much they would be willing to pay in order to avoid it.[88] That will obviously depend not only on willingness to pay but also on ability to pay, especially since there is a virtually infinite universe of potential risks against which one might, by this method, be asked to buy "insurance." The concealed switch from a selling to a buying bid is convenient: without it, a single infinite bid—unwillingness to accept a given risk at any price—defeats the economist's effort to ensure Pareto improvement (making some people better off and nobody worse off) while imposing on society a supposed benefit he thinks it should have. But it is both theoretically and morally unsound (2, 411). The only proper test of willingness to accept risk is the compensation that the acceptor demands, however seemingly unreasonable to the payer.

1.6.4 Synergies between different kinds of resources

Efficient end-use and onsite storage typically make intermittent renewable energy supply smaller, cheaper, and more effective. They can do the same for nonrenewables too, but often the benefit is bigger for renewables, suggesting important business opportunities from bundling the two together. For example:

- a more efficient showerhead enables a smaller solar water heater to provide longer showers and a larger fraction of hot-water needs (and analogously for solar process heat in industry) (52);

- comprehensively high efficiency in using and distributing hot water can make quasi-seasonal storage affordable, eliminating backup and letting a simple passive-solar downpumper meet ~100% of the family's water-heating needs (416);

- using carefully chosen, superefficient lights and appliances to reduce average household electrical load from somewhat more than 1 kW to ~0.1 kW (§ 2.3.2.11) permits a small, cheap photovoltaic array with modest storage to suffice (565);

- daylighting leverages dimmable supplementary lighting, which cuts internal heat gain, facilitating passive and alternative cooling that's largely paid for by the reduced cooling loads (429);

[88] As Karl E. Knapp points out in a personal communication, 17 October 1997, these two variables may be largely symmetric and linear for small risks, but not for large ones: "We all make tradeoffs with life risks every day, but nobody who would accept $100 to accept a one-in-a-million chance of dying will accept $100 million to be shot."

- in a cold-climate house, passive solar heating can eliminate the heating system if the envelope is superinsulated, then the saved capital cost of the heating system can pay for additional savings in water-heating and electricity, yielding the spinoff benefits noted above for solar water-heating and electricity (565);

- in a hot-climate house, high end-use efficiency can eliminate not just the heating system but also the cooling and air-handling systems, saving even more capital cost with which to leverage electric and hot-water savings that potentiate renewable supply (429);

- superefficient cars (434) can reduce the transport sector's liquid-fuel needs enough to be supportable just by converted farm and forestry wastes without requiring costly "fuel crops";

- alternatively, ultralight hybrid-electric cars can reduce fuel use, fuel requirements for range, and tractive loads so far that hydrogen fuel cells become the power source of choice—triggering in turn a rapid transition toward fuel-cell total-energy systems in buildings and toward renewable hydrogen sources (474, 758, 775).

Thus with meticulous whole-system engineering, the combination of efficiency with renewables is a natural partnership. It yields major synergies—benefits greater than the sum of the parts (288). But this synergy is actually four-way: efficiency *and* renewables *and* local energy storage *and* local energy exchanges can all help each other to yield greater benefits than any one or two of them alone. Our discussion of distributed benefits in Part Two is necessarily organized in an atomistic, reductionist fashion, but Section 2.2.6 will remind us that combining several technologies in the right way often yields much greater benefits than analyzing any of those technologies by itself.

Analyzing such synergies, however, is far from simple, partly because changes of scale can change the basic physics of the system in rather subtle and unexpected ways (464):

> For example, several analyses have found that solar district heating should be able to cut the delivered price of active solar heat roughly in half (334, 536). There are good physical reasons for this (416):
>
> - A large water tank, shared between tens or hundreds of dwellings, provides (compared to the small tank in a single house) a large ratio of volume to surface area, hence low heat losses.
> - The large tank has a favorable ratio of variable to fixed costs, and it is relatively cheap to increase the size of an already large tank.
> - One can therefore afford to use a big enough tank to provide true seasonal (summer-to-winter) heat storage.
> - This in turn provides a full summer load, improving annual collector efficiency.
> - The large tank also permits further efficiency gains by separating the storage volume into different zones with the hottest water near the center and the coolest near the periphery—this improves collector performance and further reduces heat losses from storage.
> - With true seasonal storage, collectors can face east or west with relatively little penalty, rather than only towards the Equator, so such a system would be more flexible to site, especially in a city.
>
> The net result of all these effects is a marked cost reduction....Incorporation of solar ponds or ice ponds or both would also cut costs still further, and would incorporate energy collection and energy storage into the same device.

This example illustrates how sensitively optimal scale depends on technological concept and on the proposed use. (It will certainly depend, for example, on how much heat the buildings require, and on the local climate.) It may well turn out that active solar heating is cheaper at some intermediate scale than at the scale of a single house or a whole city. And it may also very well turn out that active solar heat at any scale is uncompetitive with simpler, smaller measures to make buildings more heat-tight and to increase their passive solar gain. The question of optimal scale for a particular device is therefore not the only important question to ask; one must also determine whether that sort of device is worth building at all.

1.6.5 Smaller can be faster

Scale can also affect speed of aggregate deployment. This issue is often framed as "Even if it's much faster to build a *single* small device than a *single* large one, does that advantage persist when one must build a great *many* small devices to equal the capacity of the large one?" This is not in fact quite the right question, for three reasons:

- The big resource comes all in one gigantic lump that when first brought into service will exceed the incremental demand it was meant to serve, while the more fine-grained (in capacity) ensemble of smaller resources can more exactly match uncertain and fluctuating demand growth, avoid the "overshoot," and hence require less actual capacity to be installed by the same date. Section 2.2.2 will show how the resulting risk reduction can be quantified.

- Section 2.3.1.1 will show that to achieve the same *reliably available* (not nameplate) supply, small units permit a given amount of large-plant capacity to be replaced by a smaller amount of small-unit capacity, simply because less of the installed capacity is likely to be unavailable when needed.

- That difference becomes large, even manyfold, in isolated systems using extremely reliable distributed resources such as fuel cells, or highly reliable conventional resources such as gas turbines up to ~20 MW or backpressure steam turbines of about 40 kW to about 12 MW.

But the general issue remains fair game. To achieve a given increase in total capacity, can many small resources actually be built faster than a single large one?

Empirical data strongly suggesting that this is the case have been presented elsewhere (461). That analysis was prepared for the Pentagon at a time (~1981–82) when a recent oil-price shock and a relatively neutral Federal policy environment were combining to make many distributed resources on the supply and demand sides (chiefly the latter) collectively provide "new energy...about a hundred times as fast as all the centralized supply projects put together." This startling achievement reflected millions of individual choices: in 1980 alone, for example, Americans invested nearly $9 billion in small energy-saving devices and improvements, comparable to their expenditures on imported Japanese cars. Small-scale renewable sources were the second-fastest-growing resource, adding new supplies during 1977–80 twice as large as the simultaneous *decrease* in nonrenewable supplies. Of course, such achievements flagged later when their very success (reinforced by OPEC's indiscipline) crashed the oil price in 1986, removing both economic and psychological pressures for such vigorous deployment. Meanwhile, a largely unsympathetic U.S. Administration did its best to suppress investments in efficient end-use, renewables, and other alternatives to its centralized supply-side vision. But many international examples (461) confirm that this was no single-country fluke, and its U.S. manifestations were often impressive (see sidebar next page).

The reasons that small-scale resources can collectively be deployed so quickly through normal, civilian, market-based transactions are fundamental (461):

- Each unit takes days, weeks, or months to install, not a decade. For example, the wind turbines used in many commercial windfarms are roughly the size of a car, and can be mass-produced in much the same way, just like today's engine-generator sets. They are then installed atop standard towers, a structural-steel commodity, and hooked up to collecting cables at a rate that can be in the multi-MW-per-day range. The preceding site and utility negotiations, regulatory approvals, sitework, and infrastructure installation may take only a few months—especially now that many farmers and ranchers understand that the continuing wind royalties often match or exceed the net revenues of their farm operations. Those operations can proceed uninterrupted: the wind machines occupy a "footprint" of only a few percent of the land area and, with joint planning, need not interfere materially with operations, especially for grazing.

- Customer-installed or other non-utility capacity may also be exempt from the kinds of investment and siting regulations that apply to large utility plants.

- Because modules begin coming online almost immediately, supply can ramp up in step with demand, rather than having

---

The texture of the rapid deployment of relatively decentralized resources can be usefully illustrated by a regional case-study for New England during 1978–80 (461)—then the nation's most oil-dependent region[89]—during an oil-price shock even more severe than the 1973 Arab oil embargo. The example is summarized here not to suggest that the actual technology mix used was in any way optimal or even particularly desirable, but only to demonstrate that it was logistically successful with relatively modest effort. During those two years, the region's 12 million people, ranging from thrifty village-and-farm Yankees to ordinarily profligate suburbanites:

- increased their population by 0.7% and their real personal income by 4.6%;

- decreased their total primary energy consumption by 6.5%;

- increased the renewable fraction of their total energy supply to 6.3%, ahead of coal and just behind natural gas and nuclear fission, so that despite a drought that temporarily reduced the region's hydropower output by 22%,[90] net renewable supplies rose by 3 million barrels-equivalent-per-year to nearly 34 million bbl/y-equivalent—and hence in 1980, the region got about 46% more usable delivered energy from renewables than from nuclear power;

- decreased their total use of conventional fuels and power by 7.5%, equivalent to 46 million bbl/y, while increasing their use of coal (by 5 million bbl/y equivalent), of natural gas (by 3), and of extra Canadian power, but decreasing oil and nuclear by a total of four times that much;

- filled the resulting gap between regional energy supply and historic demand by small-scale technologies on both the supply and demand sides;

- more specifically, increased wood use by nearly 5 million bbl/y-equivalent (up 24% in two years—Vermont households' market share of wood heat rose from 22% to 56%, surpassing heating-oil consumption) to nearly three-fourths of the total renewable supply and one-fourth of northern New England's space-heating;[91]

- also increased wind, direct solar (11,000 systems installed by 1980), and municipal solid waste use by 0.5 million bbl/y-equivalent (the last of which, as of 1980, was projected to increase tenfold by 1985 and hydro-electric capacity by 30%); and, most importantly,

- through millions of highly decentralized technical improvements,[92] improved regional energy productivity by 12%—i.e., 6% per year, nearly twice the national-average rate of improvement—an achievement 14 times as large as the shift from nonrenewable to renewable resources, yet reliant on even smaller technologies, chiefly weatherstripping and insulation.

[89] Imported oil provided 80% of its total energy in 1973, 73% in 1980.

[90] Equivalent to utilities' burning an extra 2 million bbl/y of oil, made up nearly four times over by increased imports of Canadian hydropower.

[91] However, ~43% of the region's wood use was by industry, a sixth of it by diverse non-pulp-and-paper factories.

[92] Probably including also some much smaller savings from changes in lifestyle and in composition of output, though these are hard to measure, especially on a short-term regional scale. Nationwide, they account together for no more than a third of total reductions in primary-energy/GDP ratio, and probably a good deal less.

zero output throughout the entire long construction period of a monolithic single unit. This early supply not only helps to meet demand; it also provides an early income stream that reduces financing requirements for later units by permitting "bootstrapping" (§ 2.2.2.2).

- Being relatively "vernacular"—readily usable by a wide range of users with ordinary skills—the smaller units can diffuse rapidly into a large consumer-like market somewhat like cellular phones, pocket calculators, personal computers, video games, and snowmobiles, rather than requiring a slower process of "technology delivery" to a narrow and perhaps "dynamically conservative" market of a few highly specialized technical institutions, as do giant power plants (453):

> This is a function of the relative understandability, marketability, and accessibility of the technologies—of their comparative technical and managerial simplicity and the ease with which they can adapt to local conditions. These factors determine the mechanism, and hence the rate, of market penetration.

While this is hardly true of megawatt-range industrial cogeneration or commercial windfarm turbines, it is likely to be true of vernacular photovoltaic technologies such as AC-out panels and solar shingles. Of course it is also true of all but the most specialized and large-scale demand-side resource. It is more or less true of the technologies, like solar water heaters and add-on greenhouses, that are installed by ordinary contractors and do-it-yourselfers; and it should be distinctively true of the next generation of kW-to-tens-of-kW-range polymer fuel cells that are simply a black box installed by the gas or electric company in the basement right next to the gas meter, probably under a lease that includes all maintenance. It is surprisingly close to true of the ONSI 200-kWe phosphoric-acid fuel-cell package, whose manufacturer (132), no longer needs to send an installation

engineer with each unit: it's simply moved from the truck onto its pre-poured slab, connected to gas, water, and electricity, and turned on. That even individual householders can collectively act quickly when motivated is illustrated by the increase from 20% to 50% in only five years, in the 1970s, in the fraction of U.S. households trying to grow some part of their own food. At the right price, which is steadily approaching, many corporations may move just as fast to grasp opportunities for clean, super-reliable, and competitive out-of-the-box or off-the-truck electricity supplies that require no utility action or even knowledge.

- Such actions can be further accelerated by policy coordination, even on a national scale. Impressive examples include the British conversion to smokeless fuels, natural gas, and decimal coinage, the Dutch conversion to Groningen gas, the conversion of several Canadian cities and Los Angeles to 60-Hz electricity and of many Scandinavian cities and towns to district heating, and the 1967 Swedish conversion to right-hand driving. Each of these efforts took an immense number of highly decentralized actions, but all were smoothly and efficiently accomplished in a surprisingly short time. For example, many decades ago, metropolitan Montréal and Toronto retrofitted each neighborhood in turn from 25- to 60-Hz electricity, using fleets of specially equipped vans that would arrive on the appointed day (456):

> ...one van contained hundreds of clocks from which householders could choose replacements to swap for their clocks designed to run at the old frequency; another contained a machine shop for rewinding motors and rebuilding controls; all were staffed by resourceful people who had used the vans to clean up after the Normandy invasion [and were good at improvising].

Within hours, the whole capital stock of each neighborhood would be retrofitted and the vans would move on. Such technical and logistical support can quickly change very large numbers of buildings. A similar approach could be applied to the mass installation of demand-side resources or, where suitable and desired, of such supply-side resources as photovoltaics and household-scale fuel cells. Such a service could be offered by the public or the private sector as a way of conveniently achieving economies of installation scale.

- Moreover (453),

  Technologies that can be designed, made, installed, and used by a wide variety and a large number of actors can achieve deployment rates (in terms of total delivered energy) far beyond those predicted by classical market-penetration theories [for larger technologies]. For illustration, let us examine two sizes of wind machines: a unit with a peak capacity of several megawatts, which can be bought for perhaps a million dollars and installed by a heavy-engineering contractor in a few months on a specially prepared utility site; and another of a few kilowatts, which might be bought by a farmer on the Great Plains from Sears or Western Auto, brought home in a pickup truck, put up (with one helper and hand tools) in a day, then plugged into the household circuit and left alone with virtually no maintenance for twenty or thirty years. (Both these kinds of wind machines are now [in 1981–82] entering the U.S. market.) Most analysts would emphasize that it takes a thousand small machines to equal the energy output of one big one (actually less, because the small ones, being [more] dispersed, are collectively less likely to be simultaneously becalmed). But it may also be important that the small machines can be produced far faster than the big ones, since they can be made in any vocational school shop, not only in elaborate aerospace facilities, and are also probably cheaper per kilowatt. What may be most important—and is hardly ever captured in this type of comparison—is that there are thousands of times more farms than electric utilities on the Great Plains, subject to fewer institutional constraints and inertias. Likewise, California has only four main [investor-owned] electric com-

panies, but more than two hundred thousand rural wind sites that can [each] readily accommodate more than ten kilowatts of wind capacity. Not surprisingly, new megawatts of wind machines (and small hydro) [were in the early 1980s]...being ordered faster in California than new megawatts of central [thermal] power stations.

- A further reason "for suspecting that many small, simple things should be faster to do than a few big, complicated things" is that (453)

  ...the former are slowed down by diverse, temporary institutional barriers that are largely *independent of each other*. For example, passive solar may be slowed down by the need to educate architects and builders, microhydro by licensing problems, greenhouses by zoning rules. In contrast, large and complicated plants are slowed down by generic constraints everywhere at once, such as problems in siting major facilities and financing large projects. Because of their independence, dozens of small, fairly slow-growing investments can add up, by strength of numbers, to very rapid total growth, rather than being held back by universal problems. To stop the big plants takes only one [intractable]...institutional snag; to stop all the diverse kinds of small plants takes a great many. This diversity of renewable and efficiency options is not only a good insurance policy against technical failure; it also helps to guard against specialized, unforeseen social problems in implementation, offering a prospect of alternative ways to solve what problems do arise.

- It is also noteworthy that small projects (especially those using benign technologies), having typically small impacts, higher social acceptance, and often *de minimis* exemptions from the more onerous kinds of regulations, typically require no or few approvals, while approvals for large projects are not only numerous and complex but often dependent on each other, requiring an intricate dance of successive approvals and negotiations that can be stalled by one reluctant agency.

- "It may still seem counterintuitive to suggest," the Pentagon study continues, "that doing many small things can be faster than doing a few big things" (454):

  It is certainly contrary to the thrust of official energy policy [in the early 1980s]. It seems to be contradicted by one's sense of the tangible importance of a large...power plant: such a big and impressive installation must surely be the sort of thing of which our nation's industrial sinews are made, whereas a small technology—a bale of roof insulation, a cogeneration plant in a factory, a solar water heater—seemingly has only local and limited relevance. Yet in a deeper sense, the success of the free-market economic philosophy on which American private enterprise has been built depends very directly on the collective speed and efficiency of many individually small decisions and actions by sovereign consumers. It is precisely because those decisions are the fastest and most accurate means of giving practical effect to private preferences that Americans have opted for a market system—one of decentralized choice and action—rather than for a centrally planned economy....And in energy policy, recent events amply vindicate that choice.

- Although central planners and monopoly suppliers may be reluctant to rely on decentralized, small-scale actions not under their direct control (454),

  ...exactly the same mechanisms are at work in decentralized actions to increase energy efficiency [or supply] that have always been invoked as the rationale for forecasting growth in energy demand. The many small market decisions which individually constitute national demand are merely responding to a different set of signals today than they did previously. The bottom line is the proof: small, unglamorous, inconspicuous actions by individuals plugging steam leaks, weatherstripping windows, and buying more efficient cars [among other such commonplace energy-saving actions] are collectively [as of ~1981] increasing total energy capacity about a hundred times as fast as the annual investment of more than sixty billion dollars in centralized energy supply expansions with the com-

bined might of the energy industries and the federal government behind them. The hypothesis that many small actions can add up to greater speed than a few big actions is thus empirically true; there are good theoretical reasons why it should be true; and it is the approach most consistent with our national traditions.

- A few simple, back-of-the-envelope comparisons suggest that the rate at which the U.S. Energy Information Administration has forecast U.S. electric generating capability to increase during 2000–05 (199)—an average of 11.1 GW/y—could be readily matched by many plausible combinations of distributed resources. For example, it could be done on the demand side *alone* by capturing each year only 5% of the combined national lighting and drivepower efficiency potential with simple paybacks agreed by the Electric Power Research Institute in 1990 to be typically under two years (235) (for details [425, 471], current editions are available as the corresponding *Technology Atlases* from E SOURCE in hard copy or CD-ROM), with no contribution from any other demand-side resources such as improved building envelopes, HVAC, appliances and equipment, etc. The equivalent fraction of *all* cost-effective end-use-efficiency opportunities (excluding fuel-switching), across all applications and sectors, would be closer to 2%. In fact, during 1996–99, the U.S. did reduce its electricity consumption per dollar of real GDP by nearly 2% per year.

- It could be done on the supply side *alone* with just, say, 146,000 [93] or so modern 200-peak-kW wind turbines a year, or 585 per weekday, installed on the High Plains, where such machines could compete with coal-fired electricity if either externalities or distributed benefits (such as constant price) were counted. That installation rate is about equal to the number of farm tractors bought annually nationwide, so it is hardly a challenge to national manufacturing and logistical capacities.

[93] This comparison arbitrarily assumes that firm dispatchable power will be equivalent to ~38% of array capacity. Good wind sites normally have capacity factors around 0.3—somewhat higher with variable-speed, low-cut-in-speed designs; but a geographically dispersed population of wind machines can do very much better than that because so much of it is likely to receive suitable windspeeds at any given time. However, the 43% used here is a guess used only for illustration; the actual figure may vary widely, depending on machine design, dispatch economics, and geography (§§ 2.2.9–2.2.11).

- It could be done just by adding photovoltaic installations on the order of 240 million square meters a year[94]—comparable to the roof area on all the nation's new private housing starts,[95] neglecting all other kinds of building and other construction suitable for photovoltaic integration (let alone the far larger stock of *existing* structures).

In practice, of course, expanding distributed resources would involve no single kind but rather a highly diversified portfolio of many kinds in all three classes—demand-side, grid, *and* supply-side (both renewable and nonrenewable)—correspondingly reducing the burden on each class and on each specific type of resource. Viewed in this highly diversified and dispersed perspective, achieving a total supply expansion of 11 GW/y—equivalent to each American's saving and/or supplying an annual increment of about 34 delivered average (and, more or less, peak) watts,[96] or 2% of his or her average electricity use in all sectors—seems a modest challenge to the electricity industry, but a rather routine task for the country.

After all, the demand projections that some in the industry fear would require that 11 GW/y aggregated addition (659) reflect 2%/y growth caused by precisely such innocuous, incremental, individual trivial changes in consumption patterns nationwide—a total on the order of *34 watts per person per year*. If each of us can create such load growth without even thinking about it, then couldn't each of us, on average, create that much growth, just as quickly, in efficiency-plus-distributed-supplies by thinking just a little about it and doing what makes economic sense?

[94] For illustration, this assumes an onpeak availability (due almost entirely to local weather) of only 50% relative to nominal noon insolation of 1 kW/m². Nominal system efficiency is assumed to be on the order of 10% from insolation to AC output, corresponding to older or degraded monocrystalline or excellent polycrystalline silicon, or to mediocre multicolor thin-film cells, with a generous allowance for balance-of-system losses including storage.

[95] Private housing starts fluctuate between about 1 and 2 million units/y, with an average floorspace of nearly 200 square meters and an average height of just over 1.5 stories. Naturally, this would entail a 144-fold scaling-up of the photovoltaic industry, which in the U.S. shipped 77 peak MW of modules in 1999; but this is an unimportant expansion in terms of the massflow or value of many other industrial commodities such as semiconductors, float glass, or roofing materials. Realistic but impressive estimates of U.S. residential photovoltaic market potential are given in *Brittle Power* (442).

[96] Bearing in mind that at the system peak which the capacity is to meet, U.S. average grid losses are estimated by EPRI at about 14%— twice the annual-average grid loss. This correction to the 11.1 GW/y of net busbar capacity, effectively comparable to reducing it to 9.7 GW/y, was conservatively not applied to the previous comparisons with distributed resources, even though they would in fact incur less or almost no delivery loss.

1.6.6 Many littles can make a big

In some people's minds, the preceding question is not so much about speed as about raw technical capability. There is a long tradition in the electricity industry of supposing that only large, centralized plants can significantly contribute to the major level of supply required by an advanced industrial economy. Even in 2001, many commentators who should know better, having abandoned the fiction that California built no new power plants in the 1990s, changed their tune to "no major new power plants"—as if a megawatt produced by a large plant were somehow more effective than a megawatt produced by a small plant. Some historical counterexamples come to mind. When the United States in World War II was locked in mortal combat with Japan, and not always doing well, 78% of Japanese electricity was coming from small, highly dispersed hydroelectric plants (the largest single dam providing under 3% of the total, most much less), and those plants sustained only 0.3% of the bombing damage—the other 99.7% being sustained by the central thermal plants that provided 22% of the output (459).

Yet even with today's far larger demands, composed of billions of pieces of equipment run by hundreds of millions of individual people, small can become large if numerous enough. A 1982 Pentagon study noted (458):

As an analogy, in the United States today about eleven million cows, in herds averaging sixty cows each, produce fifteen billion gallons of milk per year. That is about a fifth as many gallons as the gasoline used annually by American cars, or about the same as the number of gallons that those cars would use if they were cost-effectively efficient....Yet much of that milk "is [or at least was at that time] efficiently supplied by small-scale decentralized operations"—at far lower cost than if all the milk were produced, say, in a few giant dairy farms in Texas and then shipped around the country. Likewise, "the average stripper well produces about two and eight-tenths barrels per day, which is about one-seventh of one-thousandth of a percent of what we consume in oil every day,...but...the cumulative effect of all our stripper wells [is]...twenty-one percent of continental oil [extraction]."

Some technical uncertainties remain about what fraction of an electric grid's supply could come from one or another kind of intermittent renewable source without risking instability or inadequacy of supply—though these supposed constraints are rapidly easing with more sophisticated analysis (§ 2.2.11). But no examination of the potential role of decentralized electric resources should start with a preconception that such resources cannot have a large, even a dominant, role in supply. Such assumptions guided railways' attitudes toward early cars, or mainframe computer manufacturers' attitudes toward the idea of personal computers. The resulting market lessons should by now have been well learned. It is time they were learned in the electricity industry too. Electricity demand comes in a myriad small pieces interspersed with a few bigger ones; electricity supply can do the same thing in principle, and is increasingly starting to do so in practice. We next survey why evolving the supply system in this direction can yield remarkable economic benefits.

Part Two
BENEFITS OF DISTRIBUTED RESOURCES

2.1 INTRODUCTION

For the reasons and as part of the historical processes described in Part One, market actors choosing from the electrical resource menu summarized in Section 1.2.2 are undergoing a radical shift from a short menu of the most centralized resources toward a large and diverse menu favoring more appropriate scale. A simple, though partial, explanation of this shift is the desire to minimize regret[1]—either at what one did that one wishes one hadn't done, or at what one didn't do that one wishes one had done.

In a world of increasingly rapid technological and social change, minimizing regret is greatly aided by picking options that are relatively small, fast, modular, and cheap. Sections 2.2, "System Planning," and 2.3, "Construction and Operation," describe how this way of managing risk so as to minimize regret can yield important and measurable economic benefits. Subsequent sections describe distributed benefits related to T&D (the grid); to system operation; to the quality of electrical services provided; and to social and environmental factors. Implications of these principles, barriers to their adoption, and recommendations for further action are then surveyed in Part Three.

We are now ready to explore these approximately 207 kinds of distributed benefits as systematically as current understanding and published results allow. However, three general caveats are important first:

1. *The total value of distributed benefits depends strongly on technology- and site-specific details.*

2. The total value also depends on *which benefits are counted.* In general, assessments that find relatively modest gains from counting distributed benefits, such as one 1994 survey's 4–46% gain (over central-station generation) for photovoltaics or 2–78% for wind (54), omit many significant classes of benefits. A basic lesson of Part Two will be that the harder you look, the more distributed benefits you are likely to find, and that though many of those benefits are individually small, they are so numerous that they can still be collectively large.

3. Because such limited resources have been applied to codifying and quantifying distributed benefits, the explanations and evidence we can present, especially on how much each benefit is worth, *vary widely in type* (estimates, formal calculations, field examples, etc.); in *application to particular places, systems, and times*; and in their *accuracy and rigor.*

It is not yet possible to present a neat package of analytic solutions, practical examples, lookup tables, and the rest of the toolkit that a planner would like to take off the shelf and apply. The art and science of understanding distributed benefits are far too immature for that—certainly in the open literature, and probably also even if all the proprietary literature were available. However, we have presented summary boxes and other guideposts to help clarify the relationship of the different benefits; and to avoid cluttering the narrative flow with tutorials, definitions, examples, and technical notes, we have boxed these separately as labeled sidebars.

We can only hope that this assemblage of descriptions and examples, from many disparate places and with often wildly differing

[1] This valuable phrase was coined by Group Planning at Royal Dutch/Shell in London.

levels of detail and precision, will stimulate others with greater skills and resources to expand and refine this exploration with the level of effort it merits. We trust that such improvements will focus on putting the greatest care into refining the precision of the most valuable terms, rather than seeking spurious or needless precision in unimportant terms—mindful of Aristotle's terse admonition that in addressing any problem, educated people "seek only so much precision as its nature permits or its solution requires." (13)

2.2 SYSTEM PLANNING

A noted text on corporate decision-making, *The Management of Scale: Big Organizations, Big Technologies, Big Mistakes* (138), examines case studies of disastrous large-scale blunders. Among their central causes, it identifies the adoption of inflexible technologies—those with "long lead time, large unit size, dependence upon infrastructure[,] and capital intensity." (139) Such a technology has the further attributes that:

> (a) Its development is to the direct benefit of large business organizations, able to spread some of the risk into public pockets.
>
> (b) It is likely to be an expensive failure.
>
> (c) Decision-making is highly centralized, with little debate, excluding some groups that are deeply affected by the technology.
>
> (d) The technology could have been identified as inflexible very early in its life.
>
> (e) More flexible technical alternatives exist.
>
> (f) These alternatives could be developed by organizations that are less centralized.

Many electric utilities bear extensive financial and psychological scar-tissue from their encounters with such technologies, particularly nuclear power. But as Part One described, among the key drivers of those multi-hundred-billion-dollar commitments were the perceptions that the giant plants would be necessary to keep the lights on and that they would decrease $/kW capital cost, presumed to be a surrogate for the cost of electric services. A critical part of the unraveling of this dogma was the realization that the hoped-for economies of scale were illusory and that a more sophisticated view of total cost and risk could even favor smaller units.

Rare wisps of internal criticism emanated from within the utility industry starting around 1970, but few if any squarely addressed the risks of gigantism; most, like those of Philip Sporn, dealt instead with demand forecasts and the balance between nuclear and fossil-fueled technologies (78, 297). Among the first wide cracks in the façade to be supported by rigorous analysis came in 1978, when John C. Fisher of the General Electric Company published a toned-down analysis through EPRI, and a more outspoken version in an international symposium, that was among the industry's first expert and explicit acknowledgements of diseconomies of unit scale.

Fisher presented a multiple-regression analysis of about 750 fossil-fueled steam power stations entering U.S. service during 1958–77 (238). He concluded, as he summarized in a letter (239), that

> Units with larger ratings take longer to build[,] and cost more on that account; units with larger ratings break down more often and take longer to repair and hence are out of service a larger fraction of the time. Because construction is slowed [*sic*] for larger units, the anticipated construction scale economy is diminished. Because

reliability falls off for larger units[,] the anticipated operational scale economy is reversed for units larger than an optimum size. When the cost [reductions]...associated with replication of standardized units are recognized, the optimum size shrinks to the smallest possible size consistent with maintaining full performance quality for whatever technology is being employed. For subcritical fossil steam units (the most common utility central station steam unit)[,] this size is in the neighborhood of 125 MW....

That size was only *one-tenth* the maximum then being ordered, but was consistent with British findings that estimated a 200–300-MW optimum taking fewer factors into account (1). Taking qualitative account of flexible siting, reduced reserve margin, and perhaps smaller maintenance staffs because of higher unit reliability, the conclusion drawn—heretical then, but prescient in light of GE's and other firms' later success with combined-cycle gas turbines—was:

> The replication of a series of identical generating units opens up an entirely new and profoundly different avenue for reducing the capital cost of generating capacity. The economy of scale assumes a new form, and manifests itself as the reduction of cost that can be achieved through the scale of operations in replicating large numbers of identical units. I believe that the potential for cost reduction along this new avenue is substantial.

Five years later, the *EPRI Journal* contemplated "New Capacity in Smaller Packages" (732), mainly for reasons of financial risk management. Many of its member utilities were awaking with a bad financial hangover from the combination of nuclear binge, runaway capital-cost escalation, high inflation and interest rates (amidst aftershocks of the 1979 disruption, the prime rate averaged 18.9%/y in 1981), flagging demand growth, and soaring overcapacity. The industry's flagship research journal focused less on the

engineering advantages of appropriate scale than on financial risk management, noting that "changing conditions are now prompting many utilities to take a fresh look at the matter of generating-unit size"—as if giant units were still preferable, just too risky. In particular, it noted,

> Uncertain load growth, constricted cash flow, and long lead times for large units define a new operating climate. It is risky to commit scarce capital to build a large unit that must be started many years in advance of the anticipated need....Today's financial climate requires a sharp match between capacity and demand because a major mismatch in either direction carries substantial cost. Building system capacity in small steps may be one way to optimize that match—hence, the growing interest among utilities in the concept of modular generation.

Improved system reliability (because many smaller—say, 100-MW—units were unlikely to fail simultaneously) and easier siting were also mentioned, though Fisher's inverse correlation between unit size and availability was not. EPRI's Dwain Spencer opined that:

> The concept of modular, parallel systems became a requirement and then a reality in order to achieve the high reliability required for missile and space missions. Now we have to demonstrate that this same idea can be applied to advanced power systems.

EPRI's Fritz Kalhammer saw "a broad trend toward integration of relatively small-scale, dispersed electricity sources into utility systems," and his colleague Kurt Yeager added that this trend looked durable over the long term, not a mere artifact of spiking interest rates.

Yet reflecting the ambivalence common in 1983, the article's author strongly emphasized coal combustion and coal gasification, even to run fuel cells (natural gas was then believed to be scarce and expensive). She thought the future of windpower—whose economic viability, she felt, remained to be established over the next five years—lay in gigantic 5-MW machines, which later turned out to exhibit strong technological diseconomies of scale.[2] She hoped phosphoric-acid fuel cells (the most advanced kind then contemplated—PEMFCs weren't mentioned) might "operate economically in increments as small as 10 MW"; their actual commercial scale today is 0.2 MW and falling. And she concluded, with a seeming wistfulness for the good old days, "Bigger will still be better in many applications, but as long as tight money and doubtful demand prevail, small modular units may fill a special need in prudent utility planning." As with Fisher, the overwhelming majority of the scale effects now known never got mentioned in that 1983 article; but piece by piece, the right questions were starting to be asked, even if "modular" often meant around 100–200 MW rather than much smaller.

All these themes, and many more, will emerge in the following discussion. But now, a quarter-century after John Fisher's regression analysis questioned the bigger-is-better dogma, diseconomies of scale are no longer mere tentative observations but a leading motivator of gigantic flux in the world's largest industry. Avoiding those diseconomies is increasingly emerging as a fount of quantifiable benefits that can reverse the merit order of economic choices. And making resources the right size, even if that's orders of magnitude smaller than tradition dictated, is emerging as the cornerstone of sound and profitable investments.

We begin with issues related to lead time—how long it takes to plan, site, get permits, and construct a power plant. To introduce that rich topic, we first survey the sources of uncertainty in electrical supply and demand on various timescales.

2.2.1 Many timescales, many uncertainties

The supply of electricity must be planned on a variety of timescales, ranging from a fraction of a second to decades. The reasons for this are physical, fundamental, and largely unavoidable.

Electricity is so difficult and expensive to store that except for a few special and costly large-scale installations, mostly using pumped hydroelectric storage, its supply is a real-time business (though that may change in this decade with new onsite technologies such as superflywheels and ultracapacitors (340) and even reversible fuel cells). In this respect, electricity differs from almost every other commodity. In effect, electricity is infinitely perishable—like bananas that must be eaten the very instant they are plucked, and ripened for plucking in exact coordination with the eaters' appetites. This inherent lack of inventory requires an understanding of all the diverse timescales on which those appetites may vary. We introduce this topic here in lay terms, then return to it more technically in Section 2.2.11.1 and Section 2.3.3.5 when discussing system stability and ramp rates. If you're not familiar with the operational fluctuations that electric power systems experience on a timescale ranging from milliseconds to days, please read Tutorial 1 now.

[2] By 1996 (688), commercial machines were typically rated at a few hundred kW; the largest commercial 1997 machines were 750 kW; and 1-MW machines were expected in prototype around 1998. They have since demonstrated some successes, but with caution and careful design. Earlier government-funded 2.5-MW machines, with near-supersonic tipspeeds and blades the size of jumbo-jet wings, were costly failures. Mid-1990s German engineering analyses (688) were finding cost minima around 30–40, or at most 60, meters rotor diameter, respectively corresponding to about 0.3–0.5, or at most ~1.3, MW; so on 2002 understanding of design and materials, 5-MW machines still look somewhat implausible.

Tutorial 1: Operational Fluctuations

Short-term fluctuations

Demand for electricity fluctuates from instant to instant as a myriad of users and controls unpredictably turn loads on and off. Supply may also fluctuate instantaneously as system faults, such as voltage spikes and interruptions caused by lightning or by sudden equipment failure, "shock" the grid. That shock then reverberates over distances ranging from local to vast, much like the wiggles in an enormous coupled system of weights connected by springs. Most of these fluctuations are offset by others fairly nearby, or occur on such a short timescale that they are smoothed out imperceptibly by the energy stored in the capacitance and inductance of the supply system.[3] They are the shortest of the timescales, down to microseconds, shown in Figure 2-1's graphic summary (699) of the timeframes relevant to power system management.

Longer timescales, on the order of one cycle or one "Hertz" (Hz)—in North America, 1/60th of a second or 17 milliseconds—traditionally require a specific and deliberate compensatory adjustment in supply or demand. Nowadays, transient stability on the transmission system, where even momentary glitches can cause vast quantities of power to slosh destructively around, is also requiring the evolution of new fami-

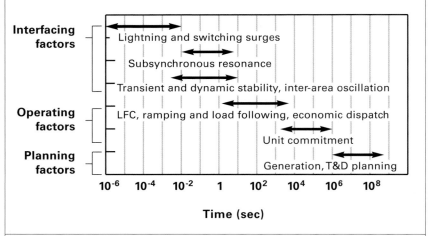

Figure 2-1: Electricity's timescales span 15 orders of magnitude
The timescales important to the planners and operators of electric supply systems span from microseconds to decades.

Source: Y. Wan and B. K. Parsons, "Factors Relevant to Utility Integration of Intermittent Renewable Technologies" (NREL, August 1993), p. 3

lies of electronic power-switching and control devices. These can extend the same control and damping capability to a timescale of milliseconds, so that the grid can eventually act much like a giant integrated circuit—about a billion times bigger than conventional chips (328). This helps to deal with not only transient instability (the voltage oscillations caused by faults) and steady-state instability (overwhelming damping forces by transferring too much power through part of a transmission system), but also small-signal or dynamic instability. That's when normally unimportant variations in generation or load, too small to be considered disturbances, nonetheless trigger low-frequency oscillations that can grow into volt-

age and frequency fluctuations large enough to spoil system stability.

On the timescale of about a second or more, uncompensated changes in demand cause changes in the speed of rotation of the large turbo-alternators at steam or hydroelectric power stations: heavier demand takes angular momentum out of the rotors, causing them to slow down, while lighter demand unburdens them so they speed up. But the frequency of the alternating-current grid, which varies directly with the speed of the rotors, must be closely controlled in order to keep different generating units synchronized (with the "top" of each rotor reaching the straight-up position at the same instant as all the others) so they are all "pulling

---

[3] Chiefly the magnetic fields of transformers and conductors, and the energy storage of capacitors located mainly at the substations.

together": otherwise they could fight each other. If not immediately disconnected ("tripped offline") by protective relays, they could suffer disastrous loss of synchrony, cascading instability, and serious equipment damage.[4]

To maintain all the rotors within an acceptable "angular shift" (difference in instantaneous shaft angle) when a given rotor starts to slow down, its operator must in the short term adjust the excitation voltage to the rotor, and in the longer term promptly adjust the flow of steam or water or (in the case of gas turbines) fuel into the turbine to restore the normal operating speed before it departs from permissible limits.[5] (Gas turbines, being aerodynamic devices, can also stall if the shaft rotation slows down too much.) In practice, this is done by automatic generation control (AGC) coordinated by a vast telecommunications network that links devices at many different levels and locations, coordinating actions on a scale of milliseconds based on sensors whose data, in modern digital versions, are sampled up to 5,000 times per second (328). Conversely, if electrical

demand decreases, the operator must correspondingly decrease the mechanical force driving the rotors, both to keep the frequency constant and to prevent them from spinning too fast (and, if that "overspeed" went uncontrolled, ultimately breaking apart—a risk if the unit isn't shut down within a fraction of a second of complete loss of its bus load [281]). The frequency must also be maintained at an average of exactly 60 Hz[6] over each 24-hour period; otherwise motor-driven electric clocks and other devices whose speed depends on grid frequency would gain or lose time. To keep this frequency rather exact, Load Frequency Control (LFC) checks and adjusts each governor's shaft speed every few seconds.

Grids currently handle these adjustments in the short term (up to a minute or so) by individual generators' shaft-speed controls, which operate automatically on a timescale of milliseconds, and by the centralized dispatch of **spinning reserve**—rotating and synchronized but not electrically loaded capacity specifically kept aside for this purpose. Additional **operating reserves** avail-

able by increasing the output of plants already operating and loaded, but not fully loaded, can also be brought online in periods ranging up to ten minutes, but often much less, since these resources are typically hydroelectric plants (which require valve-opening and rotor-spinup but no thermal warmup) and certain fast-start kinds of combustion turbines. Normally at least half of the total operating reserve is spinning, and the total operating reserve is adequate to cover the loss of the largest generating unit.

A "stability market" concept emerging first in New Zealand (303) adds a new way to meet such short-term operating requirements. Immediately interruptible loads, such as turning off an electric-resistance water heater on six seconds' notice, can be used to express the market value of offsetting other short-term increases in load, thereby stabilizing aggregate demand at significantly lower cost than could be done on the supply side (144, 399). That value is normally set by the cost of loading the spinning reserve. When the value is expressed in a two-way market, many interesting examples

---

[4] To ensure this, utility generators are almost always "synchronous" machines whose rotor current or "excitation" comes from a separate DC source or from the generator itself; with careful control, this explicit frequency control can keep all the rotors synchronized. In contrast, the induction generators used in some small-hydro and wind generators, and in many engine-driven generators, excite their rotors from an external AC source, typically the grid itself, thereby consuming reactive current (§ 2.3.2.3) so that they cannot generate without the grid's being energized.

[5] Those limits are a matter of convention, ranging from variations of less than 1 Hz to much larger values. Decades ago, frequency and phase stability limits were often said to be about an order of magnitude more stringent in North America than in Western Europe, where in turn they were about an order of magnitude more stringent than in Eastern Europe and the then Soviet Union. The lights stayed on (more or less) in all three regions across this wide range of operating philosophies: each simply dealt with the need for synchronization in different ways. In hindsight, it is not clear whether the more stringent control requirements in North American grids actually represented an economic optimum or only an unexamined assumption.

[6] The North American standard, although most of the rest of the world uses 50 Hz (50 cycles per second). Each cycle consists of a complete back-and-forth reversal of the alternating-current (AC) electric voltage "pressure" and the corresponding current flow.

of automated demand-side controls responding to real-time price signals start to emerge (515). These demand-side responses, the simplest of which are loads interruptible by underfrequency trips or by special signals, will become increasingly important and valuable in an electricity industry *dominated by its loads rather than by its generators*—a key characteristic that is already true today but not yet very widely recognized (303).[7]

On a slightly slower timescale than adjusting the steam or water valves, the power-plant operator or control system must adjust the fuel feed or combustion air, the nuclear reaction rate, or the dam's water flow. In the case of a steam plant, the steam temperature and pressure depend on the rate of combustion or nuclear reaction, requiring precise control of many interactive variables. In essence, however, all these controls are a fancy version of the old steam-locomotive boiler stoker who would shovel in coal more quickly to climb hills than to traverse level tracks. Power-station boilers, being very large metal objects, store heat and therefore have a thermal time constant that makes them respond only at a certain rate and with some delay that must be anticipated. Thermal power plants also use a large number of pumps, fans, and

other devices that can change speed only with certain mechanical delays and changes in efficiency, becoming less efficient as they depart from the ideal operating conditions for which they were designed. The resulting control optimization is quite complex—especially in the case of a nuclear plant, where, for example, the nuclear reaction creates certain neutron-absorbing fission products that later inhibit the chain reaction until they gradually decay.

Complexities mount. In addition to the ramping up and down of various units to meet or anticipate loads while maintaining constant frequency, AGC also works on a longer timescale, typically 2–10 minutes, to adjust each generator's output to optimize the system's entire generating mix against various units' thermal efficiency, fuel and operating costs, and associated transmission losses, so that the incremental production cost of each generator in different parts of the system is equal (it is then called the system lambda). And in a rolling planning process called Unit Commitment, these considerations are integrated with longer-term requirements for scheduling the various generators to allow optimal maintenance, startup and shutdown costs, and minimum fuel-burn requirements to be met at low-

est overall system cost. These criteria are typically reviewed daily and executed hourly, having regard to such longer-term considerations as seasonal availability and water storage in hydroelectric systems. But let us return to the shorter term.

Medium-term fluctuations

If a rising "ramp" of electrical demand cannot be satisfied simply by raising more steam in the plants already online, then the operator must start up additional generating capacity. In general, it takes much longer to start steam plants (like starting up a gigantic stove to get the water-kettle boiling) than to start engines or combustion turbines, so this non-operating reserve is traditionally defined as resources taking more than ten minutes to dispatch. Both for this reason and because of differing ratios of capital to operating costs, the operator typically has at her disposal a portfolio of different kinds of generating units. Based on her experience, she can "commit" (plan to start up) additional generating units in good time to meet required **ramp rates** (speed of increasing power output over time) at times of rising demand. Demand normally rises, for example, when people get to work in the mornings or come home and turn on appliances

---

[7] According to this compelling and important analysis, in future grid evolution, generators may be allowed to dispatch their output only if they provide, typically through a third-party aggregator of demand- and supply-side resources, an accompanying stability portfolio whose value is unbundled from the energy value. Otherwise they may be tempted to sell their spinning reserve margin into the profitable energy spot market rather than properly holding it back for the stability benefit of the system, and conversely, generators that provide vital stability services will not get properly compensated (303).

in the late afternoons, or when unusually hot or cold weather cause many electric heating or cooling systems to turn on more or less at the same time. Because very steep ramps may outrun the startup capabilities of the plant portfolio, utilities would be at risk of grid collapse if demand changed too quickly.[8]

This, then, is one aspect of the ever-changing operational task that utilities, running plants enormously larger than typical customers' loads, face throughout every day and night. But it is just the start of their wider planning challenge. They must carefully watch weather forecasts to ensure that, so far as possible, needed capacity will be available when severe weather causes peak system loads, rather than down for scheduled maintenance or at special risk of grid interruption by storms.

Dispatchers must plan the weekly and seasonal variations of loads—adjusted for weather, strikes, holidays, major sporting events, even flu epidemics—to coordinate with fuel deliveries and inventories, maintenance, and other factors.

And then there is system planning for supply/demand balance over the long term—a big topic to which we turn next.

[8] For this reason, when a BBC producer in the 1970s wanted to invite viewers to go turn something off and observe the collective effect of these actions as displayed on a real-time meter of demand from the National Grid, the Central Electricity Generating Board successfully implored the BBC not to do so; it was already quite challenging enough for the grid's dispatchers to cope with the fast demand ramp that routinely occurred at the end of popular evening shows when millions of Britons would simultaneously get up from watching TV and go turn on their electric kettles to make a nice cup of tea.

2.2.1.1 Long-term supply/demand balances

Amidst the "noise" of short- and medium-term fluctuations in each kind of demand from each customer on many simultaneous timescales and with fine-grained geography, utility planners must also deal with secular trends. Changes in human populations with changing ages, household structures, needs, wishes, cultures, and end-use technologies all tend to change those people's amount and time patterns of electrical consumption.

Meanwhile, similar shifts occur on the supply side. Each year, some power stations may routinely reach the end of their useful lives, when they cost more to keep running than they are worth—though that balance is an ever-shifting function of technology, market conditions, and tax and regulatory policy. Some plants, too, may change their rated capacity: upwards ("repowering") with better control technologies, better boiler- or condenser-water chemistry, or higher-quality fuels, for example, or downwards ("derating") with corrosion, warmer

Benefits

1 *Distributed resources' generally shorter construction period leaves less time for reality to diverge from expectations, thus reducing the* probability *and hence the financial risk of under- or overbuilding.*

2 *Distributed resources' smaller unit size also reduces the* consequences *of such divergence and hence reduces its financial risk.*

3 *The frequent* correlation *between distributed resources' shorter lead time and smaller unit size can create a multiplicative, not merely an additive, risk reduction.*

condenser water caused by nearby heat sources or changing climate, fouling of heat-exchange surfaces, pollution restrictions, changes in nuclear safety rules, etc. And all kinds of surprises, from local to global, may dramatically alter the portfolio of plants and fuels available for use, on notice ranging from long to little to none.

This is no simple matter. Over the very prolonged timescale—traditionally a decade or more—for building a major new power station, it becomes more like what the military calls a SWAG (scientific wild-assed guess). Despite the most sophisticated forecasting methods, few if any electric utilities in the world have a consistently accurate record. Utility planners are not amused by physicist Niels Bohr's remark that "It is difficult to make predictions—especially about the future": in this business, major planning errors can compound to multi-billion-dollar mistakes from which an especially unfortunate utility might never recover. Having many *other* utilities (let alone non-utility producers) simultaneously making similar, but not necessarily coordinated, forecasts and investments to supply the same interconnected grid does not protect against each utility's own forecasting errors, and may make them worse by reinforcing a "herd instinct."

Here, however, an obvious benefit of distributed resources reveals itself. In general, smaller resources can be planned and built more quickly than very large ones; and the longer it takes to plan, site, and build a power station, the more likely reality is to diverge from forecasts (and on the larger scale corresponding to the size of the station itself), so the greater the likelihood and scale of under- or overbuilding, so the

greater the financial risk of guessing wrong. That is (115),

> Inability to forecast precisely when power is needed involves a cost which is a function of the size *and* lead time of the units being considered and the relative flexibility provided by other units [or other resources such as demand-side management (DSM)] which the system can call on to bridge demand/supply gaps. Other things being equal, the larger the units, *and* the longer the construction lead times, the greater this cost will be, because it becomes more difficult to synchronize new power generating capacity with the growth in demand [over a larger increment and during a longer period].

Conversely, *the more closely the resource approaches the ideal of "build-as-you-need, pay-as-you-go," the lower the financial risk.*

It is important to note that this risk—of building too much or too little capacity to match demand—depends on unit size *and* on unit lead time. At least for conventional generating plants, these two variables are usually rather well correlated, so their risk-increasing effect is in principle multiplicative (though nonlinearly: only if lead time were uniformly proportional to unit size would risk rise exactly as the square of unit size). It might at first appear that the same is not true in reverse: smaller units tend to be faster (§ 1.5.7)—for much smaller distributed resources, very much faster—but they also can meet less demand, so to the extent their size and lead time are correlated (also nonlinearly), their risk-reducing advantage would be *reduced*. But this does not actually occur because small units are typically installed not singly but rather in large numbers that can *collectively* match (or more if desired) the "lumpy" capacity of the single large unit they displace. Therefore, in general, small units' risk-reducing effect is at least proportionate to their reduction in lead time, and will be even greater to the extent that large resources also take longer to build.

Chapman and Ward (115) correctly note that power planning takes place within "three separate planning horizons and processes"[9] that are "interdependent but separable, in the sense that they be considered one at a time in an iterative process, with earlier analysis in one informing the others." These three timescales, conceptually somewhat related to the scales of fluctuation described in Section 2.2.1 above, could be restated as:

- the short-term *operational* scale of keeping the grid stable, supply and deliverability robust, and the lights on, ranging from real-time dispatch to annual maintenance scheduling;

- the medium-term *planning* scale of keeping supply and demand in balance over the years through a flexible strategy of resource acquisition, conversion, movement, trading, renovation, and retirement; and

- the long-term *visionary* scale of ensuring over decades that the mix, scale, and management of energy systems are avoiding fundamental strategic errors; opening new options through farsighted RD&D and education; fostering a healthy evolutionary direction for institutional, market, and cultural structures, patterns, and rules; and sustaining foresight capabilities that will support graceful adaptation to and leadership in the unfolding future.

All three timescales are vital. So is not mixing them up. And so is seeking opportunities to serve synergistically the goals of more than one at a time, rather than creating tradeoffs between them. We therefore turn now to ways to value some specific attributes—modularity, modest scale, and short lead planning and installation times—of distributed resources that also happen to offer advantages on all three timescales and levels of responsibility.

2.2.2 Valuing modularity and short lead times

To reduce the financial risks of long-lead-time centralized resources, it is logistically feasible (§ 1.5.7) to add modular, short-lead-time distributed resources that add up to significant new capacity. But can those smaller resources create important economic benefits by virtue of being faster to plan and build? Common sense says yes, and suggests three main kinds of benefits: reducing the *forecasting risk* caused by the unavoidable uncertainty of future demand; reducing the *financial risk* caused directly by larger installations' longer construction periods; and reducing the *risk of technological or regulatory obsolescence*. Let us consider these in turn.

2.2.2.1 Forecasting risk

Nearly twenty years ago, M.F. Cantley noted that "The greater time lags required in planning [and building] giant power plants mean that forecasts [of demand for them] have to be made further ahead, with correspondingly greater uncertainty; therefore the level of spare capacity to be installed to achieve a specified level of security of supply must also increase." (90) Longer lead time actually incurs a double penalty: it increases the uncertainty of demand forecasts by having to look further ahead, *and* it increases the penalty per unit of uncertainty

[9] They add that "Additional (four or more) horizons might be usefully explored, but fewer than three will cause difficulties."

Benefits

4   *Shorter lead time further reduces forecasting errors and associated financial risks by reducing errors' amplification with the passage of time.*

5   *Even if short-lead-time units have lower thermal efficiency, their lower capital and interest costs can often offset the excess carrying charges on idle centralized capacity whose better thermal efficiency is more than offset by high capital cost.*

6   *Smaller, faster modules can be built on a "pay-as-you-go" basis with less financial strain, reducing the builder's financial risk and hence cost of capital.*

7   *Centralized capacity additions overshoot demand (absent gross underforecasting or exactly predictable step-function increments of demand) because their inherent "lumpiness" leaves substantial increments of capacity idle until demand can "grow into it." In contrast, smaller units can more exactly match gradual changes in demand without building unnecessary slack capacity ("build-as-you-need"), so their capacity additions are employed incrementally and immediately.*

8   *Smaller, more modular capacity not only ties up less idle capital (#7), but also does so for a shorter time (because the demand can "grow into" the added capacity sooner), thus reducing the cost of capital per unit of revenue.*

9   *If distributed resources are becoming cheaper with time, as most are, their small units and short lead times permit those cost reductions to be almost fully captured. This is the inverse of #8: revenue increases there, and cost reductions here, are captured incrementally and immediately by following the demand or cost curves nearly exactly.*

10 *Using short-lead-time plants reduces the risk of a "death spiral" of rising tariffs and stagnating demand.*

by making potential forecasting errors larger and more consequential. As *Business Week* put it in 1980 (83), "Utilities are becoming wary of projects with long lead times; by the time the plant is finished, demand could be much lower than expected. If you're wrong with a big one, you're really wrong.... Uncertainty over demand is the main reason for the appeal of small plants."

This forecasting risk became painfully evident in the 1970s, when the power industry consistently overestimated demand growth while lead times for large new generating plants became longer and more uncertain, the cost of capital soared, and utilities used planning models "biased toward large plants." The interaction of these four factors

created "an increased likelihood of excess capacity, unrecoverable costs and investment risk" (373) that bankrupted a few utilities and severely strained scores more. The industry therefore learned the hard way that minimizing risk "will tend to favor smaller scale projects, with shorter lead times and less exposure to economic and financial risks." (373) Specifically (373):

• An autumn 1978 *Energy Daily* review (522) of data collected by the Edison Electric Institute in autumn 1978 showed that only once in the previous 11 years had the industry underpredicted the following year's total noncoincident peak demand, and then only by 0.1 percentage point. Rather, the forecasts averaged 2.1 percentage points too high during 1968–73 and 5.1 percentage points too high after 1974. Indeed, during 1974–79, the average forecast error exceeded the average annual growth rate, and during 1975–78 the error averaged 2.5 times the actual growth—leading the editor of *Electrical World* to call for a major rethinking of traditional forecasting methods (289) (see Figure 1-41 in Part One).

- In such an uncertain forecasting environment, "The alternative to waiting 12 years to see whether demand growth did justify construction of an expensive large generator...is building smaller projects with shorter lead times." (522) For example, if a utility forecast 5.5% annual demand growth, built new generators with 12-year lead times, and actually experienced only 3.5% annual demand growth, then it would end up with 26% excess capacity. If the lead time were 6 years, however, that excess would drop to 12%; if 4 years, to 8%.

- Lead time correlated well with unit size: *e.g.*, for U.S. coal-fired plants in the 300–700-MWe range, each 100 MW of capacity required an extra year of construction. Although different analysts' values for this coefficient vary,[10] the existence of an important bigger-hence-slower correlation has long been well established (12, 557).

For these reasons, as summarized by Sutherland *et al.* (673), with emphasis added,

> The most important result is that short lead time technologies, which represent smaller units, are a defense against the serious consequences of unforeseen changes in demand. The "worst case" occurs when electric utilities build large and long lead time plants [but]...anticipated demand is unrealized. A price penalty is paid by consumers, and unfavorable

financial conditions plague the utility. Ford and Yabroff (1980, 78) concluded that the strategy of building small, short lead time plants could cut the price penalty to the consumer by 70% to 75%. *Both demand uncertainty and short lead times favor small generating units, with their synergistic effects being the most important.*

The mechanisms of that synergy become more visible when one looks more closely into the details of demand uncertainty. A lucid analysis of the tradeoffs between hoped-for power-plant economies of scale and the risk of excess capacity (75) (Figure 2-2) provides cost ratios showing how much cheaper the output from a larger unit must be, if it takes twice as long to build as a small plant, in order to justify buying the large plant under a given pattern of demand uncertainty. That pattern is expressed as the probability that during the planning period, demand will grow by one, two, or three arbitrary units, which can be interpreted as relative percentage growth rates. Those probabilities can occur in various combinations. For each, a set of ratios shows how much cheaper the large plant must be than the small plant in order to justify building the large one. In general, the assumed demand growth will justify at least one large unit. But to justify a second or third large unit, it must be modestly or dramatically cheaper than the smaller units, depending on the distribution of demand probabilities. The left-hand graph in each case shows the assumed distribution of probabilities (for example, in the first case, all three demand growth rates—*e.g.*, x, $2x$, and $3x$—are equally probable). The right-hand graph shows in the first case,

---

[10] For example (673), a RAND multiple-regression analysis by William Mooz found a correlation equivalent to ~3.5 months of construction duration per 100 MWe of net capacity (but actually a bit nonlinear), while a comparable analysis in a different algebraic form, by Charles Komanoff, found that a doubling of nuclear unit size would increase construction time by 28%. (Komanoff's capital-cost model for coal plants didn't use unit size as a variable, but unit size was the variable most significant in affecting construction duration.) A further analysis cited (673), using an EPRI database of 54 coal and nuclear plants, didn't examine unit size as an explanatory variable, but did find that 22% of the nuclear units' construction delay was deliberate in an effort not to build too far ahead of demand, implying that "the utility would have been better off with smaller and shorter lead time plants."

for instance, that a large unit is justifiable at full cost as the first unit to be built, but must be 10% cheaper than the small plant to be the right choice as the second unit, and 40% cheaper as the third unit.

Figure 2-2: Uncertain demand imposes stringent cost tests on slow-to-build resources
Long-lead-time power stations must be far cheaper than halved-lead-time smaller units in order to be an economical way to keep on meeting changing demand (unless, perhaps, demand growth is known to be accelerating).

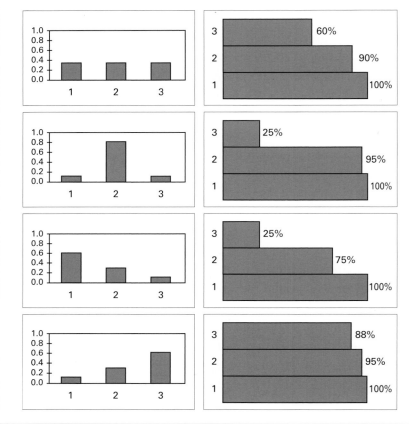

Source: E. P. Kahn, "Project Lead Times and Demand Uncertainty: Implications for Financial Risk of Electric Utilities" (Lawrence Berkeley Laboratory/University of California, 1979), p. 9, fig. 4

Thus continuing to build large plants requires them to be built at an increasingly steep cost discount even if demand growth is steady (the first case); is unlikely to be the right strategy if demand fluctuates markedly (the second case) or demand growth tapers off (the third case); and may be justifiable if demand growth is definitely and unalterably accelerating (the fourth case). This comparison—focusing only on a specific kind of investment risk, and not taking account of several dozen other effects of scale on economics—is of course a simplified illustration of planning choices that could be simulated more elaborately, typically by a Monte Carlo computer analysis. But simple though it is, the example starkly illustrates the risks of overreliance on long-lead-time plants when demand is uncertain: in the middle two cases, the third large unit could be justified only if it were *fourfold cheaper* than the competing small, halved-lead-time unit. The authors conclude (75):

> The relative cost advantage of short lead time plants can be substantial. If demand uncertainty is such that low growth rates of demand are more likely than high growth rates, or if the variance in demand growth is simply large, the capital cost of long lead time plants must be substantially decreased, under some circumstances as much as 50%[,] to make long lead time plants cheaper, even with a flat load curve. The fraction of future demand that is optimally satisfied with long lead time power plants depends on two factors. Again, the lower the probability that a given level of demand will occur, the greater the cost advantage required to make long lead time plants optimal for that level. This conclusion is modified by the existing mix of short lead time—high [fuel] cost plants and long lead time—low fuel cost plants. The more short lead time plants in the existing mix[,] the smaller the cost advantage of long lead time plants needs to be. In general[,] unless long lead time plants have a substantial cost advantage or the probability of the demand['s] growing at the maximum rate is large, it is rarely optimal to supply all the projected demand with long lead time plants.

In summary: if too many large, long-lead-time units are built, they are likely to overshoot demand. Paying for that idle capacity will then raise electricity prices, further dampening demand growth or even absolute levels of demand, and increasing pressure for even further price increases to cover the revenue shortfall. This way lies financial crisis, as the industry found to its cost in the 1970s and 1980s.

Of course, forecasting errors go both ways: you can build capacity that you turn out not to need, or you can fail to build a plant that you *do* turn out to need. Are those risks symmetrical? In the 1970s, when power-plant (especially nuclear) vendors were trying to justify their seemingly risky GW-range products, they cited studies purporting to show that underbuilding incurred a greater financial penalty than overbuilding (100, 671). However, those studies' recommendation—to overbuild big thermal plants as a sort of "insurance" against uncertain demand—turned out to result from artifactual flaws in their models (243, 249, 417).[11] More sophisticated simulations, on the contrary, showed that (at least for utilities that don't start charging customers for power plants until they're all built and put into service) if demand is uncertain, financial risk will be minimized by deliberately *under*building large, long-lead-time plants (75, 243–4, 246–7, 249).

For example, given an illustratively irregular pattern of demand growth characteristic of normal fluctuations in weather and business conditions, excessive reserve margins and electricity prices can be reduced by preferring short-lead-time plants (Figure 2-3):

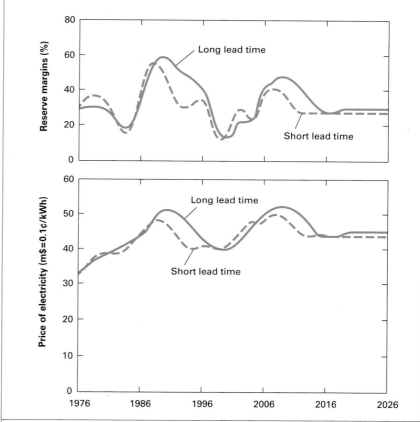

Figure 2-3: Faster-to-build resources help avoid capacity and price overshoot
Short-lead-time plants help to avoid excessive reserve margins and tariffs under uncertain demand.

Source: A. Ford and A. Youngblood, "Simulating the Planning Advantages of Shorter Lead Time Generating Technologies" (*Energy Systems and Policy* 6, 1982), p. 360, figs. 7 and 8

[11] The EPRI models assumed that all forms of generating capacity are expanded at the same rate, so that baseload shortages automatically incur [large] outage costs rather than extending the capacity or load factor of peaking or intermediate-load-factor plants. (This assumption means that the plant-mix questions at issue simply cannot be examined, because plants are treated as homogeneous.) Furthermore, the use of planning reserve margin as the key independent variable obscured the choice between plants of differing lead times. Capital costs were assumed to be low, so that even huge overcapacity didn't greatly increase fixed costs. Outage costs were treated as homogeneous, even though it would make more sense to market interruptible power to users with low outage costs. Uncertainties were assumed to be symmetrical with respect to under- or overprediction. And the opportunity costs of over- or underbuilding were ignored, whereas in fact, overbuilding ties up capital and hence foregoes the opportunity to invest in end-use efficiency or alternative supplies, while underbuilding means one still has the capital and can invest it in ways that will hedge the risk. For further comparative discussion of conflicting studies, see (249).

There are four reasons for this:

- operating short-lead-time, lower-thermal-efficiency, low-capital-cost stopgap plants (such as combustion turbines fueled with petroleum distillate or natural gas) more than expected, and paying their fuel-cost penalty, is cheaper than paying the carrying charges on giant, high-capital-cost power plants that are standing idle;[12]

- even if this means having to build new short-lead-time power stations such as combustion turbines, their shorter forecasting horizon greatly increases the certainty that they'll actually be needed, reducing the investment's "dry-hole" risk;

- smaller, faster modules will strain a utility's financial capacity far less (for example, adding one more unit to 100 similar small ones, rather than to two similar big ones, causes an incremental capitalization burden of 1%, not 33%); and

- short-lead-time plants can be built modularly in smaller blocks (301), matching need more exactly.

This last point is so obvious that it is often overlooked: big, "lumpy" capacity additions *invariably* overshoot demand (absent gross underforecasting of rapidly growing demand), leaving substantial amounts of the newly added capacity idle until demand can "grow into it" (Figure 2-4).[13]

Thus adding smaller modules saves three different kinds of costs: the increased lead time (and possibly increased total cost) of central resources; the cost of idle capacity that exceeds actual load; and overbuilt capacity that remains idle. Both curves maintain sufficient capacity to serve the erratically growing load, but the small-module strategy does so more exactly in both

[12] Naturally, this sort of conclusion is not immutable, but rather depends on interest rates, fuel costs, and other factors that change over time.

[13] This is quite an old and familiar problem in mathematical economics (588, 657). The latter paper concludes that "efficient production when there is uncertainty of demand forces the supplier to sacrifice economies of [unit] scale in order to achieve greater flexibility through a larger number of plants. Equally important is the result that full efficiency requires a set of plants of different sizes. Thus there is no optimal scale of plant or minimum efficient scale and in fact such a concept is meaningless in the present context. Only the collection of all plants is efficient."

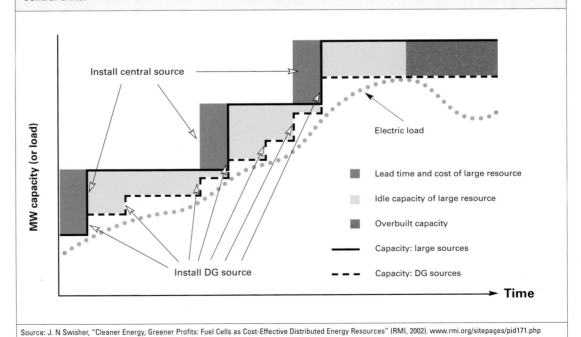

Figure 2-4: Slow, lumpy capacity overshoots demand in three ways
The yellow areas show the extra capacity that big, lumpy units require to be installed before they can be used. Small distributed-generation (DG) modules don't overshoot much; they can be added more closely in step with demand. The blue areas show the extra construction and financing time required by the longer-lead-time central units.

Source: J. N Swisher, "Cleaner Energy, Greener Profits: Fuel Cells as Cost-Effective Distributed Energy Resources" (RMI, 2002), www.rmi.org/sitepages/pid171.php

quantity and timing, and hence incurs far lower cost.

This load-tracking ability has value unless demand growth not only is known in advance with complete certainty, but also occurs in step-functions exactly matching large capacity increments. If that is not the case—if the growth graph is diagonal rather than in vertical steps, even if it is completely smooth—then smaller, more modular capacity will tie up less idle capital for a shorter period.

If demand grows steadily, the value of avoiding lumps of temporarily unused capacity can be estimated by a simplified method modified by Hoff, Wenger, and Farmer (324) from a 1989 proposal by Ren Orans. The extra value of full capacity utilization is proportional to:

$$\frac{T(d - c)}{1 - e^{-T(d-c)}}$$

where d is the [positive] real discount rate, c is the real rate at which capacity cost escalates, and T is years between investments. This approximation yielded reasonable agreement with PG&E's estimate (§ 2.3.2.6) for deferring Kerman transformer upgrades (324).

This analysis also provides a closed-form analytic solution for the case where the distributed resource is becoming cheaper with time, so even if it's not cost-effective now, it is expected to become so shortly. If the relative rates of cost change between the distributed and traditional resources are known, due allowance can be made. The equations provided (324) can also use option theory (§ 2.2.2.5) to account for uncertainties in the cost of the distributed resource. Such uncer-

tainty may create additional advantage by suitably structuring the option so that the manager is entitled but not obliged to buy, depending on price. For these reasons, in an actual situation examined, a distributed resource costing $5,000/kW can be a cost-effective way to displace generating investments that would otherwise be made annually, plus transmission investments that would otherwise be made every 30 years—largely because the lumpiness of the latter investment means paying for much capacity that will stand idle for many years.[14]

In any actual planning situation, depending on the fluctuating pattern of demand growth, the extra cost of carrying the lumpy idle capacity can be calculated from the detailed assumptions, and then interpreted as a financial risk. Some tools for this calculation are described below. In principle, but not in most models, such a calculation should take into account an important economic feedback loop—the likelihood that the higher electricity tariffs needed to pay that extra cost will make demand growth both less buoyant and less certain, further heightening the financial risks (247–8). This sort of feedback is probably best captured by system dynamics models (248). Those models broadly confirm the "death spiral" scenario characteristic of plants that take longer to build than it takes customers to respond to early price signals from the costly construction—especially if demand is as sensitive to price as many econometric analyses suggest.[15] Avoiding the risk of the "death spiral" is an important potential benefit.

[14] It's important for the analytic tools used in this situation to capture declining costs incrementally and immediately, so that no cost reduction is delayed or lost through stepwise capture at longer intervals.

[15] Econometric studies collected by Ford and Youngblood (248) found long-run own-price elasticities of demand as large as −1.5 in the residential and commercial sectors and −2.5 in the industrial sector, with widely varying time constants. In general, elasticities with an absolute value larger than unity can lead to trouble; many of the values cited, including most of the industrial ones, are in this range. (An elasticity of −1.5 means that each 1% increase in price leads to a 1.5% decrease in demand. "Own-price" refers to the price of the same commodity whose demand is being measured; that differs from "cross-price" elasticities, which describe substitution of one resource for another as their relative prices change. "Long-run" typically refers to a period of years.)

Benefits

11 Shorter lead time and smaller unit size both reduce the accumulation of interest during construction—an important benefit in both accounting and cashflow terms.

12 Where the multiplicative effect of faster-and-smaller units reduces financial risk (#3) and hence the cost of project capital, the correlated effects—of that cheaper capital, less of it (#11), and needing it over a shorter construction period (#11)—can be triply multiplicative. This can in turn improve the enterprise's financial performance, gaining it access to still cheaper capital. This is the opposite of the effect often observed with large-scale, long-lead-time projects, whose enhanced financial risks not only raise the cost of project capital but may cause general deterioration of the developer's financial indicators, raising its cost of capital and making it even less competitive.

13 For utilities that use such accrual accounting mechanisms as AFUDC (Allowance for Funds Used During Construction), shorter lead time's reduced absolute and fractional interest burden can improve the quality of earnings, hence investors' perceptions and willingness to invest.

14 Distributed resources' modularity increases the developer's financial freedom by tying up only enough working capital to complete one segment at a time.

15 Shorter lead time and smaller unit size both decrease construction's burden on the developer's cashflow, improving financial indicators and hence reducing the cost of capital.

16 Shorter-lead-time plants can also improve cashflow by starting to earn revenue sooner—through operational revenue-earning or regulatory rate-basing as soon as each module is built—rather than waiting for the entire total capacity to be completed.

17 The high velocity of capital (#16) may permit self-financing of subsequent units from early operating revenues.

18 Where external finance is required, early operation of an initial unit gives investors an early demonstration of the developer's capability, reducing the perceived risk of subsequent units and hence the cost of capital to build them.

19 Short lead time allows companies a longer "breathing spell" after the startup of each generating unit, so that they can better recover from the financial strain of construction.

20 Shorter lead time and smaller unit size may decrease the incentive, and the bargaining power, of some workers or unions whose critical skills may otherwise give them the leverage to demand extremely high wages or to stretch out construction still further on large, lumpy, long-lead-time projects that can yield no revenue until completed.

21 Smaller plants' lower local impacts may qualify them for regulatory exemptions or streamlined approvals processes, further reducing construction time and hence financing costs.

22 Where smaller plants' lower local impacts qualify them for regulatory exemptions or streamlined approvals processes, the risk of project failure and lost investment due to regulatory rejection or onerous condition decreases, so investors may demand a smaller risk premium.

23 Smaller plants have less obtrusive siting impacts, avoiding the risk of a vicious circle of public response that makes siting ever more difficult.

2.2.2.2 Financial risk

For all the reasons described in Section 2.2.2.1, shorter lead time and smaller, more modular capacity additions can reduce the builder's financial risk and hence market cost of capital (371, 417–8). But there are even more causes for the same conclusion (675):

1. Shorter lead time means less accumulation of AFUDC, a lower absolute and fractional burden of interest payments during construction (140), higher-quality earnings that reflect more cash and less fictitious "regulatory IOU" book income, and lower cost escalation during the construction interval (384, 493). One manifestation of these effects is that with highly modular projects, the developer "only needs enough working capital to finance one segment at a time. Once the first segment is completed, the unit can be fully financed, and the proceeds used to finance the next segment" (Figure 2-5).

This is analogous (317) to building houses that are sold as they're completed, rather than tying up much more capital in an apartment building that can't yield any rental revenue until it's all finished.

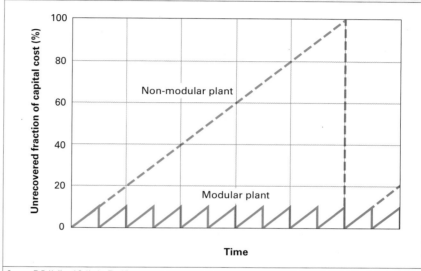

Figure 2-5: Modular plants reduce need for working capital
Modular plants can need 10+ times less working capital than lumpy plants, reducing default risk and perhaps therefore the modular units' cost of capital.

Source: T. E. Hoff and C. Herig, *The Virtual Utility: Accounting, Technology and Competitive Aspects of Emerging Industry* (Kluwer Academic Press, 1997), p. 26, fig. 9

Allowance for Funds Used During Construction (AFUDC) is a U.S. utility accounting practice virtually unknown in most countries and baffling to non-utility businesspeople. Especially during the nuclear construction boom of the 1970s, many state utility commissions issued a sort of "regulatory IOU" by permitting utilities to reflect on their books a fictitious, noncash income item representing the *cost* of capital (both debt and equity) tied up in the construction project but not yet ready to generate electricity and hence to earn revenue. The principle was that the utility's financial reports would then look as healthy (superficially) as they would actually become when the project was completed, electricity flowed to customers, and real revenues were earned. Unfortunately, some utilities became so dependent on this unreal revenue that it came to provide a substantial fraction of their book income. If the project were then abandoned, as sometimes occurred, then the gap between reported and actual cash income would become painfully apparent. The alternative regulatory treatment—including **CWIP (Construction Work in Progress)** in the commission-approved **rate base** of assets on which utilities were authorized to earn a return on and of capital—allowed the utilities to start charging customers for money spent on projects not yet completed. This method defied the normal principle that ratebased assets must be "used and useful," and it had:

- the economic advantage of providing a more nearly correct marginal price signal early enough that customers could value the electricity more appropriately and presumably use it more judiciously—possibly making the plant largely or wholly unnecessary;

- the economic disadvantage that this price signal did no good because the utility had no intention of canceling the project even if demand growth slackened or reversed;

- the political advantage of placating the utility and its investors; and

- the political disadvantage of infuriating customers who were having to pay for an asset that was doing them no good and might never operate at all.

The resulting regulatory and legal wars are now history, and the wholesale competition begun in 1992 has largely transformed the structure that created them, but even a few decades later, their scars persist on some utilities' financial and political balance sheets.

2. Shorter lead time means that the utility does not have to keep as much capacity under construction, costing money and increasing financial risk, to meet expected load growth in a timely fashion.

3. Shorter lead time means that units get into the rate base[16] earlier, or, in the case of a privately owned plant, can start earning revenue earlier—as soon as each module is built rather than waiting for the entire total capacity to be completed. This benefit has been quantified (317), with an example of a 500-MW plant built in one segment over five years *vs.* ten 50-MW modules with 6-month lead times (Figure 2-6). If each asset runs for 20 years, then under either plan, the same capacity operates identically for the middle 15 years—but the modular plant has higher revenue-earning capacity in the first five years, and conversely in the last five years as the modular units retire. But because of discounting, the early operation is worth much more today. Using a 10%/y discount rate and $200/MWy revenues, the modular solution will have an astonishing 31% higher present-valued revenue. If the modular plant were infinitely divisible and had zero lead time, then regardless of the life of the plant, the ratio of present-valued revenues would be $(e^{Ld} - 1)/Ld$, where L is the number of years it takes to complete the nonmodular plant and d is the annual real discount rate (317).

4. Short lead time allows the companies a longer "breathing spell" after the eventual startup of the large units that are currently under construction (so that they can better recover from the financial strain of those very costly and prolonged projects). This is analogous to a mother's stretching out the spacing of her bearing children.

5. These four advantages allow the company to avoid poor financial performance. Thus, the short-lead-time unit allows the company to avoid the increase in financing costs that can occur when a firm misses its financial goals.

These conclusions are also reinforced by four other factors that affect financial cost and risk, notably:

6. Shorter lead time decreases the burden on utility cashflow as expressed by such indicators as self-financing ratio, debt/equity ratio, and interest coverage

[16] Under traditional U.S. (and most other) rate-of-return regulation, utilities are entitled to charge customers approved tariffs expected to yield "revenue requirements" that consist of two kinds of prudently incurred costs: operating expenses, and a fair and reasonable return on and of capital employed to provide "used and useful" assets. The "rate base" on which the utility has the opportunity to earn that regulated return is thus the sum of those used and useful assets. Therefore, the sooner a power station enters service, the sooner it starts earning returns.

Figure 2-6: Modular resources' early operation increases their present value
Modular plants can start yielding revenue while big, slow, lumpy plants are still under construction.

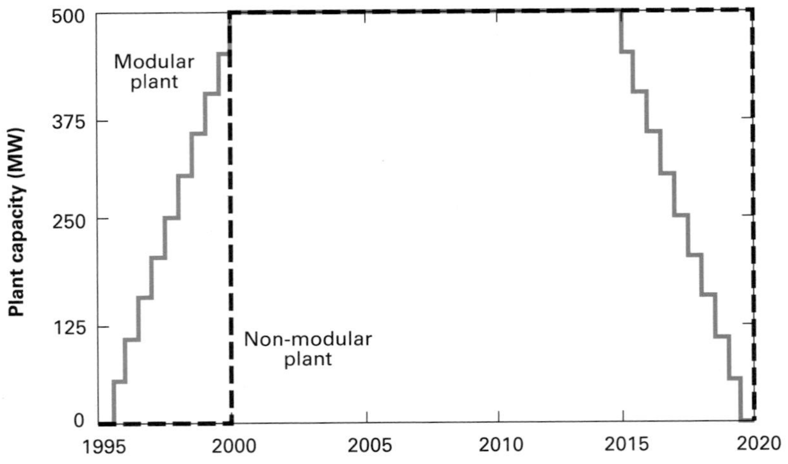

Source: T. E. Hoff and C. Herig, *The Virtual Utility: Accounting, Technology and Competitive Aspects of Emerging Industry* (1997), p. 22, fig. 7

ratios—all used by financial analysts to assess risk for such purposes as bond ratings and equity buy/sell recommendations (375, 757).

7. Shorter lead time may decrease the incentive, and the bargaining power, of some workers or unions. Otherwise their indispensable skills may give them the leverage to demand extremely high wages or to stretch out construction still further, as occurred on the Trans-Alaska Pipeline System and many of the later U.S. nuclear power plants.

8. Smaller plants may have less obtrusive siting impacts (250). This can avoid the vicious circle, pointed out by H.R. Holt, in which utilities seeking to minimize siting hassles may maximize capacity per site, making the project so big and problematical that the plant is perceived as a worse neighbor, hence increasing political resistance to such projects and making the next site that much harder and slower to find, and so on.

9. Shorter lead time reduces the risk of building an asset that is already obsolete—a point important enough to merit extended discussion in the next section.

The first five of these benefits emerged strikingly from a Los Alamos National Laboratory system dynamics study in 1985 (677). The analysts used a Northern California case study for Pacific Gas and Electric Company under the regulatory policies prevailing in the early 1980s. They examined how both the "lead time" to plan, license, and build a generic power station and the financial or accounting cost of that lead time (due to real cost escalation and interest on tied-up capital) would affect its economic value over a 20-year planning horizon. However, to clarify choices, they inverted the calculation: Rather than modeling longer-lead-time plants as riskier or

costlier (in present-valued revenue requirements), they simulated the utility's financial behavior and asked how much "overnight" (zero-lead-time) construction cost could be paid for the plant as a function of its actual lead time in order to achieve the same financial objectives.

Adding also a similar analysis for a coal-fired utility (677) and another for Southern California Edison Company (245), the Los Alamos team found that shorter lead times justified paying about one-third to two-thirds more per kW for a plant with a 10- instead of a 15-year lead time; that a 5-year lead time would justify paying about *three times* as much per kW; and that a 2.5-year lead time (analyzed only for SCE) would justify paying *nearly five times* as much per kW. In each case, these far costlier but shorter-lead-time plants would achieve exactly the same financial performance as their 15-year-lead-time competitors under the same exogenous uncertainties, for the first five reasons listed above. Shown all on the same graph, the results look like this:

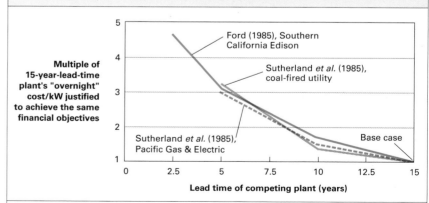

Figure 2-7: Power-plant financial feasibility *vs.* lead time
To achieve the same financial performance and risk, power plants with severalfold shorter lead time can compete even at severalfold higher construction costs.

Source: W. R. Meade and D. F. Teitelbaum, "A Guide to Renewable Energy and Least Cost Planning" (Interstate Solar Coordination Council, 1989), p. 11, ex. 8; R. J. Sutherland *et al.*, "The Future Market for Electric Generating Capacity: Technical Documentation" (Los Alamos National Laboratory, 1985), pp. 145–146

Figure 2-8: Slow construction multiplies its costs
Construction costs spiral with the combination of lead times, interest rates, and cost escalation rates.

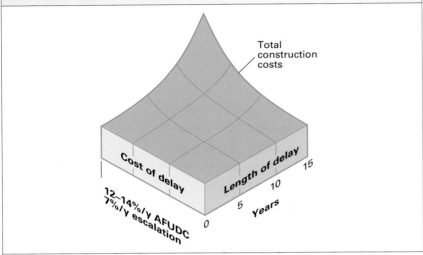

Source: R. J. Sutherland *et al.*, "The Future Market for Electric Generating Capacity: Technical Documentation" (Los Alamos National Laboratory, 1985), p. 114

These Los Alamos simulations show that plants with a 3–4-fold shorter lead time can cost (in "overnight" $/kW terms) about three times as much per kW, yet still yield the same—or, taking account of resilience under surprises, better—financial performance. Yet most distributed resources have lead times considerably shorter than the smallest value analyzed, 2.5 years; some take more like 2.5 months, weeks, or days to install. As construction time converges toward the theoretical "overnight" ideal, wouldn't distributed resources earn an even *larger* tolerance of higher overnight cost? Moreover, wouldn't similar considerations apply not just to generating but also to *grid* investments? If so, mightn't it be worth even more to avoid grid investments, since

- U.S. utilities have lately been investing more than twice as much on grid as on generating assets. As recently as 1978, during the nuclear boom, U.S. utilities invested only one-third as much in the grid as in generating capacity. However, as Figure 2-9 shows, since the mid-1980s, investments in the grid have become dominant, even before much new generating capacity began to be financed and owned by non-utilities;

- emerging pure-distribution companies have almost no investments *but* the grid; and

- it is even more difficult to forecast demand accurately for a small area (which has less load diversity and is more subject to the vagaries of individual large customers, sectors, or neighborhoods) than for a whole utility system (which tends to average out random differences between customers, sectors, or regions)?

These findings clearly show that the longer or costlier the actual lead time, the greater its cost, and hence the costlier the short-lead-time plant that could compete with it:

However, that analysis (678) is conservative—it *understates* the benefits of short lead time—because it

> ...assumes a surprise-free, predictable future. There are no unexpected changes in regional economic growth, fuel prices, lead times, or [competing private generation] activity that might lead to adverse ratepayer or stockholder impacts when implementing the...resource plan. Thus, the fourfold cost advantage identified for short lead time plants...does *not* depend on the flexibility that shorter lead time plants offer in the face of uncertainty.

Sensitivity tests of the effect of a surprise (a ±100% change in demand growth rate halfway through), under a variety of other assumptions, confirmed that in most cases, short-lead-time plants would substantially increase the benefits or reduce the penalties of surprises, further increasing the value of short lead times (674).

Until 1997, no answer to these questions had been published. But in that year, energy economist and systems analyst Thomas Hoff

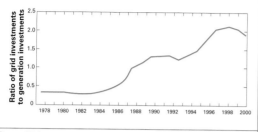

Figure 2-9: Utility investments are now dominated by the grid
U.S. investor-owned utilities are now devoting more than twice as much capital expenditure to the grid as to generation.

Source: EEI (Edison Electric Institute), *Statistical Review of the Electric Power Industry 2001* (EEI, 2002)

released a closed-form analytic solution (315) for the simplified case where demand growth fluctuates according to stochastic binary steps, in much the way others analyzed using decision theory (§ 2.2.2.6). This can make distributed resources cheaper than lumpy grid upgrades or generation expansions—the opposite of the conclusion reached when demand is viewed statically (via low, medium, and high growth scenarios) rather than dynamically as an unfolding process. For example, because the longer the lead time, the greater the demand uncertainty, if in any year there is a 50% probability that demand will increase (assumed to occur at a rate that uses up system reserve margin in one year), then at a 10%/y real discount rate, a $1,000 plant has a lower expected value—the longer its lead time, the less valuable it becomes. That is especially true if demand growth is considered as a dynamic process (Figure 2-10) based on those assumptions. The message of the graph— more fully explained by Hoff (315)—is that the dynamic unfolding of demand over time increases the risk reduction offered by short-lead-time plants; and the longer the difference of lead time (or the smaller the probability of rapid demand growth), the more dramatic this value advantage becomes.

Hoff's analytic approach (315) is illustratively applied to a system with equal probability of 0- or 5-MW demand growth each year; five years' worth of grid capacity remaining before the maximum rate (5 MW/y) of demand growth would require either expansion or distributed-resource reinforcement; and a 10%/y discount rate. Grid expansion is assumed to cost $25 million ($500/kW) and have a 5-year lead time, while distributed PV capacity would come with 1-year lead time and in 5-MW increments, each costing $15 million but returning $5 million in system benefits for a net per-unit cost of $2,000/kW. Thus ten increments of PV expansion would provide the same total capacity as the single 50-MW lump of grid upgrade. On these assumptions, *the expected present-valued cost is lower ($24 million) for the PV than for the grid-expansion ($25 million) choice, even though per kW the PV choice is four times as costly.*

Figure 2-10: Counting the dynamic nature of demand growth increases the value of short-lead-time plants
Considering demand growth as a dynamically unfolding process makes longer-lead-time plants even less valuable because so much more uncertainty accumulates about whether and when they might be needed.

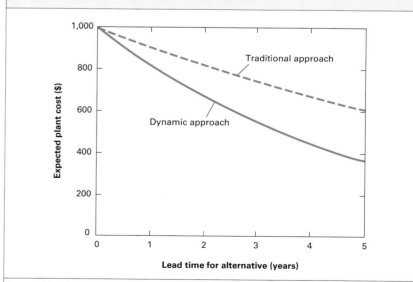

Source: T. E. Hoff, *Integrating Renewable Energy Technologies in the Electric Supply Industry: A Risk Management Approach* (NREL, March 1997), p. 39, fig. 5-5. www.clean-power.com/research/riskmanagement/iret.pdf

Thus "highly modular, short lead time technologies can have a much higher per unit cost than the non-modular, long lead time T&D upgrade and still be cost-effective." The analytic solution shows the following variation of breakeven PV net cost with both module size and lead time, based on the grid-displacement benefits flowing from the assumptions in the previous paragraph:

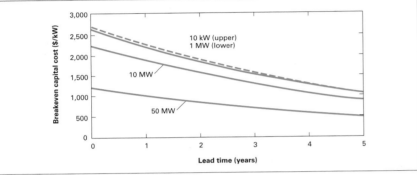

Figure 2-11: Smaller, faster grid-support investments are worth more
For a typical grid-reinforcement application, smaller and faster distributed-resource modules can compete with a lumpy grid expansion even if they cost manyfold more per kW. Please see text for assumptions.

Source: RMI analysis from Eq. 5.6 in T. E. Hoff, "Using Distributed Resources to Manage Risks Caused by Demand Uncertainty" (PEG, 1997)

Thus Hoff shows that the value of short lead time, shown by the Los Alamos studies for generating plants down to 2.5-year lead times, *also continues all the way down to zero lead time*, and is equally valid for analogous grid applications. Moreover, Hoff quantifies the additional value of small modules that better respond to fluctuating demand growth. (That value can also be assessed using option or decision theory, as discussed below in Section 2.2.2.5 and Section 2.2.2.6 respectively.) The analytic solution is (311):

$$E = I \left\{ 1 + \left(\frac{p}{d} \right) \left[1 - \frac{1}{\left(1 + \frac{d}{p}\right)^{N-1}} \right] \right\} \left(\frac{1}{1 + \frac{d}{p}} \right)^{T-L}$$

where

E = expected present-value cost

I = total investment cost of all plant increments

L = lead time (in years) of units, which are assumed to differ only in this respect and in cost, not in capacity

p = probability that demand will increase at a given step

d = real discount rate (in decimal format, per year)

T = number of years before demand growth at the highest possible rate (by growing at all possible steps) will use up available capacity (assuming $T>L$)

N = number of units needed to achieve desired increase in capacity

The term before the multiplication sign expresses the benefit of modularity; the second term shows the benefit of short lead time. Of course, as noted earlier, these two values are especially powerful in combination. That will occur when smaller modules *also* have shorter lead time, so that these two attributes are associated rather than unrelated. This will frequently occur in practice.

Moreover, Hoff's graphed results for the illustrative assumptions listed above (Figure 2-11) assume that the distributed resource has a real price that doesn't change over time. But in fact, PV prices have been declining at about 9%/y (311). If that continues, then "PV could have a current price of more than $6,000/kW [excluding non-grid benefits such as generating capacity, energy, energy loss savings, externalities, etc.] and still be a lower cost alternative than the T&D upgrade. This is because [if the grid upgrade takes five years but the PV installation only one year] there will be no investment in PV for at least four years (when its cost will be reduced to about $4,000/kW)." (311)

In fact, the Sacramento Municipal Utility District's turnkey bid price for complete residential PV systems was $5,060/kW in 1998, $3,950/kW in 2000, and $3,400 per installed kW of alternating-current ouput in 2002—a decrease of nearly 10% per year in nominal terms—so it appears that in the reasonable illustrative case offered by Hoff (especially bearing in mind that a substation application offers greater economies of scale than a residential one), actual market conditions for a decision-maker *already* meet these cost targets. Such dramatic price decreases are both a benefit to distributed resources and a competitive threat to centralized resources, as described in the following section.

As a final illustration of the importance of fast, granular resources, consider a perfect distributed generation resource that can be built in exactly the increments needed to meet annual load growth, with a one-year lead time—shorter than that of a larger central station. On those assumptions, the following table shows the percentage increase in the net-present-value cost of the central source compared with a distributed source *with the same unit capital cost* ($/kW). For example, if the central source has a capacity increment equivalent to six times the annual load growth, and a four-year lead time, it carries an effective 45% cost premium compared with a same-$/kW distributed source. Conversely, in this situation the distributed generator could cost 45% more per kW and still yield the same net-present-value capital charge as the central source. The only difference is in their lead time and their "lumpiness": the central resource costs more because it must be built earlier and because it has excess capacity until load growth catches up, as illustrated earlier in Figure 2-4. This calculation, however, is not as flexible and

Table 2-1: Smaller can cost more but can make more money
Net-present-value increase in benefit (percent) of a small resource with a 1-year lead time, compared to a large resource whose incremental capacity is the "size ratio" times annual incremental load growth.

Size ratio	Large resource lead time (years)				
	1	**2**	**3**	**4**	**5**
1	0%	5%	10%	16%	22%
2	5%	10%	16%	22%	28%
3	10%	15%	21%	27%	34%
4	15%	20%	27%	33%	40%
5	20%	26%	32%	39%	46%
6	25%	32%	38%	45%	53%
7	31%	37%	44%	52%	60%
8	36%	43%	50%	58%	66%
9	42%	49%	57%	65%	73%
10	48%	55%	63%	72%	81%

Source: J.N. Swisher, "Cleaner Energy, Greener Profits: Fuel Cells as Cost-Effective Distributed Energy Resources" (RMI, 2002). www.rmi.org/sitepages/pid171.php

inclusive as Hoff's analytic solution above, as illustrated in Figure 2-11, so that form is recommended for practical calculations.

2.2.2.3 Technological obsolescence

Technological change is very rapid. During the 1990s, the aeroderivative gas turbine, an offshoot of military jet engine R&D, halved the long-run marginal cost of fossil-fueled power generation, captured most of the market for new capacity, and triggered industry restructuring by making more acutely visible the spread between cheap new power and costly old power. What might happen next? Mature backpressure turbines, new microturbines, and emerging fuel cells promise still cheaper power (134), especially when their waste heat is harnessed. The whole proton-exchange-membrane fuel-cell revolution is based largely on better membranes, lower pressures, higher performance, and much lower cost (largely via an order-of-magnitude reduction in catalyst loadings, plus design for

Benefits

24   *Small units with short lead times reduce the risk of buying a technology that is or becomes obsolete even before it's installed, or soon thereafter.*

25   *Smaller units with short development and production times and quick installation can better exploit rapid learning: many generations of product development can be compressed into the time it would take simply to build a single giant unit, let alone operate it and gain experience with it.*

26   *Lessons learned during that rapid evolution can be applied incrementally and immediately in current production, not filed away for the next huge plant a decade or two later.*

27   *Distributed resources move labor from field worksites, where productivity gains are sparse, to the factory, where they're huge.*

28   *Distributed resources' construction tends to be far simpler, not requiring an expensively scarce level of construction management talent.*

29   *Faster construction means less workforce turnover, less retraining, and more craft and management continuity than would be possible on a decade-long project.*

30   *Distributed resources exploit modern and agile manufacturing techniques, highly competitive innovation, standardized parts, and commonly available production equipment shared with many other industries. All of these tend to reduce costs and delays.*

manufacturing and assembly). Many of these developments were unforeseen a decade ago. Similar breakthroughs seem possible in manufacturing high-temperature molten-carbonate and solid-oxide fuel cells. Completely new kinds of photovoltaics based on inherently cheap materials are also emerging, based, for example, on sulfur, polymers, self-assembling structures, synthetic organic molecules, or chlorophyll analogs. Many other technological surprises are increasingly likely as more and smarter technologies are fused into new combinations. Even the possibility of wholly new energy sources, based on an improved understanding of basic physics, cannot be excluded.

Amid such flux, the smaller and faster the units ordered, the less the risk of large capital commitments to technologies that are obsolete and uncompetitive even before they're installed. Sinking less capital in costly, slow-to-mature, slow-to-build projects, and inflexible infrastructure reduces financial regret, and may also shrink the institutional time constant for getting and acting on new information. Thus less capital is tied up at any given time in a particular technology at risk of rapid obsolescence; a larger fraction of capacity at any time can use the latest and most competitive designs; and the associated organizations can learn faster.

The value of the resulting risk reduction may be hard to quantify, because the nature and size of the technological risk is by definition unknowable. Yet that value features prominently in the thinking of strategists in such industries as telecommunications and information systems. It should be no less a core element of strategic planning for electricity. There is also a link between unit scale, the pace of technological improvement, and economics. Smaller units with short development and production times and quick installation can better exploit rapid learning—many generations of product development can be compressed into the time it would take simply to build a single giant unit, let alone operate it and gain experience with it. As with electronics, then, the lessons learned drive continuous improvements that can be rolled incrementally and immediately into successive modules—not filed away for the next generation of engineers (if they remember) to apply to the next giant unit.

Obviously such agile technologies also offer far greater economies of mass production—less like giant bridges, more like computers. They move labor from field worksites to factories, offering far greater scope for productivity gains—like building cars, not cathedrals. They exploit modern and agile manufacturing techniques, highly competitive innovation, standardized parts,[17] and commonly available production equipment shared with many other industries. Their short construction cycles minimize the big-project headaches of workforce turnover and retraining. Their far less complex construction management draws on a deeper and cheaper talent pool.

All these attributes interact. They also increase the likelihood that more ponderous competing technologies may become obsolete and need to be written off before the end of their planned amortization lifetimes. The displacement, already underway, of operating and unamortized nuclear plants by combined-cycle gas turbines (which can

be built and run more cheaply than just operating and repairing the average nuclear plant) offers a sobering lesson. Such lessons in turn make the capital markets wary of nuclear-like assets whose fair market value may depend far less on how far along they are in their projected engineering or accounting lifetimes than on the pace of technological evolution among competing technologies. Wary capital markets mean higher discount rates, costlier capital, and reduced competitiveness.

In general, too, central thermal power stations have neoclassical supply curves—the more units you build, the *more* each one costs—for reasons fundamental to democratic societies (§ 1.2.2, Figure 1-8). In contrast, efficiency and dispersed renewables perceived as benign have experience curves. For PVs, for example, each doubling of cumulative production has *cut* real marginal cost by nearly one-fifth. In any long-run competition between these two types of technologies, with their fundamentally different processes of both technical innovation and public acceptance, the more ponderous and unpopular ones are likely to lose. We return to this issue in Section 2.4.10.

2.2.2.4 Regulatory obsolescence

The cost, siting, and even practical availability of technologies depends on regulatory requirements, tax rules, and other public policy. Continuous conflicts between various groups amidst a swirling and ever-changing mass of environmental, social, and economic concerns make the regulatory process often unpredictable in detail (though often rather predictable in general trend), and hence a source of risk just as

[17] *Business Week* (84) reports that the U.S. military's wider adoption of standard commercial parts has reduced availability lags from months to hours and cut costs by fourfold or more.

Benefits

31 *Shorter lead time reduces exposure to changes in regulatory rules during construction.*

32 *Technologies that can be built quickly before the rules change and are modular so they can "learn faster" and embody continuous improvement are less exposed to regulatory risks.*

33 *Distributed technologies that are inherently benign (renewables) are less likely to suffer from regulatory restrictions.*

34 *Distributed resources may be small enough per unit to be considered* de minimis[18] *and avoid certain kinds of regulation.*

35 *Smaller, faster modules offer some risk-reducing degree of protection from interest-rate fluctuations, which could be considered a regulatory risk if attributed to the Federal Reserve or similar national monetary authorities.*

important as technological obsolescence. For example (317), PG&E's 1994 *Annual Report* discloses that bringing its existing power stations into compliance with current NO_x emission rules could require investments of up to $355 million over a decade—costs probably not anticipated when the plants were bought. (They were sold soon thereafter as part of restructuring.) Similarly, plants ordered today could require costly retrofits, operational restrictions, or fuel-price changes in another five or ten years because of greater understanding of fine-particulate or carbon emissions.

Obviously, technologies that can be built quickly before the rules change, and are modular so they can "learn faster" and embody continuous improvement, are less exposed to such regulatory risks (384). Still less exposed are plants that are inherently benign, so they are less likely to suffer from regulatory restrictions, or simply small so they may be considered *de minimis*.[18] Smaller, faster modules may also offer some protection from interest-rate fluctuations, which could be considered a regulatory risk if attributed to the Federal Reserve.

[18] This phrase, effectively meaning "too small to worry about," comes from the old legal maxim *De minimis non curat lex*, "The law isn't concerned with trifles."

2.2.2.5 Flexibility/modularity value assessed by option theory

"Flexibility," in a management context (117),

> ...is generally used to describe the ability to do something other than that which was originally intended....Similar terms...are 'adaptable' and 'versatile' (defined respectively by the *Concise Oxford Dictionary* as 'capable of modification' and 'able to turn readily from one activity to another').*** Other things being equal, one position is more flexible than another if:
> (1) It leaves available a larger set of future positions....
> (2) It allows the attainment of new positions in a shorter period of time....
> (3) It requires less additional cost to move to another position.

There are many potential tradeoffs between these dimensions. Obviously, flexibility is not desirable *per se*, but only insofar as its benefits exceed its costs. That is, "Flexibility is valuable in so far as it is able to reduce the cost of inflexibility." (116) Until recently, however, flexibility's benefits were qualitative and abstract while its costs seemed quantitative and concrete, so big investment decisions tended to default to the inflexible but measurable. Now new tools from financial economics are starting to shift that balance, encouraging the purchasing of flexibil-

Benefits

36 *The flexibility of distributed resources allows managers to adjust capital investments continuously and incrementally, more exactly tracking the unfolding future, with continuously available options for modification or exit to avoid trapped equity.*

37 *Small, short-lead-time resources incur less carrying-charge penalty if suspended to await better information, or even if abandoned.*

38 *Distributed resources typically offer greater flexibility in accelerating completion if this becomes a valuable outcome.*

39 *Distributed resources allow capacity expansion decisions to become more routine and hence lower in transaction costs and overheads.*

40 *Distributed generation allows more learning before deciding, and makes learning a continuous process as experience expands rather than episodic with each lumpy, all-or-nothing decision.*

41 *Smaller, shorter-lead-time, more modular units tend to offer cheaper and more flexible options to planners seeking to minimize regret, because such resources can better adapt to and more cheaply guard against uncertainty about how the future will unfold.*

---

ity where it is worthwhile.[19] This may result in preferred decisions that appear to traditional utility planners to be counterintuitive, such as deliberately building *less* conventional capacity than required to meet expected demand; but that is the right answer if flexible resources create an asymmetry in the over/undercapacity "penalty function." "Essentially, if being late [in building enough central plants to meet demand] is less expensive than being early, it pays to be late, although there are costs involved." (116)

The theoretical future envisaged by traditional power-station planners is deterministic, making such choices seem invisible and therefore unnecessary to consider. But the *actual* future inhabited by electricity providers and users is not deterministic at all. Rather, it *gradually unfolds* in unpredictable ways. The inevitability of uncertainty in how that future unfolds makes modular resources especially valuable. Why? Because, as financial-economics con-

sultant Dr. Shimon Awerbuch (34) correctly notes, modular resources create

> ...valuable *flexibility* options since managers can install capacity slowly, over time, to match load increases. Moreover, capacity expansion decisions become more routine—like the installation of additional telephone central office capacity—and hence less costly. Recent work on flexibility suggests that when valued in a traditional manner, inflexible projects are comparable to [*i.e.*, potentially competitive with] flexible ones only if their present value is considerably greater.

Understanding the dynamic nature of the demand-growth process is important to reducing financial risk by choosing the right investments. Hoff and Herig (317) have shown that on reasonable assumptions, the cost premium worth paying for a modular resource can easily double using a dynamic rather than a static model of demand—for reasons similar to those described in Section 2.2.2.2.

[19] Many other sources of flexibility for utility planners, such as extending the retirement of old units or trading wholesale power with other utilities, are important (115) but are beyond the scope of this book. So, largely, is demand-side management—one of the greatest sources of flexibility in the electricty industry, and one of the highest-return investments available in the entire economy.

From a slightly different perspective (494), deferring major resource commitments has a direct economic value (saved carrying charges and opportunity costs), and that deferral can be achieved by substituting small investments that, while allowing an option to revisit the big investment decision later, meanwhile allow more learning before deciding. Smaller, faster modules can therefore allow an intelligent response to the uncertain-load dilemma that Applied Decision Analysis consultant Peter Morris draws in Figure 2-12:

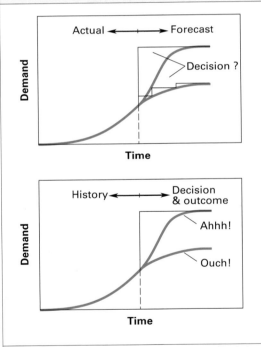

Figure 2-12: Minimizing regret as the uncertain future unfolds
When demand is uncertain, smaller modules of supply can minimize regret.

Source: P. Morris, "Optimal Strategies for Distribution Investment Planning" (EPRI, 1996)

What tools can measure how well different decisions can maximize the "ahhh-to-ouch" ratio?

As described in Tutorial 2 below, option theory—a tool widely used in sophisticated financial investment management—is one way to put an economic value on the way modular resources create managerial options that, if exercised *in the future*, are beneficial even though they do not affect short-term accounting costs (except by paying for the option itself—if it *has* a cost, which for many distributed resources is often negative anyway). That is, rather than comparing the net present values of decisions envisaged *now*, option theory assesses the additional value of flexible *choices* now to delay a decision until more is known.

Some of the option benefits described in Awerbuch's quotation above have already been considered above, though more in the engineering and planning than the financial and option-theory metaphors. But some other benefits, such as making expansion decisions more routine and hence less costly, were not previously described. Also, the option description of portability or salvage value, described in Section 2.2.2.8, may capture some additional element of flexibility.

Tutorial 2: Option Theory

Option theory helps to recognize and value opportunities where "the range of potential outcomes presents an upside potential that can be quite high. The downside risk is only the cost of procuring the option, which is much more limited than the possible loss resulting from a sunk investment in an uncompetitive resource." Capturing that spread yields "a 'just in time' resource commitment philosophy" in which "shorter lead time resources possess value beyond what is indicated by a standard [net-present-value] calculation because they allow a utility to wait for better information and thereby eliminate some uncertainty prior to commitment." (377)

Rigorous applications of option theory to modular utility resources are few and early, but highly suggestive. For example, the same article cites New England Power Company case studies in which, in certain specific circumstances,

- a resource with an option value is worth paying up to $167,500 more to acquire if lead time is one year but not if it is two years (*i.e.*, the flexibility of the shorter lead time is worth up to that much);

- a hydro repowering project, because of exogenous uncertainties, was worth $5 million more if deferred than if bought immediately;

- shutdown of two old, small coal-fired units should be deferred as long as economically possible to await better information on NO$_x$-emission upgrade requirements; and

- option theory was used to optimize buyout provisions in independent generators' contracts.

A fuller case study is provided by a Harvard Business School paper (684) that includes in its option pricing model of asset value

> ...descriptive factors frequently ignored,...including lead time, lumpy and sequential cost outlays, irreversibility of expenditures, and uncertainty about regulatory outcomes for completed projects. The analysis shows the value of shorter lead time technologies, the value of flexibility to delay or abandon construction, [and] the incentive to delay construction under uncertain regulation....

Under basic option valuation theory, "the value of an option increases as future uncertainty increases. Since exercise of the option is never required, managers are not forced to incur losses; however, they have the opportunity to take advantage of good outcomes by exercising the option if they choose." (684) That opportunity can be extremely valuable, because managers retain future choices, rather than being locked into one present choice and no future choices by the inflexibility of a large, lumpy, irreversible, long-lead-time investment chosen now.

The Harvard analysis considers a hypothetical project similar to a 500-MW coal-fired steam plant in the highly uncertain environment of the 1970s, whose costs might or might not be recoverable through regulatory decisions—somewhat akin to today's market-structure and competitive uncertainties. The project exhibits option values of flexibility, short lead time, and modularity that increase with future uncertainty. Specifically, flexibility to phase the construction in parts according to need, rather than all-or-nothing construction of a single monolithic resource, greatly increases option value, as shown in Figure 2-13:

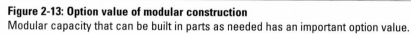

Figure 2-13: Option value of modular construction
Modular capacity that can be built in parts as needed has an important option value.

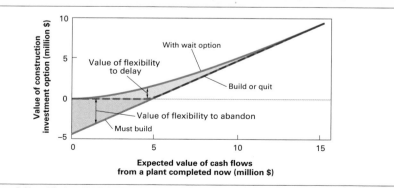

Source: E. Olmsted Tiesberg, "An Option Value Analysis of Investment Choices by a Regulated Firm" (*Management Science*, April 1994), p. 542, fig. 2

Moreover, "a shorter lead time increases the firm's flexibility to complete the project more quickly. Thus, the value of a shorter lead time is another example of the option value of decision flexibility, and[,] in general, flexibility adds value to the project when the future is uncertain." In particular, "the value of the project is higher when the minimum possible lead time is shorter" under different regulatory outcomes (Figure 2-14).

The importance of such effects was also found to increase with less or no regulation, since the firm will then face greater uncertainties in the marketplace, making flexibility even more valuable. This is now true of most countries.

Option theory is well established and widely used. Like the Capital Asset Pricing Model (§ 2.2.3.2), it uses, of necessity, certain idealized assumptions that may hide essential aspects of actual markets (287, 501, 585).[20] But it is certainly better than the alternative of deterministically ignoring option values. Utilities can, after all, acquire tangible options to mitigate their risks. These options may include (121)

> ...(depending on the underlying resource) identifying sites, testing technologies, training installers, determining market potential, developing commercial relationships with suppliers, and perhaps reserving some generation or construction capacity. A small investment today may [ensure]...the availability of resources (existing generation, photovoltaic cells, accelerated DSM, new central generation) in the future, at the time they turn out to be most valuable.

Such valuable options are more likely to be bought if their option value is explicitly known.[21]

Figure 2-14: Shorter lead time increases option value
The value of reducing the lead time from 4 years to 2 years is $16 million in the full cost allowance example.

Source: E. Olmsted Tiesberg, "An Option Value Analysis of Investment Choices by a Regulated Firm" (*Management Science*, April 1994), p. 544, fig. 5

[20] For example, option pricing models assume that financial markets are frictionless (with full and free access to perfect information) and that assets may be sold short without restriction. Some particular models have further restrictive assumptions discussed in financial texts.

[21] The fundamental reasons that small resources can be faster to deploy *en masse* (§ 1.6.5) can also be considered aspects of their flexibility.

2.2.2.6 Flexibility/modularity value assessed by decision analysis

Another important approach to valuing flexibility and modularity uses decision analysis (Tutorial 3)—a quantitative technique rooted in operations research, management theory, and financial economics. If you know what choices you will have at a series of future times, what you'll *then* know (and with what certainty) to help make those choices, and how much each outcome is worth, then decision analysis uses an elaborate simulation of millions of possible decision trees to determine the optimal decision under each set of possible

Tutorial 3: Decision Analysis

Decision analysis explicitly describes all the key uncertainties and produces a flexible series of investment decisions that respond to the way uncertainties actually develop and resolve over time (as Morris reminds us, we don't learn the whole future in one year). This typically yields much better—more profitable or valuable—decisions than the traditional methods: high returns to investment are more probable and low returns are much less probable, both at the cost of significant analytic effort. The technique has been applied by the Northwest Power Planning Council, New England Electric System, and others. In other words, as Resource Insight analyst Paul Chernick reminds us (122),

> Traditional approaches to reflecting uncertainty in utility planning, which essentially define a strategy as a fixed mix of resources, are too limited. A strategy includes options for responding to changing circumstances. Utilities should devise plans based not on simple expected values of future events, but on decision analysis by modeling a sequence of event, decision, event, decision. In more complex situations, Monte Carlo simulations, in which random events alternate with realistic utility reactions, can replace formal decision analysis.

> The lack of realism in the traditional resource-modeling process creates the impression of in-depth analysis without teaching the utility about the relative flexibility and risk-mitigating value of various resource plans and capability building. As a result, these approaches cannot reflect the major advantages of DSM, renewable resources, and distributed generation over conventional supply: small increments, short lead time, security of continued supply, protection from fuel-price fluctuations, lack of environmental risk, and load-following in installation and operation. These methods also cannot reflect the advantages of 100-MW additions over 300-MW additions, determine the cost-effectiveness of building [combustion turbines]... with provisions for conversion to combined-cycle operation or coal gasification, determine the cost-effectiveness of pre-licensing potential additions to reduce lead time, or otherwise [reveal] the costs and benefits of alternatives under uncertainty.

In a particular site and a well-characterized set of circumstances, decision analysis can fully model the economic value of distributed resources' short lead times, small and modular units, and hence ability to install exactly the amount of capacity where and when it is needed. How can this work in practice?

A proprietary EPRI study (496) applied decision analysis to Pacific Gas and Electric Company's photovoltaic installation to reinforce the fully loaded Kerman substation—a case study summarized in Section 3.3.5.5. In that particular case, the flexible policy based on the modular resource and its ability to adapt to unfolding circumstances turned a large cost into a significant net benefit: the difference had a net present value equivalent to about one-eighth of the PV project's total value. If added (as it should be) to the separately evaluated other benefits of the project, this additional and previously uncounted benefit would rank fourth, just behind improved reliability ($225/kWy), saved energy ($194/kWy), and the avoided cost of upgrading the substation ($115/kWy).

Unfortunately, no decision analysis of this or any other distributed-resource applications appears to be publicly available in full detail: it is a very active and profitable field for many consulting firms, which are naturally reluctant to make their methods, models, and findings public. But this example persuasively illustrates the important economic value of modularity, and of the flexibility it provides.

Section 2.2.7 will mention a further value of these attributes when demand is uncertain. Distributed-resource investments can be arranged so that their costs are most likely to be incurred during periods of demand growth when the firm is more profitable and hence better able to afford those costs. This makes profits more reliable and hence the distributed resource more valuable.

A final example akin to decision-theory valuation is offered by Hoff (314). He describes the hypothetical case of a developer who wants the utility to extend its grid to a greenfield site so he can build the first five of a planned 50-home development, but the utility isn't sure the project will succeed, so it's reluctant to pay $200,000 for the grid extension. On reasonable assumptions, the utility will lose an expected value of $45,000 if it extends the grid immediately. However, suppose instead it installs PV generation and storage for the first five houses, and then for five more houses if warranted. This gives the developer much greater confidence that the project will succeed. Only then would the grid-extension investment be committed for all 50 houses. The original 10 PV systems would be removed for resale or reuse elsewhere, recovering ~95% of their value (§ 2.2.2.8). This yields an expected gain of $72,000 in expected value. The difference between the expected value of these two strategies pays 59% of the grid-extension cost.

conditions, and thence the optimal decision policy to pursue under the assumed uncertainties.

This theoretical framework tends to be more deterministic than the decision-maker's real world, which is full of all kinds of wild cards, including changes in the structure of the whole problem. It also requires specific assumptions to be made about the value and probability of future choices—a requirement that may not be a great deal easier to satisfy than knowing the future itself. However, decision analysis does differ fundamentally from, and can yield better decisions than, the traditional engineering-oriented methods of dealing with uncertainties: namely, to

- ignore them, or
- recognize a number of possible outcomes and assign judgmental probabilities to them, but still pursue a single probability-weighted course of action,[22] or
- recognize them and develop a different course of action depending on how each key uncertainty unfolds.

2.2.2.7 Project off-ramps

Benefit

42 *Modular plants have off-ramps so that stopping the project is not a total loss: value can still be recovered from whatever modules were completed before the stop.*

Hoff and Herig (317) point out that managers can gain valuable options not only in deciding when to buy resources but also in deciding when to *stop* buying them: "Modular plants have off-ramps so that stopping the project is not a total loss." Suppose that a series of units is being built, their cost is uncertain, and this uncertainty will be largely resolved when the actual cost of the first unit is known because subsequent units will have similar costs. If the actual cost turns out to be excessive and managers want to cut their losses, then (assuming no salvage value) more value can be recovered if whatever has already been built can operate and yield revenue. "Thus, while modularity provides value to utilities [or other developers] who want to control demand uncertainty, it is also of value to investors who are funding an [independent

---

[22] This approach is still being adapted by some analysts to distributed-utility planning (495), but seems attractive more for its similarity to familiar planning tools for centralized systems than for its recognition of the unfolding-future context of making decisions to minimize regret.

power-producing project]...and are unsatis-
fied with the project's progress." Even if
investors pull the plug on financing partway through a modular project, they can
still get some value from whatever modules
were already finished, rather than being
stuck with an inoperable piece of an
uncompleted large plant.

2.2.2.8 The extra value of modules' portability and reversibility

Once a power plant is sited and constructed, it's conventionally presumed to be there
forever, at least until demolished. However,
many short-lead-time, small-scale technologies are "sited" only temporarily, because
they are inherently portable. As Awerbuch
remarks (34),

> ...although renewables such as [photo-
> voltaics]...are generally quite capital-inten-
> sive, and thus often thought of as inflexi-
> ble on the basis of their supposedly high
> 'sunk cost,' the proportion of *sunk cost* is
> probably lower for PV as compared to[,]
> say, a large coal plant or even a gas tur-
> bine. This is evidenced by the fact that PV
> installations can be (and have been) unin-
> stalled at some future time and sold for a
> reasonable fraction of their original cost.
> This managerial option likewise creates
> flexibility and hence has significant value.

That value arises because the resource
remains flexible in use throughout its engineering life; it can be physically redeployed
to a different site or even a different utility
system. Thus if, for example, a photovoltaic
array is sited at a particular substation to
support expected demand growth that fails
to occur there, then the array can be disconnected and unbolted (leaving behind only a
very small fraction, perhaps nominally
around 5%, of its value in footers, cables,
etc.). It can then be loaded onto a truck and

Benefits

43 *Distributed resources' physical portability will typically achieve a higher expected value than an otherwise comparable non-portable resource, because if circumstances change, a portable resource can be physically redeployed to a more advantageous location.*

44 *Portability also merits a more favorable discount rate because it is less likely that the anticipated value will not be realized—even though it may be realized in a different location than originally expected.*

45 *A service provider or third-party contractor whose market reflects a diverse range of temporary or uncertain-duration service needs can maintain a "lending library" of portable distributed resources that can achieve high collective utilization, yet at each deployment avoid inflexible fixed investments that lack assurance of long-term revenue.*

46 *Modular, standardized, distributed, portable units can more readily be resold as commodities in a secondary market, so they have a higher residual or salvage value than corresponding monolithic, specialized, centralized, nonportable units that have mainly a demolition cost at the end of their useful lives.*

47 *The value of the resale option for distributed resources is further enhanced by their divisibility into modules, of which as many as desired may be resold and the rest retained to a degree closely matched to new needs.*

48 *Distributed resources typically do little or no damage to their sites, and hence minimize or avoid site remediation costs if redeployed, salvaged, or decommissioned.*

reinstalled at another "hot spot" where its
output will be worth more.[23]

On the logic illustrated in the sidebar on
p. 142, a large utility may well wish to
maintain a sort of internal "lending

---

[23] Sometimes this is a deliberate design feature. For example, when Robert Sardinsky was designing a photovoltaic system to power a house being built in a sensitive mountain site, he made the PV system first ground-mounted, to run the construction tools (thus avoiding a smelly and noisy portable generator), then simply installed the PV array on the roof afterwards.

Examples: Portable resources

As part of a 1996–2000 demonstration project under the Rural Electric Research program of the National Rural Electric Cooperative Association, a 200-kWe ONSI phosphoric-acid fuel cell made in Connecticut and mounted in Texas on an ordinary truck-towable trailer was scheduled for successive tests in Georgia, Colorado, mainland Alaska, and the Aleutians (507).

A rural utility even deployed a portable photovoltaic generator to serve a customer whose continued presence was uncertain (507, 669). Plains Electric G&T [generation and transmission], later merged, was a sprawling rural cooperative serving 20% of New Mexico's population but over 60% of its land area; it had only 5 customers per mile of distribution line (3.1/km). Its 14.4/24.9-kV line serving the town of Cibecue, 36 miles (58 km) from the substation, had a radial feeder with over 95% of its load in town and over half the peak demand due to a lumber mill with unreliable load. Projected peak demand would soon create a voltage drop exceeding Rural Electrification Administration guidelines, but the circuit would then be at only 25% of capacity. A 69-kV upgrade and substation would cost $4.2 million; but since the mill's load was not considered reliable, that major fixed investment would be at risk. A 100-kW photovoltaic generator, providing sufficient load match 85% of the time (equivalent to a 3-year deferral), or 100% with less than an hour's storage, was therefore ideal. It would provide peak-load voltage support, requiring less investment than capacity support would, but if the mill load vanished, the asset could be redeployed elsewhere in the sprawling system. A conservative assessment of a few distributed benefits—energy and capacity value, grid construction deferral, and line and transformer loss reductions, $12/kWy worth of externalities, but no reactive support—found a breakeven cost equal to the then-current PV cost of $9/W for installed systems. However, the value increased to $11.11/W when the system was assumed to be redeployable four times, at a moving cost of $1/W, achieving at each successive site a 3-year deferral with only a minimal 30% of the benefits of the original site. Both that 30% and the implicitly assumed system life look conservative; more like 10+ deployments would be possible in a highly likely 30-year system life if the system didn't need to wait for the next new site to become ready to receive it.

library" of flexible, portable resources redeployable at need, just as at least one major East Coast utility reportedly maintains (or at least holds an option on) barge-mounted gas-turbine capacity that can be connected to its grid from dockside if a major generator should fail. To be sure, this option is not available for all kinds of renewables—it is less suitable for a windfarm than for a fuel cell or a PV array—nor do fuel cells necessarily enjoy inherent flexibility advantages over skid-mounted gas turbines in this respect. Nonetheless, the concept can be an important risk-reducer for utility planners who want to match temporary or uncertain-duration resources to similar revenue streams, rather than sinking inflexible costs to serve potentially ephemeral loads. Since the dominant benefits are usually to the distribution system, a competitive industry structure in which power is readily wheelable should not greatly alter this conclusion. The value of optimal siting of distributed resources within the network may also be dramatically increased as new software permits nearly instantaneous power-flow optimization calculations on portable computers.[24]

In cases where site-specific benefits *somewhere* in the system are expected to remain available and high throughout the distributed resource's life, the economic value of its portability can be approximated by counting zero or very low forecasting uncertainty for the realization of site-specific benefits throughout that life. If the planned benefits don't materialize in that place as expected, or prove smaller or briefer there than expected, then they can probably be achieved elsewhere instead. Following the logic of financial economics (§ 2.2.3), therefore, a favorable discount rate should be

---

[24] This capability is claimed by Optimal Technologies (www.otii.com).

applied to that stream of benefits. The same is true (subject to adjustments for certain infrastructure) to engine generators, wind turbines, fuel cells, etc.

For this reason, in the proprietary EPRI study discussed in Tutorial 3 above (496), where a flexible (PV-based) investment's modularity changes a negative into a positive expected net present value, any comparison of the two investments should reflect their respective risk-adjusted discount rates (§ 2.2.3). These will differ—the discount rate should be more favorable (approaching riskless) for the modular resource—*because it is portable* and therefore incurs little or no risk that its modularity benefit will not in fact be realized. That is, a resource that is both modular/short-lead-time *and* portable will typically achieve a higher expected value than an otherwise comparable inflexible resource (such as a substation upgrade or central power station) because of its modularity and short lead time, but will *also*, separately, discount that expected value at a more favorable rate because its portability virtually eliminates the risk of not achieving those benefits. This "double-dipping" concept does not appear to be widely reflected in the literature, which therefore understates this kind of distributed benefit.

For a private owner unconcerned with the distribution utility's distributed benefits and lacking another site to which to redeploy a portable resource, the equivalent value is realized by the ability to resell (salvage) the installation at will rather than being stuck with an entirely and permanently sunk cost: that is, most of the investment is reversible. Moreover, its value is divisible, so all or any part can be liquidated as desired. As noted in Section 2.2.2.5, this value can be calculated using option theory or other financial-economics techniques.[25] Hoff and Herig (317) offer two illustrations:

- Hypothetically, two developers propose 50-MW battery storage facilities with identical prices and performance, but one is a single 50-MW battery while the other comprises 50,000 individual car batteries. If future storage-technology breakthroughs made these approaches obsolete, then at least the car-battery project could still be salvaged for resale and use in cars (so long as car technology doesn't change too much) or small PV systems, while the single 50-MW battery would probably have to be scrapped.

- In practice, when Arco Solar, for strategic reasons, resold its 6-MW Carissa Plains PV plant and the new buyer dismantled it, the used modules were in fact resold at retail prices of about $4,000–5,000/kW at a time when new modules were selling for about $6,500–7,000/kW. That is, the used modular assets could be marked to market and lucratively resold—and were.

Hoff presents "the value of the option to abandon a plant as a percent[age] of current market value versus the plant's salvage value." It assumes a 10 percent risk-free rate, 5 percent dividend rate (value of plant output), infinite plant life, constant salvage value over time, and standard

[25] Hoff and Herig (317), and in detail Hoff (306), analyze the value of investment reversibility using techniques analogous to those used for valuing American "put" options on dividend-paying stocks.

deviations of the plant's value of 10%, 25%, and 50%:

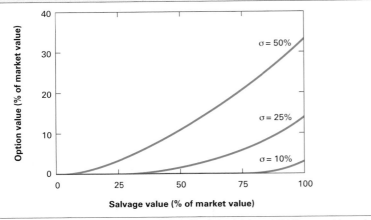

Figure 2-15: Salvage option value
The more uncertain a plant's value, the greater the option value of being able to salvage much of its original investment.

Source: T. E. Hoff, *Integrating Renewable Energy Technologies in the Electric Supply Industry: A Risk Management Approach* (NREL, March 1997), p. 65, fig. 5-20. www.clean-power.com/research/riskmanagement/iret.pdf

Thus, for example, if the standard deviation of uncertainty in the plant's value is 50%, and salvaging the plant can yield 75% of its current market value, then the value of the abandonment-and-salvage option is equivalent to adding 20% to current market value.

Moreover, and not captured in that example, in either of the cases bulleted above it would be possible for the proprietor to sell only part—*any* desired part—of the capacity and keep the rest to meet a reduced but nonzero need. That valuable decremental option is available only because modular technologies are divisible.

Another aspect of reversibility is that most renewable sources have small and relatively benign impacts on the site where they are installed. In contrast, nuclear units may permanently "sterilize" other land-uses, while most fossil-fuel plants entail substantial civil works and some may risk long-term soil and

[26] Paul Chernick of Resource Insight states that he developed a similar method of risk-adjusted discounting in 1987 (120, 698). Chernick disapproves of some of Awerbuch's risk-adjustment methodology, including an alleged difference between tracking absolute *vs.* percentage price changes, and other issues considered below. We have not been able to resolve this dispute nor to establish intellectual priority in developing the application of risk-adjusted discounting to energy projects.

water contamination. These differences affect residual site value and the flexibility of later reuse. Often they give coal plants a small or even negative salvage value (45).

2.2.3 Avoiding fuel-price volatility risks

This and the following three sections (2.2.4–2.2.6) draw heavily on the pioneering work of independent financial-economics analyst Dr. Shimon Awerbuch, now at the International Energy Agency in Paris, and particularly on his March 1996 first-draft *How to Value Renewable Energy: A Handbook for State Energy Officials* (29).[26] Unfortunately, its sponsor, the Interstate Renewable Energy Council, lacked funding to complete this work, though an IEA follow-up study is due in 2002 (44). Many important sections therefore remain only in outline, not fleshed out. However, the following description seeks to sketch the basic concepts in sufficient detail to establish that they are often, along with short lead times, the most important single source of distributed benefits—especially those that can often make renewables highly competitive against fossil-fueled power generation.

Many distributed resources happen to be renewable, and hence use no depletable fuel. Some others, such as fuel cells and cogenerators, may use depletable fuels but at higher thermal efficiency. Such resources can thus eliminate or reduce exposure to the financial risks of fluctuating fuel prices.

Market prices for fossil (or for that matter nuclear or biomass) fuels fluctuate in the same manner as for any other freely traded commodity. For example, the 115-year

Benefits

49 *Volatile fuel prices set by fluctuating market conditions represent a financial risk. Many distributed resources do not use fuels and thus avoid that costly risk.*

50 *Even distributed resources that do use fuels, but use them more efficiently or dilute their cost impact by a higher ratio of fixed to variable costs, can reduce the financial risk of volatile fuel prices.*

(1881–1995) record of movements in the real price of crude oil on the world market is a Brownian random walk,[27] as revealed by statistical analysis of the following graphic data (from Worldwatch, British Petroleum, and USEIA):

Figure 2-16: The random walk of world real crude-oil price, 1881–1995

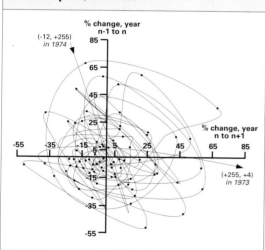

Source: RMI graphic in format by H.R. Holt (USDOE), based on Worldwatch Institute compilation of data from British Petroleum, *BP Statistical Review of World Energy* (BP, 1993); 1995 update *id.*; EIA, *Annual Energy Review 2000* (EIA, 2001)

The popular notion that the Organization of Petroleum Exporting Countries has somehow stabilized oil prices is incorrect. Though OPEC's market power and initial cohesion did permit it to raise general oil-price levels for some years, the main effect of OPEC in the longer run has been not to make world oil prices more consistent (nor even lastingly higher) but rather to treble their volatility.[28] This in turn has helped to elicit oil futures and options markets that are deep for about the next decade, then more shallowly traded, but still permit one to buy today, at a fixed price, oil for delivery 20 years hence. (The same can also be done for natural gas, and there is arbitrage between the two forward markets.) Naturally, such a contract embodies not only expected price trends, but also "price-risk insurance" for which an insurance premium must be paid: the underwriter will demand compensation for accepting the risk of unknown future prices—which could go either way—by selling future oil today at a fixed price. Typically the long-term oil mar-

---

[27] In such a series, distance from the origin (0,0) increases as the square root of the number of steps. If the series contained systematic trends or autocorrelations—*e.g.*, if an increase in price in one year made price more likely to increase also in the following year—then its shape around the central point would be distorted, favoring one sector over the others. The originator of this style of graph, H. Richard Holt, recently retired from a senior policy-analysis position at the U.S. Department of Energy, where he created a similar graph in an unpublished internal memo more than a decade ago (335). He found that for 1901–87, year-to-year price swings could be nearly 60% up or down, with a one-in-five chance of 20%; prices are jagged, so "the smoother the path, the less likely it is to occur"; 12 years with no price spikes bigger than 10% (up or down) is about as likely as nine reds in a row in roulette; there are no significant year-to-year price correlations; "extremely 'noisy' short-term fluctuations are imposed on a long-term price trend that is rather flat"; and "the price-time series has the properties of a random walk as tested by three statistical measures." These included the observation that "runs" of successive price increases or decreases occurred 42 times in 87 years, almost exactly as expected from a normal distribution. The implications of such mathematical behavior for economic value are described by Manne (273).

[28] Holt (335) found that the average real price during 1974–87 was more than twice that of the previous 71 years, the variance was 18 times higher, and the mean deviation was up fivefold higher. This counts only variance between average full-year prices, but volatility was also high within some years: during 1986, for example, the price dropped 56% and rebounded 35%. Many observers now consider there to be a hidden stabilizing mechanism: the "dirty little secret of the oil business," as one put it, "is that neither producers nor consumers want oil to trade outside a reasonable price range, because either too high or too low prices cause too much pain to both parties, and they have a shared interest in pain avoidance." However, this mechanism does little to damp normal commodity-price volatility on many timescales, especially given political instability in the Gulf, Venezuela, etc.

ket shows flat or declining underlying real prices, for many fundamental technical and economic reasons. However, the size of the risk premium depends somewhat on the sagacity, size, and financial solidity of the underwriter and on a wide range of technical, political, and other conditions that the market continuously reassesses.

Arbitrage is the practice, in markets not fully in equilibrium (*i.e.*, essentially all actual markets), of locking in a riskless profit by buying and selling an identical asset simultaneously in different markets at different prices. (The actual form of the transactions can also be more complex than simple purchase and sale.) Arbitrageurs help to equilibrate markets by ironing out minor differences between different market-makers' prices. For most traded commodities, large trading houses have huge rooms full of traders, or their automated equivalents, constantly scanning global markets for arbitrage opportunities, which often make up on volume what they lack in price margin per unit.

The existence of such market mechanisms for unbundling the real-time price of a commodity from the cost of insuring against the volatility of that price underscores how *the inherent unpredictability of fuel prices set by fluctuating market conditions represents a financial risk*. Such risks are common in everyday life, and aversion to those risks affects our buying choices. Why, for example, do about 80% of Americans, given a choice, prefer a fixed-rate mortgage over an adjustable-rate mortgage, even though the ARM has an initial interest rate typically at least two percentage points lower? (37) Simply because they either guess or calculate that the risk of losing their financial

In financial theory, **risk** means the variation, over time, of a particular cost stream value around its expected value. Similarly, **riskless** describes financial instruments whose yield does not vary at all with general market conditions. For laypeople, "riskless" is perhaps a misleading term: a risk with the attributes of insurance, such as a mortgage whose payments varied with your income, would be "much less risky than riskless," which sounds contradictory. In financial theory, it makes sense because insurance, by definition, reverses risk—in effect, it represents "negative risk" that offsets undesired positive risk.

flexibility, or even their home, if the interest rate spikes up isn't worth the initial discount—or in other words, that the *risk-adjusted* present value of a volatile interest rate isn't as good a deal as the present value of a fixed rate. This is a logical conclusion in an efficient capital market, where holders of large, highly diversified investment portfolios can absorb the interest-rate risk at far lower cost than can the recipients of single, undiversified home mortgage loans.

Another version of this choice is presented whenever an investor decides whether to purchase, for example, riskless Treasury debt that yields, for illustration, 6% annual interest or a junk bond at perhaps 12%. Why will most investors tend to prefer the Treasury debt? Because the extra six percentage points' yield on the junk bond represents a "risk premium" to compensate them for the possibility of not being paid interest—perhaps even losing their principal too. They know that $1,000 invested in Treasury debt will have a much lower nominal yield, but it will also eliminate the risk of losing their money. Only investors with a high appetite for risk and the ability to lose their principal will find the junk bond's premium yield worth the risk.[29]

Incredible though it may seem, *such a commonplace balancing of risk against reward has never been customary for utility managers*: the historically standard way to compare different utility investments was to compare all alternatives *as if they bore exactly the same risk.* Specifically, the streams of expected or forecast costs and benefits of different kinds of central power stations were all discounted to

[29] Awerbuch (26) shows that if a $1,000 16%-coupon junk bond trading at half its face value is compared with a $1,000 7%-coupon Treasury bill, discounting both at a 10% WACC (see p. 147) yields a net present value of $956 for the junk bond and $228 for the Treasury. However, using appropriate risk-adjusted discount rates—7% for the Treasury bill and 32.5% for the junk bond—shows that the Treasury is indeed more valuable per dollar invested, which is why it sells for twice as much. More precisely, in a perfect market, securities sell at zero net present value: the market discounts the junk bond by half to make its net present value equal to that of the Treasury bill selling at face value.

present value[30] *using the same discount rate.* This procedure "improperly converts [an]...uncertain stream of future fuel costs into a stream of certain costs without accounting for uncertainty." (317) Yet almost all analysts schooled in the "engineering economics" method of cost comparison insist on using the same discount rate. Even in a recent and widely used textbook on distributed generation, the authors specifically criticize comparing different options at different discount rates as a sign of biased analysis (765)—apparently unaware that sound financial economics usually *requires* it.

> **Weighted Average Cost of Capital (WACC)** is the melded cost (calculated either at the margin or, more usually, as an historic average of outstanding financings) of debt capital, usually from long-term bonds, and shareholder (equity) capital compensated by dividends. Both these costs of capital are explicitly corrected for tax effects, because at least in the United States, interest on debt is a tax-deductible expense but dividends to shareholders are not. Therefore (307), a utility 50% financed with 8%/y debt and 50% with equity at 12%/y, and with a combined all-jurisdictions effective marginal tax rate of 40%, has an 8.4%/y aftertax WACC.

The single uniform discount rate at which U.S. utilities traditionally compared all cashflows of all projects when buying about a trillion dollars' worth of assets was essentially equal to each utility's Weighted Average Cost of Capital (WACC). That's natural for organizations run largely by engineers and accountants, dealing with investment alternatives that all have broadly similar attributes. But any business student will appreciate that its rear-view-mirror accounting perspective is an unsound guide to future choices among diverse and very different resources, in two ways:

- WACC shows the returns that bond- and stockholders require (or required in the past) to risk their capital in a given utility. It therefore reflects—more or less accurately depending on the quality of investors' information and decisions—the *aggregate* of the utility's entire set of activities, risks, and rewards. It reflects investors' assessment of the aggregate risk of *all the firm's net cashflows.*[31] But it says nothing whatever about the risk of the *cost* streams of the *individual new project* investments now being proposed. These are only costs and not revenues; are diluted by all other old and new investments; may be of an entirely different character than previous utility activities; and may even be unknown to utility investors. Obviously, WACC is highly likely to be the wrong discount rate for expressing the unique risks of a *particular* project's costs. The proper discount rate for a given project's specific cost stream is the profit we would have to pay an informed and competitive investor to undertake the obligation to pay that cost stream. If that rate happened to equal the WACC, it would be purely by coincidence. Nor will the discount rates for various cashflows of a particular new project, or probably even for the whole portfolio of new projects, average out to equal the WACC except by coincidence. Investors' historic per-

---

[30] As described in Section 1.2.3, p. 13 box, discounting expresses the time value of money. If $1 deposited in the bank today earns 5% annual interest, it will be worth $1.05 next year and $1.11 the year after that. Conversely, $1 next year ("future value") is worth only $0.95 today ("present value"). The present value of a stream of future values is the sum of their values, each discounted to today's value by applying the chosen "discount rate" of compound interest for however many years in the future each value arises.

[31] "Unregulated firms typically seek projects that maximize the present value of *net cash flows* (NCF). NCF is the difference between inflows and outflows; it is the cash flow stream that investors see. In contrast the [Revenue Requirements Method (RRM) used by regulated utilities]...identifies resources that minimizes outflows or costs....The WACC is the investor's discount rate; it is appropriate to use to project the firm's net cash flows. It reflects the full measure of operating risks coupled with the financing risks. As such, applying it to the revenue requirements of a particular project is incorrect...because...while the RRM examines only the costs (or outflows), the WACC reflects the risk of the *net* cash flows. The two cost streams can have very different risks, hence should be evaluated at very different discount rates." (25) In fact, regulated utilities traditionally behave as if all their customers had signed an unforeseeable long-term power purchase contract, but they haven't. The risk that customers will buy efficiency or self-generation instead, or in a competitive environment even switch suppliers, has long been ignored or underestimated, and is increasing.

ceptions of the risk of investing in a given utility are simply *unrelated* to the risk of a particular cost stream for a particular future project.

- Using a one-size-fits-all discount rate for all new investments presumes that they have identical risk profiles, but from a financial perspective, they definitely don't. A gas-fired power plant is completely exposed (to a degree inversely proportional to its thermal efficiency) to the financial risk of fluctuating gas prices over the next few decades, while a wind-power or solar plant isn't. For simplicity, consider two utility projects: operating a zero-capital-cost fossil-fueled resource with a certain fuel cost, or operating a zero-capital cost solar resource with an identical fixed maintenance cost. If both projects have the same operating life, and both are discounted at the same aftertax WACC (say, 10%/y), the utility would say they have identical present-valued costs, so it doesn't matter which is selected. But because a stream of unpredictable fuel costs is inherently much riskier than a stream of fixed maintenance costs, using the proper (different) discount rate for each would show the solar project to be much superior in this respect.

Awerbuch correctly states that "Engineering cost approaches that ignore risk will always indicate that riskier, lower cost alternatives such as gas-fired turbines are the most economic, a result that is equivalent to arguing that junk bonds are a better investment than U.S. Treasury bills because they promise a higher annual payment stream for each $1,000 invested and are hence 'cheaper.'" That error obviously misallocates resources. Tutorials 4–6 show why by considering the concept of risk and the methods of calculating utility costs.

Levelized cost, applied each year for a specified period (usually the life of a project), has the same present value as an actual stream of costs that may vary year by year. It is utilities' standard way of expressing a stream of time-varying costs as a single number that also reflects the time value of money. Since the levelization computation applies the discount factor appropriate to each year, levelized cost multiplied by project lifetime equals the present value of that cost stream. **Busbar cost** is cost measured at the output terminals (the "busbar") of the generator; it does not count grid losses downstream, and it usually but not always includes power consumed by the plant itself.

Beta is a measure of the volatility of prices in a market. The equity market, as measured by the *Standard & Poor's 500 Index* or the *MCSI Europe Index*, has by definition a beta of 1.0. A value lower than 1.0 indicates a less risky (less volatile) stream of returns than such a broad market will earn. A beta of zero indicates riskless, nonvolatile securities like Treasury debt, or those that vary completely independently of the market; a value greater than one connotes a stream more volatile than the brand market in equities; and a value less than zero indicates a stream that varies in the opposite direction from the market. The calculational method for beta is described in such financial economics texts as (585); formally, beta is the ratio of the covariance of the investment's return with that of the market, to the variance in the market return.

Beta can be expressed not only for stocks and bonds but also for cashflows, such as a stream of expenditures to buy fuel. It can therefore be used to estimate a market-based discount rate—"the rate at which an investor would willingly undertake the risk of owning or underwriting a particular cost stream." For example, if we're uncertain about the maintenance costs of a wind turbine over the next 20 years, we could pay a lump sum now to an investor in return for a promise to do the maintenance. The investor will accept that sum only if it's at least as attractive as a broad market investment, having due regard to their relative risks. The investor could prudently diversify any unsystematic risk (like a particular turbine that turns out to be a lemon—see Tutorial 5 below) by owning many such maintenance contracts.

Tutorial 4: Utility Accounting *vs.* Financial Cost Valuation

Traditional electric utilities have well-established, relatively complex, but at root quite unsophisticated ways of analyzing and comparing the costs of different generating resources (25). These methods, enshrined in generations of courses and handbooks, are based on a rough-and-ready non-financial technique called "engineering economics." They compare technologies based on an imaginary direct accounting cost—the *levelized busbar cost* of producing electricity, or more precisely, the present-valued revenue requirements to produce electricity at the busbar. Most industries would call this quantity not a cost at all, but a *price*, because if they are paid it, they will recover not only their cost of production but also the regulated utility's authorized profit. Nonetheless, the term "cost" is commonly and confusingly used, as if it were equivalent to other industries' Cost of Goods Sold. It is calculated by the following method:

1. project annual direct operating costs over an assumed operating life (whose value may be only very loosely related to actual engineering life and may be more related to arbitrary accounting conventions about amortization life),

2. assume an imaginary linear recovery of the capital investment,

3. add up the projected operating and capital costs for each year, and discount them to present value using an arbitrary discount rate, typically WACC, to obtain the Present-Value Revenue Requirement (PVRR),

4. optionally, convert PVRR into a levelized cost—an imaginary constant tariff (cents per kWh) which, if charged every year for the plant's life, would have a present value (discounted at the WACC) equal to the PVRR, and

5. optionally, test the sensitivity of the results to modest variations in assumed inputs.

This methodology's main defects are:

- Electricity is actually paid for at the wholesale node or retail meter, not at the busbar, so busbar-cost comparisons don't properly count different delivery costs that vary over space and time, according to delivery voltage, and that may also reflect different reliability and other attributes.

- The direct costs calculated by the utilities' traditional method overlook important overhead and indirect costs that are probably larger for nonrenewable than for renewable resources, so their omission, or their later addition as a constant percentage markup of each option's capital cost, biases the result against renewables. (See Technical Note 2-2, pp. 161–162. However, distinguishing between different resources' overhead and indirect costs usually requires advanced Activity-Based Costing, which few if any utilities have adopted [40].) The engineering economics approach conceals important

costs, such as reserve margin and spinning reserve, inside opaque and aggregated accounting categories such as "plant in service" or "fuel." (40) Being invisible in the accounting for a particular resource or activity, such hidden costs tend to persist because there is no incentive to save them. This is analogous to traditional cost accounting in manufacturing, where "there is no manufacturing cost category for 'producing defective parts.'" (28) Traditional utility cost accounting doesn't properly categorize "most transaction...costs including the negotiation, purchase, movement and storage of fuel and other supplies or the activities associated with meter-reading and billing, which may be significant in the case of small accounts." (43) Proper Activity-Based Costing could well reveal that in such small accounts, the avoidable transaction costs can tip the balance in favor of supposedly uneconomic renewables (43).

- The procedure doesn't credit renewables for such significant capabilities as the ability to issue a long-term fixed-price contract. It doesn't credit any modular technologies for such attributes as short lead times and ability to adapt to rapidly changing requirements (29). In financial language, such alternative resources may "create valuable managerial or strategic options[32] which can be 'exercised' at a later time," making the resource more valuable without reducing its immediate annual accounting costs (27). As described above, option theory (§ 2.2.2.5) or decision analysis (§ 2.2.2.6) can be used to value these attributes.

- Sensitivity testing cannot reveal, correct, or make up for incorrect prior treatment of expected values or discount rates (25).[33]

- Discounting all costs at the uniform and arbitrary WACC rate is not correct for any particular stream of utility costs, and does not properly adjust for differences in risk between different resources. It falsely makes a risky annual cost stream look as if it had the same present value as a safe cost stream with the same annual expected values (25). This flies in the face of basic finance theory, which holds, obviously enough, that "dollar for dollar, a risky cost stream, such as future outlays for fuel, must have a *higher* present value since it is less desirable than a safe cost stream." (29) (Why higher? Because it's a cost. Higher costs are less attractive. If it were an income stream, a lower discount rate would make it bigger and hence more attractive.[34])

Let's start with the last of these effects because it is often the most important, then return to the others. For all of them, we will use not the utility industry's engineering economics approach, but the very different philosophy of financial economics. In financial economics, *all values are fair market values*. Thus the price of a share of stock is simply the market's perception of the present value of its stream of future dividends. It equals the probability-weighted present value of the dividend payment in each future year, discounted at the risk-adjusted discount rate appropriate to that stock, plus the discounted expected terminal value of the share itself.

In the case of fuel prices, fair market value is not an abstraction. The cost of a futures contract indicates the present value of the fuel delivery at a future date, discounted at the market-determined discount rate *reflecting the perceived risk* associated with that fuel price. In other words, if you want to know what the fuel-price-volatility risk is worth, just ask

a provider of price-risk insurance, such as any large energy trading firm, how much more that firm will charge you for fixed-price gas than for floating-market-price gas. The difference is the compensation that the trader requires to take the price risk off your hands. That difference—plus any risk associated with the possibility of default on the constant-price contract[35]—can be directly reflected in a cost comparison with, say, a windfarm; or the "base," not risk-adjusted, price of gas could be used in such a comparison if discounted at the appropriate risk-adjusted rate, as described next. But if neither of these adjustments is made—if risk is reflected in neither fuel price *nor* fuel-cost discount rate—then an important financial fact is being improperly omitted. That is precisely what traditional utility accounting-cost procedures have done for decades, causing serious misallocations.

---

[32] Awerbuch gives the example that manufacturers adopting numerically controlled process technology, such as machining, in the 1970s were easily able to adopt computer-controlled manufacturing about a decade later; this opportunity could not have been exactly foreseen at the time of the first change, but created valuable strategic capabilities as the technological future unfolded. Traditional discounted-cashflow analyses obviously cannot anticipate such outcomes, and hence incorrectly value them at zero. Awerbuch plausibly conjectures that certain distributed resources may create "opportunities to serve new customers, or provide different levels of quality and reliability as different types of services." A few such examples are given in Section 2.3.3.8.

[33] For example, in Awerbuch's Treasury bill/junk-bond comparison above (25), reducing cash inflows for both investments by some arbitrary and equal amount, say 10%, makes yield look more volatile from the Treasury bill, even though it is in fact a riskless investment, so such a comparison is invalid to start with. Sensitivity testing also has many well-known but often overlooked pitfalls (25), such as trying to change one variable in isolation when it is in fact linked to others, using modal (most likely) rather than expected values, multiplying rather than combining expected values, choosing unhelpful sensitivity ranges, not knowing how to interpret them, and obscuring the requirement to evaluate projects on a marginal basis.

[34] Awerbuch also notes (26, 32) a more elaborate argument that is important for theorists but not for our purposes here.

[35] Enron was considered a very large and financially strong company, but its collapse in 2001 proved, as earlier drafts of this book had remarked, that the risk of its defaulting on a constant-price gas contract was not zero. Any long-term contract "is only as secure as the risk of default of parties on both sides of the contract....[W]hile contracts may be fixed price, they are not necessarily risk-free." (309) In the post-Enron climate, counterparty creditworthiness is the key factor.

Tutorial 5: Financial Risk

Risk—defined (p. 146) as the variation, over time, of a particular cost stream around its expected value—must be properly reflected by applying the discount rate corresponding to the level of risk; otherwise the wrong asset will be bought. But there are really *two different kinds of risk*.

The classic distinction between the two classes of risk is that an individual oil exploration firm's managers may worry a lot about their firm's significant risk of drilling many "dry holes" in a given year. Yet investors care a lot less, because they can diversify that risk by owning stock in multiple exploration firms. One firm or another may do poorly in a given year, but on average, a certain amount of oil will be found collectively by the firms in the portfolio. For a single drilling firm, the dry-hole risk is called "systematic" or "undiversifiable."

Capital markets compensate shareholders only for undiversifiable risk. Why not for diversifiable risk too? Because it can easily be eliminated by diversifying the investment, even with quite a small number of stocks—usually a half-dozen or so.[36]

Collectively, the stocks of oil-exploration firms tend to be less volatile than the entire universe of equities: seeking oil discoveries may be risky for a given firm (though markedly less so with the latest technologies), but most of that risk is "unsystematic" or "diversifiable" (also called "random" because it is not correlated with economic events) and hence is not reflected in shareholder-required discount rates. That is, a given oil exploration firm may have a *high random risk*, but its industry as a whole has a *low systematic risk*. (Your horse may not win the race today, but *some* horse will.) Some other kinds of firms have a higher-than-usual systematic risk because their value rises and falls with, but more than, that of the equities market generally, thus increasing investors' risk and meriting a risk premium. Some kinds of firms, such as gold mines, may have or claim to have an opposite, countercyclical, quality that is valued by investors because it helps protect them from market downturns.

It follows that in proper resource valuation, discount rates are not adjusted for such unsystematic risks as the risk that a particular turbine rotor, wind-turbine blade, or photovoltaic inverter will prematurely fail. Such risks can be avoided simply by diversifying the portfolio of such technologies—or, better still, the portfolio of different *kinds* of technologies. The value of the risks should be reflected instead by cash-flow estimates that include probability-weighted outcomes for technological failures.[37]

Conversely, even if smoothed long-term averages of fuel prices could be accurately predicted, fuel would still be *risky in a financial sense*, because its price tends to vary in step with changes with other asset values in the economy, making its variations hard to hedge against. (Note how this financial use of the term "risk" differs from most engineers' use of the same word—a source of endless confusion.) Thus a portfolio containing all oil companies still bears the systematic risk that their revenues and profits will tend to rise and fall with other asset values in the marketplace.

Awerbuch (44) quotes Stewart Myers of MIT on another helpful example:

---

[36] As a rule of thumb, five stocks are often enough to reduce diversifiable risk by about 95%.

[37] Thus, using an example from Awerbuch (35), if you expected that 80% of PV modules would last for their rated 30 years while 20% would fail a decade earlier, then the expected life of a large array would be the weighted average, or 28 years. (This probability-weighted expected value is the right number to use; the "modal" or "most likely" value of 30 years is not.) Alternatively, you could expect a certain failure rate per year somewhere in the array, establish a reserve fund to replace failed modules, and fold that into the project's operating cost. Either way, you have established an expected cost of keeping the project working for an expected lifetime. No adjustment in discount rate is appropriate, since the failure risk has been converted into a known cost—much as buying constant-price (price-insured) natural gas converts its volatility risk into a known cost.

The owner of a roulette wheel is exposed to considerable business risk; fortunes can be made or lost by the "house" in any one night. But this business risk is random or unsystematic and the owner can easily diversify it by owning many roulette wheels so that on any given night some make money while others lose.

Having diversified the random risk, the owner is exposed only to the remaining, non-diversifiable, systematic risk: when the economy is good[,] more tourists show up to play than when the economy is poor. This remaining systematic risk (which is usually measured using the financial "beta") is impossible to diversify or hedge since there are few (if any) investments that provide a counter-cyclical stream of returns.

Only systematic risk—the risk that cannot be diversified—can be properly handled by discount rates. Unsystematic or random risk that can be handled by diversification, such as random fluctuations in fuel price or random failures of individual wind machines, must be handled by correctly estimating expected costs. If these two steps are not properly done, no amount of later sensitivity testing can rehabilitate the risk valuation; and in any event, sensitivity testing can help only with unsystematic risk. (That is, "Planners cannot perform analyses at an arbitrary discount rate and then expect sensitivity [testing] to demonstrate the riskiness of a particular technology.") Sensitivity testing is an engineering method of finding out which variables most sensitively affect outcomes, and can help to evaluate unsystematic

risks, but it is often a poor technique in economic and financial problems where many variables are correlated.

To estimate present-value costs for energy resources, therefore, requires two steps:

- estimating expected values (probability-weighted average outcomes) for each cost stream, such as the expected revenues of an oil firm (which depend on discoveries, volumes, and prices) or the expected life and output of a wind turbine, and

- applying market-based (risk-adjusted) discount rates appropriate to each cost stream.

Cost streams typically come in four flavors:

- fuel, for which a financial tool called the Capital Asset Pricing Model is usually the best practical way to find the right discount rate;[38]

- fixed operating outlays (debt-equivalent);

- tax-shelter benefits (riskless); and

- variable operating-and-maintenance (O&M) costs.

Power purchase contracts should be evaluated like financial leases. Some "cost" streams commonly used in *accounting* analyses have no place in such *financial* analyses because they do not affect present value—for example, depreciation plus allowed earnings (which must add up to the original outlay) and tax normalization. Instead, one simply adds up the initial outlay and the several individually discounted cost streams; no modeling of accounting fictions such as depreciation is required, so one can devote more effort to properly estimating costs.

[38] Standard texts such as Seitz (585) describe the CAPM and its application. The CAPM-based discount rate for a particular cost stream is the riskless rate of return plus the product of that cost stream's beta times the difference between return to a widely diversified portfolio and the riskless return. Seitz's Chapter 11 cites the following as typical CAPM assumptions: wealth-maximizing single-period decision-makers choosing portfolios for expected return and its standard deviation; universal agreement on all assets' expected returns, standard deviations, and covariances; unlimited capital at the risk-free interest rate; no taxation; no transaction costs; completely divisible and fungible investments; markets unaffected by single investors' trades; and fixed quantities for all investments. These assumptions clearly differ from actual market behavior, but this does not prevent the CAPM from being a useful and widely applied approximation. In practice, hundreds of tests have shown (586) that the CAPM does explain *much* of market assets' observed risk/return correlations—but not *all*, since other factors are also at work, including the omission of many classes of potentially tradable assets from normal financial markets. (There are endless debates about whether beta should reflect the entire universe of risky assets, from racehorses to real-estate and from stamps to beer-steins.) Models more accurate than the CAPM are available, such as Arbitrage Pricing Theory (501), but since they specifically correct for each asset's sensitivity to a variety of risk factors, each of which bears a certain risk premium, they are more complex and harder to use. The CAPM is therefore widely used for its practicality and simplicity, and "appears to be the model of choice in practice" (585); it may not give the right answer for a specific asset that may be affected differently by some special kind of risk than are other assets, but it will be reasonably accurate for a portfolio. As Stanford economist Prof. William Sharpe, who shared a Nobel Prize for Economics for his development of the CAPM model, remarks, "The Arbitrage Pricing Theory uses fewer assumptions [than CAPM] about investor utility and actually obtains a less powerful result, but it is extremely difficult to implement in practice." (589)

Tutorial 6: Valuing Risk

If volatile fuel prices increase financial risk, what is that increased risk worth? Its value can be estimated using tools that were developed by financial analysts to measure the risk of stock portfolios, but can be applied to any other cash-flow too. These tools are the basis of the modern financial system, and several of their developers received the 1990 Nobel Prize in Economic Science.

The basic principles of capital asset valuation are straightforward. Each stock has a certain level of historic price volatility. That volatility can be compared with the price volatility of the entire stock market using the "beta" measure.[39] In round numbers, the U.S. stock market during 1928–2001 had a volatility on the order of 20% per year,[40] and trends upward at an arithematic average rate of 12% per year. For comparison, historic real prices for fossil fuels seem relatively stable in the long run, but as of the mid-1990s, had exhibited annual volatilities around 15–30% (232). Specifically, the standard deviation of gas prices was about 38% (so that 66% of the time, the gas price will be in a range of ±19% around the mean), while the standard deviation of coal prices was about 20% (26); so coal price is about as "safe" as stocks, but gas price is about twice as volatile. (The 2000–01 gas price spike may have increased these values.) A formula from capital market theory (using the Capital Asset Pricing Model[41] in some but not all cases) can then be used to determine the discount rate that is appropriate to each cashflow's or asset's value of beta. Applying the resulting risk-adjusted discount rate to the expected returns from or values of each cashflow or asset will fully adjust for their different financial risks. This permits investments with different degrees or patterns of price volatility to be fairly compared, just as one would do when choosing between a junk bond and Treasury debt.

[39] For simplicity, we ignore here the refinement of adjusting beta for increased leverage if the asset is debt-financed and moves the particular firm away from its optimal capital structure (313). This consideration would not apply to a debt-financed publicly owned entity.

[40] The standard deviation of returns for the S&P 500 was 20.1% during 1928–2001 (777). Note that this is for an entire stock portfolio: the 1987–91 standard deviation of an *individual* common stock averaged 50% (574). We have not analyzed whether equity returns are becoming more volatile.

[41] The CAPM assumes that investors in a given asset will demand a return equal to the riskless rate they could earn from, say, Treasury debt, plus the product of two terms: the asset's sensitivity to market trends (beta), times the market risk premium (*i.e.*, the difference between the expected market return and the riskless rate). Thus if the riskless return expectation is 4%/y and the general market return expectation is 12%/y, then an asset with a beta of 0.8 (20% less volatile than the general market) would be fairly priced at a 10.4%/y return; one with a beta of 1.2 (20% more volatile than the general market), at 13.6%/y.

Applying standard risk valuation techniques to fueled power plants yields strikingly different financial rankings than traditional utility accounting perspectives that ignore differences of risk. For example, historic betas for U.S. natural-gas prices would suggest an appropriate aftertax discount rate of around 4%/y (31). Suppose, for illustration, that you expected the real price of natural gas over your planning horizon might escalate at either 2%/y or 4%/y, with equal probability. Each stream of costs for gas to run your proposed power plant could then be discounted at nominal rates of 4%/y, and the totals weighted by probability and summed. This will yield a very different answer—over (say) 20 years, *2.9-fold different*—than discounting the same cost streams at a typical WACC, say around 9%/y. That is, using WACC would in this case tacitly suppose that gas-price volatility represents a financial risk worth 2.9-fold less than it actually is, and could therefore fool you into buying a gas-fired power plant whose fuel-price risk is unjustifiable. This is a special case of the general proposition that even if you don't know exactly the right discount rate to use for fuel that has a volatile price, "a rough approximation...is far better than using the WACC, which is generally much too high as a fuel discount rate, and hence significantly biases outcomes in favor of fuel-intensive technologies." (25)

To pick a broader example, suppose that a utility is comparing four power-plant investments—nuclear, coal, geothermal, and wind. Assume that in that utility's circumstances, the traditional comparison, applying the same 15%/y discount rate to all four technologies, makes nuclear look cheapest. But applying to each technology a risk premium associated with its particular attributes (for illustration, a 15%/y risk premium or 30%/y total discount rate for nuclear if one considers it a speculative investment, a 9%/y risk premium for coal, 6%/y for geothermal, and 4.5%/y for wind, reflecting judgments of their respective exposure to political risks such as carbon taxes, technical disappointments, or risk of poor financial performance based on uncertainty of demand) changes their ranking, as in Figure 2-17:

Figure 2-17: The importance of risk-adjusted discount rates
Risk-adjusting plant comparisons can change their economic priority.

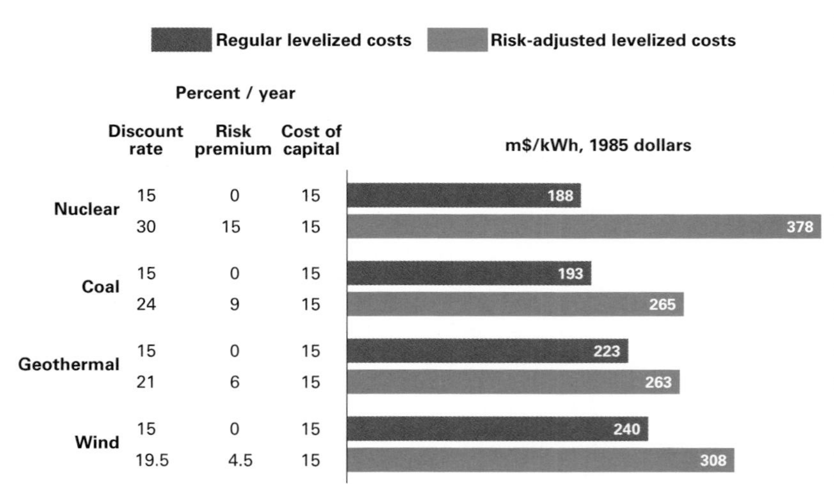
Regular levelized costs
Risk-adjusted levelized costs
Percent / year
Discount rate
Risk premium
Cost of capital
m$/kWh, 1985 dollars
Nuclear
15 0 15 188
30 15 15 378
Coal
15 0 15 193
24 9 15 265
Geothermal
15 0 15 223
21 6 15 263
Wind
15 0 15 240
19.5 4.5 15 308

Source: W. R. Meade and D. F. Teitelbaum, "A Guide to Renewable Energy and Least Cost Planning" (Interstate Solar Coordination Council, 1989), p. 40, ex. 29

On these illustrative assumptions, buying a coal instead of a nuclear plant would reduce the variability of financial performance, under the same exogenous uncertainties, by 40%; geothermal, by 60%; and wind, by 70%. Naturally, other choices of discount rate could yield different rankings; or sensitivity testing could reveal what relative discount rates would be required to change the rankings. But although the appropriate risk premium for a given technology or project depends on many factors, especially including exposure to fuel-price volatility, the correct value is certainly not zero; and as Peter Bradford remarked when Chairman of the New York Public Service Commission, it is better to be approximately right than precisely wrong.

Another and even simpler example of how using risk-adjusted discount rates can change outcomes is Awerbuch's comparison of a CAPM analysis (assuming the full range of observed 1982–91 fuel betas, 7%/y riskless return, and 14%/y expected market return) with a conventional 1991 utility-style comparison prepared by the Finance and Technology Committee of the National Association of Regulatory Utility Commissioners (25). The NARUC analysis uses a uniform 10.4% WACC discount rate (Figure 2-18):

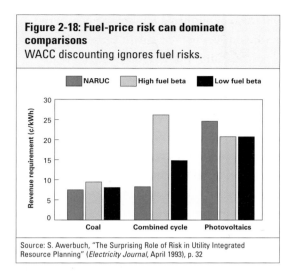

Figure 2-18: Fuel-price risk can dominate comparisons
WACC discounting ignores fuel risks.

Source: S. Awerbuch, "The Surprising Role of Risk in Utility Integrated Resource Planning" (*Electricity Journal*, April 1993), p. 32

The uniform WACC discount rate—implicitly assuming that all fuel cost streams (or none) are equally risky—makes coal-fired electricity look slightly cheaper than it is on a risk-adjusted basis, combined-cycle gas-fired electricity about 2–3-fold cheaper than if risk-adjusted, and photovoltaics, with no fuel cost, *more* costly than if risk-adjusted. (This is because such sunk capital costs deserve essentially riskless treatment, especially with a portable and fungible technology.) If the less favorable end of the natural-gas beta range is used, the combined-cycle electricity actually has a *higher* risk-adjusted price than the photovoltaic electricity, conventionally thought of as costing around $0.25/kWh! Moreover, the photovoltaic example illustrates how WACC-based discounting gives the wrong answer even with no-fuel technologies: for example, it underestimates the "true, market-based value of...fixed maintenance outlays" by about 20% (26).

Awerbuch points out (26) that if the coal plant's overall revenue requirements (income) were discounted at a notional "composite" rate of 6.5%/y, combined-cycle gas at 4.5%/y, and photovoltaics at 11.5%/y, their present values would match those obtained by using the correct risk-adjusted discount rates for each cost stream individually. The seven-percentage-point spread happens to echo the spread of yields between riskless Treasury debt and common stocks.

Similarly, in the appendix to a 2001 paper, Awerbuch finds that under market-based financial criteria, a 50-MW photovoltaic plant in Hawai'i (at $4,810/kW in 1996) can produce cheaper levelized power than a 200-MW combined-cycle plant in the northeastern U.S., thanks to Hawai'i's generous solar tax credit and current depreciation rules. He also presents the same comparison using WACC, which overstates the photovoltaic project's net aftertax outlays by 57%—illustrating how engineering-economics methodology is biased against capital-intensive options like renewables (43).

A generalizable way of illustrating the sensitivity of power-plant economics to assumed discount rate is to graph a levelized avoided cost—say, the power-supply cost from fossil-fueled plants that a new renewable source could avoid—as a function of the risk adjustment applied to those fossil-fueled plants relative to an assumed 10%/y base-case discount rate (Figure 2-19):

Figure 2-19: Effects of discounting avoided costs at risk-adjusted discount rates
Risk-adjusting levelized costs can change their value by about 50–500% compared with assuming that they all have equal risks.

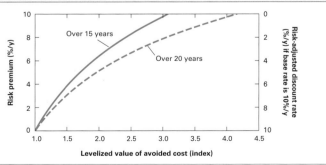

Source: P. Chernick, PLC Inc., "Quantifying the Economic Benefits of Risk Reduction: Solar Energy Supply Versus Fossil Fuels" (ASES Solar Conference, June 1998)

Thus over, say, a 20-year planning horizon, a 10%/y risk premium increases levelized costs by a factor of 4.15. A 4%/y risk premium (applied to, say, natural gas as compared with riskless renewables) means that the levelized accounting cost of gas should be multiplied by about 1.5–1.7 (for 15 or 20 years' planning horizon, respectively) to yield the risk-adjusted cost. An Enron analyst confirmed in early 2001 that the firm's market-dominating gas trading normally used a 5–6-percentage-point risk premium for gas price risk—the range depending on individual traders' temperaments and trading positions. This implies a multiple of ~1.7–2.3 for 15–20-y gas cost streams.

Even a one- or two-percentage-point risk premium, which is probably far too low for a gas/renewables comparison, yields a cost difference an order of magnitude greater than ~1999 Pacific Northwest cost differences (around $0.001/kWh or around 3%) between, say, a windfarm and a combined-cycle gas power plant. The gas plant was in fact bought because the gas-price risk was not taken properly into account—clearly an economic blunder. The first methodologically correct such solicitation apparently occurred in the U.S. in July 2001, when the investor-owned utility Xcel Energy required fuel-indexed bids to come with a ≥10-year fixed-fuel-price bid (770).

2.2.3.1 Valuing electricity price volatility

So far we have discussed the greater value of constant fuel costs as an advantage of renewable resources, which use no fuel, or of extremely efficient resources, such as fuel cells or some kinds of co- or trigeneration, which use little fuel per unit of service. (At a larger scale, combined-cycle gas turbines

will also exhibit this advantage compared with simple-cycle or steam plants.) That is, resources with a low ratio of variable to fixed costs incur less cost volatility and hence merit more favorable discount rates. This is important because the ratio of variable to fixed costs is about 40-fold different for gas-combined-cycle than for photovoltaic plants, as shown in Figure 2-20.

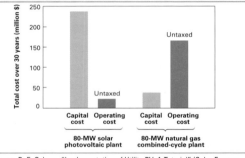

Figure 2-20: Solar and fossil-fuel technologies have opposite cost structures
Solar and gas plants have ~40-fold different ratios of fixed to variable cost—and the only variable costs are tax-deductible as business expenses (the investments must be amortized).
This table assumes 1995 combined-cycle and 1997 PV central-station technologies, each rated at 28% capacity factor.

Source: D. E. Osborn, "Implementation of Utility PV: A Tutorial" (Solar Energy International, March 1995), p. 22

This ratio is often effectively increased by tax distortions and further exacerbated by accounting and financial-market distortions (see Technical Note 2-2, pp. 161–2). On the other hand, large sunk costs, though stable, are also inflexible, especially if the asset is too large or rooted to its site to be portable—the point of our previous discussion of option values (§ 2.2.2.5), decision analysis (§ 2.2.2.6), and portability or reversibility (§ 2.2.2.8). Resources that are both modular *and* zero-fuel (renewables or efficiency) thus offer the best of both worlds—flexible capital investment to meet evolving needs, and perhaps redeployability too, but no exposure to volatile fuel costs later.

Benefit

51

Resources with a low ratio of variable to fixed costs, such as renewables and end-use efficiency, incur less cost volatility and hence merit more favorable discount rates.

The same financial-economics logic that values avoided volatile fuel costs applies also to the market value of *electricity* whose price is constant or nearly so. Just as four-fifths of American homebuyers are willing to pay at least a two-percentage-point premium for a fixed-rate mortgage, and many natural-gas buyers are willing to pay an energy trading firm a 5–6-percentage-point premium for fixed-price gas, so too many electricity buyers are willing to pay a premium for constant-price electricity, or equivalently, to buy a separate price-insurance policy. So, to a degree, does the Sacramento Municipal Utility District's green pricing policy, which guarantees buyers of photovoltaic power that for their 15% price premium, they will get their electricity at a constant price until other customers' prices rise above the green price, then be partly insulated from any further increases. In Texas, Austin Energy's Green Choice program guarantees a constant electricity price for 10 years, charging an unvarying $0.0285/kWh premium *instead of* the fluctuating fuel charge, which may therefore exceed the green power charge. The fixed price won a purchase subscription (doubled to 24 GWh/y) by chipmaker AMD.

In short, the more the price of electricity is unbundled from its other attributes, and the more the contractual and physical flows are distinguished, the more possible it will become for risk-averse customers to choose higher but constant prices over initially lower but volatile ones, just as they now do with mortgages (37). This elimination of fuel-price risk adds another dimension to the value of renewable resources.[49] Proper discounting of *each* cost stream for each resource should in principle capture this value, but only if careful attention is paid

not only to the volatility of fuel costs but also to highly uncertain *non*-fuel items. But that is an important condition. Much of the disappointment in nuclear power's economic performance was caused by using conventional and comfortable *assumptions* about the expected values and volatilities of such cost streams as O&M and net capital additions (major repairs), rather than examining the historic data, or from ignoring the possibility that new regulatory requirements could increase costs above historic levels.[50] The common current practice of ignoring possible restrictions on or taxation of carbon emissions is an analogous bias against renewable and efficiency resources, and could be interpreted and quantified as a financial price or risk bias. And the nearly universal practice among regulated utilities of passing through fuel-price volatility risks to utility customers through a "fuel adjustment clause"—removing from utility shareholders any reward or penalty for sound management of those risks—is clearly disadvantageous to renewables that avoid the risks altogether.[51]

One word of caution. While it is important and essential to discount future cost streams, of whatever kind, at the appropriate risk-adjusted discount rate, this procedure is too static to account for operating flexibility. That is, if managers have the opportunity to change the direction of a project as they learn new information over time, then mere risk-adjusted discount rates may not fully capture the benefit of that managerial discretion. This benefit is more explicitly captured by such tools as option theory (§ 2.2.2.5) and decision theory (§ 2.2.2.6) and by portability (§ 2.2.2.8).

[42] His analysis assumes, for illustration, a 4%/y riskless rate, a 12%/y expected market return, an 8%/y marginal cost of debt, a 39% combined federal/state income-tax rate, and (for variable O&M) the firm's aftertax WACC if the pre-tax WACC is 10.4%/y with 50% debt and 10% preferred stock. His 2002 update for IEA (44) suggests smaller but still generally negative ß for oil and gas in IEA-Europe economies, *vs.* the U.S. estimate of −0.5 to −1.25.

[43] Year-over-year changes in EIA real producer prices expressed in chained 1992 $ (186) compared with year-over-year changes in the annual average of monthly changes in the *S&P 500* annual return (with dividends reinvested monthly to 1980 and then daily). (231) Awerbuch uses somewhat different data, but ours are close enough for a first-cut reality check.

[44] And that of other analysts such as Chernick (119), who found a beta of −1.11 for December residual-oil prices during 1971–86. The beta of oil price with respect to the average retail residential electricity prices of Western Massachusetts Electric Company was −2.1 for this period: "oil prices varied in the same direction as rates (and hence in the opposite direction from ratepayer welfare), and twice as much." Awerbuch's 2002 EIA compilation (44) further documents negative energy-price ß values from recent econometric literature, including −0.78 (± 0.27 standard error) for U.S. sopt wellhead gas *vs.* S&P 500 for 1980–mid-1992.

Technical Note 2-1:
Valuing empirical fuel-price volatility

How realistic are the risk premia suggested by Awerbuch? He offers (39) the following nominal aftertax discount rates for components of U.S. project cashflow: [42]

Countercyclical risk (econometric/beta estimate)	
oil and gas ($\beta = -0.5$ to -1.25)	0.0%/y to 4.0%/y
coal ($\beta = -0.2$ to -0.4)	2.0–3.0%/y
Debt-equivalent cashflows (by convention)	
property taxes, insurance, fixed O&M, working capital	5%/y
Riskless cashflows (by convention)	
depreciation tax shields, tax credits	4.0%/y
Procyclical risk (judgmental estimate)	
variable O&M	8.5%/y

The proper procedure is then simply to discount, over project life, each cost stream at its own discount rate, then add up their present values. This is not complex, since fossil-fueled projects typically have only three or four cost streams, and most renewables, lacking fuel, have even fewer.

But are these suggested fuel discount rates reasonable? At first glance, one might suppose, with Hoff (308), that discounting fuel cost at a lower rate than the risk-free rate, resulting in "very high present value fuel costs," may be too severe a treatment, because "while the economy may slow during large oil price shocks, it is not clear that fuel prices are negatively correlated with the market in general...." This question does require further analysis, as Hoff suggests. However, as a quick check on Awerbuch's 1982–91 beta for utilities' fuel purchases (−0.20 for coal, about −1.25 for gas and heavy fuel oil), (25) RMI analyzed real producer prices for the post-embargo period 1974–96. We obtained: [43]

1974–96 U.S. producer prices of:	coefficient of correlation with S&P500 returns	β
crude oil	−0.335	−3.1
natural gas	−0.569	−2.2
coal	−0.638	−0.84

These results are only a first approximation, using different data series and less sophisticated analysis. Nonetheless, they appear to justify Awerbuch's contention [44] *that post-embargo fuel prices are countercyclical—they go up when general equity markets go down and vice versa. (Although one can argue about whether fuel prices are big enough to move the macroeconomy, it is certainly true qualitatively over the past few decades that rising fuel prices tended to presage if not trigger economic declines while falling fuel prices provided some economic boost.)*

Cost streams with this countercyclical property "are risky because their price rises when the firm's revenues are declining" and vice versa (25) , so such cost streams deserve low (unfavorable) discount rates. That is, "A cost stream, such as

fuel, that co-varied negatively with the economy produces the worst possible set of expectations for the [utility] firm and its ratepayers since this cost will be at its highest when the economy is doing poorly, and ratepayers are feeling the pressures of recession—low incomes and depressed home values." (26) They not only fluctuate over time, but "do so in a negative systematic manner relative to the economy and the returns on other assets." (This is the flip side of the increased value enjoyed by countercyclical income-earning assets.)

Hoff further states that while negative betas "are theoretically possible, they are rare. For example,...every one of the 1,700 stocks listed in the Value Line Investment Survey (1995) had positive betas." True, but only in a limited sense: some classes of equities, such as gold-mining stocks, routinely exhibit negative betas, and some other particular equities do so occasionally. One energy economics expert notes that negative betas are rarely encountered, not so much because they're actually rare as because they're typically ignored or removed from databases as "anomalies" that few analysts know how to interpret (578). Moreover, there is no obvious reason to extrapolate from equity returns to other kinds of cashflows such as pure fuel prices, which can have any sort of cyclical or countercyclical behavior that arises from their nature. It will take much more than this analogy to cast much doubt on the observed data.

There is no theoretical or practical reason why fuel prices, or any other cashflow, should not vary countercyclically with market values and hence have a negative beta. This is well accepted by such eminent economists as Robert Lind (400), Robert Wilson (403), and Hayne Leland (402): since often "what is most relevant for determining the riskiness of an investment project is the covariance of its return with the returns to the economy as a whole and not the variance of its own return...[, t]he development of an energy technology with very uncertain future returns may not constitute a risky project. If it will have a high payoff under just those conditions when the rest of the economy will do poorly, it will reduce the overall variability of national income and therefore reduce risk. Such an investment has the characteristics of insurance." (404) In such cases, "rather than reducing the value of net benefits to reflect the cost of risk, the value of net benefits should be increased by the amount of their insurance value." (401)

In a more recent publication (26), Awerbuch presents a less dramatically negative beta (−0.5 to 0.0) for U.S. gas outlays, zero for coal outlays, and corresponding discount rates of 1–3%/y and 3%/y if the riskless rate is 4.7%/y—i.e., risk premia of about 1.7%/y for coal and 1.7–3.7%/y for gas. In the past few years he has even been prepared to accept arguendo that perhaps fuel-price betas might ultimately approach zero rather than negative values (42). That is, like many other commodities, fuel prices might continue to fluctuate but not move systematically with or against the market. But that's still not as safe as generating with, say, photovoltaics, whose cost is essentially all sunk up front, leaving no fluctuating future cost stream of any significance.

Awerbuch then further suggests that based on Enron's 1995 fixed-price gas contracts ($3.50/million BTU) and then-preva-

[45] It could also be argued that a fixed-price gas purchase contract might deserve a slightly higher (debt-equivalent) discount rate to the degree that it's fungible by means of gas swaps, or can be resold if one goes bankrupt or under other contingencies.

[46] Those big swings, in a particular equity or a whole portfolio, are exactly what makes investors think of the stock market as risky: they know it's reasonable to have a long-run return expectation of 12%/y or so, but they lose sleep over a 30% market crash tomorrow.

[47] For example, many power stations have dual oil/gas capability, but they can still burn only certain grades of those fuels (just as coal plants are generally limited to only certain kinds of coal), and certainly not coal or uranium.

[48] For example, different stock portfolios can easily be ~0.6 correlated, and it is hard to get much worse than ~0.4 (41). These values are lower than the ~0.85 correlation between fuel prices, but not vastly so; and most utilities insulate their market performance from their fuel prices, not only by the fuel-cost adjustment clauses (risk passthroughs) most of them traditionally enjoy, but also by the more durable means of having largely fixed costs.

2.2.4 **Reduced overheads**

Distributed resources commonly reduce overhead costs in several ways that are not commonly considered because traditional accounting systems don't properly identify them. These distortions are described in Technical Note 2-2.

2.2.5 **Planning resource portfolios**

Another defect of the traditional accounting-cost comparison is that it evaluates technological choices in isolation, rather than for their effect on the resource *portfolio*. Such evaluation cannot detect options that reduce risk (cost variability) more than they raise the average cost per kWh. Awerbuch properly notes that "At any given time[,] some alternatives in the portfolio may have high costs while others have lower costs, yet over time, the astute combination of alternatives serves to minimize overall generation cost relative to the risk." Just as investors use financial portfolios to provide over the years a consistent risk-managed performance under unpredictably varying economic conditions, so electricity providers should evaluate technologies not simply in isolation but for their effect on total portfolio cost or return. Awerbuch continues:

> Financial investors understand that the future is unpredictable; therefore, rather than emphasizing fortune telling, investors focus on building robust portfolios that are expected to maximize return for the given level of risk undertaken. Portfolio theory is well-developed. Its principles suggest that the important measure for valuing alternative resource options is *how a particular option affects the generating costs of the portfolio of resource options relative to how it affects the risk of that portfolio.*

Thus the objective should be not just to find least-cost technologies but to evolve optimized portfolios that may combine technological, financial, and contractual resources. To understand the economics of such portfolios, it is necessary to consider fuel diversity.

2.2.6 **Fuel diversification**

2.2.6.1 **Engineering perspective: diversify fuels and sources**

Most utilities interpret fuel diversification in engineering terms—contracting for fuel supply with providers in different regions, and using a mix of fuels. In both cases, the aim is to make a particular kind of physical or market disruption less likely to interrupt supply. Disruptions could include a failed pipeline, a rail or coal strike, frozen coal barges or coal stockpiles, or interruption of affordable fuel shipments from a particular part of the world. Such events do occur, often in related clusters. For example, a cold wave in January 1994 across the Northeast and Mid-Atlantic states (175)

> ...created record high electrical demands and caused fuel-related problems and mechanical failures resulting in unexpected generating capacity outages. The electric utility system in the eastern two-thirds of the United States was strained to the point that demand could not be met, as many utilities experienced their winter peak demand at the same time....Voltage reductions were instituted in many regions and public appeals were issued to conserve electricity....Rolling blackouts were required in the Pennsylvania-New-Jersey-Maryland Interconnection (PJM) and Virginia Electric & Power Company...control areas to maintain a balance between available capacity and demand....PJM had almost 19 [GW]...unavailable...and could not meet demand. The majority of the [nearly 14 GW of] unplanned outages were due to fuel-related and equipment failure problems, 35 percent and 48 percent, respectively. Fuel availability was interrupted due to delivery problem caused by icy roads and rivers, frozen coal and load-

[49] And probably also of end-use efficiency resources, since the effective price of their "negawatts" varies exactly with the price of the electricity they save, yielding a constant net price. The more electricity costs, the more saved electricity is worth.

[50] Awerbuch and Preston (47) give the example that if a $500-million outlay in 10 years has only a 10% chance of being mandated, it still generates "a relatively significant present value of $32 million" (discounting the $50-million expectation at the riskless aftertax rate of ~4.5%/y).

[51] Transferring this risk to customers does not make it go away—it is only a redistribution of risk—and becomes less likely to persist in a more competitive environment when firms will be expected to absorb their own risks and reflect those risks in their prices. Otherwise they cannot be properly rewarded or penalized for the quality of their decisions.

Benefits

52   *Fewer staff may be needed to manage and maintain distributed generation plants: contrary to the widespread assumption of higher per-capita overheads, the small organizations required can actually be leaner than large ones.*

53   *Meter-reading and other operational overheads may be quite different for renewable and distributed resources than for classical power plants.*

54   *Distributed resources tend to have lower administrative overheads than centralized ones because they do not require the same large organizations with broad capabilities nor, perhaps, more complex legally mandated administrative and reporting requirements.*

55   *Compared with central power stations, mass-produced modular resources should have lower maintenance equipment and training costs, lower carrying charges on spare-parts inventories, and much lower unit costs for spare parts made in higher production runs.*

ing docks, and a loss of natural gas interruptible supply because of increased heating needs. Equipment problems occurred mainly at coal plants from frozen conveyor belts and frozen mine equipment, as well as derating of scrubber[s] and precipitators that were affected by the cold weather. For PJM's 22 million customers on 19 January 1994, net peak demand was only 2 GW higher than expected, but net supply was about one-fourth lower than expected (176), due largely to these kinds of logistical problems and to power plants with frozen exposed pipes, tanks, pumps, fuel stores, and fuel and ash conveyor systems. Another example, developing more slowly, was the November 1991–June 1992 drought in New Zealand, when lake inflows were the lowest in 60 years of record-keeping, and the complex hydroelectric system, which generates three-fourths of the nation's power, nearly ran dry. Distribution companies used a variety of rationing, curtailment, and price methods to cope. One South Island distributor achieved nearly 20 percent voluntary demand reductions through price signals alone as the grid's spot price rose by sevenfold (516).

2.2.6.2 Financial-economic perspective: guard against systematic price risk

Diversification has a completely different meaning in financial-economic than in engineering terms. Both are important, but they're complements, not substitutes. Engineering diversification keeps the lights on; financial-economic diversification saves money. Awerbuch (36) explains this as follows.

Geographic diversification is meant to guard against random (unsystematic) risks; it is like buying multiple roulette wheels. Using different fuels is usually directed mainly at random risk too. The trouble with both strategies is that in financial terms, they cannot effectively protect against systematic risk, like the 1973 and 1979 oil-price shocks,

Technical Note 2-2:
Tax, accounting, and financial-market distortions

Tax distortions

Systematic bias is introduced by the current, but not necessarily permanent, practice of allowing current tax-deductibility for operating costs, which are about four-fifths of the total lifecycle costs of a gas combined-cycle plant, but not for capital costs,[52] which are over nine-tenths of the total lifecycle costs of a photovoltaic plant (524)—amplifying the pretax cost-structure disparity shown in Figure 2-20.

Indeed, this is but the tip of a sizable iceberg of tax-related disincentives to renewable sources, at least in the United States (355). Although input (e.g., sales and employment) taxes and state income taxes are unlikely to distort renewable/nonrenewable investment choices (282),

- Most of the energy subsidies remaining after the 1986 U.S. tax reform favor the more capital-intensive options (387), but not all are available to non-utility or small-scale buyers, introducing a potential bias against distributed resources.

- Tax and other subsidies to fuels (387) disfavor renewables correspondingly (302).

- Distributed resources that are capital-intensive but have little or no fuel and O&M cost may be more burdened by local property taxes than are nonrenewable resources (282), especially if sited in an urbanized area with higher property tax rates than rural areas. For investor-owned utilities, local property taxes can increase U.S. levelized costs by anywhere from 7–9% for conventional resources to 8–31% for renewables (wind being at the top end), while for non-utility generators, local property taxes disadvantage only a few renewables, notably wind. (However, 18 states exempt solar property from property taxes, and six more offer localities that option [156]).

Against these disadvantages must be set the tax credits and other subsidies available in some jurisdictions to some renewables. For example, most renewables in the U.S. have relatively short (favorable) tax depreciation lives. Wind and dedicated-plantation biomass investments—plus most solar-electric resources since 1995—also receive a $0.015/kWh federal production credit, increasing with inflation. Non-utility generators can further benefit from solar and geothermal investment tax credits (reduced proportionally if the owner is subject to Alternative Minimum Tax). These tax bonds may be meant to offset larger subsidies to nonrenewables, to act as a surrogate for avoided externalities, or both; and the analysis, changing as it does over time and space, can become quite complex. Some of the United States now offer quite substantial tax credits, buydowns, and other incentives, especially for solar electricity (156, 168).

An indicative Congressionally mandated 1993 U.S. review by the U.S. Department of Energy (282) found that the net effect of all U.S. taxes and credits, for an investor-owned utility, was to make windpower 9% and plantation biomass 3% cheaper, to make other renewables 6% (geothermal) to 40%

(hydropower) costlier, and to make the electricity from conventional resources 18–22% costlier too. Of the seven classes of renewables considered, the net effect of all these public-policy interventions was to disadvantage three or four kinds of renewables (hydro, PV, solar-thermal, and perhaps waste biomass) while advantaging three (geothermal, plantation biomass, and wind). For non-utility generators, however, the analysis found a net favorable effect for all renewables, especially for plantation biomass and wind because of their specific tax credits. These disparities appear to reflect the results of political lobbying more than any rational weighing of relative societal benefits. The analysis is probably also incomplete.

Accounting distortions

Utilities' traditional accounting-basis comparisons are incorrect even in their own terms, because they assume that all technologies incur the same indirect or overhead costs and the same transaction costs. In fact, they don't. For example, an operator of a fossil-fuel power station must maintain staffs for environmental compliance, fuel logistics and purchasing, fuel inventory management and accounting, facility engineering, etc. These incur significant overhead costs that have no analogues for such distributed renewables as rooftop photovoltaics. Similarly, spinning reserves, reserve margin, meter-reading, and other operational overheads may be quite different for renewable and distributed resources than for classical power plants. Modular, mass-produced resources should have lower maintenance equipment and training costs, lower carrying charges on spare-parts inventories, and much lower unit costs of spare parts made in higher production runs than do central power stations, whose parts are often highly specialized (dropping a rotor could cost many millions of dollars in an instant). (463) Properly reflecting these differences may require more sophisticated accounting systems, such as Activity-Based Costing.

More broadly (45), the traditional technology suite based on central power stations and extensive grids "cannot be operated outside of large, hierarchical organizations" with "broad capabilities which can provide the needed support and agglomeration economies," and such organizational support "therefore consumes significant overhead and transactions cost." Mass-produced, small-scale, modular, and renewable or demand-side resources do not, in general, have any of those institutional requirements, nor, in principle, their inherently large overheads. Though such overheads "are fixed in the short-run, they are clearly avoidable in the long-run and hence should not be ignored in utility planning" (46) as a source of economic difference between these resources. Small or informal organizations can often also avoid legally mandated administrative, reporting, personnel, and other requirements that can add considerable overhead cost; many U.S. employment-law provisions, for example, only apply to firms with at least 50 employees.

Another bias arises from the tacit assumption that all supply technologies provide a service of the same quality. Again, they don't. As will be shown in Section 2.3.3.8, for example, many

[52] These are instead typically amortized, although in some circumstances they may be eligible for investment tax credits or other capital-based subsidies. The figures given are over 30 years and reflect roughly 1990 technology.

distributed renewable sources have far better inherent reliability and power quality than any grid-transmitted resource. Applications that need these attributes may find that their value outweighs all other cost considerations. Even more importantly, distributed resources in alliance with demand-side technologies (the two may even blur indistinguishably together, as in daylighting) can often offer benefits of amenity and productivity that add customer value one or two orders of magnitude more important than the entire energy bill (§ 2.4.1.3).

Awerbuch notes an even more fundamental problem (46): the tools of the accounting profession are not yet fully adequate for comparing distributed, capital-intensive, "passive," low-operating-cost assets with centralized, less capital-intensive, actively managed, high-operating-cost assets. The previous section (2.2.3.1) noted that the tax-deductibility of operating costs such as fuel creates a bias against renewables and efficiency; but long before that, the very way we think about assets and investment choices is distorted by accounting language embedded in utilities' central-station tradition. For example, utility managers are accustomed to marginal-cost functions based largely on operating costs. But a resource with essentially no operating costs has a flat (roughly zero) marginal cost over its operating range, then an infinite cost at its capacity limit—an idea alien to thermal-plant practice. Similarly, levelized costs are a convenient shorthand, but they mask important issues of intergenerational equity and cost burdens.

Accounting can accurately allocate operating costs but has much more trouble allocating capital costs. Indeed, while "accountants view depreciation as an allocation of historic (sunk) cost in an 'arbitrary but systematic' manner," economists "always view depreciation as a measure of changing economic value" that is related not to sunk costs but only to real-time competitive market conditions dependent on, among other things, competing new technologies (48). Accounting categories and principles are just as bad at decision support for distributed non-fueled resources as they are for, say, fax machines (48). How can you use the accounting cost of avoided stamps and envelopes to convince a bottom-line-driven manager to buy a fax machine when its big benefits are probably in efficiency, throughput, speed, and better decisions—and how, having done so, could you then use accountancy to decide when to replace the fax machine with an improved one? Accountancy is an important tool for understanding what you've done, but it's not an instrument for navigating through future uncertainties and innovations. It's like a rear-view mirror, not a windshield. Technologies that change the topology, the architecture, the basic structure of how a service is delivered cannot be compared with the technologies they replace by using accounting costs—because the most important effects of changing the whole way you do business will be, as Robert Frost said of poetry, "lost in translation." For distributed resources, most of which have largely or entirely fixed costs and low or no operating costs, these distortions of capital value are especially burdensome—and the actual value of their speed and modularity, though not fully recognized by standard accounting concepts, becomes especially important.

Such simple examples suggest an important research agenda for accountants, economists, and management theorists to support the transition to distributed utilities, because "Given our limited accounting vocabulary, the task of understanding renewable technologies is roughly equivalent to trying to appreciate Shakespeare by 'listening' to a rendition in Morse Code." (48)

Financial-market distortions

Modular, short-lead-time, fuel-less generators might be presumed to find favor among financiers. But on the contrary, Awerbuch (43) argues that the opposite is true, because of widespread misunderstandings of finance theory among lenders. For example, lenders worry about how debt service obligates much of renewable projects' cashflow, apparently without noticing that variable costs and systematic risks are almost zero, so "high loan-to-value makes a lot of sense. Everyone seems to understand that. For example, when you put up [T]reasuries against your margin loan, your broker might lend you 90 percent of their value. But you might only get 50 or 75 percent of value if you put up risky stocks." (The confusion is also linked to another misunderstanding: lenders traditionally include loan payments when calculating operating leverage—contrary to finance theory.) Just as lenders cheerfully finance shopping centers, office towers, and other real-estate projects where most operating cashflow goes to service the debt, they should do the same for renewable projects. The real question for both is the creditworthiness of the tenants or electricity buyers.

Another class of distortions is equity investors' frequent expectation that renewable projects will yield returns as high as risky startup ventures. They needn't, because they're almost riskless (if they have creditworthy power buyers)—"a simple, clean business with controllable costs and little risk." So the real issue isn't the technology, but rather, outmoded ways of perceiving its value.

because the prices of all fossil fuels are highly correlated (historically, around 0.85): when one rises, the rest tend to follow in due course. An additional systematic risk could come from carbon taxes, which would affect all fossil fuels, albeit unequally (coal twice as much as natural gas). Thus PG&E, before divesting its fossil-fueled generating capacity, had the least carbon-exposed portfolio in the United States (519), its supply portfolio was highly diversified and strongly renewable, yet it was not well hedged against financial risk from fuel price, because ~70% of its 1992 fuels were indexed to gas price, and the rest, other than nuclear, were well correlated with gas price. Its generating business then eliminated this problem by retaining only renewable and nuclear capacity, but its distribution business remained at risk for incomplete recovery of high costs for power purchased in the market—costs still strongly correlated with natural gas prices, as earlier drafts of this book noted. When gas prices soared in 2000–01, the company went bankrupt.

Standard fuel diversification, Awerbuch notes, is therefore like owning a stock portfolio consisting just of automakers: you're protected against competitive fluctuations among them, but not against a general downturn in the car business. But portfolio theory instructs us that "it is possible to develop a [far more diversified] portfolio that has a higher expected return than the all-automobile portfolio [but] with no added risk." (29) This requires simply adding shares of firms whose returns correlate poorly with those of automakers. Even if those firms' shares are as risky as automakers' and offer no better return, the portfolio as a whole will then be less risky for the same return.[53] Better still, add some percentage of riskless U.S. Treasury obligations, up

to the level that makes the whole portfolio efficient—unable to yield more without increasing risk, or to provide less risk without lowering expected return.

This financial-economics philosophy—utterly alien to most utility managers—implies that true risk diversification and efficient generating portfolios must include resources whose cost streams correlate badly with each other (*e.g.*, renewables with fossil fuels), and must also include some element of technologies with no systematic risk components (*e.g.*, efficiency or most renewables).[54] Portfolio theory, concludes Awerbuch, "yields a basis for [rigorously] quantifying the value of 'fuel-diversity,' which is now generally treated as a 'soft' benefit."

He helpfully illustrates this concept by showing that a coal/gas portfolio provides no obvious optimal point to choose: "Adding gas to an all-coal portfolio reduces cost, but does so at an almost linear increase in risk so that there is no portfolio effect" on overall value. This is because gas costs less than coal but has a more volatile price (larger standard deviation of historic prices), and the two effects offset each other because their prices are 84% correlated, graphed as the heavy curve in Figure 2-21.

Suppose hypothetically (and incorrectly) that the prices of gas and coal were perfectly *anti*correlated: when each rose, the other fell equally. Then a perfect portfolio effect could be created: we could shift through point *B* (each point represents a 5% gas addition to the portfolio) to point *A*. At that point, 35% gas and 65% coal, the cost is about one-fourth lower than the original all-coal portfolio, but the risk has fallen to zero. Point *C* is even better than *B* because it's

Benefit

56
Unlike different fossil fuels, whose prices are highly correlated with each other, non-fueled resources (efficiency and renewables) have constant, uncorrelated prices that reduce the financial risk of an energy supply portfolio.

equally risky but lower in cost. Unfortunately, a resource perfectly *anti*correlated with coal price is hard to imagine. But as a first step, renewables and efficiency have prices that should be more or less perfectly *un*correlated with coal. Introducing such a riskless technology, even in modest proportions, provides a striking improvement over the futile substitution between coal and gas shown above. For example (Figure 2-22), point M represents a mix of 70% coal, 30% gas, and no renewables. But adding a little renewable component moves to a point such as Q—about 15% renewables, 60% coal, and 25% gas—with a lower cost *and* a lower risk than the all-coal portfolio. Many other tradeoffs are of course available that cost less than the riskless but putatively costly 100%-renewable portfolio at the upper left corner.

Moreover, while the price and risk reductions in going from P to Q are small, the renewables will also provide other valuable benefits, such as modularity, flexibility, redeployability, short lead times, and protection from possible future carbon taxes or supply interruptions. Moreover, further contracts or other mechanisms should be devisable that could move from points such as P or R to points such as S. The diversity value of the renewable component will then be the difference between the risk/cost ratio of the portfolio with and without it.

Awerbuch points out that this result—the portfolio desirability of adding renewables even though they may cost more—is exactly like "the textbook result for financial portfolios, which show that every optimal portfolio must include some riskless U.S. Treasury bills even though they are the lowest yielding, and hence the most *expensive*[,] investment alter-

native." In this sense, S is "the analog of an optimal financial portfolio when borrowing is permitted," and could result from "some type of option under which customers purchase (or sell) riskless renewable electricity or riskless capacity." (29) Green pricing and other unbundled brokerage options should offer exactly such an opportunity in restructuring electricity markets.

Another way to represent Awerbuch's illustrative example is to use the same assumptions:
This would lead to the risk/cost relationship shown in Figure 2-23 as a function of the percentage of photovoltaics added to the coal-and-gas portfolio. The curves are relatively close together because of the 84% correlation between price movements of gas and coal. Yet adding the riskless photovoltaic resource to the gas-coal portfolio can materially reduce risk at the same cost, as represented by horizontal movements between the curves.

In a practical application of this concept, the Sacramento Municipal Utility District has tariffs significantly sensitive to natural-gas price. A doubling of the gas price could, for example, increase residential tariffs by ~10% for residential customers, whose tariffs are about 25% energy costs, or by ~30% for industrial customers, whose tariffs are ~75% energy costs. However, SMUD could hedge fuel-price risks by focusing new renewable-resource acquisitions on those customers whose tariffs depend most heavily on energy costs. If fuel prices rose sharply, this reallocation of risk could yield major savings for those most fuel-price-sensitive customers at only a slight cost to the least fuel-price-sensitive customers. This suggests that "it may be beneficial to manage fuel price risks for those customers most sensitive to price changes

[53] For such countercyclic behavior it may even be worth paying a premium in the form of a lower return.

[54] "Most" because biofuels may in some instances, such as regions dependent on forest-products industries, have prices correlated with local economic activity. On the whole, Awerbuch suggests, pending further study, that "biomass price risk may be largely unsystematic and diversifiable across multiple geographic areas." (38) He and Martin Berger drafted in July 2002 an illustrative practical application of the whole fuel-diversification thesis to the supply portfolio of the European Union.

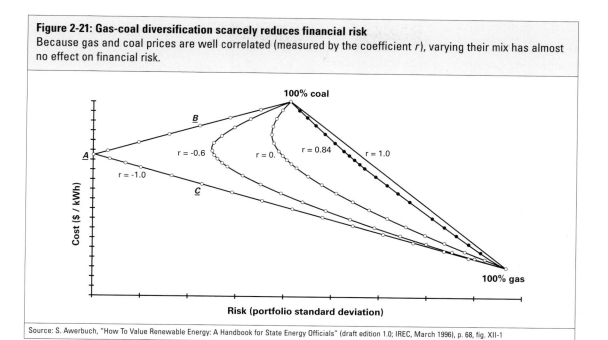

Figure 2-21: Gas-coal diversification scarcely reduces financial risk
Because gas and coal prices are well correlated (measured by the coefficient *r*), varying their mix has almost no effect on financial risk.

Source: S. Awerbuch, "How To Value Renewable Energy: A Handbook for State Energy Officials" (draft edition 1.0; IREC, March 1996), p. 68, fig. XII-1

rather than for the utility as a whole." (745)

2.2.7 Load-growth insurance

A special risk-reducing advantage of two kinds of distributed resources—end-use efficiency and cogeneration—is that they provide automatic "insurance" against uncertainties in load growth. This is because their output expands in proportion to the activities that create demand. If a factory using these resources adds an extra shift, or a building using them stays open longer hours, then the resulting increase in demand will be moderated by expanded end-use savings or cogeneration output. For example, if a given luminaire (lighting fixture) that has been equipped with technical improvements reducing its watts per lumen by 50% is then run for twice as many hours, it will achieve the same *percentage* saving but twice as large an *absolute* saving as on the original operating schedule. Similarly, a cogeneration plant whose operation depends on demand for

coproduced process steam or fuel-cell waste heat will also produce more electricity when it needs more heat; normally both rise more or less proportionately.[55]

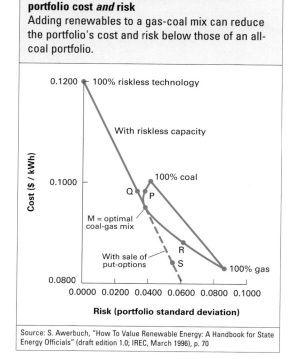

Figure 2-22: Adding renewables can reduce portfolio cost *and* risk
Adding renewables to a gas-coal mix can reduce the portfolio's cost and risk below those of an all-coal portfolio.

Source: S. Awerbuch, "How To Value Renewable Energy: A Handbook for State Energy Officials" (draft edition 1.0; IREC, March 1996), p. 70

Such resources differ from conventional generating capacity in two ways. First, the beneficial saving or provision of electricity is not fixed but expandable. Second, *it expands in precisely those conditions in which it has a higher economic value*, because rapid load growth places the greatest stress on existing resources. In those respects, these two kinds of distributed resources have the essential features of an insurance policy—insurance against load growth. The value of that zero-premium insurance can be estimated by calculating the present value of building and operating the avoided supply-side resources, adjusted for the extent and probability of the activity growth that might occur.

Hoff (315) has identified a generalized supply-side analogy to such load-growth insurance. In essence, he finds that when uncertainty in demand directly affects a firm's profits, then modular, short-lead-time resources are worth more because their quick and flexible response to fluctuating demand will be worth the most in exactly the condi-

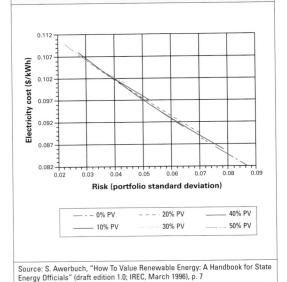

Figure 2-23: Costly renewables improve portfolios
Adding a riskless solar resource to a gas-coal portfolio can yield lower risk at the same cost, just like adding Treasuries to an investment portfolio.

Source: S. Awerbuch, "How To Value Renewable Energy: A Handbook for State Energy Officials" (draft edition 1.0; IREC, March 1996), p. 7

Table 2-2		
Characteristics		
Fuel	**Risk (std dev)**	**Cost ($/kWh)**
Coal	0.045	0.1
Gas	0.087	0.082
PV	0	0.12
Correlation matrix		
	Coal	Gas	PV
---	---	---	---
Coal	1	0.84	0
Gas		1	0
PV			1

Source: S. Awerbuch, "How To Value Renewable Energy: A Handbook for State Energy Officials" (draft edition 1.0; IREC, March 1996), p. 70

tions when it is most needed. That in turn is because the costs of the distributed resources can be made highly correlated with the firm's profits under certain market condi-

tions, notably that the distributed resources are part of a larger portfolio that cannot be diversified. This approach goes beyond normal decision theory (§ 2.2.2.6) by taking explicit account of managers' risk aversion. It can therefore explicitly evaluate how "distributed resources offer utilities [or other investors] an important tool in managing the risks associated with demand uncertainty."

2.2.8 Matching loadshape

So far we have discussed load growth as if it affected only *how much* electricity is used, but not *when* it is used, especially on a daily scale. However, different patterns of load growth do affect loadshape. For example, most heavy industries add steady loads over two or three shifts a day, most commercial loads are heavy in the daytime and small at night, and most residential loads have peaks related to household schedules and space-conditioning.

Although every utility has a different pattern, and the patterns all shift over the years, Southern California Edison Company's 1990 peak-day loadshapes by rate class offer a fairly typical illustration of these effects (Figure 2-24). (782) Conceptually, diverse distributed resources could nicely match aggregated loadshapes, as in Hoff's example of how PVs help meet summer, and cogeneration winter, loads (Figure 2-24a).

But for optimal integration of all kinds of distributed resources, and to anticipate how time-of-use or real-time pricing might affect loadshapes, it is useful to disaggregate sectoral loadshapes by end-use, as in new Lawrence Berkeley National Laboratory analyses of California's statewide electricity demand on the summer peak day:

For any type of load, future demand is not fate but choice, and can be chosen with great flexibility by using a balanced portfolio of demand- and supply-side resources. Careful investment in end-use efficiency, load management, and electric-thermal integration (such as cogeneration or thermal storage) can alter the size and timing of demand from almost any load over a very wide range in order to achieve the desired service quality at least cost. It can generally turn load growth into load stability or shrinkage, at any desired time or overall, for any customer or class of customers, on any desired geographic scale, if that is the cheapest way to meet customers' service needs.

The loadshapes in Figures 2-24 and 2-25 are so smooth because they are highly aggregated. They reflect the diverse timing of loads across a large utility's entire service territory, as different customers do the same things at somewhat different times and different things at the same times. But as one

examines ever smaller portions of the utility system—distribution planning areas, then the areas served by particular substations, then those by particular distribution lines— the curves become more jagged because fewer customers' loads are being aggregated, and fewer means less diverse. This is nicely illustrated by residential loadshapes presented below in Section 2.3.2.12.

Moreover, as one travels from the biggest power stations and transmission lines out through the ever finer branches of the distribution system, costs rise steeply. For example, Detroit Edison's Murray Davis estimates that whereas transmission capacity typically costs about \$100–150/kW to build, adding distribution investment brings the grid investment up to ~\$400–500/kW (and makes it even more site-dependent). Yet by definition, distribution capacity *must reach each and every customer*. This means that the costliest (and, as we'll see in Sections 2.3.2.2 and 2.3.2.3, the highest-electrical-loss and worst-power-factor) part of the power system inherently suffers from the lowest load diversity and the worst load factors (*i.e.*, the lowest capacity utilization). But that customer end of the distribution system is precisely where distributed resources are often easiest to install and can create the greatest value.

Small units obviously allow greater flexibility in matching supply with demand, both systemwide and locally—the more fine-grained and localized the resources, the better the match. Demand-side resources, the most tailored and local kind, specifically decouple a specific customer's service delivery from electric loadshape (by providing the same service with less electricity or with electricity in a different time pattern). They

---

Benefit

57
Efficiency and cogeneration can provide insurance against uncertainties in load growth because their output increases with electricity demand, providing extra capacity in exactly the conditions in which it is most valuable, both to the customer and to the electric service provider.

---

[55] Naturally, the expansion in both cases is a function of more hours run, and cannot exceed the physical capability of the resources in terms of total potential annual output.

Benefits

58 *Distributed resources are typically sited at the downstream (customer) end of the traditional distribution system, where they can most directly improve the system's lowest load factors, worst losses, and highest marginal grid capital costs—thus creating the greatest value.*

59 *The more fine-grained the distributed resource—the closer it is in location and scale to customer load—the more exactly it can match the temporal and spatial pattern of the load, thus maximizing the avoidance of costs, losses, and idle capacity.*

60 *Distributed resources matched to customer loads can displace the least utilized grid assets.*

61 *Distributed resource matched to customer loads can displace the part of the grid that has the highest losses.*

62 *Distributed resources matched to customer loads can displace the part of the grid that typically has the biggest and costliest requirements for reactive power control.*

63 *Distributed resources matched to customer loads can displace the part of the grid that has the highest capital costs.*

---

can be complemented by distributed supply-side resources on the scale that will best harness load diversity so as to share capacity among multiple customers' or uses' needs, so as to take advantage of not everyone's wanting to do the same thing at the same time.

Figure 2-24: Loadshapes are diverse
Loadshapes differ between customer classes.

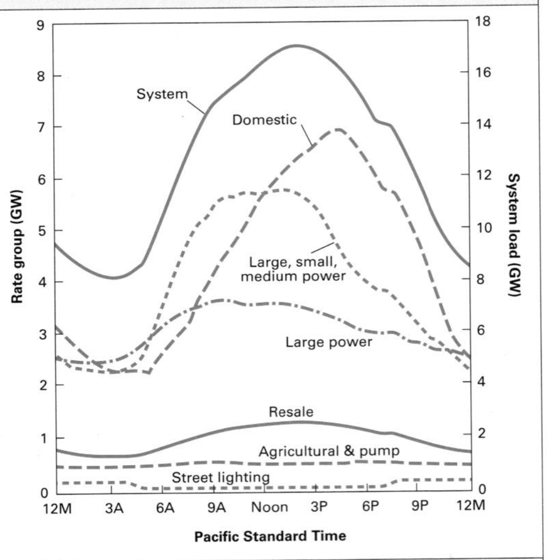

Source: SCE, 1990 summer peak day data from H. W. Zaininger, "Distributed Renewables Project" (Distributed Utility—Is This the Future? Conference 1992)

Currently prevalent loadshapes should be used only with caution as a basis for system

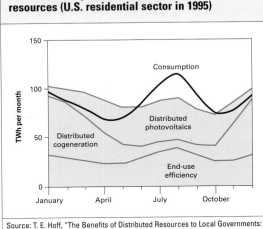

Figure 2-24a: Measured electricity consumption and estimated production using distributed resources (U.S. residential sector in 1995)

Source: T. E. Hoff, "The Benefits of Distributed Resources to Local Governments: An Introduction" (draft report to NREL, 12 September 2000), p. 8, fig. 4

planning, because they are a consequence of price signals and end-use technologies that are often far from optimal for both the customer and the system. Emerging real-time pricing is likely to reduce or suppress many peaky loads that never before had to pay their way. That suppression will probably be less behavioral than technological—*e.g.*,

Figure 2-25: End-use structure of 1999 California summer-peak-day statewide load
Note that all but the bottom two segments are building loads. The residual "other" term shows differences between loads reported to the FERC and the structure of the California Energy Commission forecasting model; the differences are probably due mainly to small utilities that don't report to CEC, and are of little consequence for understanding end-use structure.

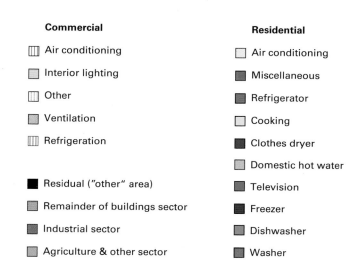

Commercial	Residential
Air conditioning	Air conditioning
Interior lighting	Miscellaneous
Other	Refrigerator
Ventilation	Cooking
Refrigeration	Clothes dryer
	Domestic hot water
Residual ("other" area)	Television
Remainder of buildings sector	Freezer
Industrial sector	Dishwasher
Agriculture & other sector	Washer

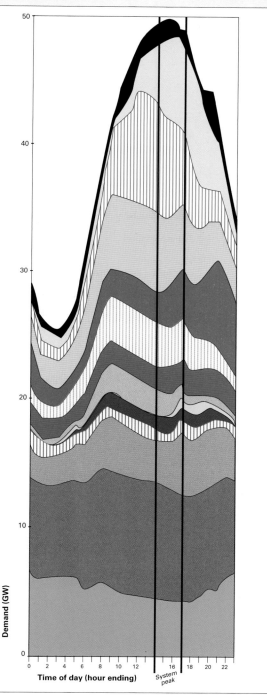

Demand (GW)

Time of day (hour ending)

System peak

Source: R. E. Brown and J. G. Koomey, "Electricity Use in California: Past Trends and Present Usage Patterns," LBL-47992 (forthcoming in *Energy Policy*, 2002)

Figure 2-26: End-use structure of 1999 California summer-peak-day residential load
"Miscellaneous" includes lights, pools, spas, waterbeds, and small appliances. The sequence is the same as in the legend. The previous figure's residual "other" term is not included.

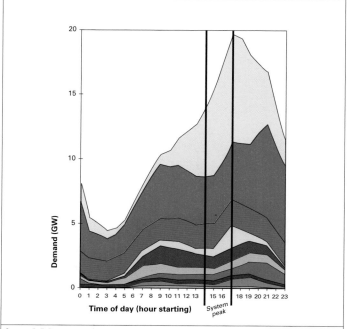

Demand (GW)

Time of day (hour starting)

System peak

Source: R. E. Brown and J. G. Koomey, "Electricity Use in California: Past Trends and Present Usage Patterns," LBL-47992 (forthcoming in *Energy Policy*, 2002)

Commercial

▥ Air conditioning

▢ Interior lighting

▥ Other

▢ Ventilation

▥ Refrigeration

▪ Office equipment

▢ Domestic hot water

▥ Exterior lighting

▢ Cooking

**Industrial, Agricultural,
and Public Service**

■ Assembly industry

▪ Agriculture

▥ Process industry

▫ Transportation

▥ Water pumping
(California Department of Water Resources)

▫ Other industry

▪ Street lighting

Figure 2-27: End-use structure of 1999 California summer-peak-day loadshape for the commercial sector, excluding the residual "other" term shown in the statewide total graph

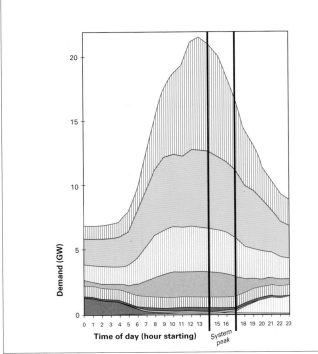

Source: R. E. Brown and J. G. Koomey, "Electricity Use in California: Past Trends and Present Usage Patterns," LBL-47992 (forthcoming in *Energy Policy*, 2002)

Figure 2-28: End-use/sectoral structure of California summer-peak-day industrial, agricultural, and public-service load, excluding the residual "other" term shown in the statewide total graph

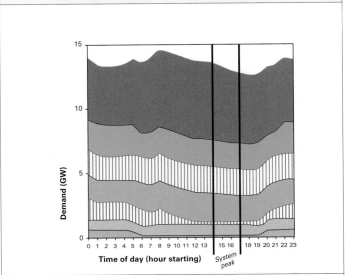

Source: R. E. Brown and J. G. Koomey, "Electricity Use in California: Past Trends and Present Usage Patterns," LBL-47992 (forthcoming in *Energy Policy*, 2002)

using more efficient building envelopes, lights, equipment, etc. to provide more comfort with less cooling on hot afternoons or with less heating on cold nights. Nonetheless, load-matching is of practical economic interest, because unlike conventional generators, "the capacity value of intermittent renewable energy strongly depends on the correlation between the utility load and the pattern of resource availability" (701)—more than on any other factor (720). And if loadshapes do change, making distributed resources less valuable where they are, they can be relocated to preserve or enhance that value to the system (§ 2.2.2.8), so changing loadshape need not be considered an unmitigated source of additional investment risk.

Those renewables' energy value will also depend on when they can be dispatched and in what portion of the system load-duration curve, because that will determine the value of the fuel and other operating costs that they can displace (124–5). It is therefore necessary to consider the output patterns, and their matching with typical loadshapes, for distributed resources whose output depends mainly on the vagaries of weather.

2.2.8.1 Evaluating field data for renewables

To the extent that traditional loadshapes do persist after real-time pricing is introduced, some renewable resources happen to fit them very well. For example, at a seasonal level, northern Europe tends to have both high electric demand and high windspeed during the winter (as do San Francisco and the nearby Altamont Pass windfarm area in the summer [701]). This means there is more windpower (which varies as the cube of

windspeed) just when it is most valuable, as shown in Figure 2-29. What degree of capacity credit should be given to renewables that are intermittent but generally match utility loadshapes?

This question is normally asked only for individual types of renewables by themselves, and we shall address it in a moment. But more interesting and far less studied is the potential for *combinations* of renewables to work under complementary kinds of conditions. For example, the combination of windfarms and PVs turns out to match almost perfectly the typical Southern California loadshapes. That's partly because afternoon valley heating draws the wind through the turbine-equipped passes late in the day, when a secondary load peak occurs after the peak PV output has passed.

The combination is especially valuable, of course, if it includes hydropower with water-storage capacity. In one combined wind-solar-hydro system (8), for example, "very small generation reserves would be needed because of the energy storage capabilities of the hydro facilities." (722) A combined wind/hydroelectric proposal for Medicine Bow, Wyoming showed similar advantages (80). Unfortunately, few multi-resource integrated systems seem to have been studied, so the rest of the field data discussed next are only for single renewable technologies in single or multiple sites. We return in Section 2.2.10.1 to opportunities for technological and siting diversity. Until then, we consider further the value of load-shape-matching for individual renewable technologies.

At the level of both seasonal and daily load-shape, Effective Load Carrying Capacity

Benefits

64 *Many renewable resources closely fit traditional utility seasonal and daily loadshapes, maximizing their "capacity credit"—the extent to which each kW of renewable resource can reliably displace dispatchable generating resources and their associated grid capacity.*

65 *The same loadshape-matching enables certain renewable sources (such as photovoltaics in hot, sunny climates) to produce the most energy at the times when it is most valuable—an attribute that can be enhanced by design.*

66 *Reversible-fuel-cell storage of photovoltaic electricity can not only make the PVs a dispatchable electrical resource, but can also yield useful fuel-cell byproduct heat at night when it is most useful and when solar heat is least available.*

67 *Combinations of various renewable resources can complement each other under various weather conditions, increasing their collective reliability.*

A **load-duration curve** (introduced on p. 80) conventionally plots the percentage of maximum experienced load that a generating or grid resource experiences in a typical year against the number or percentage of hours in the year. Examples are given in Figures 1-35–1-37. Such a curve typically has a shoulder of always-on load, tapering down through intermediate load factors to rare peak loads (conventionally shown at the upper left corner).

(ELCC)—the fraction of the plant's rated capacity credited as being fully dispatchable on a given utility system—ranged in early studies from 5% to nearly 50% of installed wind capacity for seven U.S. utility systems, depending on their weather and load patterns, other capacity, and degree of wind-power saturation (assumed to range from 5% to 20% of system capacity).[56] (240) A recent Canadian analysis using the Hydro-Québec/Canadian Electrical Association model found 42–43% ELCC for 3.3–9.9 MW windfarms on Prince Edward Island (24). European analysts were meanwhile finding ELCCs of about 100% at low penetration in North Germany (284), and in the Netherlands, 26% at modest or 7% at high penetration (31% of total installed capacity). (135) Low figures generally resulted from particular assumptions about how the nonrenewable grid would be operated, and did not necessarily represent a practical or economic limit in light of knowledge gained later, as noted in Sections 2.2.10.1–2 below.

In many areas, photovoltaics can have even higher ELCCs, largely because of the "better irradiance-demand [than wind-demand] correlation: human activities tend to follow the sunlight cycle" (721), and at least in California, "the combination of air conditioning and commercial load follows insolation very closely," (610) whereas the wind often blows when most people are asleep. A study of PV potential for 20 diverse utilities across the entire United States, ranging from hundreds of MW to tens of GW and totaling 100 GW of peak load, found that a 10% market penetration of fixed PVs would yield matches to loadshape typically above 50% and ranging from 36% to 70%, depending on location; two-axis tracking PVs would do 5–15 percentage points better (range 38–80%). (546) This is largely because many U.S. utilities have big air-conditioning loads on the sunny summer afternoons when PVs produce the most power.

Largely for this reason, the empirical ELCC of the Carissa Plains PV plant on the PG&E system was 79% (305, 319); that of PG&E's Kerman substation PV, 77% (735); that of simulated PV capacity in New York, 62% (falling to 50% as capacity reached 500 MW) (547), and similarly in New Jersey (391). As would be expected, the higher the utility's ratio of summer to winter peak loads, in general, the higher its PV ELCC (546).

Figure 2-29: Correlation of wind and electricity demand in England
English winds are strongest in the winter when electricity demand is highest (mainly because of electric space heating) and the power is therefore worth the most.

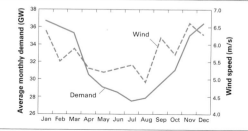

Source: R. W. Thresher, "Wind as a Distributed Resource" (EPRI 2nd DR Conference, 6 November 1996)

Such results are not confined to sunny climes. In relatively cloudy Massachusetts, the Gardner Project, installed in 1985–86, used 30 home and five commercial sites (the latter rated at 1.8–7.3 kW each) to achieve a "high concentration of PV systems (53%) on a single distribution feeder" which therefore often fed back net power to the utility: in effect, 28 PV-equipped homes met their own loads plus the loads of 25 other homes, with as much as 56% of some homes' total power output being "exported" to the feeder during summer months. A 2-kW home-roof PV system saved 1.2 kW of capacity at the summer peak hour (the six-year range was from over 1.1 to 1.5 peak kW)—a 60% average ELCC (85). For a relatively poor solar climate, this is most encouraging.

Another indication of PVs' often good match to loadshape is the following comparison of PG&E's system annual load-duration curve and the annual output of its PV sites, shown first at full scale (Figure 2-30) and then magnified (Figure 2-31) for the top 25 hours of annual load duration, when of course generating and delivering electricity is most costly

and any shortfall would require correction by demand- or supply-side investment: SMUD finds a similarly close match with tracking PVs for its top 25 load hours (Figure 2-32) and for its top five load days (one of which shows a 20% PV output loss from passing clouds at one hour, but all of which generally show the predictability of daily PV output) (Figure 2-33).

A spreadsheet tool is available for modeling the effect of any desired photovoltaic output on the California peak load for June 2000, and hence for valuing the resulting peak load reduction.[57]

2.2.8.2 Improving loadshape match by technical design

There are three important ways to stretch photovoltaic output later into the afternoon to match many utilities' peak loads better. One is to use a two-axis tracking mount—often based on highly reliable satellite-dish technology using just a few watts—to keep pointing the array directly at the sun as its azimuth changes. For the 14 July 1994 system peak of the Sacramento Municipal Utility District, a fixed PV is generating only about 50% of its rated power at the peak time (1800), while the tracker is generating 80% (Figure 2-34).

Another option, simpler but less widely appreciated, is to point a fixed collector more towards the west rather than the due south normally assumed (in the Northern Hemisphere). For example, pointing a 20°-tilt Sacramento rooftop PV 30° west of due south increases its capacity credit by ~25% while reducing its annual energy production by only 1%—a clear economic win (Figure 2-35). (316)

[56] For many existing installations, total nameplate capacity is not the correct denominator. For example, the Altamont Pass windfarms (over 0.7 GW at that time) experienced an ELCC equivalent to 22% of their nameplate rating in 1987 but only 14% in 1988, because of different wind and load patterns. However, compared instead with maximum actual *output* so as to account for any nonoperating or overrated machines, these ELCCs would increase to about 40% and 20% respectively, and PG&E's Solano 2.5-MW experimental MOD-2 turbine (since decommissioned) achieved not 74% but 80% in 1987 (240).

A due-west orientation increases the capacity credit by 48% while reducing annual energy production (compared with due south) by 12%. This may or may not be preferable, depending on who owns the system and on the relative value of capacity and energy. Under the conditions of the SMUD PV Pioneer project, the actual average orientation, ~30° west of due south, maximizes societal benefits (Figure 2-36):
A third method, seldom considered but likely to become attractive later in this decade, is to store offpeak electricity locally and use or resell it in peak-load periods when it is most valuable. Traditionally, storage entailed costly chemical batteries, but emerging methods—superflywheels, ultracapacitors, flow batteries,[58] and reversible fuel cells—promise superior economics and efficiencies. Reversible fuel cells, which can with equal ease convert hydrogen into electricity or vice versa, are especially interesting because they scale up or down to virtually any size, incur little efficiency penalty as compared with a one-way

fuel cell, can produce pressurized hydrogen directly for tank storage at low cost, and add little cost to that of an ordinary fuel-cell stack, combining a fuel cell and an electrolyzer (a fuel cell run backwards) into the same dual-function equipment. This approach means one can not only can (say) store photovoltaic electricity as hydrogen for nighttime use, but also convert that hydrogen back into electricity plus useful byproduct heat from the fuel cell, just when the most space (and, often, water) heating is required and when solar heat is least available.

2.2.8.3 Prospecting to maximize loadshape-matching's economic value

Close matching between loadshape and output maximizes many benefits of distributed

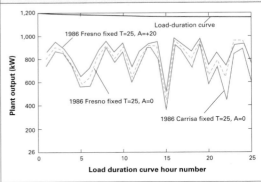

Figure 2-31:
...especially in the top 25 peak-load hours
The match is even better on the hottest days when the solar generation is most valuable.

Source: D. Shugar et al., "Benefits of Distributed Generation in PG&E's Transmission and Distribution System: A Case Study of Photovoltaics Serving Kerman Substation" (PG&E, November 1992), p. 3–4

generation, including increased grid conductor capacity, voltage support, loss savings, and enhanced reliability (592). Good source-to-load matching can be maximized by screening tools. At PG&E, these revealed a frequently excellent match for most of its 201 Distribution Planning Areas—if the optimal

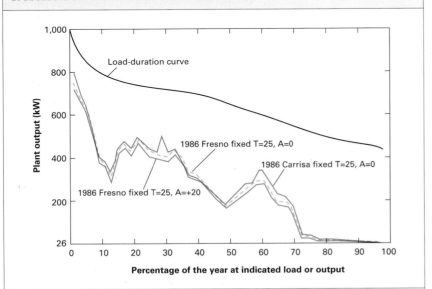

Figure 2-30: PVs well match PG&E's annual load-duration curve...
Photovoltaics can produce the most electricity at about the same times when PG&E most needs it.

Source: D. Shugar et al., "Benefits of Distributed Generation in PG&E's Transmission and Distribution System: A Case Study of Photovoltaics Serving Kerman Substation" (PG&E, November 1992), p. 3–4

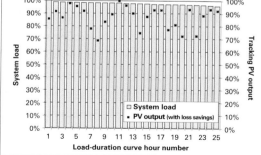

Figure 2-32: Peak-load/PV match in Sacramento
The hottest afternoons maximize both air-conditioning loads and photovoltaic output.

Source: H. Wenger *et al.*, *Photovoltaic Economics and Markets: The Sacramento Municipal Utility District as a Case Study* (SMUD, CEC, and USDOE PV Compact Program via NCSC; 1996), p. 8-8, fig. 8-4.

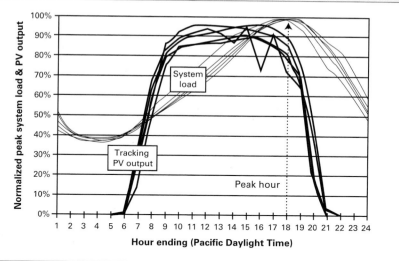

Figure 2-33: Peak-load/tracking-PV match in Sacramento
Tracking photovoltaics fit all but the evening tail of Sacramento's peak load.

Source: H. Wenger *et al.*, *Photovoltaic Economics and Markets: The Sacramento Municipal Utility District as a Case Study* (SMUD, CEC, and USDOE PV Compact Program via NCSC; 1996), p. 8-8, fig. 8-5. www.energy.ca.gov/development/solar/SMUD.pdf

kind of PV mounting were chosen in each case to improve the match to the timing of loads in that area (Figure 2-37) (593):

Such screening can help to determine where tracking mounts are worth their extra cost and where cheaper, simpler, lower-maintenance fixed mounts are a better buy. It also shows that in the foggier, milder coastal areas, Planning Area Load Carrying Capacity (PALCC) tends to be lower because loads are more constant, while in the central valley, clear anticyclonic summer weather and high commercial air-conditioning loads make demand peaky but also make PV output fit it well, yielding much higher PALCC values. Overall, the graph shows that for more than 8 GW of peak load (assuming 10% of it were met by PV) at the PG&E Distribution Planning Area level, a north-south-axis tracker will provide 60–100% Planning Area Load Carrying Capacity, meaning that 60–100% "of the PV system's rated capacity can be counted upon by planners to shave the Distribution Planning Area's load peak...an excellent match...." Excellent PALCCs were found for more than 40% of PG&E's planning areas. These matches were then overlaid on

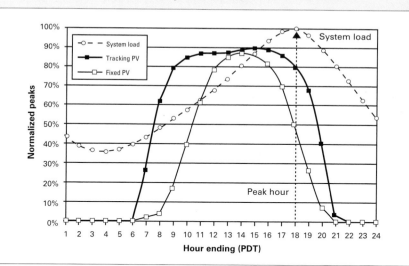

Figure 2-34: Tracking PVs prolong late-afternoon output
Tracking PVs match late-afternoon peak loads better than do fixed PVs...

Source: H. Wenger *et al.*, *Photovoltaic Economics and Markets: The Sacramento Municipal Utility District as a Case Study* (SMUD, CEC, and USDOE PV Compact Program via NCSC; 1996), p. 8-7, fig. 8-3. www.energy.ca.gov/development/solar/SMUD.pdf

graphs (594) of marginal grid costs to disclose the most promising initial sites for detailed engineering study.

2.2.8.4 Fine-grained prospecting

[57] Janice Lin (jlin@powerlight.com), personal communication, 13 March 2001.

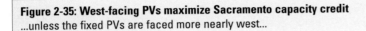

Figure 2-35: West-facing PVs maximize Sacramento capacity credit
...unless the fixed PVs are faced more nearly west...

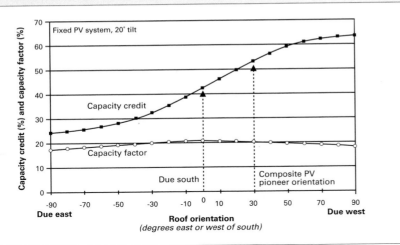

Source: H. Wenger *et al.*, *Photovoltaic Economics and Markets: The Sacramento Municipal Utility District as a Case Study* (SMUD, CEC, and USDOE PV Compact Program via NCSC; 1996), p. 8-9, fig. 8-6. www.energy.ca.gov/development/solar/SMUD.pdf

in time and space

Such "loadmatch prospecting" can then focus on time-dependent behavior at the micro-level of the individual substation to start quantifying the load-matching provided by onsite PVs. This yielded particularly striking results for PG&E's 498-kW[59] Kerman PV plant (Figure 2-38).

To this must be added the benefit from

Figure 2-36: Intermediate azimuth maximizes PVs' economic value to SMUD
...which can maximize their economic value.

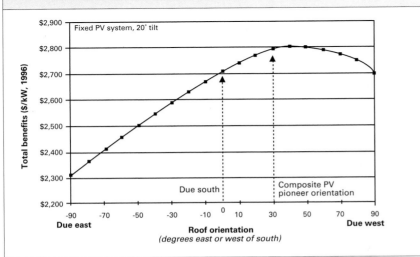

Source: H. Wenger *et al.*, *Photovoltaic Economics and Markets: The Sacramento Municipal Utility District as a Case Study* (SMUD, CEC, and USDOE PV Compact Program via NCSC; 1996), p. 8-11, fig. 8-7. www.energy.ca.gov/development/solar/SMUD.pdf

adjusting PV output for avoided system losses (Figure 2-39).

This fine-grained level of detail is important because "load profiles of different feeders vary dramatically across the...system.... In addition to occurring at different hours than system peak, feeder peaks are typically sharper and more pronounced. The task of meeting demand in the distribution system may thus focus on a few hours of the year." (608)

Focusing on the specific substation transformer, and doing so dynamically rather than statically so as to reflect important time-dependent behavior, reveals in this case that relieving peak load *before* the absolute peak "pre-cools the transformer bank. The two effects of pre-cooling and delivering power during the absolute peak combine to lower the maximum transformer temperature by 4 [C°]...and boost transformer capacity by about 410 kW" (735)—capacity worth on the order of $30,000 incrementally (§ 2.3.2.5). (Conversely, this dynamic behavior means that the time-matching between PV output and peak transformer need not be as tight to achieve capacity-saving and life-extension benefits for the transformer as for conductors. [592])

Moreover, while the PV plant yielded a 77% peak capacity availability and ELCC at the

---

[58] A flow battery is like a cross between a battery and a fuel cell. It combines a split battery (two half-cells separated by an ion-exchange membrane) with extra storage of electrolyte whose chemical energy can be separately regenerated by electricity. Its power output, often well into the MW range and potentially approaching GW, is then determined by plate area, but its energy capacity is determined by tank volume. The two can be independently chosen within a very wide range to fit the use. This emerging technology is versatile, scalable, durable, and relatively simple. It could have important benefits in grid and load management (409).

system level, at the *transmission* level the ELCC was an even higher, hence more valuable, 90%, increasing transmission capacity by about 450 kW (736). Capacity factor in the first year and in the final assessment averaged 25% and Performance Index[60] a gratifying 91%, the shortfall being due to normal teething problems at startup. The plant's output provides an excellent season-

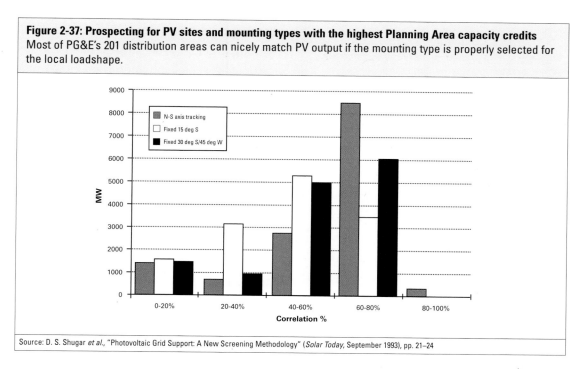

Figure 2-37: Prospecting for PV sites and mounting types with the highest Planning Area capacity credits
Most of PG&E's 201 distribution areas can nicely match PV output if the mounting type is properly selected for the local loadshape.

Source: D. S. Shugar *et al.*, "Photovoltaic Grid Support: A New Screening Methodology" (*Solar Today*, September 1993), pp. 21–24

al fit to the utility's residential loads in March and in August through November, but falls short by about one-third in April through July and by about one-half in December through February (17).

Another nice illustration of photovoltaics' ability to save grid capacity—and, by unloading hot equipment, to extend its life—is the equal kW savings for the top ten hours' or for the top one hour's load duration on a specific Arizona Public Service Company feeder (Figure 2-40). Thus PVs' match to air-conditioning-driven peak loads is valid not only at the system level

(Figure 2-31 for PG&E, Figure 2-32 for SMUD) but also down to the even peakier-load feeder level.

2.2.9 Reliability of distributed generators

Utilities have a long tradition of valuing capacity to see if it's worth installing to meet a predicted demand. In emerging competitive markets, this valuation may come to be based increasingly on market parameters, but historically it has been based instead on the *cost* of installing capacity of some conventional kind

Benefit

68

Distributed resources such as photovoltaics that are well matched to substation peak load can precool the transfomer— even if peak load lasts longer than peak PV output— thus boosting substation capacity, reducing losses, and extending equipment life.

instead. Typically generating capacity is modeled as a least-capital-cost proxy such as a simple- or combined-cycle gas-turbine plant or a steam plant—in any event, a plant of broadly similar operational role to the unit being evaluated, with comparable load-carrying capability. This makes the arithmetic simple. A combustion turbine can provide available capacity essentially anywhere for ~$40–50/kWy, which trans-

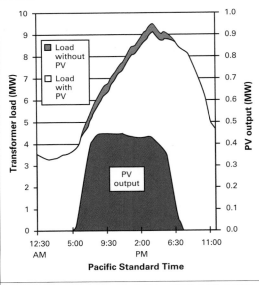

Figure 2-38: PVs precool distribution transformers
At the substation level, PV output that peaks before the load can nonetheless provide extra value by precooling the transformer.

Source: H. J. Wenger *et al.*, "Measuring the Value of Distributed Photovoltaic Generation: Final Results of the Kerman Grid-Support Project" (First World Conference on Photovoltaic Energy Conversion, December 1994)

lates to a capacity value of ~$0.004/kWh (187). (This is why firm and non-firm onpeak energy at, say, the California-

Oregon border traditionally differ in price by that amount.)

However, the less the potential new resource resembles the traditional one being used as a proxy, the more problematic the calculation of "capacity credit" becomes. How can a wind turbine or a solar cell be properly compared with a conventional power station, which is "dispatchable when available"— able to send out power whenever it is in working order, has fuel, and is called upon? Fortunately, such evaluation is possible by applying to nontraditional resources the same kinds of availability, loadshape-correlation, and other technical/economic concepts long used to evaluate traditional resources. Indeed, these techniques differ only in degree, not in kind, from those long used to forecast the availability of hydroelectric resources, which depend on rainfall, nonelectric demands for flow or storage, and many other technical and non-technical parameters.

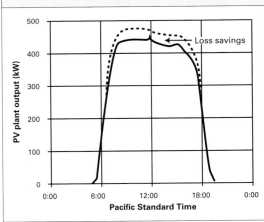

Figure 2-39: PV loss savings are like extra kW
System losses saved by PV output can be interpreted as equivalent to additional PV output.

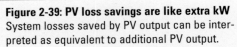

Source: Hoff, T. E., "Distributed Generation: An Alternative to Electric Utility Investments in System Capacity" (*Energy Policy* 24, no. 2, 1996), fig. 11

[59] PVUSA AC rating at 1000 W/m² irradiance, 20ºC ambient temperature, and 1 m/s windspeed. The PV installation was actually not *at* the 10.5-MVA substation as often described, but on a semi-rural 12-kV distribution feeder about 8 circuit-miles downstream. Where a place to mount the PVs is available, such flexibility in offsite location is both common and valuable.

Section 2.2.8 above considered one part of the capacity-credit calculation—whether the renewable source is likely to be operating when customers want its output. But that ELCC value actually mixes together two quite different effects—first, *when* the wind blows or the sun shines (compared to customers' peak-load times), and second, *whether*, at the time when that renewable energy flux exists, the equipment installed to capture it is ready to do so. The first effect expresses the "availability" of the weather needed for the renewable source to generate electricity, while the second expresses the availability of the source at those times. Unless distributed renewable resources like solar and windpower come with electric or (for thermal sources) thermal storage,[61] or with fueled backup firing,[62] traditional utility analysis does not consider their power "firm." To see the flaw in that reasoning, we must consider in turn how a renewable source's reliability depends on each of three factors: the renewable energy flow itself, how reliably the hardware works to capture that flow, and whether the energy is needed at that time. Since Section 2.2.8 already evaluated the statistics of the renewable energy flow, we add to that discussion only some broad observations.

2.2.9.1 Renewable energy intermittency

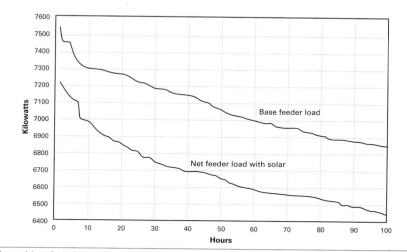

Figure 2-40: PV/peak-load match (Figures 2-31–32) remain valid at feeder level
On an Arizona feeder, PV support saves the same amount of load at ten- as at one-hour-peak load duration; the better the match at shorter load durations, the greater the value in avoiding capacity investments.

Source: Arizona Public Service Company, "Creating a Bright Future" (Distributed Utility—Is This the Future? EPRI, PG&E, and NREL conference; December 1992)

Renewable energy flows are less intermittent than one might suppose. In an illustrative Pacific Coast *array* of wind machines, a lull "which reduced power to about a third of the summer mean would last for [no more than] fifteen hours with ninety-five percent probability," and only 1% of the time would the windpower fall below one-sixth of the summer mean for as long as ten hours (372). In contrast, major outages in light-water reactors in the late 1970s were lasting for an average of about 300 hours at zero output.

Moreover, even a modest amount of storage can make many intermittent renewables a great deal firmer. Analyses in the 1970s showed that a mere ten hours' storage would make a typical *single* wind machine in Denmark as reliable as a typical light-water reactor of that period (663–4). Recently, a Great Plains utility-scale windpower array with the right number of turbines, supplemented by 20 hours of com-

[60] Performance Index was used in PG&E's Kerman studies to mean actual photovoltaic output relative to long-term expected output at a 27% capacity factor. It is a simple figure of merit for comparing different siting and orientation choices for PVs.

pressed-air energy storage (at a discharge rate of 150 MW per 225-peak-MW wind turbine), was simulated to raise the wind system's effective capacity factor from 34% to 93% at a delivered electricity cost premium of only about 15% (97). This simulation apparently did not fully account for the load-following and spinning-reserve benefits of storage, which are often quite valuable (227, 781). This extra value appears to merit more attention, although we don't count it here among distributed benefits because this book focuses on generation resources rather than on storage resources.

The fluctuations in renewable energy flows are also better understood and more predictable than those in the supply of conventional fuels and power. Year-to-year variations are quite small; the standard deviation for Danish windpower, for example, is about 9–10% (389). The methods used to forecast the weather in a few days (as used by hydroelectric power dispatchers and short-term load forecasters), or the movements of the sun in a century, are considerably more reliable than those used to try to predict such other critical drivers of energy policy as reactor accidents or Saudi politics: "One can have greater confidence that the sun will rise tomorrow than that someone will not blow up Ras Tanura [the main Saudi oil-loading port] tomorrow." (455) Clouds may persist for days or even weeks, but not for months—analogously to a complete cutoff of oil imports or of major pipeline systems.[63] (442)

2.2.9.2 Distributed resources' technical availability reduces reserve-margin requirements

Some distributed resources can clearly be highly reliable:

- End-use efficiency resources tend to

operate whenever their associated energy-using systems do (except for bad control systems that are bypassed or turned off). From the perspective of the electricity supplier, some end-use efficiency resources can even have an equivalent availability in excess of 100%. That is because, say, a failed efficient lighting ballast increases the "savings" to an even larger value, namely the ballast's entire consumption—albeit at the temporary expense of no light output. However, some other kinds of efficiency technologies could in principle be less than 100% available, if their failure causes higher rather than lower energy use.

- Phosphoric-acid fuel-cell generators have demonstrated 98+% uptime (508) with clean fuel and suitable care. UTC Fuel Cell (formerly IFC) has logged over 5 million unit-hours on several hundred units. As of 2 April 2002, the North American fleet had achieved unit uptime averaging 93.2% over the previous 30 days and 88.9% over the previous year in standby applications without special uptime attention. However, a premium service stocking spare parts and guaranteeing two-hour service achieved 99.4% and 96.8% respectively (779). Commercial proton-exchange-membrane fuel cells, which operate at a much lower temperature, promise even greater reliability when matured for volume production. The alkaline fuel cells used in aerospace are the most reliable dispatchable power source known. British submarines have run more than 13 million fuel-cell-hours with zero failures. Even a relatively complex molten-carbonate fuel cell a decade ago was expected to have a 98% technical and 97% equivalent availability, with a 1,385-h mean time between failures of 25-h mean duration (285). This promise has been borne out in practice. A commercial system that combines UTC phosphoric-acid units with a crossbus, switchgear, and switch-transient eliminator guarantees 99.99% to 99.9999% avail-

[61] Some interesting solar thermal electric generators use molten salt, hot rocks, or other ways of storing high-temperature heat to raise steam long after the sun has set, or even, if desired, round-the-clock.

[62] For example, the 355 MWe of LUZ parabolic-trough solar thermal electric generators installed in Southern California are allowed by FERC rules to provide up to one-fourth of their annual output by gas firing. This time- and weather-independent heat source makes them dispatchable.

Benefits

69 *In general, interruptions of renewable energy flows due to weather can be predicted earlier and with higher confidence than interruptions of fossil-fueled or nuclear energy flows due to malfunction or other mishap.*

70 *Such weather-related interruptions of renewable sources also generally last for a much shorter time than major failures of central thermal stations.*

ability at the customer's option (672).

- Most renewable energy *technologies* are highly reliable technically (*i.e.*, in working order and ready to send out energy when the renewable energy flux is present)—most of all, good photovoltaics. For example, in seven years' operation of the Gardner project mentioned in Section 2.2.8.1 (85), even using 1985–86 technology, none of the 332 1.59-m² modules failed in any way, while two of the 120 smaller (0.91-m²) modules experienced cracked glass, one possibly from vandalism. The electronics also proved highly reliable. With isolated exceptions, "Most of the systems have been operational every single day since installation...."

- Similarly, the 300-kW passive-tracker Austin demonstration PV plant was 99.8% available and 99.7% equivalently available (during the daytime) in 1990. It experienced only 18.7 hours' total downtime, distributed over 11 incidents—all under an hour except one 16-hour daytime interruption due to a grid fault, and all but one other planned. Planned maintenance person-hours, three-fourths of it simply meter-reading, accounted for 83% of all maintenance person-hours. Several other PV plants without self-resetting inverters and routine inspections experienced higher downtime, due mainly to manual resets of inverter trips, but could still be 97% available with no onsite staff (572). PG&E's Carissa Plains plant showed typical availabilities of 99.12% in 1987 and 99.44% in 1988 (609).

- Very high technical availability is also

observed in properly designed and maintained wind machines despite their mechanical stresses and moving parts.

Benefits

71 *Some distributed resources are the most reliable known sources of electricity, and in general, their technical availability is improving more and faster than that of centralized resources. (End-use efficiency resources are by definition 100% available—effectively, even more.)*

72 *Certain distributed generators' high technical availability is an inherent per-unit attribute—not achieved through the extra system costs of reserve margin, interconnection, dispersion, and unit and technological diversity required for less reliable central units to achieve the equivalent supply reliability.*

73 *In general, given reasonably reliable units, a large number of small units will have greater collective reliability than a small number of large units, thus favoring distributed resources.*

For example, the Danish utility ELSAM's 43-MW windfarm availability through 1991 averaged 97.6%, and since then, the best Danish makers have reported availabilities consistently over 98% (145); lost energy is well under 2% because maintenance is done only in low winds. The availability of the 4-MW Delabole windfarm in Cornwall, England was 97.9%, with outages due to lightning (48%), breakdown (37%), routine maintenance (13%), and grid failure (2%). (688) As for long-term durability, a 40-kW Enertech machine in Texas is reported to have run

[63] A celestial collision, huge volcanic eruption, or nuclear war could hide the sun for months or even years; but if that happened, energy supply would be the least of our worries: we'd first run out of food and perhaps of breathable air.

at ~97% availability for 15 years—equivalent to ~52,000 hours' operation, or nearly twice the engine life observed in the record 950,000-mile Mercedes car (23). And Bergey wind machines have even withstood tornadoes (61).

- Even in the extreme conditions of St. Paul Island (one of Alaska's Pribilof Islands in the Bering Sea), a 500-kW standalone wind/diesel cogeneration system (225 kW wind + 2×150 kW diesel) commissioned in June 1999 achieved combined availability of 99.88% and 99.93% in its first two years, even though teething problems held the wind turbine's initial availability to 83% and 70%. (499) This illustrates how combining just two technologies can yield extremely high overall availability.

For comparison, all U.S. fossil-fueled power stations of all sizes during 1989–93 (those in the GW range did worse) averaged only 85% available;[64] nuclear, 73%; gas-turbine, 90%; combined-cycle, 88%; and even hydropower, 91% (511).

Summarizing these and the previous illustrative data for well-designed systems shows the contemporaneous distributed resources' higher technical availability (Figure 2-41). In general, later improvements tended to favor the distributed resources more than central ones—except for U.S. nuclear plants, some of the least reliable of which have been abandoned; the long-term availability of the rest remains to be seen.

The 80s-of-percent availability of all conventional power stations' collective output is achieved only at the expense of unit and technological diversity, geographic dispersion, and costly reserve margin. These precautions are so common in normal utility practice that their cost and even their existence are often overlooked, as if distributed resources incurred such burdens but central stations didn't. Yet an *isolated* utility that can rely only on reserve margin, not interconnection, makes starkly clear the value of those other attributes. The Rural Electric Research program of the National Rural Electric Cooperatives Association found (507) that for a small, isolated rural utility to achieve the same 99.95% generation availability (4.4 hours' outage per year) from a 200-kW, 98%-available ONSI fuel cell would require reserve capacity consisting of *one* extra fuel-cell package, or approximately *four* 93%-available internal-combustion engines, or about *six* 89%-available gas turbines—or, of course, large numbers of smaller backup units, whose diversity would then reduce the required reserve margin, just as distributed resources routinely do.

In this situation, "the higher cost per kW of the fuel-cell power plant is not as significant as one might initially think"—if, of course, that level of generation availability is actually required and cannot be provided in other, cheaper ways, such as onsite backup for critical uses, or dispatchable load management.

Similarly, binomial probability distribution analysis shows that an isolated system seeking 100 kW of firm capacity from dispatchable units with an assumed 5% forced outage rate, to serve a constant load from homogeneous customers, can get that capacity from five 50-kW units, twenty-five 5.26-kW units, or one hundred 1.16-kW units. These three alternative plans have total capacities of 250 kW, 131.5 kW, and 116 kW respectively, so going to the smallest units reduces the required total capacity by 54% compared to the larger (50-kW) units (323).

Of course, the degree to which "a system composed of a large number of small plants is more reliable [than]...a system with a small number of large plants" depends also on how reliable the plants are. An empirically derived formula allows this to be taken into account too (323): the ratio of capacity of the generating system to the load is

$$\exp [A(\ln N)^B]$$

where

$A = 1.20 - 0.212 \ln D + [14.40 - 2.139 \ln D] \times$ (forced outage rate)

$B = -1.159 + 0.1024 \ln D + [0.1689 - 0.00512 \ln D] \ln$ (forced outage rate)

D = number of days when demand is expected to exceed capacity in a 10-year period, *i.e.*, the established loss-of-load probability target

N = number of generating units

From this formula, families of curves can be plotted showing the capacity savings from smaller units or more reliable units or both.

Another useful way to think about distributed generators is that they can flexibly achieve a wide range of combinations of low cost and high availability. Where extremely high-availability resources like fuel cells are not used, redundancy and technological diversity can be used in many combinations to tailor the cost/reliability result to the requirement "to a degree [that] the electric utility often cannot match." This often gives distributed resources a competitive edge where permissible cost and desired reliability are *both* either high or low (767).

2.2.9.3 Modular resources' reduced variance of availability further reduces reserve margin

So far we have considered the availability of individual distributed generators to be a point value, rather than having an uncertainty and a probability distribution. However, Hoff (317) points out that:

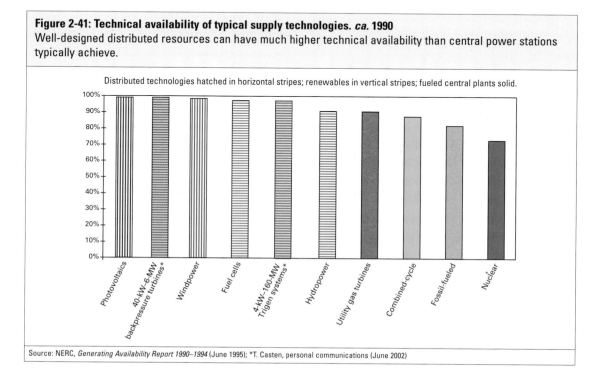

Figure 2-41: Technical availability of typical supply technologies. *ca.* **1990**
Well-designed distributed resources can have much higher technical availability than central power stations typically achieve.

Distributed technologies hatched in horizontal stripes; renewables in vertical stripes; fueled central plants solid.

Source: NERC, *Generating Availability Report 1990–1994* (June 1995); *T. Casten, personal communications (June 2002)

[64] Not corrected for annual or seasonal deratings, but those equivalent availabilities are even worse, by about three percentage points for the steam plants, five for gas turbines, and seven for combined-cycle.

Figure 2-42: Isolated systems' units must be very reliable if large
Even modestly less reliable units enormously increase required reserve margins to maintain the same reliability in isolated power systems.

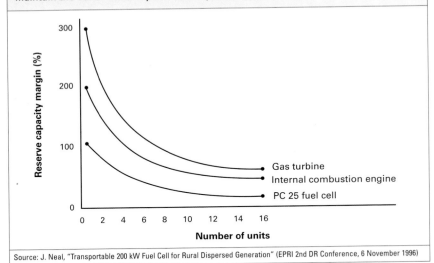

Source: J. Neal, "Transportable 200 kW Fuel Cell for Rural Dispersed Generation" (EPRI 2nd DR Conference, 6 November 1996)

...[M]odular plants have less variance in their equipment availability than non-modular plants when equipment plants in the modular plant are independently distributed. A non-modular plant can be considered to be either operating or not operating. If its forced outage rate is $(1-p)$, it has full availability with probability p and is unavailable with a probability of $(1-p)$. Modular plants, by contrast, can have partial [collective] availability. For example, a modular plant with two identical segments has three possibility levels of availability... the plant is 100 percent available if both segments are functional; it is 50 percent available if either the first or the second segment is functional...; and it is unavailable if both segments are non-functional.

This means that if the non-modular plant *and* the segments of the modular plants all have identical 10% forced outage rates, but the modular plant has ten segments, then the variance of availability for the non-modular plant is 9% (standard deviation is 30%), but the variance for the modular plant is less than 1% (standard deviation is 10%). Lower variance means higher confidence in using the availability figure to plan for a desired reliability of supply.

More generally, for a plant with n independent identical modules, the variance of availability equals $p(1-p)/n$, producing a graph of the shape in Figure 2-43 (the specific numbers shown assume a 15% forced outage rate):

This greater predictability of availability has an economic value that can be calculated from the total cost of the avoided reserve capacity. We shall revisit this idea in Section 2.3.1.1.

2.2.9.4 Outage durations and ease of repair

Most distributed resources, especially renewables (being free of the chemical and thermal stresses of combustion), tend not only to fail less than centralized plants, but also to be easier and faster to fix when they do break. Mending a broken wind turbine is more like fixing a car, or at worst more like re-masting a sailboat, than it is like fixing a major turbine or boiler failure in a thermal power station and providing costly replacement power in large blocks during lengthy repairs. Mending a broken PV panel is not very different than fixing a broken window. Mending a failed PV inverter is rather like servicing any other kind of electronics, usually requiring a plug-in replacement module or an adjustment of setting or software—not at all like fixing failed utility generators or switchgear. Parts are standardized and off-the-shelf, not unique and made-to-order. They arrive by courier pack, not by barge or railcar. And in the repair itself, quite aside from the obvious differences of scale and complexity, the distributed renewable resources are more physically accessible, needing at most a ladder or mobile lift. Other than normal precautions, they are also quick and safe to work with: little or no post-shutdown thermal cooling,[65] let alone radioactive decay, need be waited out before repairs can begin. Most importantly, while the failed individual module, tracker, inverter, or turbine is being fixed, all the rest in the array continue to operate.

Similar reparability advantages apply to

> **Benefit**
>
> **74**   *Modular distributed generators have not only a higher collective availability but also a narrower potential range of availability than large, nonmodular units, so there is less uncertainty in relying on their availability for planning purposes.*

modular microturbines, where a replacement unit the size of a large watermelon can be hot-swapped into a "ten-pack" mounting frame, and to modularly designed elements of a fuel-cell package. The turbine does use combustion, but cools quickly because it has little thermal mass. The fuel cell is an electrochemical device with no combustion, so its stack can be swapped with simple dis- or reconnections of the pipes and wires; moreover, if it's a proton-exchange-membrane model, it runs at only ~70–80°C, hence requires fewer and simpler procedures. And if a PEM fuel-cell stack is inadvertently poisoned with, say, carbon monoxide beyond the level that can be regenerated by a whiff of oxygen, the poisoned stack can simply be sent back to the factory, disassembled into its constituent layers, re-membraned, bolted or (for low-pressure designs) glued back together, and returned to service. This is far cry from trying to repair a large steam tur-

bine with a broken blade or a bent shaft.

2.2.9.5 Renewable capacity credit is real and valuable

For both meteorological and technological reasons, therefore, it is simply incorrect, as was fashionable in the 1970s and is still occasionally proposed, to suppose that renewables are only fuel-savers, merit no capacity credit, and require complete nonrenewable backup to ensure dispatchability. On the same argument, any nonrenewable power source with less than 100% availability would also merit only a partial capacity credit.

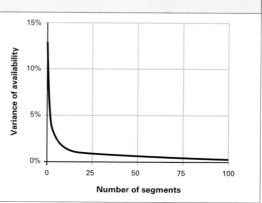

Figure 2-43: Multiple small modules dramatically reduce variance of collective availability
The shape of the curve is similar for individual-unit forced outage rates different than the 15% assumed here.

Source: T. E. Hoff, *Integrating Renewable Energy Technologies in the Electric Supply Industry: A Risk Management Approach* (NREL, March 1997). www.clean-power.com/research/riskmanagement/iret.pdf

Detailed analyses in ever greater detail over the past two decades have shown that variations in renewable energy flows even in a confined geographic area tend to affect different technologies in opposite ways (§ 2.2.10.6), and to cause shorter, more predictable interruptions, measured in smaller increments, than conventional systems'

Benefits

75   *Most distributed resources, especially renewables, tend not only to fail less than centralized plants, but also to be easier and faster to fix when they do fail.*

76   *Repairs of distributed resources tend to require less exotic skills, unique parts, special equipment, difficult access, and awkward delivery logistics than repairs of centralized resources.*

77   *Repairs of distributed resources do not require costly, hard-to-find large blocks of replacement power, nor require them for long periods.*

78   *When a failed individual module, tracker, inverter, or turbine is being fixed, all the rest in the array continue to operate.*

79   *Distributed generation resources are quick and safe to work with: no post-shutdown thermal cooling of a huge thermal mass, let alone radioactive decay, need be waited out before repairs can begin.*

80   *Many distributed resources operate at low or ambient temperatures, fundamentally increasing safety and simplicity of repair.*

---

forced outages, grid collapses, oil embargoes, strikes, or freeze-ups (152, 370, 373, 663–4, 682). It is on the contrary the centralized, *non*renewable systems that have a serious reliability and energy storage problem. For example, the late Astronomer Royal, Sir Martin Ryle, a distinguished Cambridge scientist, showed that a large-scale British system of wind generators would yield a substantially more reliable and less storage-intensive power supply than a grid similarly reliant on nuclear power stations, based on the empirical performance of both (577).

Often the key to such valuable results is to optimize the wind turbines' technical design—especially their low-speed cut-in, so that they run more hours even at a slight penalty in high-windspeed efficiency—and to put more turbines in a windfarm so as to produce more power during the more numerous hours when windspeed is below the turbines' top rated speed. This strategy can yield a Kansas windfarm with only slightly higher delivered electricity price at a combined capacity factor of 0.55–0.62 than of

0.36 (96)—or, as noted above, it can achieve around 0.95 at slightly higher prices using compressed-air energy storage onsite (97). Storage may be far more economically interesting in this application than for mere load-leveling in conventional utility operations.

In short, "The common views that intermittent renewable energy technologies only provide replacement energy to the utility systems but no capacity value and that they require significantly lower costs to be competitive in the utility industry are misleading. The studies have shown that intermittent renewable energy technologies can have capacity values. Renewable technologies are competitive today in many situations. The renewable technologies were at a disadvantage because their unique attributes...were not always considered by the conventional planning process." (723)

It is also important to note that just as utility restructuring permits contract and settlement paths to be unbundled from physical delivery of electrons, so it also permits stor-

[65] As is also required for high-temperature fuel cells, though for small modules this is little different than waiting for a car engine to cool before attempting repairs.

age, or the technological and geographical diversity options described next, to be unbundled from the intermittency of a particular renewable resource. A renewable energy aggregator (§ 1.2.12.3), for example, could separately contract for an intermittent renewable resource, a suite of complementary renewables of other kinds or in other places, perhaps some storage, and contractually bundle them together into a firm source. The combination of technological and physical diversity would presumably decrease the amount, if any, of storage required, improving the economics of the resource bundle. As such transactions become more convenient and commonplace, it will become increasingly fallacious to consider the intermittency of single renewable resources and sites in isolation.

2.2.9.6 Geographic dispersion and technological diversity

We have seen how many renewable energy flows, though intermittent, can still yield a substantial proportion of firm dispatchable power. That capacity credit can be increased by paying attention not only to details of technical design (§ 2.2.8.2 and § 2.2.10.1) but also to the energy resources' geographic dispersion and their technological diversity. These issues affect the dynamic stability of the grid and its ability to ensure a reliable fit between customer or system needs and renewable generation, so that fluctuations in wind or sun do not excessively burden thermal stations' load-following capabilities. We return in Section 2.2.10 to the implications of unusual short-term fluctuations in renewable energy flows for what fraction of the grid's supply can safely come from intermittent renewables. Here we focus instead on a

> **Benefit**
>
> **81** *A small amount of energy storage, or simple changes in design, can disproportionately increase the capacity credit due to intermittent renewable resources.*

related but separate question: how capacity credits can be increased by spreading more *kinds* of renewable generators over a larger physical *area*.

Over a diversified geographic scale, local weather-related fluctuations are quickly smeared out. Even on the ~300-km scale of the Netherlands, for example, adding 1 GW of windpower to the nation's ~9 GW of fossil-fueled generation would ensure that "hour-to-hour variations of total wind power output are never greater than 40% of installed [wind] capacity, while an hourly wind power output decrease of 30%–40% of installed [wind] capacity—equivalent to the loss of a single thermal unit of only 300–400 MW, equivalent to 3–4% of total generating capacity—might occur only four times in 10 years." (135) Similarly, a North German study (67) found (715) that

> ...the coherence in wind speed fluctuations was very low at high frequency from a large number of wind turbine sites dispersed over distances of some tens of kilometers. Wind farm power fluctuations with frequencies higher than 10^{-2} [Hz] were in fact leveled out. Thus...the output could be treated as smooth in time scales of several minutes. The low correlation of fluctuations in wind power over long distances, observed from these studies, gave hope that the negative impact of intermittent wind power on the operation of an electric power system might be less severe than previously thought. Halberg (283) simulated wind generation in the Dutch electric system using wind data recorded

at six sites spread across the coastal area of the Netherlands. He concluded that although wide variations in wind power could be expected to occur for longer periods (several hours), the frequency of severe variations...per 1,000 MW of wind capacity appeared to be comparable with the frequency of forced outages of large thermal generating units.

On a still larger scale, deploying the same kind, or more than one kind, of renewable resources interconnected on a scale of hundreds of km or more can often "jump over" particular weather systems (374). On this mesoscale, for example, the wind is virtually always blowing *someplace*. Such widely dispersed turbines are very unlikely to be all becalmed simultaneously, so the British integrated grid *could economically generate more than half of its total power requirements from wind turbines* dispersed over many attractive wind resource regions, and could therefore start to displace thermal baseload generation (279).

Weather variations can be huge locally and are random (more precisely, mathematically chaotic) in detail, but are statistically predictable in general pattern (664). Those variations are quite well understood (subject to potential shifts as the global climate experiment proceeds and weather volatility increases). A properly designed renewable energy system can therefore cope with variability by using the combination of sources, design parameters,[66] and demand-side resources best suited to each site and application (486). In general, and specifically with windpower (280), more refined recent understanding of temporal and spatial diversity, and of how intermittent renewables can be integrated into largely thermal power systems, has overturned more pessimistic early assessments (716).

Even more reliable output can be achieved by combining geographic dispersion with technological diversity, so conditions bad for one resource are good for another. For example, storms are bad for PV but good for wind and small hydro; calm sunny periods are bad for wind but good for PV. Indeed, a grid that combines such diverse and dispersed renewables can actually yield a *more* reliable supply than one using fossil- or nuclear-fueled central plants. This often surprises the plants' designers, but it shouldn't, because *all power supply systems are unreliable—just in different ways, for different reasons, for different durations, with different probability and predictability.* The need to design for fluctuating or intermittent output is nothing new. Today's power systems fluctuate too, because of forced outages, grid faults, and (far less predictable by statistical techniques) embargoes, strikes, sabotage, war, etc. Those kinds of fluctuations and interruptions, too, must be guarded against by design, and are, at great cost (442). We shall return to this theme in Sections 2.3.1.2 and 2.3.3.5 when discussing certain distributed resources' special ability to displace conventional spinning reserve.

Important system reliability benefits also come from distributed resources' small unit size, because less capacity can fail at a time and it is unlikely that many units will fail simultaneously. This separate effect, hinted at in Section 2.2.9.3, is discussed in Section 2.3.1.1 below, where it is credited for corresponding reductions in reserve margin and spinning reserve.

One other qualification of the geographic dispersion thesis is important. Dispersion may not be worthwhile if it means installing resources in sites where renewable energy fluxes are poor. For example,

PG&E found (644) that "the energy and capacity values of a single PV plant at the best insolation site within PG&E's service territory are greater than the sum of the energy and capacity values of many smaller plants scattered throughout the area." However, as Section 2.2.10.1 elaborates,

> ...there is a statistical benefit of multiple dispersed sites: since the instantaneous amounts of insolation at each site are not perfectly correlated, variations in the collective output of the plants become smoothed. Given these complications, the valuation of PV plants require[s] the simulation of hourly insolation and PV output at many sites over the period of a year (644).

While such "results suggest that the correlation of weather at sites within PG&E's service territory is strong enough to make the dispersion benefit comparatively small and the central site superior in terms of both energy and capacity values," (644) this may not be true elsewhere. And even for PG&E, we shall see in Section 2.3.2.6 that reasonable estimates of the best site's extra energy and capacity credits for a 50-MW central PV plant compared with 50 dispersed 1-MW sites—$23/kWy—are outweighed nearly 7:1 by the distributed plants' offsetting benefits, chiefly in the avoided costs and losses of the grid. A proper comparison of different degrees of dispersion, hence implying unit scale, therefore requires far more than a comparison of the renewable resource fluxes at dispersed *vs.* central sites.

2.2.9.7 Generating reliability and grid reliability

We have seen in the past few pages that the technical availability of many distributed resources is extremely high. Section 2.2.9.2 above cited 97–100% for distributed photovoltaics, wind turbines, and fuel cells. (The first two are subject also to correlations

between irradiance or windspeed and load-shape, and the fuel cells to any potential unavailability of fuel supplies, though, with suitable design, pipeline gas can be backed up by a cheap onsite bottled-gas reserve.) However, even the most reliable remote generator cannot deliver its reliable supply if "bottlenecked" by a less reliable grid.

In the United States, virtually all power interruptions are caused by the grid, and most of those grid failures are in distribution, not transmission (330).[67] Indeed, one source states that "the distribution system is responsible for 95% of all outages, power quality problems, and other drivers of customer [dis]satisfaction." (112)[68]

There is a lot to go wrong in far-flung networks of aerial wires. For example, for the extensive pre-2000 PG&E system in Northern California, embracing the full range of conditions from coastal to mountainous, desert to rainforest to marine to alpine, and urban to near-wilderness (93),

- the average customer experienced about 2.5 hours' outage per year (99.971% availability),

- at least 80% of customer outage hours originated in the distribution system,

- the ~100,000 miles of distribution lines experienced ~20,000 sustained outages per year, and

- the ~20,000 miles of transmission lines experienced ~100–200 sustained outages per year.

This difference between transmission and distribution outages is attributed to three causes: different distances between lines, and between lines and the ground; that transmission poles are steel, not wood; and that transmission lines tend to be in more remote areas

[66] For example, wind machines can easily be designed to cut in at lower windspeeds (§ 2.2.9.5). Even if this sacrifices a little efficiency at high windspeeds, it may be the economically better solution in certain wind regimes, and can considerably expand the zones in which windpower is competitive.

Benefits

82   *Distributed resources have an exceptionally high grid reliability value if they can be sited at or near the customer's premises, thus risking less "electron haul length" where supply could be interrupted.*

83   *Distributed resources tend to avoid the high voltages and currents and the complex delivery systems that are conducive to grid failures.*

less prone to interact with people, vehicles, etc. For PG&E's distribution system in 1995, the main causes of the *distribution* outages were 24% equipment failure, 15% trees, 12% animals, 5% car crashing into equipment, 9% weather, and 25% unknown external causes. Most distribution failures involved connectors and conductors, not fixed equipment like poles and transformers (93).

This might seem at first glance to suggest that

• more centralized resources increase the risk of unreliable ultimate supply because they rely on longer distances, higher voltages and currents, and greater grid complexity to reach the customer, while conversely,

• distributed resources have an exceptionally high grid reliability value if they can be sited at or near the customer's premises, thus risking less "electron haul length" where supply could be interrupted.

However, whether that is true depends on design details: for example, a distributed resource designed to turn off when the grid to which it is connected fails has thereby forfeited its potential "resilience benefit" for the customer. This is no longer necessary (§ 2.3.2.10.6), but remains a common design practice. Moreover, the more widely the resources are distributed, the more numerous their links if they're interconnected; so

exposure to disruption, though reduced in length, may be increased in number of links. Though a richer topology of links also provides more options for rerouting and backup supplies, the absolute number of outages among those more numerous links may also increase, although each would affect far fewer customers, and probably for a shorter time. For these and other reasons, reliability cannot be simply compared in safe generalizations between highly centralized and highly dispersed configurations.

2.2.9.8 Diversity, complexity, and resilience

As Section 1.2.9 mentioned, an elaborately developed and documented argument (442) shows that not only naturally caused[69] but also deliberate disruptions of supply can be made local, brief, and unlikely if electric power and other energy (and nonenergy) systems are carefully designed to be more efficient, diverse, dispersed, and renewable. Such design applies the principles that underpin the resilience of biological systems (378, 442), where "resilience" means not mere ability to keep working despite disruptions, but an active "learning" quality that adapts the system to become even more resilient next time (see Technical Note 2-3).

[67] In many developing countries, and even in some (chiefly rural) areas of the United States, both generation and the grid are far less reliable, so considerations and conclusions may be different.

[68] These two statements may be consistent, since the latter counts power-quality issues as well as outages, and PG&E may be more reliant than average on long, remote transmission lines such as the Pacific Intertie. EPRI's Hoffman (328) states that "approximately 90%...of customer outages in the United States stem from problems with distribution system equipment...."

The two book-length arguments (summarized in Technical Note 2-3) are far too detailed to treat further here. Suffice it to say that the *economic* benefits of and arguments for distributed resources are analogous to *structural* benefits and arguments more familiar to the designers of extremely reliable technical systems or of institutional arrangements to ensure mission-critical industrial or military security. Importantly, too, to the extent that potential disruptions of supply are maliciously caused, resilient design using distributed resources—the strategy of the diverse ecosystem, not the monoculture—will not only blunt those disruptions' effect; it will also thereby reduce the motivation to cause them in the first place, because the difficulty and risk will seem less worthwhile when the effect is so much smaller. We return to this theme later, in Section 2.4.10.1.

2.2.10 Permissible saturation of renewable generators

Section 2.2.8.1 referred to the successful Gardner operation of a normal utility distribution feeder with up to 53% of its supply coming from roof-mounted residential photovoltaics (85)—a level most engineers a decade earlier would have considered imprudent or impossible. Similarly, high local saturations of photovoltaics in SMUD's residential PV project are having no negative impacts and indeed seem to be improving rather than degrading system stability (529). The gratifying Gardner results were achieved using 1985 inverters, but today's SMUD installations use inverters many generations more evolved. Modern solid-state inverters have very reliable, fast, and sophisticated protective devices built-

in, are digitally controlled and hence flexibly programmable, and produce very clean waveforms. These attributes will be discussed further below in Section 2.3 in the context of interconnections and power quality. This experience and others like it confirm that "most interfacing issues are resolved or resolvable with state-of-the-art hardware and design," though often case-specific, and that the literature "does not reveal any unsolvable technical problems," so "In the near-term, it appears that there are no technical constraints that impede the integration of intermittent renewable technologies into...utility systems." (723)

A wider issue sometimes raised, however, is on the scale not of the individual feeder but of the whole area or regional utility system: namely, what degree of saturation could intermittent renewables, such as PV or wind, achieve before endangering system stability by increasing the load-following requirements on traditional turbogenerators? This question has often been answered with very small numbers—that practical engineering economic constraints would confine intermittent renewables to only perhaps 5%, or 10%, or at most 20%, of electrical capacity or of load (definitions vary).[71] In fact, such an artificial generic constraint is among the most widespread, durable, and most misunderstood canards about renewables, and should finally be laid to rest. It is *not* an engineering or economic requirement, but rather an artifact of unrealistic assumptions made about those requirements or about how they can be overcome by sensible adaptations of operating procedures or equipment without compromising reliable operation.

The following brief review of the history

[69] For example, major earthquakes or weather events.

Benefits

84   *Deliberate disruptions of supply can be made local, brief, and unlikely if electric systems are carefully designed to be more efficient, diverse, dispersed, and renewable.*

85   *By blunting the effect of deliberate disruptions, distributed resources reduce the motivation to cause such disruptions in the first place.*

and resolution of this issue, drawing heavily on an excellent survey by NREL (727), shows why such saturation concerns are generally unwarranted—for the same reasons already discussed in Sections 2.2.8.1 and 2.2.9.6 in the context of capacity credits for intermittent renewables. A central finding is that even where very high renewable penetration is undesirable for *economic* reasons, that threshold is likely to be far lower than any *technical* limit posed by reliability, stability, or other operating requirements. This is partly because distributed generation in a large, far-flung grid can help to change its basic transient-response dynamics from unstable to stable (154)—the more so as the distributed resources become smaller, more widespread, faster-responding, and smarter.

[70] Similar principles emerge in many other contexts. For example, lessons about how large, hierarchical organizations can avoid costly mistakes, such as Collingridge's penetrating analysis (§ 2.2), often emphasize (138) such ideas as doing minor trials with low cost of failure, making marginal changes, achieving trial results rapidly, focusing the energy of critical scrutiny proportionately to the cost of mistakes, involving many diverse stakeholders in decisionmaking and sharing power among them, and coordinating choices by mutual interaction rather than central planning. Collingridge summarizes: "Since error is unavoidable, it makes sense to make minor mistakes rather than major ones. Intelligent choosers will exploit the mistakes they inevitably make, learning from them as they go." Thus the emphasis is on trial-and-error learning—just as Kelly remarks that "Even the most brilliant act of human genius, in the final analysis, is an act of trial and error," and that biological "evolution can be thought of as systematic error management." (380)

Technical Note 2-3: Resilience

The basic system design principles for resilient energy (or other) systems are (445):

- *A resilient system is made of relatively small modules, dispersed in space, and each having a low cost of failure.*

- *Failed components can be detected and isolated early.*

- *Modules are richly interconnected so that failed nodes or links can be bypassed and heavy dependence on particular nodes or links is avoided.*

- *Links are as short as possible (consistent with the dispersion of the modules) so as to minimize their exposure to hazard.*

- *Numerically or functionally redundant modules can substitute for failed ones, and modules isolated by failed links can continue to work autonomously until reconnected.*

- *Components are diverse (to combat common-mode and common-cause failures), but compatible with each other and with varying working conditions.*

- *Components are organized in a hierarchy so that each successive level of function is little affected by failures or substitutions among components at lower levels.*

- *Buffer storage makes failures occur gradually rather than abruptly: components are coupled loosely in time, not tightly.*

- *Components are simple, understandable, maintainable, reproducible, capable of rapid evolution, and socially compatible.*

Or in summary: more efficient and renewably based energy systems—reliant on relatively fine-grained, richly interconnected, redundant, cooperative, loosely coupled modules that are diverse, have low failure costs, are easily repaired, fail gracefully, and are so organized that failures at one level have little effect on another—can make large-scale or long-term failures of supply impossible.

These principles[70] *are strikingly parallel to, though less inclusive than, those articulated in a richer biological context as Kevin Kelly's "The Nine Laws of God" (379)—the essential design principles observable in the results of some 3.8 billion years of evolution. We offer them only in Kelly's tantalizing summary form, to encourage readers to consult his insightful original:*

- *Distribute being*

- *Control from the bottom up*

- *Cultivate increasing returns*

- *Grow by chunking*

- *Maximize the fringes*

- *Honor your errors*

- *Pursue no optima; have multiple goals*

- *Seek persistent disequilibrium*

- *Change changes itself*

> **Benefit**
>
> **86** *Distributed generation in a large, far-flung grid may change its fundamental transient-response dynamics from unstable to stable— especially as the distributed resources become smaller, more widespread, faster-responding, and more intelligently controlled.*

2.2.10.1 Simulated penetration limits and available responses

A widely cited 1981 analysis (395) performed a highly pessimistic simulation of permissible renewable penetrations by *simultaneously* assuming the loss of the largest conventional unit, the maximum probable drop in renewable output, and the maximum probable ramp-up in system load, while not properly crediting the renewables for spatial or technological diversity. This analysis, deriving a practical intermittent-source limit of 5% of system peak demand, was widely quoted and adopted by those unfamiliar with its assumptions (or at least comfortable with its conclusions), creating a false impression that took over a decade to undo.[72] Those who accepted the conclusion at face value should have known better, since *every* source is intermittent for one reason or another (§ 2.2.9.6) and hence the simulation is really rather like conventional rules-of-thumb limiting the largest unit to 5% of system demand. However, more modern and realistic studies of potential intermittent-renewable penetration have yielded far more encouraging answers that are finally starting to get the attention they deserve.

As usual, there is a small grain of truth in the middle of the hairball. On a local scale, fluctuations in intermittent renewable sources can indeed be quite sudden. For example, during the summer peak season, passing clouds can cause PV generation to drop suddenly, especially if it is concentrated at a single site (9). For a particular shaded array, capacity value then vanishes. But this risk (and its control complications, since older regulators and tapchangers take on the order of a minute to respond to the change in system voltage) can be diversified—most simply and restrictively, through interconnection with other photovoltaic arrays that are not shaded at the same time.

Depending on weather patterns, those arrays could be just down the street or might have to be km or even tens of km away. For example, a squall line can cause total loss of PV generation over a 1,000-square-km area in 17.6 minutes (363). However, this depends on scale. The PV generation can be lost over 100,000 sq. km in 176 minutes, but over 10 sq. km in only 1.8 minutes (363). Thus the bigger the area, the slower the change in output and the more easily it can be handled just like any other changing load or supply. The same diversification also occurs in microcosm when scaling up from the single PV-equipped house to the entire feeder;[73] but the more fine-grained the scale, the more such fluctuations look like normal load fluctuations and hence "For feeder sized areas...can be easily regulated using standard voltage regulators." This means that "conventional feeder designs and voltage regulation techniques deal adequately with the photovoltaic induced load flow fluctuations." (648)

[71] On this basis, even an excellent five-National-Laboratory study in 1990 (662) provided the option of artificially constraining to 20% all renewables' potential long-term total contribution to national electricity supply— the sort of assumption that invites reductions in R&D budgets and otherwise fundamentally distorts policy.

Much better results are also obtained if clouds are simulated stochastically rather than deterministically, as is appropriate for normal, high-probability fluctuations. For example, fairweather cumulus clouds are simulated to cause a maximum one-minute PV output change of 16% over a 10-sq.-km area, but only about 3% for 1,000-sq.-km or larger areas. Thus while the local effect can be dramatic (worst at the level of a single house), that is no worse than normal demand-side fluctuations such as turning an electric HVAC system on or off. (At the single-PV-house level, voltage flicker from normal cloud variations "would not be perceived even if the excursions in generation were abrupt [infinite ramp rate] rather than gradual. For the full feeder circuit, one percent voltage fluctuations occurring over two to ten minute periods can be expected during extreme partly cloudy conditions." [648]) Thus with deterministic treatment, two largely coal-fired Arizona utilities could violate Area Control Error criteria[74]—an instantaneous indicator of supply/demand imbalance—under worst-case autumn squall-line simulations with PV output equivalent to 1.5–16.3% of peak demand. But much better performance would result under stochastic conditions even with 17–24% PV fractions, and ACE would then stay well within allowable limits, leading the authors (9) to conclude that random PV output was "not a serious problem for the power system." (713)

Similar issues of both up- and down-ramping windpower output arise from local gusts and from larger-scale rapid changes in windspeed, *e.g.*, from a squall line. For this reason, some analysts have suggested (581) that wind arrays susceptible to a single storm front be limited to about 5% of a utility's system generating capacity—but have also agreed that this first-order requirement could be relaxed with better wind forecasting and control strategies (582), or presumably with other mitigation options mentioned below for PV fluctuations.

These squall-line simulations, though important, represent exceptional conditions, which, like other potentially harmful conditions, are infrequent. Infrequent conditions can be handled just like other utility planning contingencies (major and multiple generating outages or grid faults, etc.): *i.e.*, "proper design and operational changes can be made to deal with such occurrences." (717) Thus the 1–16%-of-system-load PV stability limits found in the Arizona simulations (9) reflect established average ACE criteria, worst-case conditions, and the special circumstances of those two utilities, whose generators responded relatively slowly. But "With prudent generation dispatch and operating practice, a power system can generally accommodate PV generations up to 5% of its system load in its generation mix." (713) PG&E's Kerman team concurs (632), "[W]e can state generally that experiments to date have offered no indication of voltage regulation problems due to PV generation." Moreover, that nominal 5%-of-load limit is far from the actually achievable limit, because

[72] Similarly erroneous rules-of-thumb can come from other methodologies. For example (358), a common British practice for "deciding if a [distributed] generator may be connected is to require that the three-phase short-circuit level (fault level) at the point of connection is a minimum multiple of the embedded generator rating. Multiples as high as 20 or 25 have been required for wind turbines/wind farms in come countries, but again these simple approaches are very conservative. Large wind farms have been successfully operated on distribution networks with a ratio of fault level to rated capacity as low as 6 with no difficulties."

[73] In the Gardner system, "(1) measurements at a single two kilowatt photovoltaic system on a partly cloudy day produced ramp rates of about 200 watts per second for excursions of 1500 watts, (2) measurements for thirty systems spread over fifty acres produced ramp rates of 1470 watts per second for excursions of 33 kilowatts, (3) simulations of five hundred six-kilowatt-capacity photovoltaic systems (three megawatts) dispersed over an entire 5.4 mile diameter feeder area yielded ramp rates of 14 kilowatts per second for excursions of 1380 kilowatts." (649)

even large transient losses of renewable generating capacity, which appear to violate NERC-OC[75] operating guidelines, "may not be much different from [those]...induced from a large load change that frequently is caused by large fluctuating industrial loads" such as the trip or startup of a smelter or an electric steel minimill (717)—demand-side shocks that often reverberate throughout an entire regional grid. That point is important both for symmetrical policies and for realism, since the industrial transients are routinely handled by existing utility systems.

This analogy suggests an alternative approach that can in fact support stable generator operation (though perhaps also requiring more careful transmission operation to stabilize voltage swings) at much higher PV penetrations, via either or both of two independent changes. The first is simply what any prudent utility would do if a large fluctuating *load* such as an arc-furnace were joined to its system (713):

> "...[C]orrective measures such as assigning more generating units to regulating duty or installing fast-response combined-cycle generators are available. These measures are effective if carefully planned [and] ...show higher [intermittent renewables] penetration limits...."

PV system design and operation (713) also offer further options for relieving these supposed constraints by at least two methods:

- Disperse the PV generators from the assumed central-plant configuration; in a Kansas simulation (364), the same 1%-per-minute ramp rate would support *nearly five times* as much dispersed (over 10 sq. km) as centralized PV capacity without

causing unscheduled tieline flows. Moreover, this is advantageous anyhow: as a case study cited in Section 2.3.2.1 below suggests, dispersing the PV capacity may also increase its distributed benefits, unrelated to this issue, by severalfold.

- Failing that, at periods of peak demand, PV output could be curtailed, sacrificing its valuable output in order to reduce its potential fluctuation.

Of these, the latter appears costly—PV output is most valuable precisely at the system peak load—but the former appears profitable at all times. Moreover, system ramp rates could be advantageously increased (§2.3.3.5) by invertor-driven resources that provide other operational, economic, and reliability advantages. And unscheduled fluctuations in renewable output could be offset by dispatchable load management that instantly drops *or* adds such loads as water heaters.

Still another adaptive option suggested by NREL (717) would be for NERC-OC to "establish different operating criteria for large penetrations of intermittent" generators. Since Area Control Error "caused by intermittent generation may not be much different from the ACE induced from a large load change that frequently is caused by large fluctuating industrial loads," and this is a daily fact of life for utilities that welcome the opportunity to serve such loads, such intermittence should not be used to penalize renewables that may experience output fluctuations of comparable magnitude and frequency.

---

[74] ACE "measures a combination of frequency deviation and net tie-line power flow." Under NERC operating standards, it "must equal zero at least once and most not vary beyond a certain range during each 10-minute interval." (532) ACE is not easy to measure in the field, since the same conditions that cause intermittency in renewable sources (fast-moving thunderstorms or weather fronts) can also cause rapid changes in loads, and may also be obscured by external conditions such as sudden changes in power flows from or to a neighboring utility across a transmission tieline (532).

Operational problems at high PV penetrations were also studied in the Jewell study cited above with reference to large-area PV integration. For an Oklahoma utility, 15%-of-load penetration was found to reverse some flows on subtransmission lines under certain fluctuating cloud and load conditions. However, while this reversal *could* "cause operational problems to an electric utility's protective equipment" not designed for reverse flows, it is routinely accepted and welcomed by the Massachusetts utility's Gardner-experiment distribution feeder actually operated without difficulty at 53%-of-load PV penetration (85), where protective equipment has been designed to accommodate this condition. Moreover, 30%-of-load Oklahoma PV penetration could reverse flow in transmission lines, but that "is not a problem for most utilities because transmission lines are designed to transfer power from either direction." (714)

A strong hint that distribution-level load-flows are not an issue even with high PV penetrations, assuming that protective equipment is designed to permit bidirectional distribution flows, comes from the Gardner study by New England Power Service Company (648). It found that at the feeder level where voltage fluctuations would typically be of greatest potential concern, with careful but ordinary system operation,

> Voltage profiles with the projected three megawatts of [PV]...capacity on a ten megawatt feeder circuit are only slightly changed. Rather than posing a problem, photovoltaic generation reduces the voltage drop during periods of heavy loading by reducing the net circuit loading. Even with...three times the projected 3 MW photovoltaic generation, the voltage remains within standard normal bounds unless the feeder circuit is lightly loaded and VAR compensating capacitors are connected. ...The extension [to simulated long-run

high PV penetration] of the findings of this [experimental] study suggest no limits on penetration exist, from a transient response standpoint, if the [inverter]...controls are equal to those of the [1985 inverters] used in the Gardner experiment.

In practice, no instability issues arose in the Gardner Project, nor in the very dense 1.25-MW mainly residential photovoltaic installation in Amersfoort (Netherlands), nor in the Sydney Olympic Village, Newington, Australia, which meets the needs of 665 homes with a 1-kW PV system integrated into each roof. (The only technical issue in Newington was interference between the anti-islanding circuits of inverters connected to the same point.) Other dense residential PV projects are installed or being expanded without instability problems in Nieuw-Sloten (Amsterdam), San Diego, Sacramento, Japan, and elsewhere (74).

Another study for a Virginia location typical of the southeastern U.S. found that system cost would fall as central-station PV penetration rose to 13.3% of total system *capacity* (hence a large fraction of system *load*, depending on the reserve margin), and would then rise again as load-following and spinning-reserve costs dominated—but without using the mitigations just suggested (126). This and similar studies consider only fuel cost to conventional generators, and do not count higher O&M costs for those generators if they take on increased regulating duties. However, the studies also do not count PV economics—neither capital costs *nor* distributed benefits. Until all these factors are taken into account, it's premature to assume that higher PV penetration at any particular level will increase system costs. On the contrary, the distributed benefits described elsewhere in this book are so numerous, diverse, and

[75] The Operating Committee of the North American Electric Reliability Council, the governing body of the industry on that continent. It is based in Princeton, New Jersey, and operates through distinct but coordinated regional power pools. NERC-OC sets minimum guidelines for both daily operation and long-range planning.

often significant that high PV penetrations, using the measures required to ensure stable and reliable electric system operation under fluctuating insolation, may well *reduce* total net system cost.

Results of many studies of windpower penetration were similar. Restrictive assumptions—such as no fast-response generators, isolated grids, little or no geographic diversity, inflexible operating procedures designed for large thermal plants, and no distributed benefits—typically suggested that windpower capacity be limited to about 5–15% of system load depending on local circumstances (728). However, as with PVs, these restrictive assumptions are neither necessary nor optimal—as can be inferred from the successful, routine, and economical operation of windpower to produce about 8% of some California utilities' entire offpeak electrical supply in the early 1990s (723), 18% of Denmark's in 2002,[76] and reportedly upwards of 20% in some regions of windpower-intensive countries such as Denmark, northern Germany, and northern Spain. In fact, at some times and places, those regions can produce windpower exceeding 100% of local loads, apparently without difficulty.

Studies partly described in Section 2.2.8.1 above found that appropriate mitigation strategies can yield much higher permissible wind penetration, such as 20–25% of total annual generation in Britain (more with more realistic wind forecasting, as is common practice in Denmark today—akin to the Scandinavian practice of using weather forecasts to predict heat loads on district-heating cogeneration systems). (11, 685) More than 50% also works if spatial diversity is properly counted and if nonrenewable units are allowed to cycle (but nuclear units not to cycle to below 40% of rated output)—

albeit without fully detailed simulation of short-term system operability). (279)

Mindful of these potentials and building on its sophisticated experience of the world's highest windpower fraction, Denmark in the late 1990s was officially projecting windpower to provide by 2005 nearly all of the country's minimum demand and half of its maximum demand, and by 2015, more than its minimum demand and about two-thirds of its maximum demand—with nearly all the rest to come from distributed cogeneration (357). Indeed, the entire European Union officially expects to get 22% of its electricity from renewable sources, including hydroelectricity, by 2010—nearly the current Danish level (27% expected in 2003) or twice the current U.S. level—with no expectation of grid instability or other technical problems.

2.2.10.2 A temporary issue?

Thus the still-widespread assumption that renewable sources' intermittence seriously limits their potential contribution is not technically valid. Even in 1993, this was already clear to leading researchers (716):

> Later studies seem to point to higher allowable penetrations [of windpower] than the earlier reports. This can be attributed to better knowledge of wind speed and its spatial and temporal correlations.[77] Wind data collected worldwide over the past few years indicate that aggregate wind power output from a wind farm is less variable than previously thought.[78] Some pessimistic assumptions of wind behavior, which result in projections of low wind penetration levels, have been shown to be unrealistic. Grubb studied the problem (280) and tried to explain the wide difference in results obtained by different researchers. He concludes that the difference in study assumptions can adequately explain the different results. Changing just a few basic parameters, par-

ticularly those relating to system operating reserve allocation, limits of thermal units['] partial loading, and wind diversity and predictability, can have a dramatic effect on the computed value of the wind energy output. Simplifying assumptions on these factors often lead...to substantially overestimating operating penalties of wind generation at higher penetration levels.

Similar conclusions apply to photovoltaics. Dispersing PV systems over an area of 1,000 sq. km or more, for example, could raise PV penetration to 36% of system load, not 5–10% (717), and presumably to even more with better mitigation, such as fast-response fuel cells. "The studies are strongly influenced by assumptions made on resource intermittency and [utility] system modeling. More recent studies usually suggest higher penetration limits than earlier studies." (717) And for all kinds of distributed intermittent generation, "Operational experiences and several recent studies with factual weather data indicate that hour-to-hour variations of...output are much less than early studies suggested."

In any event, intermittent renewables will actually be deployed gradually, during marked improvements in both understanding and hardware elsewhere on the grid. EPRI's Flexible AC Transmission System, for example, and analogous distribution automation—in many ways the best friend of distributed resources—are meanwhile likely to go from concept to installed reality.[79] "A complete microcomputer-based protection scheme could be *integrated into future distribution automation systems* [at little

or no extra cost and with many other benefits]. The redesigned...systems *should not pose any penetration limits* on intermittent generators, except for the capacity of the lines." (709) "Concentrating on finding feasible penetration levels with today's knowledge and system structure"—which will no longer exist by then—therefore "may not be a worthy research topic in the near future. Instead, the effort should be directed toward finding feasible technical solutions to facilitate the integration of intermittent renewable energy technologies." (718)

As that decades-long coevolution of supply, grid, and end-uses unfolds, at some point the unspoken question will need to be asked: why are we assuming a utility system where central steam plants operate perpetually by divine right, and anything not perfectly matched to their requirements is penalized by the notional costs of adapting the steam plants—instead of allowing *all* available, feasible, and cost-effective options to enter the grid in fair competition, ascribing to each the costs and benefits found from disinterested comparisons? Will we even reach the point where the grid is dominated by distributed, often intermittent, but highly diversified renewables, and it is the central steam plants instead that must justify their existence because of the high cost of their incompatibility with the next-generation technologies?

And there is one more important wild-card: local electricity storage. If, as many analysts expect, superflywheels or ultracapacitors,

---

[76] Normalized to an average wind year; 21% is expected in 2003 (146). Interestingly, Denmark's extensive use of windpower has developed within the context of a vertically integrated utility system (360). In early 2002, however, a hostile new Energy Minister cancelled a further 450 MW of offshore windfarms due online in 2004–08, on the peculiar basis that Danish renewables in 2003 should produce 9.2 TWh, 35% above the 6.8-TWh target.

[77] "Analyses of actual wind-speed data have concluded that there is a high degree of spatial diversity in wind resources. Some early assumptions on wind-speed distribution and spatial correlation appear too simplistic and pessimistic. Exploiting spatial diversity of the wind resource may result in a higher allowable penetration limit." (716)

[78] "Wind speed can change rapidly, but these changes are found to be bounded and can be represented statistically. Power output from a wind farm actually fluctuates less than previously assumed; therefore, the electric system should be able to integrate more wind power into the system." (717)

with smart controls responsive to real-time price of energy or stability, and dispatchable on remote command, enter the market at affordable prices (341), than all bets are off. If that occurred, or if (as seems even more likely) reversible fuel cells with minimal efficiency and cost penalties for their reversibility enter the market at a variety of unit sizes, then intermittent renewables associated with or able to call upon or interact with such storage could well become *more* firmly dispatchable than central power stations. They could then provide valuable distributed benefits with no capacity discounts or operational penalties for intermittence. This is not merely possible; it is quite likely. And real-time pricing in a more competitive environment makes it a good bet over a period much shorter than the likely deployment of renewables on a collective scale that could approach the penetration limits discussed in the previously cited studies. Thus once again, the generals may be refighting the previous war. By the time we have PV and wind capacity widely deployed on a large scale, its intermittence may be just an historical footnote.

2.2.11 Buying time

Before we conclude this discussion of how distributed resources can minimize regret in system planning, and move on to how they can reduce the costs of system construction and operation, one more point deserves mention. It is related to and consistent with, but different than, the philosophy behind option and decision theory, and is best illustrated by a story.

Around 1984, Royal Dutch/Shell's engi-

neers were designing hardware to bring ashore the oil from Kittiwake, a deepwater North Sea field, intending to sell it for $20 per barrel. But Group Planning had an unpleasant surprise for them: by the time the oil landed in 1986, the oil price was likely to have crashed, so Kittiwake oil could only be sold for $12. It would not be possible to lose money on every barrel and make it up on volume. Either figure out how to cut costs another 40%, the engineers were told, or find another job and the oil would stay where it is.

This shock—the engineers were by then sweating out the last percent of their cost budgets—soon turned into a challenge. They met it in just over a year by designing completely new technology. Why hadn't they done so earlier? Because, it turned out, they'd been asked how to bring the oil ashore as quickly as possible, no matter what it cost, rather than bringing it in as cheaply as possible even if that took a little longer. From this new question followed many new answers.

From those new answers, in turn, followed more new questions: Where else had we been asking the wrong question and getting the wrong answer? (All over the place, as it turned out.) And if the new technology could turn $20 oil into $12 oil, couldn't it also turn $30 oil into roughly $18 oil? (Yes.) And if so, wouldn't the whole supply curve of oil be pushed down toward the lower right, substantially postponing economic depletion? (Yes.) And wouldn't we then have more *time*—in which we could develop and deploy still better techniques, on both the supply side and the demand side, for postponing depletion still further, thus

---

[79] Some analysts muse whether it is worthwhile, or possible, to buy both of these options since both are in a sense different solutions to the same problem. However, they are also both worthwhile for different reasons, and just happen to work especially well together, so it isn't clear that they should seriously compete for resources.

buying still more time, and so on?

Yes, indeed. That is the big lesson of this story—a lesson all but ignored in the toolkits used by most energy and utility policymakers. The most precious thing we can buy is *time*. The highest leverage comes from wisely reinvesting that time to buy still more time.

As option theory teaches, it is worth paying for *time in which to learn more* about how to do better. With the passage of time, as the future unfolds into the present, many problems will solve themselves, others will emerge from the shadows, and we will gain much better information about which are important and how best to address them. We will also gain much better and cheaper technologies: a few years can turn a laboratory experiment into a commercially available product.

These things have long been true, but never more than now, when the pace of technological advance, social change, and industry restructuring seems to be rapidly accelerating. In such turbulent times, the ability of modular, short-lead-time technologies to temporize—to do the job while we buy more time—gains a special strategic value whose fundamental importance not even the most elaborate financial-economic theories can properly capture.

Benefit

87 Modular, short-lead-time technologies valuably temporize: they buy time, in a self-reinforcing fashion, to develop and deploy better technologies, learn more, avoid more decisions, and make better decisions. The faster the technological and institutional change, and the greater the turbulence, the more valuable this time-buying ability becomes. The more the bought time is used to do things that buy still more time, the greater the leverage in avoided regret.

2.3 CONSTRUCTION AND OPERATION

Distributed resources can directly displace the construction of new power-system assets, *and* can advantageously change the operating patterns of existing assets. We discuss these two benefits together because they are so closely related to each other, as well as to the system planning issues just discussed in Section 2.2. For example, operational improvements that help grid equipment to last longer will also reduce the need to build replacement equipment and the associated planning and financial risks. Similarly, avoided grid losses are primarily an operational improvement, but also reduce the need for capacity. Our somewhat artifi-

cial division of distributed benefits into the categories of planning, construction, and operation is thus for taxonomic convenience only. Regardless of taxonomy, all types of benefits should be considered as an integrated and interactive whole; otherwise important synergies between them may be lost.

2.3.1 Generation

We begin with generation on the assumption that most readers think, traditionally, in the direction in which the electrons flow. However, although this will serve for narrative purposes, it is conceptually more help-ful to *think* from downstream to upstream—in the direction in which market demands are expressed, savings from avoided losses compound, and money flows.

This is the conceptual revolution of Local Integrated Resource Planning (§ 1.4.1): it starts with the service the customer wants, then asks how much electricity (and mix of other inputs) is needed to do that task in the best and cheapest way, then assesses distri-bution needs for that electricity, then trans-mission, and finally generation. Generally, LIRP experience shows that end-use effi-ciency, load management, and better wires management turn out to be enough to do the job at least cost, and that if generation is needed, it will typically be local, not central. The bigger the savings downstream, the more the avoided conversion and wire loss-es compound back upstream into savings of capacity and energy. This is the new para-digm of modern utilities and a key to their market success. However, for the conven-

ience of readers more comfortable with tra-ditional top-down, generation-based plan-ning methods, we temporarily adopt the generation-centric mental model, start there, and compare different generating options by unit scale.

2.3.1.1 Reserve margin

The amount of generating capacity needed to meet load was shown in Section 2.2 to be a sensitive function of how well other invest-ments, such as end-use efficiency and bad management, are allowed to compete and of how sophisticated a planning strategy is used to minimize regret. However, whatever the increment of generation that may ulti-mately be needed, it exacts a toll—a "sur-charge" equivalent to an extra ~10–30% of capacity—for the reserve margin whose role is described in Sections 1.2.2 and 2.2.9. The size of that toll, however, is not fixed, but depends on the size of the generating units relative to the size of the grid they supply.

Benefits

88 *Smaller units, which are often distributed, tend to have a lower forced outage rate and a higher equivalent availability factor than larger units, thus decreasing reserve margin and spinning reserve requirements.*

89 *Multiple small units are far less likely to fail simultaneously than a single large unit.*

90 *The consequences of failure are far smaller for a small than for a large unit.*

91 *Smaller generating units have fewer and generally briefer scheduled or forced maintenance intervals, further reducing reserve requirements.*

92 *Distributed generators tend to have less extreme technical conditions (temperature, pressure, chemistry, etc.) than giant plants, so they tend not to incur the inherent reliability problems of more exotic materials pushed closer to their limits—thus increasing availability.*

93 *Smaller units tend to require less stringent technical reliability performance (e.g., failures per meter of boiler tubing per year) than very large units in order to achieve the same reliability (in this instance, because each small unit has fewer meters of boiler tubing)—thus again increasing unit availability and reducing reserves.*

Figures 2-42 and 2-43 in Section 2.2.9.2 showed how the amount of reserve margin required could vary by severalfold, depending both on how many units there are (*i.e.,* the scale of each unit relative to the whole system) and on how reliable each unit is. The costliest units to back up are the biggest ones, both because they need a higher percentage of system capacity to be set aside as their reserve and because reserves are bought by the kilowatt. In short, when a big generating unit dies, it's like having an elephant die in your living room. You need a second elephant, equally big, to haul the carcass away. Those standby elephants are expensive and eat a lot.

Power-system reserve margin requirements, then, rise with large unit size: "Larger units impose a more substantial burden of reserve capacity on the system." (252) How much so? The canonical formulation was for decades, and still qualitatively remains, that of the following 1958 graph, typical of U.S. utility systems and unit reliabilities of that era (266):

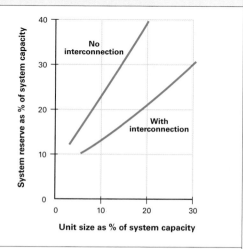

Figure 2-44: Late-1950s view of reserve margin *vs.* unit size
Even in interconnected systems, reserve margin rises steeply with big, lumpy units.

Source: A. Ford and T. Flaim, "An Economic and Environmental Analysis of Large and Small Electric Power Stations in the Rocky Mountain West" (Los Alamos National Laboratory, October 1979), p. 29, fig. 4–3

Larger units "require a larger reserve margin quite independent of differences in performance of large and small units" as measured by forced outage ratio (256). Simply because they are big, they need more reserve to make up for their potential loss. Unless that loss is so unlikely that the unit's forced outage probability is comparable to the desired system loss-of-load probability, the potential loss of the unit will degrade system reliability, so backup is absolutely required, and the only question is how much. You can't count on substituting a mule-team for a potentially dead elephant, no matter how healthy the elephant seems. If you have no spare elephants, then buying an elephant means you also need to buy a sufficient number of ox-teams so you can be confident of having an elephant's worth of hauling power in reserve. But if you had, say, mules instead of an elephant in the first place, then it would be extremely unlikely that a whole elephant's worth of mules would fail at the same time, and each sick mule could be routinely replaced by another. That is, with enough good mules, you needn't maintain many spare mules—just as with small generating units, at the left-hand edge of Figure 2-44, the required system reserve margins become very small.

The size and cost of electrical generating reserves can be estimated by an analogous process. In a given utility system, adding smaller units can provide the same amount of reliably dispatchable supply as adding a *greater* capacity of large units, even if all clustered at the same site. (Details then depend on how many, large, and reliable the system's generating units are and how reliably and diversely they are interconnected.) This is because:

- each small unit tends to have a lower forced outage rate and a higher equivalent availability factor than each large unit, and

- because there are many small units, it is far less likely that one large unit's worth of them will simultaneously fail than that one large unit by itself will fail.

The former point is illustrated by Figures 1-9 and 1-10 (§ 1.2.3) and Figure 2-41 (§ 2.2.9.2).[80] Its effect in combination with the second point has been both assessed in theory (369, 383) and illustrated by detailed analysis of specific cases.

Of these case studies, perhaps the clearest is a noted comparison (250) of two ways to add a large coal-fired station to a hypothetical 7.6-GW Western U.S. power system which already had a small (1%-of-system, or 76-MW) average unit size. A new 250-MW coal unit, the Los Alamos team found, would have a 5.7% forced outage rate, compared with 13.8% for a 750-MW unit burning the same coal at the same site. Because of this crucial 59% lower forced outage rate, a new station comprising nine 250-MW units could provide the same Effective Load-Carrying Capacity (ELCC, § 2.2.8.1)[81] as four 750-MW *units—a decrease in station capacity from 3,000 to 2,250 MW, or 25%,* to do the same task just as reliably.[82]

Deeper analyses then revealed that when the diseconomy of large unit scale was counted, the balance between economies and diseconomies of unit scale tipped the *overall* economic advantage to the smaller units:

- the smaller-unit plant would have a 15% higher capital cost per kW—assuming economies of scale four times as strong as were actually observed in the United States—but 25% fewer kW, and 44% shorter construction time (5 rather than 9 years, with the first unit coming online at the end of year 4 rather than of year 5), so its present-valued generating capital cost would be 17% lower for the same firm output capacity;

- the smaller units would burn 12% more coal per kWh (11,500 rather than 10,000 BTU/kWh) and pay slightly more to transport coal and electricity, but would achieve a higher capacity factor (68% *vs.* 57%) and incur 3% less O&M cost, hence 1% lower total operating cost;

- the smaller-unit plant would therefore provide 6% cheaper electricity; and

- the smaller-unit plant would enjoy important additional advantages because of its shorter permitting time, greater financial and perhaps regulatory flexibility, shorter lead time, reduced forecasting risk, better load-matching, and reduced financial stress described in earlier sections (§§ 2.2.2–2.2.10).

---

[80] Figure 1-10 is the more useful, not only because it is up-to-date through 1993, but also because it separates unit-size effects from unit-maturation (or -senescence) effects. The distinction between maturation of a given unit's performance and of the design going into units of that type and size is discussed by Ford & Flaim (257); it was important during the 1970s, when bigger units had immature designs while smaller, older units were outliving their most reliable years. Without graphing by unit age, as Figure 1-10 in Section 1.2.3 does, it is impossible to distinguish between trends related to unit performance at different sizes and to trends due to the diverging age structure caused by shifts in investment trends.

[81] A closed-form analytic solution for ELCC is provided by Ford & Flaim (255) for adding to a grid (with a known number and size of units) a new unit with known size and forced outage rate. In principle, a simulation that took account of smaller units' potentially shorter downtime—not just their smaller capacity loss per unit outage and their lower forced outage rate per unit—could yield even more favorable results. So might explicit allowance for any forms of technological, fuel, or other diversity that might mitigate the forced outage rate—*e.g.*, if conditions likely to force an outage at one unit were actually favorable to another.

[82] As Ford & Flaim note (253), "A word of caution is in order. A direct comparison of the energy output (kWh) of the two plans [*i.e.*, a large- or small-unit plan for the station] by finding the product of rated capacity, capacity factor, and period hours is not appropriate. The correct intepretation is that the expected energy output of the rated capacities is equivalent. The equivalent ELCCs of the two plans reflect this expectation."

This example is striking enough. But how could its logic extend to units far smaller than the still relatively large[83] 250-MW units—say, three to five orders of magnitude smaller? At such small scale, there could be qualitative as well as quantitative differences. But as vernacular units at kilowatt and sub-kilowatt scale proliferate, their comparison with five- or six-order-of-magnitude larger steam plants will become ever more important to understand. And a key question will be whether small renewables can beat big *non*renewables, where not only scale but also all other attributes differ profoundly.

It may also become relevant that such distributed resources as fuel cells, photovoltaics, and end-use efficiency can have very small units, extremely high availability (§ 2.2.9.2), and a further availability advantage—few and generally brief scheduled or forced maintenance intervals. Reserve margin is meant to cope with *all* sources of uncertainty in the supply/demand balance—severe weather, unusual customer activities, plant outages, transmission faults,[84] scheduled maintenance, or whatever. A reduction in scheduled maintenance requirements and in the duration of forced outages should therefore, in principle, contribute to some marginal reduction in reserve requirements.

Benefit

94   *"Virtual spinning reserve" provided by distributed resources can replace traditional central-station spinning reserve at far lower cost.*

2.3.1.2 Spinning reserve

Spinning reserve, as explained in Tutorial 1, is a special subset of operating reserve, which is in turn the quickly available portion of reserve margin. Spinning reserve is the generating capacity kept synchronously spinning under load, ready to take up the slack instantly if a major generator or transmission link fails or if a massive new load, like an electric steel mini-mill, suddenly comes onto the grid.[85] NERC-OC guidelines (§ 2.2.10.1) require each region or subregion to maintain operating reserve, at least half of it spinning reserve, large enough to provide a normal regulating margin and to cover the most severe single contingency—normally the sudden loss of the largest generating unit. In our earlier metaphor, spinning reserve is thus like a full-sized spare elephant that is not just lying there asleep but is standing by, alert to instructions and poised for immediate service. As that analogy would suggest, such an elephant costs the same to buy as an elephant kept asleep, but it also eats more. That is, spinning reserve costs the same in capital but also uses slightly more fuel (§ 2.3.3.3).

---

[83] A 500-MW plant containing two such units would typically have a 15-story generator building, two 200-foot-long (61-m) cooling-tower blocks, and two 500-foot-high (152-m) smokestacks (about one-third the height of the Empire State Building). It would also require about 500 acres (208 ha) of land (254).

[84] A fault is the interruption of function in a powerline or other electrical device. Common causes for line faults include lightning strike, vehicle/pole collisions, downing of a line by trees or wind, etc., and technical malfunction.

[85] As discussed below in Section 2.3.2.10.3, spinning reserve is sometimes thought to be required for reversible and rapidly unloadable generation in case a large block of renewable generation suddenly stops working, *e.g.* because of squall-line cloud cast on a central PV plant (§ 2.2.10.1). However, this is not a correct perception, and would be an unfair burden to impute to an intermittent renewable source, for two reasons: (1) rapidly variable customer loads may already incur the same requirement, and (2) renewable dropouts are typically predictable at least ten minutes in advance (often much more) from weather observations, permitting orderly scheduling (702). This is not a new idea or requirement. Utilities already forecast hourly loads, and even transmission-line capabilities, by carefully watching ambient temperature, sun, and other weather conditions (§ 2.3.2.4).

How much is spinning reserve worth? A rare calculation is provided by an analysis (212) of a Los Angeles 2-MW molten-carbonate fuel-cell proposal, which found a benefit of $0.0011 (1991 $) per kWh of fuel-cell output. This was calculated as the investment plus operating costs of a combustion turbine that would otherwise have provided spinning reserve equivalent to that of an unloaded but synchronized 0.25 net MW of surplus fuel-cell capacity on which the system could firmly rely—"stretch" capability amounting in this case to one-eighth of expected normal output.[86] That value is equivalent to a present value of $192,543 over the plant's assumed 30-year life at the assumed 65% capacity factor. For each of the 250 kW of *spinning reserve capacity* provided, that's equivalent to $770/kW (present-valued 1991 $).[87] Note that the capital cost of reserve margin can be counted as spinning reserve, as it was in the Los Angeles study, *or* as other operational reserve, but not both: it should be counted once and only once.

Spinning reserve happens to be traditionally provided by synchronized rotating machines because they are what dominate the present generating system. But this is not necessary and may not be optimal. Spinning reserve's *function* can be provided instead by inertialess, electronically con-trolled, hence instantly-responding resources ("virtual spinning reserve"), whether supply-side (§ 2.3.3.5) or demand-side (§ 1.2.11). "Is it possible," ask two PG&E analysts (276), "managing all system resources, to remove the need for spinning reserve" in its literal angular-momentum sense?[88] Technically, there seems no reason why not; economically, the question is empirical, and first indications are that the demand-side methods, at least, can yield a ~30% lower cost (§ 1.2.11). The *value* of virtual spinning reserve should be comparable to that assessed for the Los Angeles fuel cell, as long as the marginal spinning-reserve resource being displaced is a conventional rotating machine. However, if cheaper resources such as fast load management become the recognized marginal resource, their lower *cost* may redefine the proxy. There is no engineering or economic principle—only century-old tradition—that requires the functionality traditionally provided by the spinning reserve of a standby electric generating machine to be provided in that way if it can be provided more cheaply and just as reliably by, say, a radio signal that instantaneously turns off thousands of electric water heaters. Such distributed demand-side resources are already used, *e.g.*, in New Zealand, to provide grid stability on timescales as short as six sec-

---

[86] This capacity would be available because the normal operating point (1.77 MW nameplate capacity) is less than the 2.0-MW nominal capacity and the 2.25-MW maximum capacity. Such elastic output is not unusual for fuel cells, depending on thermal and other conditions; in this case the 2.0-MW output could be sustained for substantial periods, and the 2.25 MW maximum capability for up to about four hours a day, without reducing stack life (218). (Presumably a tradeoff could also be calculated between fuller loading at maximum output and the economic value of the reduced stack life.) The assumed spinning-reserve value conservatively counted only 0.25 MW of this 0.47 MW of output flexibility. The analysis counted the spinning-reserve value *or* the value of optional peaking generation, whichever was greater; obviously the unit's spare capacity cannot be allocated to both roles at the same time.

[87] The split between operating and capital cost is not stated, but is probably about 2:1. This compares with nearly 9:1 for combustion turbines highly loaded in an operational role: ~88% operating cost can be inferred from using the study's assumptions for a combined-cycle plant and, to first order, adjusting to a simple-cycle turbine's heat rate (13,090 instead of 8,000 BTU/kWh) and plant cost ($620 instead of $737/kW according to EPRI's 1993 supply *TAG*™ for a similar timeframe). The difference is because the combustion-turbine proxy in the spinning-reserve role is unloaded, which uses less fuel than if it were fully loaded (220).

[88] They add that operability is traditionally defined as "the ability of a power generation unit to be started, to be brought to desired load, to be maneuvered to participate in the changes of served load, to support real and reactive power voltage regulation, [and] to respond to mitigate emergency conditions"—but that these definitions might not be appropriate for the distributed utility. Stability—"the ability of the generation-transmission-load system to remain in synchronous operation under steady-state operating and transient fault conditions"—is still essential, but could be achieved by very different means, as described below.

onds. These techniques' technology, reliability, and economic soundness are thoroughly established. In a world that allows demand- and supply-side resources to compete fairly for all roles—energy, capacity, reserves, stability, etc.—there should be no policy or analytic bias against demand-side solutions in any role.

2.3.1.3 Life extension

It is obvious that using distributed instead of central-plant resources to provide the functionality of traditional spinning reserve can extend the life of the central plants that will therefore be kept hot and spinning for fewer hours. It is less obvious, but also important, that most machinery operates most reliably if run steadily. To the extent that distributed resources reduce cycling, turn-on/shutdown, and low-load "idling" operation of generating units, they can reduce mechanical wear, thermal stress, corrosion, and other processes that shorten the life of expensive, slow-to-build, and hard-to-repair central generating equipment, thus incurring more and sooner the costs and risks of replacing it. Published analyses do not appear to quantify this effect. Yet when capital-averse or -short owners are seeking to "milk" old capacity for as long as possible rather than having to replace it, extending the engineering life—ideally, far beyond the amortization life—can be very attractive. Its economic value could be measured by changes in the present value of new equipment investments otherwise required to

Benefit
95   *Distributed substitutes for traditional spinning reserve capacity can reduce its operating hours—hence the mechanical wear, thermal stress, corrosion, and other gradual processes that shorten the life of expensive, slow-to-build, and hard-to-repair central generating equipment.*
96   *When distributed resources provide "virtual spinning reserve," they can reduce cycling, turn-on/shutdown, and low-load "idling" operation of central generating units, thereby increasing their lifetime.*
97   *Such life extension generally incurs a lower risk than supply expansion, and hence merits a more favorable risk-adjusted discount rate, further increasing its economic advantage.*

replace the old capacity. That present value should be adjusted as necessary for any differences of operating cost (*e.g.*, because the new equipment might be more efficient, use different fuels, or need to meet newer emission requirements). It will also depend on differences in risk (§ 2.2.2.2). In general, life extension carries lower risk than building new capacity, so proper risk-adjusted discount rates will give life extension a considerable further advantage.

2.3.2 Grid

The U.S. electricity industry invests in assets other than nuclear fuel a sum on the order of $30 billion per year, over half of it for transmission and distribution (collectively called here the "grid,"[89] contrary to some usages that use that term for trans-

---

[89] The distinction between the two levels of the grid varies. PG&E, for example, considers transmission to be 60 kV or more and distribution to be 21 kV or less, and has nothing in between. Many utilities consider distribution to be 13 kV or less; some, anything under 69 kV. Some utilities define a third level of Extra High Voltage transmission used for bulk power transfers over substantial distances. However, emerging FERC practice, jointly proposed by PG&E and Natural Resources Defense Council and adopted in the 1996 "mega-NOPR" ruling on operating the grid as a common carrier, uses a functional rather than a voltage-level definition. In essence, transmission moves power for resale to someone other than its end-user, whereas distribution moves power to its end-user for use rather than for resale. This approach sensibly avoids the likelihood that any voltage-based definition would be gamed during industry restructuring. A useful tutorial on typical transmission and distribution voltages, equipment counts, etc. is available at Willis & Scott , 2000 (762). In most grids, the boundary between transmission and distribution traditionally occurs at the substation, which is often fed by multiple power sources but feeds each neighborhood with a single radial line (763).

Benefits

98   *Distributed resources can help reduce the reliability and capacity problems to which an aging or overstressed grid is liable.*

99   *Distributed resources offer greater business opportunities for profiting from hot spots and price spikes, because time- and location-specific costs are typically more variable within the distribution system than in bulk generation.*

100   *Strategically, distributed resources make it possible to position and dispatch generating and demand-side resources optimally so as to maximize the entire range of distributed benefits.*

mission alone). Most of that—by one reckoning, as much as 70% of it (494), and in 1998–2000 for investor-owned utilities, 79% (166)—is spent on distribution.[90] Thus most North American utilities' investment needs, and most of their corresponding appetite for capital, are dominated by distribution. (In developing countries, or those whose power systems are driven by bureaucratic momentum rather than by market discipline, this may not be true.) At the larger end of the scale, distribution investments are relatively large and lumpy—not as much so as GW-scale power stations, but certainly enough to present significant risks of excessive or premature capacity, analogous to those discussed earlier in the context of generation.

New distribution investment is undertaken not only to serve areas with load growth but also to replace equipment nearing the end of its operating life. Many long-established utilities with mature markets have portfolios dominated by such equipment, and find it a major source of cascading problems—inadequate capacity, overheating, quicker and more widespread failures, more voltage drops and outages, more customer complaints—and therefore a serious threat to providers' business success in both traditionally regulated and competitive

markets. In cases where customers perceive that grid upkeep is being neglected to their detriment, the political reaction can even be strong enough to endanger the utility's whole business, as nearly occurred in the Chicago franchise renewal discussions in the 1990s.

Distribution, like the rest of the power system, is traditionally planned by forecasting demand and building to meet it. The methodology is essentially the same as for generation or transmission. The difference is that the forecast is based on highly localized conditions like the age of a specific transformer or the capacity of a specific feeder, rather than on aggregated system loads.

In a more competitive environment, however, three distinct market functions emerge: markets for energy, for its delivery, and for the grid's operational stability. Of these, the one reflecting local system constraints, and hence most likely to dominate dispatch decisions for distributed generation, will be the prices discovered in the *delivery* market (303). In New Zealand transmission by TransPower in 1996–97, for example,

- price differences ranged up to about 30% between different system nodes (303) at the same time;

[90] The share shown in Figure 2-50 below , based on EEI data for investor-owned utilities, is lower than that, but the graph does not include public utilities, one-fourth of the national system, which tend to serve lower customer densities. Aggregated grid construction expenditures for public utilities are not publicly available.

- order-of-magnitude price spikes could occur if a market actor erred (thereby incurring a salutary direct financial liability); and

- price differences and fluctuations at the more fine-grained distribution level were typically larger than in the more homogeneous and diversified transmission system.

The delivery market at both transmission and distribution levels—especially the latter, closest to distributed resources—typically exhibits more volatile prices than energy markets. It therefore offers greater business opportunities for distributed resources that can profit from hot spots and price spikes. That is, the scale of distributed resources, far down in the distribution system near the customers, is precisely the scale that offers those resources the greatest profit opportunities by mitigating real-time delivery constraints.

Distributed resources' ability to capture that profit depends on being first *deployed in the right place* and then *dispatched at the right time*. Proper deployment depends on careful assessment of avoidable grid costs, losses, reliability needs, and other technical attributes. There are broadly three siting alternatives: at the distribution substation, relieving transformer loading and perhaps somewhat improving reliability; on the distribution circuit, deferring local circuit upgrades while improving reliability and voltage profile;[91] or on the customer's premises, achieving the greatest com-

pounding savings in grid capacity and losses while offering further potential for riding through outages by using the local generation alone (perhaps shared with nearby customers in an "island" of isolated load [§ 2.3.2.10.6]).[92] That is, the most dispersed resources may save the most money by providing the greatest variety and intensity of distributed benefits.[93]

Some applications, not all of them remote, warrant standalone applications (§§ 2.2.9.2, 2.3.2.11)—either to avoid remote generation, grid, and connection costs altogether or to improve power quality and reliability beyond the levels obtainable from grid power. However, the most common function of most distributed resources is neither to displace the grid nor to displace generating capacity, but rather *to use the grid optimally to locate and integrate generating resources and end-uses* (304).

This is certainly true for most if not all generating technologies except photovoltaics. However, for photovoltaics it is rapidly becoming truer than it was a few years ago. In 1990, only 3% of U.S. photovoltaic shipments were installed in grid-connected applications (746). But that share then rose by nearly tenfold in nine years as interfaces became easier and cheaper to obtain and as interconnection barriers were lowered (Figure 2-45).

---

[91] Except perhaps (562) in underground cables, whose high capacitive reactance may cause lower loads to produce higher voltages.

[92] This scheme, in principle, might be more prone to harmonic-related power-quality problems (§ 2.3.3.8.1) because the system impedance is higher looking back upstream from the customer's service transformer (108). However, recent experience, including RMI's (where the PV power we sell back to the grid has lower harmonic content than what the grid sells us from a rural feeder), suggests that this issue can readily be handled by modern inverter design (85, 699).

[93] SMUD's assessments do show greater benefits at secondary than at primary voltage, but we are suggesting the possibility of a more sweeping conclusion based on the full range of benefits.

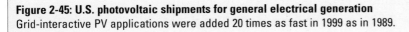

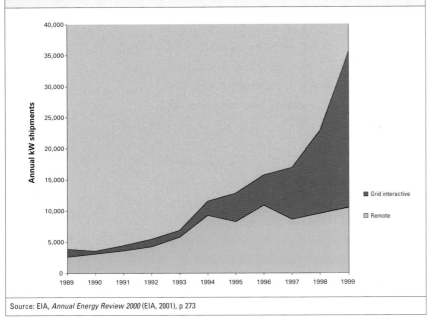

Figure 2-45: U.S. photovoltaic shipments for general electrical generation
Grid-interactive PV applications were added 20 times as fast in 1999 as in 1989.

Source: EIA, *Annual Energy Review 2000* (EIA, 2001), p 273

The statistical classification may omit some further uses that are certainly "remote" but intended for one specific use: the two categories shown in Figure 2-45 are only the crosshatched portions of a much larger set of shipments and end-uses (Figure 2-46).

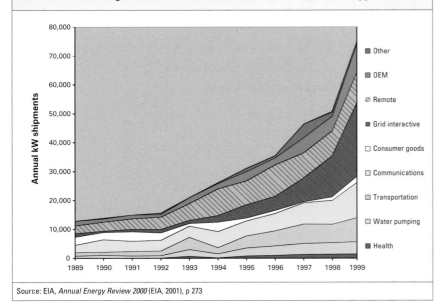

Figure 2-46: U.S. photovoltaic shipments by end-use, 1989–1999
The "remote" and "grid-interactive" uses totaled less than half of all applications.

Source: EIA, *Annual Energy Review 2000* (EIA, 2001), p 273

But by any measure, an important expansion of grid-connected applications is clearly underway. Much of it is designed to take advantage of the distributed benefits of grid support.

Many of the case-study results discussed in this section use system-average estimates for the cost of transmission and even of distribution, much as utilities traditionally use standardized proxies for the value of generating capacity. In principle, however, all of these quantities should be dynamic, not static; they should change with the load that the distributed resource affects. Greater experience with market pricing of grid services should improve understanding of the true economic value (as opposed to the accounting cost) of grid assets.

Grid capacity must be adequate to deliver the desired energy and power, net of grid losses. Grid investment also includes reactive support (§ 2.3.2.3). Since delivering real and reactive power has capital as well as operating components, we discuss the grid in an integrated fashion in this section, rather than breaking out the operating-cost (notably energy) part of grid losses into Section 2.3.3.1 on operations.

We begin this integrated discussion by more closely examining the grid's inherent losses of electricity, mainly because losses determine the grid capacity and the generating capacity required to meet customers' delivery needs, and because the closer a resource is to the user, the smaller are the losses incurred en route.

First, therefore, we review some historic context about the grid's losses and costs; then we explore how to reduce both.

2.3.2.1 The mysterious grid

Under traditional rate-of-return regulation, grid operators had no incentive to find out much about how their grids worked; they got paid whether they built and ran the grid efficiently or not. Measurement was therefore held to the minimum necessary to keep the lights on. For example, in 1990, a multi-million-customer utility with 3,000 feeders did not maintain time-series records of loads for more than a half-dozen of those feeders, all dominated by major customers (554). This illustrates a pervasive, disturbing, and fundamental ignorance about grid operations and economics that we illustrate next by considering total grid losses and costs.

2.3.2.1.1 Losses

One might suppose that a commodity as universally metered as U.S. electricity could not be lost in large quantities without someone's noticing. But for whatever reason, the authoritatively reported U.S. grid losses[94] for, say, 1998 diverge by an amount equivalent (at the average utility retail price) to nearly $8 billion worth.[95] The reported losses in that year ranged from about $15 billion to $23 billion in retail value:

Table 2-3

1994 TWh	Edison Electric Institute EEI, 1997 (163)	Energy Information Administration EIA, 1995 (173)	Energy Information Administration EIA, 1996 (178)	Energy Information Administration EIA, 1996 (177)	Energy Information Administration EIA, 1996 (179)
Lost & unaccounted for	216.706	220.948	230	~275	–
Net utility generation	2,920.712	2,924.961	2,911	~2,910	–
Purchases from nonutility generators	203.189	208.778	209	~208	–
Imports	50.520	52.230	52	~53	–
Exports	6.328	7.592	8	~11	–
% Losses[96]	6.86 %	6.95 %	7.27 %	~8.7 %	"approximately …9 %"

[94] "Lost and unaccounted for" electricity, counting directly only real power, but reflecting its extra losses due to poor power factor (§ 2.3.2.3).

[95] Lest the truth be presumed inevitably to lie fall somewhere between these authoritative sets of values, a 1990 Electric Power Research Institute study (222) cites transmission losses *alone*—excluding the larger distribution losses—as 6.1% baseload, 12% intermediate-load-factor, and 12.9% peaking—values clearly inconsistent with the published industry statistics. Industry sources, including EPRI, have been unable to clarify the origin of those 1990 figures or to confirm or deny their validity, although they sound implausibly high.

[96] Calculated as lost and unaccounted for, divided by the sum of: net utility generation, plus utility purchases from nonutility generators, plus imports, minus exports. We omit here electricity that was generated by nonutilities for their own use rather than for sale to utilities, since it typically never enters the grid. We also omit electricity that was accounted for but not sold, consisting of energy furnished without charge plus energy used by the electric utility department (but not inside the generating station; such uses are already debited from gross generation to yield net generation). This omitted term of electricity accounted for but not charged for is reported by EEI (preliminary 1994 data) to total 11,324 GWh; by EIA (final 1994 data) (171), 15,495 GWh. As a reality check, the then-largest investor-owned utility, PG&E, reported in its 1995 FERC-1 form "total energy losses" of 7.37% of total supplies, close to EIA's statistical national average of 7.27%. Loss allocation between utility and nonutility generators is poorly understood.

Figure 2-47: Lost and unaccounted-for U.S. electricity (utility plus nonutility), 1989–2000

Lost and unaccounted-for electricity as a percentage of total generated and net-imported electricity in the U.S.[97] The government data show a downward trend but are poorly correlated with the investor-owned utility industry's data, especially in 1998. Hidden in both data sets are substantial deficiencies in measurement and accounting.

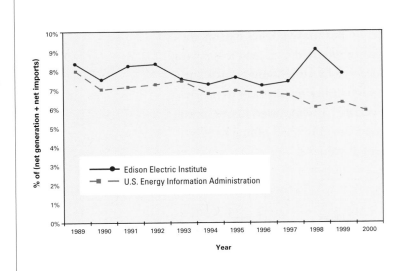

Source: EIA, *Annual Energy Review 2000* (EIA, 2001); EEI, *Statistical Review of the Electric Power Industry 2001* (EEI, 2002)

Figure 2-48: USEIA lost and unaccounted-for electricity *vs.* cooling-degree days

Even using the more stable EIA data on lost-and-unaccounted-for electricity, that quantity is anticorrelated with populated-weighted U.S. average cooling degree-days, both measured by calendar year.

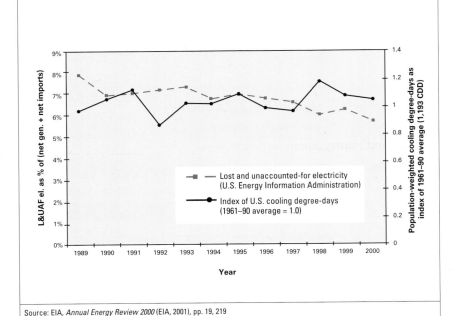

Source: EIA, *Annual Energy Review 2000* (EIA, 2001), pp. 19, 219

The Energy Information Administration's 1996 statistics (184) estimated utilities' 1996 grid losses at 9% of gross generation, corresponding to 9.47% of net generation—higher than any of the values explicitly reported. However, EIA privately noted in August 2000 that the 9% figure is only a rough internal estimate with little analytic basis. This estimate was repeated a year later in the 2000 *Annual Energy Review*, which stated (203) that "of electricity generated [apparently by both utilities and nonutilities], approximately 5 percent is lost in [power] plant use and 9 percent is lost in transmission and distribution." Yet EIA's statistics for 2000 (201) show "lost and unaccounted for" electricity totaling only 5.8% of net generation, continuing the recent downward trend shown in at least the government if not the industry data (Figure 2-47).

These two authoritative data sets are essentially uncorrelated (coefficient –0.054, 1989–99). Their difference ballooned to an implausible 2.91 percentage points, nearly one-third, in 1998—worth $2.3 billion at a nominal short-run marginal wholesale cost of $0.02/kWh. Just their year-to-year variability, and the lack of the same general trend in both data sets, causes concern over data quality. And even for the seemingly plausible EIA data, losses are negatively correlated (coefficient –0.485, 1989–2000) with cooling degree-days, as shown in Figure 2-48. One might presume that the correlation of electric load with space-cooling needs, and of those needs with hot weather, would add

---

[97] The EEI data include (and presumably consist largely of) line losses, but exclude utility use and free service, which are shown separately. The EIA data, based on statistical sampling, include "losses that occur between the point of generation and delivery to the customer, and data timeframe differences and nonsampling error." Both data sets exclude use at the power plant, which are already reflected in net generation. As explained in Section 1.1, EIA lacks data on the disposition of nonutility electric generation before 1989.

to the heating of conductors and transformers (heat increases their resistivity)[98] to make losses correlate well with summer heat, but surprisingly, this is not the case.

Many of the differences may well arise from nonuniform definitions and universes, data-collection-frame differences and nonsampling error (*i.e.*, erroneous data) (178), and noise in statistical sampling techniques. Indeed, EIA's *Electric Power Annual 1995* (171) states frankly that grid-losses-and-unaccounted-for kWh *are not measured*, but are only the residual term required to make the electricity books balance. The same source tactfully states (172) that "Due to the complexity of electric power transactions that involve specifics of contracts, simultaneous energy transactions, the unintended receipt and delivery of energy (inadvertent flow), and losses, uniformity in reporting the classification and quantity of each transaction among utilities may not exist."

Moreover, in the early 1990s, most U.S. utilities did not even have fully metered distribution systems (611)—metered even with one number a month in arrears downstream of main substations, let alone detailed real-time data on a finer timescale. (It was presumed that since the customers would be billed for whatever their retail meters showed, losses further upstream were simply an overhead that the regulators would pass through whether they were measured or not, so the meters were just a needless expense.) There is still a great deal of room for improvement in distribution-system metering. And though electric meters are fairly accurate and reliable in reading real

power drawn by resistive loads, they can be spoofed by nonsinusoidal waveforms from highly nonlinear loads—so much so that a Bonneville customer was reportedly found to have a meter spinning *backwards* because of a bizarre fifth-harmonic injection from end-use equipment.

Fair enough: it's not so easy to measure electrical flows or, therefore, losses. But do those practical difficulties justify discrepancies of up to billions of dollars per year among the final lost-and-unaccounted-for statistics published for the same year *by the same agency*? Surely this suggests that under traditional incentives and mindsets, there is little incentive to measure losses carefully; nobody is responsible for them, and whatever they are, the customers simply absorb them as an ineluctable overhead cost. In contrast, in competitive transmission and distribution systems like New Zealand's today, each party is responsible for a quantity carefully measured at an exactly defined delivery point, and all losses are explicitly allocated costs whose reduction is a business opportunity. This simple and transparent incentive creates thorough and unremitting efforts to find, measure, and reduce losses (394). Losses decrease when someone owns them. If the EIA (but not the EEI) lost-and-unaccounted-for data are correct, there may indeed be an encouraging recent trend in this direction as competitive pressures increase and management attention gets more focused on this issue (Figure 2-47).

2.3.2.1.2 Costs

If grid losses are so ill-defined, what about grid *costs*? Here again we find signs of pervasive inattention and opacity.

---

[98] Heating a conductor from 300 to 400 K (80.3 to 260.3°F), for example, through a combination of weather and load, increases its resistivity by 42% if it's aluminum and 39% if it's copper.

> **Benefit**
>
> **101**
> *Distributed resources (always on the demand side and often on the supply side) can largely or wholly avoid every category of grid costs on the margin by being already at or near the customer and hence requiring no further delivery.*

Figure 2-49: U.S. investor-owned utilities' construction expenditures, 1950–2000
The unprecedented ~1966–87 spurt in power-plant (mainly nuclear) construction expenditures nearly broke the industry. Recent spending has been reduced by capital risk aversion and by outsourcing of new power supplies to nonutilities.

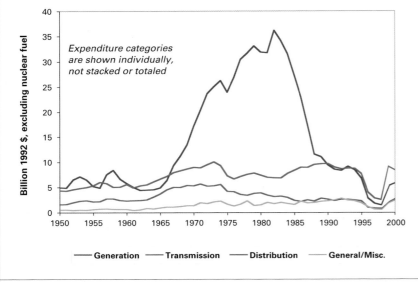

Source: EEI, *Statistical Yearbook of the Electric Utility Industry 2001* (EEI, 2002), p. 70

About four-fifths of the total costs of the existing U.S. electric grid are capital costs, the rest operating costs. *Marginal* capital costs are extremely utility- and site-dependent, but the aggregate importance of grid investments to the industry as a whole is evident from investor-owned utilities' *historic* allocation of construction expenditures. The bulge of construction during the boom period of big nuclear and coal plants is its most striking feature (Figure 2-49).

Clearly, the ~1966–87 power-station boom was an extraordinarily anomalous period— so big that it nearly bankrupted many utilities, and strained capital formation nationwide.[99] Indeed, its magnitude substantially distorted the totals for the entire half-century: the period after 1987 has been much nearer to the pre-1966 norm than to the boom period, and indeed looks quite like 1925–65, as the following summary figures show (ending in 1998 to avoid distortion by the major shift of investment from utilities to nonutilities):

Table 2-4			
Shares of investor-owned utilities' year-by-year construction expenditures (undiscounted)	Generation	Transmission and distribution	General and miscellaneous
1925–40 (data only every five years)	32.5%	58.5%	9.0%
1945–65	20.6%	57.4%	21.9%
1966–87	66.0%	29.5%	4.5%
1988–98	37.2%	41.3%	10.5%
Total 1945–98	**54.7%**	**38.8%**	**6.4%**
Total 1945–98 except 1966–87	**34.0%**	**51.4%**	**14.5%**

[99] Investor-owned utilities' investments in the early 1980s (167) peaked at 66% of all durable-goods manufacturing industries' investments (692) before retreating after 1987 to a more normal level of ~28%.

With due caution instilled by knowledge of that anomalous period of power-plant construction, we can now interpret the history of different assets' *shares* of utility investment:

This graph confirms that both before and after the power-station bulge, *more construction budget went into grid than generating facilities*. Indeed, since 1989, distribution invest-

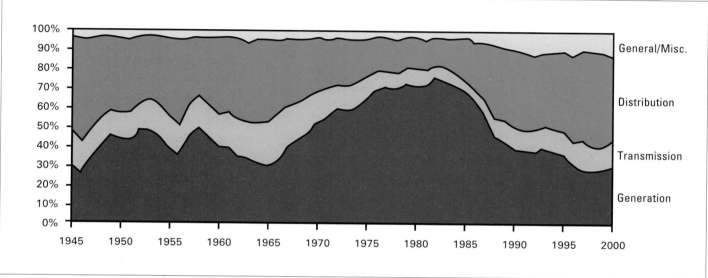

Figure 2-50: Allocation of U.S. investor-owned utilities' construction expenditures, 1945–2000, excluding nuclear fuel
Except for the 1970s power-station binge, grid investments have dominated private utilities' expenditures for more than half a century.

Source: EEI, *Statistical Yearbook of the Electric Utility Industry 2001* (EEI, 2002), p. 76

ments alone, not counting transmission investments, have about matched—and since 1990 (except in 1993) have exceeded—utilities' generation investments. To be sure, many distribution-oriented utilities have recently been shifting part, and some have shifted all, of their generating investments to nonutility providers, while continuing the grid investments needed to serve their own customers. Yet the trend is unmistakable: throughout nearly the entire history of the electricity industry other than the nuclear binge, *grid investments have dominated total investments*.

Furthermore, these national data mask important marginal effects by averaging slower- with faster-growing utilities. Many of the latter are in the Western Systems Coordinating Council, whose utilities' capital

expenditures in 1990 were over 75% for transmission and distribution (630)—one-third above the national average fraction. Especially now that new generating capacity is becoming dominated by cheap combined-cycle gas turbines, this 3:1 investment ratio (grid: generation) is probably a good surrogate, and may even be conservative, for regions with strong load growth.

Of course, the grid costs less to *operate* per kWh than do power stations, which consume fuel and, being full of moving parts and high temperatures, tend to be maintenance-intensive. But since the grid dominates total utility investment, shouldn't grid costs be an important part of electricity's *total* delivered price?

Certainly—but that is hardly clear from standard industry reporting. The significance of grid costs was revealed instead in a pioneering 1976 study (59) in which graduate students laboriously examined by hand the individual Federal reports filed on paper by a 48-state sample of investor-owned utilities serving ~80% of the total U.S. utility business.[100] The 1976 study—apparently to this day the only thorough examination of this subject, at least in the United States—found that

> The costs derived from the transmission and distribution (T&D) system have historically comprised about 2/3 the costs of producing and delivering electricity to residential-commercial customers, and over 1/3 the total costs [of] supplying electricity to large industrial customers.

Focusing on major terms that accounted for ~80% of total T&D costs (and may have neglected T&D *losses* as an equivalent cost), the study found that for the smaller customers (average load 1.04 kW, only 15% below the average U.S. household in 2000), who accounted for ~55% of the electricity sales of the utilities analyzed,

- the average dollar spent on electricity went ~19% to transmission equipment, 24% to distribution equipment, 21% to all that equipment's operation, maintenance, metering, and billing, ~6% to profit and to arithmetic discrepancies in the analysis (largely because different costs were escalating at different rates), and *only 29% to producing or acquiring electricity;*

- thus *delivering the electricity to these smaller customers in 1972 cost 2.2 times as much as generating it;* and

- their 1972 T&D costs ranged from $0.010

to $0.023/kWh between different regions of the country.

For large customers with average loads of 177 kW, delivering the electricity cost 1.2 times as much as making it, and the regional averages of T&D costs ranged from $0.0037 to $0.0082/kWh.

Such large grid costs are relevant because distributed resources (always on the demand side and often on the supply side) can largely or wholly avoid them on the margin by being already at or near the customer and hence requiring no further delivery. Yet that analysis for 1972 has never been updated,[101] and utility statistics are kept in such a form that doing so would entail extensive effort. Although marginal costs, as Section 1.4.2 showed, are extremely site-specific within each utility, and are not normally published anyhow, we were therefore curious about what might have changed in the aggregated historic cost structures. We therefore performed our own analysis, reported in Technical Note 2-4, using 1995–96 data to predate the distorting effects of selling major utility assets to nonutilities.

The analysis in Technical Note 2-4 shows that for low-voltage (chiefly residential) customers, whose delivery costs are much

[100] Including public utilities would in general have strengthened the results, since the IOUs tend to serve higher-density, less rural loads: as noted in Section 1.3, rural cooperatives own and run about 43% of all U.S. distribution-line mileage (330), even though they sell only 305 TWh or 9% of the utility industry's electricity output (506).

[101] To the knowledge of its senior author (M. Baughman, personal communication, 17 February 1997) and ourselves.

Technical Note 2-4:
1996 U.S. Electricity Delivery Costs

The Energy Information Administration's *Financial Statistics of Major U.S. Investor-Owned Electric Utilities 1996*,[102] *in financial statistics' Tables 27 and 28, classifies the accounting value of utility plant in service (in undiscounted, undeflated mixed current dollars) as shown in Figure 2-51:*

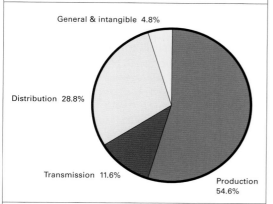

Figure 2-51: U.S. utility plant in service 31 December 1996 (major investor-owned utilities)
Two-fifths of utilities' historic investments are for the grid.

General & intangible 4.8%

Distribution 28.8%

Transmission 11.6%

Production 54.6%

Source: EIA, *Financial Statistics of Major U.S. Investor-Owned Electric Utilities 1996* (EIA, December 1997)

That is, neglecting the niceties of discounting and deflation of various years' dollars, historic production investments[103] are only 40% larger than historic grid investments.

Table 23 classifies the reported disposition of these same utilities' electric revenues in the conventional way, namely by accounting categories (using the more complete EIA data, which differ immaterially from EEI data) (Figure 2-52).

With due allowance for minor distortions, such as the fraction of expenses (7.3%) and revenues (8.1%) due to non-electric utility operations (some electric utilities also sell gas, steam, etc.), total revenues clearly go mainly to pay for fuel and purchased power—the vertically striped wedges at the upper right of Figure 2-52—and for returns on and of capital—the horizontally striped items. But these accounts say nothing about what that capital, or other operating expenses, got used for.

To estimate that activity-based allocation, we reallocated to five functional categories—production, transmission, distribution, general and administrative, and customer service, sales, and information—according to their respective shares of embedded total asset value (Figure 2-51), the utilities'

- *capital charges (depreciation, amortization, interest, common and preferred dividends, and retained earnings),*

- *taxes (income taxes because the assets generate the income, sales and franchise taxes likewise, and property taxes on the presumption that they broadly reflect the asset values) and*

- *nonfuel operating and maintenance costs (using the utilities' own reported allocation to the same five functional categories). (180)*

Using the simplifying assumptions that 3% of total generated electricity is lost in transmission and 4% (of the same original generated base) in distribution—reasonable nominal values consistent with 1996 data—we then allocated those fractions of the production assets' capital and operating costs to the transmission and distribution functions.[104] The result, though approximate and completely aggregated, seems to be the near-

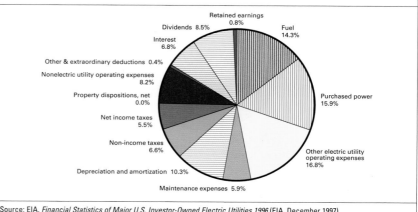

Figure 2-52: Conventional accounting allocation of the 1996 electric revenues of large investor-owned U.S. electric utilities
(average ultimate-customer revenue = $0.07105/kWh)
Accountants' cost allocation says nothing about functional uses.

Retained earnings 0.8%
Dividends 8.5%
Interest 6.8%
Other & extraordinary deductions 0.4%
Nonelectric utility operating expenses 8.2%
Property dispositions, net 0.0%
Net income taxes 5.5%
Non-income taxes 6.6%
Depreciation and amortization 10.3%
Maintenance expenses 5.9%
Fuel 14.3%
Purchased power 15.9%
Other electric utility operating expenses 16.8%

Source: EIA, *Financial Statistics of Major U.S. Investor-Owned Electric Utilities 1996* (EIA, December 1997)

[102] That group is identical, within much less than 1%, to the universe of all investor-owned utilities, which were responsible for 77% of all U.S. net utility generation. As noted earlier, this group probably underrepresents nationwide grid costs and losses, because investor-owned utilities tend to serve more built-up areas with higher load densities than public utilities do.

[103] In highly aggregated mixed current dollars total, having no regard to timing, inflation, or differing tax treatment; depreciation or amortization patterns; or asset lives. These factors, especially the last, probably account for differences between Figure 2-51 and the historic totals of construction expenditures.

[104] Conservatively, however, we did not try to assess the value of any transmission losses already built into purchased power at the point of transfer.

est thing available to an update of the 1976 study of 1972 data—a functional allocation of where electricity dollars paid to major investor-owned utilities went in 1996:

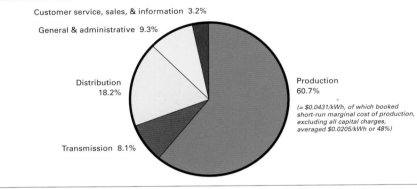

Figure 2-53: Functional allocation of the 1996 electric revenues of large U.S. investor-owned electric utilities
(average ultimate-customer revenue = $0.07105/kWh)
Accounting costs of electricity reallocated by function

Customer service, sales, & information 3.2%
General & administrative 9.3%
Distribution 18.2%
Transmission 8.1%
Production 60.7%
(= $0.0431/kWh, of which booked short-run marginal cost of production, excluding all capital charges, averaged $0.0205/kWh or 48%)

Source: RMI analysis from EIA, *Financial Statistics of Major U.S. Investor-Owned Electric Utilities 1996* (EIA, December 1997)

Finally, for an even more realistic (though necessarily approximate) picture, the General & Administrative overheads should be allocated to the functional categories they support. Doing this simply pro rata on the other costs' share of the non-G&A costs (a reasonable rough-and-ready method, absent better data) yields:

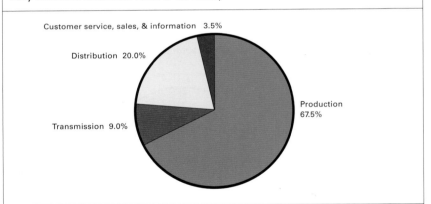

Figure 2-54: Functional allocation (G&A allocated out) of the 1996 electric revenues of large investor-owned U.S. electric utilities
Fully allocated functional costs of electricity

Customer service, sales, & information 3.5%
Distribution 20.0%
Transmission 9.0%
Production 67.5%

Source: RMI analysis from EIA, *Financial Statistics of Major U.S. Investor-Owned Electric Utilities 1996* (EIA, December 1997)

or for the mythical "average" customer of those large investor-owned utilities in 1996, the values shown in Figure 2-55.

Many interesting comparisons can be made between the disaggregated 1974 and the highly aggregated 1996 results. For example, the study of the 1972 data found that nonproduction costs were 2.23 times production costs for residential and commercial customers, 0.83 times for industrial—an energy-weighted average of 1.66 times for both together. After the

1980s construction bulge, the 1996 data (undeflated, undiscounted, and hence somewhat weighted for the more recent generating investments) had changed this ratio to a combined figure of 0.53 (or 0.65 if we didn't allocate costs to the grid losses) before G&A costs are allocated to the four functional categories, or 0.49 afterwards. Nonetheless, even at embedded historic values, before the G&A allocation, 26.4% of all 1996 electrical revenues to large investor-owned utilities went to grid-related costs, only 60.4% (2.3 times as much) to production costs.

After G&A allocation, 29% of the 1996 customer dollar went to grid costs, 4% to other costs of retail service (also part of the delivery function)—i.e., one-third to delivery, two-thirds to production. For the average ultimate-customer revenue (i.e., excluding sales for resale) of $0.0713/kWh received by investor-owned utilities in 1996 (167), this implies total average delivery-related costs of $0.0235/kWh. That was about one-seventh more than the marginal cost of operating these utilities' power stations, or 90-odd percent of the total busbar production costs from a new combined-cycle gas plant (§ 1.2.4). Thus to build and run a new combined-cycle plant costs scarcely more than the embedded cost of just delivering its output to the average customer. Delivery to residential customers, or new ones, typically costs substantially more than such marginal generation.

Figure 2-55: Where the $0.07105/kWh of 1996 ultimate-customer revenue to large investor-owned U.S. utilities went
For the average kWh sold by large investor-owned utilities in 1996 (public utilities probably have higher delivery costs), the fully allocated delivery costs slightly exceeded the reported accounting cost of running the existing power stations.

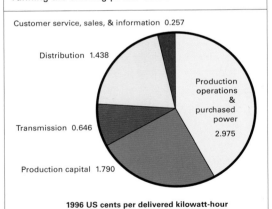

Customer service, sales, & information 0.257
Distribution 1.438
Transmission 0.646
Production capital 1.790
Production operations & purchased power 2.975

1996 US cents per delivered kilowatt-hour

Note: The reported accounting allocations differ (*e.g.*, 2.054¢/kWh production operations, 4.061¢/kWh total production), partly because they do not allocate out grid losses, the costs of producing lost electricity, or general and administrative costs; production is considered at the busbar, not delivered to the retail meter. Grid losses are assumed to be 3% transmission, 4% distribution.

Source: RMI analysis from EIA, *Financial Statistics of Major U.S. Investor-Owned Electric Utilities 1996* (EIA, December 1997)

higher than the average for all customers, embedded T&D costs already considerably exceed long-run marginal production costs.[105] For the average customer, embedded T&D costs even exceed the short-run marginal production cost, and exceed by ~1% the average total operating cost including all O&M—the quantities around which utilities' dispatch decisions and their mental universe revolve. This is such an important conclusion that we repeat it more plainly:

Of course, the costs already paid to build the existing grid, and the costs of that part of its operation and maintenance that

> For a typical customer of a large U.S. investor-owned utility in 1996, the utility paid more to *deliver* electricity to its customers than just to *produce* that electricity in existing stations. This industry is engaged in cutthroat competition (and turned inside-out by restructuring to encourage such competition) over tiny differences in the marginal cost of producing electricity. Yet over the long run, a cost even greater than the average of the power plants' *total* non-capital generating cost *could be entirely avoided by distributed resources* that require no delivery to the customer because they're already there.[106]

depend on time rather than on throughput, cannot be avoided, because one can make decisions only about the future, not about the past. But distributed resources often can avoid the *future* costs of expanding the grid's capacity, and such expansion typically costs even more than the old grid did cost. For example, as Section 1.4.2 noted, adding to Pacific Gas and Electric Company's grid capacity would cost an average of at least \$230/kW[107] and a maximum of \$1,173/kW, or five times as much. For comparison, PG&E's *embedded* book cost for old grid capacity was probably closer to the national average for investor-owned utilities—on the order of \$214/kW.[108]

PG&E's worst-case "hot-spot" T&D marginal cost, which is far from the worst in the industry, carried an avoidable T&D investment (at an illustrative 10%/y real fixed charge rate) equivalent to a capital charge alone of 3¢/kWh if the circuit had the PG&E system-average distribution load factor of 45%—or proportionately higher if it's worse than that, as many obviously are. As Section 1.4.2 noted, PG&E's actual worst-case total marginal cost[109] was around *100 times* that large—around \$3/kWh! Distributed resources' potential to avoid marginal T&D costs by being already delivered to the customer therefore become even more impor-

---

[105] For example, although the New Hampshire pilot project on retail access featured (around 1996) advertisements of *energy* commodity costs around \$0.01–0.03/kWh, they didn't mention other cost components. For a typical residential customer paying \$0.035/kWh for energy, the total charge would be \$0.105/kWh—51% for stranded assets and acquisition premium, 3% for transmission, and 23% for distribution. For a typical large business customer paying \$0.031/kWh for energy, the total charge of \$0.061/kWh would be 64% stranded assets, 2% transmission, and 4% distribution (260). These figures confirm both the importance of delivery costs to small customers and the incentive for all, especially large, customers to leave the grid altogether if that can profitably avoid stranded-asset and delivery costs.

[106] This advantage would be diminished by the transaction costs of marketing, designing, and installing the resource in both cases.

[107] By another estimate, \$282/kW for transmission alone in 1992 \$. (626)

[108] For major investor-owned utilities in 1995, total net utility plant less nuclear fuel had a book value of \$378 billion (181), divided by ~704 GW of capacity (706 GW for all IOUs times this subset's 99.7% of sales to ultimate consumers), yielding \$537/kW of embedded net system costs. The national-average T&D fraction of those costs, according to our analysis from the EIA financial statistics as described above, was 39.8%, or \$214/kW. PG&E's average cost of marginal T&D was probably higher than for most systems nationwide because of higher land costs and relatively rapid demand growth that used up much of the older surplus grid capacity. (The company's 1995 FERC-1 filing implies a higher value, on the order of \$376 per kW of peak load sent out in 1995—considerably less than the grid's peak capacity—but this included many longer-lived assets with considerable excess capacity that would not be used up for quite a while.)

[109] Including generation, which is a relatively minor part of such a large total cost.

tant than average embedded costs would suggest.

Yet none of this shows up in utilities' operating statistics. The standard reports—external and, often, internal too—say nothing about the total costs of the T&D functions, the energy flows in between the power-plant busbar and the retail meter, or the T&D-to-generation total-cost ratios. These blank spots on the mental map reinforce utilities' historic tendency to compare generating options *only in busbar cost*, as if all options were simply alternative black boxes that got plugged into the same grid at the same place and hence incurred the same delivery costs. That was more or less true when all plants were of GW scale. But it is certainly not true of more distributed options that derive major value from being closer to customers and hence reducing the capital and operating costs of delivery. With generation, as with real estate, competitive advantage can depend on "location, location, location."

This persistent underemphasis on grid costs is all the more surprising when one recalls that the classical rationale for treating electric utilities as regulated franchise monopolies has always been the supposedly prohibitive cost of duplicating the grid's infrastructure![110] If the cost of duplicating grid infrastructure is so enormous (as indeed it is—40% of investor-owned utilities' entire embedded investment of every kind as of 1995), then why isn't it considered important enough to feature at least a transparent mention in industry statistics as a component of retail electricity prices?

Fortunately, as Section 1.4.1 noted, a few utilities *have* lately started to practice "Local Integrated Resource Planning" that pays very careful attention to the fine-grained geographic structure of grid costs. These smart utilities often achieve striking financial benefits, because although distributed resources cannot avoid the already sunk capital costs of the existing grid, they often *can* reduce, defer, or avoid the [often much higher] *marginal* costs of grid capacity that does not yet exist. Their reward is to turn marginal consumption from a major money-loser into an opportunity for both operating profit and customer satisfaction.

2.3.2.2 Grid losses: potential reductions

The sequence of computing losses or avoided losses matters because "reduced load has a compounding effect, [so] it is important to start [the evaluation] with the distribution system, determine loss savings through the station transformer based on [distributed]... generation and feeder loss savings, and finally determine transmission loss savings based on all of the above." (605) That is, the calculation should start downstream and work upstream. Since distributed resources are all the way downstream, or nearly so, their location maximizes the compounding of the upstream grid losses that they reduce.

This is illustrated by a 1993 experiment in which Pacific Gas and Electric Company sited a 500-kWAC photovoltaic array on a 12-kV distribution feeder eight circuit-miles downstream of a Fresno-area 10.5-MVA substation. (Many results from PG&E's pathfinding analyses of this Kerman project are cited

---

[110] An electric system as large and diverse as America's has an example of almost any anomaly, and in fact, at least 17 U.S. jurisdictions, some rather sizable, have long had duplicate or checkerboarded distribution facilities enabling customers to choose a private or a public utility. Some observers of this oddity have the impression that, contrary to a commonly assumed doctrine about "natural monopoly," this supposedly wasteful duplication of infrastructure actually raises costs less than competition reduces them.

Tutorial 7: Grid Losses

Grid losses compound—multiply—as they occur successively. Setting aside for the moment the reactive power losses discussed in Section 2.3.2.3 below, the grid losses in conductors are of three main kinds:

- **Corona-discharge** losses from high-voltage transmission lines can be much reduced by good design but not eliminated except by using lower voltages, which in turn would require fatter, heavier, and costlier conductors to avoid incurring more resistive loss per unit of power transmitted.

- **Radiative** losses refer to the radiation of electric and magnetic fields that are absorbed by surrounding objects or media. These too depend on the voltage and current being conveyed and on technical design details. They are related to the reactance of the line; reactive losses are discussed below in Section 2.3.2.3.

- **Resistive** losses dominate total losses. In a simple conductor (wire), they are proportional directly to its length and to its resistivity,[111] inversely to its cross-sectional area (which of course increases as the square of diameter), and to the square of the carried current (conventionally denoted as I). They are therefore often called I^2R losses. Resistivity is a function of material and temperature. The bigger the losses, the hotter the conductor becomes, increasing its resistivity and—if the same current continues to be delivered—thereby increasing losses and heating still further, subject to equilibrium between how fast the heat is added and how fast it is given up to the environment. For this reason, the thermal capacity[112] of a conductor such as a transmission line is increased significantly by any breeze that may help remove heat, but is reduced by sunlight that heats the conductor (unless the solar warming also stimulates winds that more than offset the solar heating). The most severe thermal conditions occur on the hottest days, when peak air-conditioning loads require the lines to carry more current just as they are hottest and have the highest resistance. That is why peak grid losses in the United States are traditionally estimated by EPRI to be about twice the nominal average loss of ~7%. Similarly, SMUD reckons that the losses from its system gateway (downstream of the long-distance transmission) to secondary distribution voltage peak at 7.94% on system-peak summer days, but average only 5.81% year-round, and drop to 5.35% at the winter offpeak (744). This relationship of grid losses to time-of-day and time-of-year is linked to a further distributed benefit discussed in Section 2.3.2.5.

In addition, transformers and other inductors (such as inductive "reactors") typically have iron cores that lose energy through eddy current within the laminations and through hysteresis as the iron's magnetic domain walls shift. These "no-load" or "iron" losses are largely independent of both temperature and load: a good 50-kVA transformer will use nearly 1 kW to heat its core even at zero load.[113] Iron losses traditionally range from about 50% of total losses with small, low-quality dry-type transformers to under 10% of total losses in large, high-quality units. However, iron losses can be further reduced by careful choice of materials and geometry. Most new transformers should use amorphous iron, which costs more but is worth it because it reduces iron loss by at least fivefold.[114] Where loads are very peaky, as in many buildings, a second-best solution may be to share the load between two or more stepdown transformers (as is often done anyhow for reliability) and then to de-energize one or more of those units during periods of light loading in order to avoid iron loss.[115]

[111] Resistivity is electrical resistance R per unit of cross-sectional area and length.

[112] Thermal capacity is how much power the conductor can carry in given conditions without overheating. Actual capacity may be lower than this because of stability limits.

[113] However, the transformer's *copper* losses are proportional to the conductor's resistance and to the square of current, and again produce heat that further increases resistance, and so on until thermal equilibrium with the surroundings is reached.

[114] For example, in a standard oil-filled cylindrical 25-kVA distribution transformer of the type found on most U.S. utility poles, a good 99.33%-efficient amorphous-iron model with doubled copper content and other premium features can cut total real-power losses by 57% or 588 kWh/y with a 14%/y real aftertax return on investment, compared with an already respectable 98.44%-efficient oil-filled standard model (many are nearer 96%). (339) To

These physical loss mechanisms imply that distributed resources can reduce grid losses in four main ways:

- shorter haul length from the more localized (less remote) source to the load, hence less *R*;

- lower current if the resource is end-use efficiency or local generation that reduces required net inflow from the grid, hence less *I²*;

- effective increases in conductor cross-section per unit of current if an unchanged conductor is carrying less current, hence less *R*;

- less conductor and transformer heating (hence less *R*) if current is reduced by more efficient use, by load management or peak-shaving that reduce onpeak coincidence, by better management of existing transmission assets, or by better distribution circuit management that better shares loads among parallel distribution capacity.[116]

achieve this, 150 W of *full*-load copper loss is avoided plus 40 W of iron loss, but of course full load is infrequent while iron loss is continuous. The premium unit costs 112% more ($680 *vs.* $320), but at 50% load factor, $0.06/kWh, 0.95 power factor, and a 5%/y real discount rate, it saves energy at $0.024/kWh, about the utilities' short-run marginal cost of generating and transmitting it. Its present-valued 20-year savings are 109% of its total and 2.1 times its marginal cost. The superefficient model also offers a temporary overload capacity around 30%, *vs.* about zero for the standard model, providing valuable service flexibility. Alternatively, it should last much longer under either normal or excessive loads before its insulation fails. Yet since most utilities carefully analyze transformer efficiency only at or above the ~2-MVA level, and use lowest-first-cost criteria for

routine "small" purchases, amorphous-iron distribution transformers had only a ~10% U.S. market share in 1993. (New England Electric System, at 75%, was a notable exception.) The U.S. has about 35 million distribution transformers in service; American utilities buy another million, and their customers another half-million, each year. These purchases thereby waste every year a third of a peak GW plus 3 TWh/y—for the next 20–30 years. At marginal costs (say, $0.02/kWh plus $700/kW delivered to the distribution pole), that's a $1-billion-a-year misallocation of U.S. capital.

[115] For the dry-type transformers widely used in buildings and equipped with standard silicon-steel rather than amorphous cores, many of the subtleties of choosing models that yield top efficiency at the desired load range and temperature are discussed

by Howe (342). Part-load efficiency is especially important, since the average such transformer of 500 kVA or less is so oversized that it experiences an average load only 35% of its rating. Poor specification of such transformers in the U.S. currently wastes upwards of $1 billion a year, or ~2–6% of a typical U.S. building's electricity costs (342).

[116] For example, a 900-MHz communications and automation system at the TVA-fed Joe Wheeler Power Company, which has 4,000 miles of distribution lines and 23 substations, paid for itself in under 1.5 years, partly by halving grid losses (from 10% to 5%) through selective line upgrades and better load balancing between circuits. These upgrade opportunities could not be identified or analyzed before the data were acquired by the new communications system (118).

in various sections of this book.) The losses avoided by the PV output compounded to a 1.08 multiplier for avoided transmission capacity[117] and 1.12 for avoided generating capacity (618). The annual loss savings were 92 kWh/kWy on the feeder circuit (625), 39.5 kW and 227 kVAR in the transformer (624), and still more in transmission, for a total loss reduction (616) of 300 kWh/kWy. The *total* losses avoided were worth $21/kWy (613), including no correction for heating effects (625). These were worth $0.0092 per kWh of

PV output[118] (1992 $), consistent with $0.003–0.017/kWh (1991 $) for fuel-cell output in Los Angeles (216).

Naturally, the losses avoided by a PV generator, whose output peaks fairly coincidently with the system load (§ 2.2.8), are greatest around the system peak when they are also most valuable (Figure 2-56).

The actual losses that distributed resources can avoid are thus quite complex, and depend not only on the grid load displaced

[117] On the simplifying assumption (617) of similar transmission and generation loads: "If the transmission loads were significantly different, the analyst should estimate a separate peak plant availability for the transmission system."

[118] Converted not at the expected output of 2,766 kWh/kWy plus expected avoided losses of 300 kWh/kWy, or 3,066 kWh/kWy (616), but at the 1,080 MWh of actual output plus 58.5 MWh of actual loss avoidance found in the final evaluation (735). The final evaluation used the exact PVUSA rating of 498 rather than the nominal 500 kW, and the 25% capacity factor observed in the evaluation year rather than the 32% assumed in the original analysis or the 27% that would have been achieved if the final design had met all expectations in the evaluation year (735).

Benefits

102   *Distributed resources have a shorter haul length from the more localized (less remote) source to the load, hence less electric resistance in the grid.*

103   *Distributed resources reduce required net inflow from the grid, reducing grid current and hence grid losses.*

104   *Distributed resources cause effective increases in conductor cross-section per unit of current (thereby decreasing resistance) if an unchanged conductor is carrying less current.*

105   *Distributed resources result in less conductor and transformer heating, hence less resistance.*

106   *Distributed resources' ability to decrease grid losses is increased because they are close to customers, maximizing the sequential compounding of the different losses that they avoid.*

107   *Distributed photovoltaics particularly reduce grid loss load because their output is greatest at peak hours (in a summer-peaking system), disproportionately reducing the heating of grid equipment.*

108   *Such onpeak generation also reduces losses precisely when the reductions are most valuable.*

109   *Since grid losses avoided by distributed resources are worth the product of the number times the value of each avoided kWh of losses, their value can multiply rapidly when using area- and time-specific costs.*

but also on the time, weather, load conditions, loadshapes, and—especially—physical placement in the grid. For example, a fuel-cell analysis for Los Angeles (212) found that supply directly into the 4.8-kV distribution system would reduce losses all the way from the central-plant generator busbar to the local distribution feeder (power-plant substation transformer, transmission, transmission substation transformer, subtransmission, distribution substation transformer, and distribution feeder to the point where

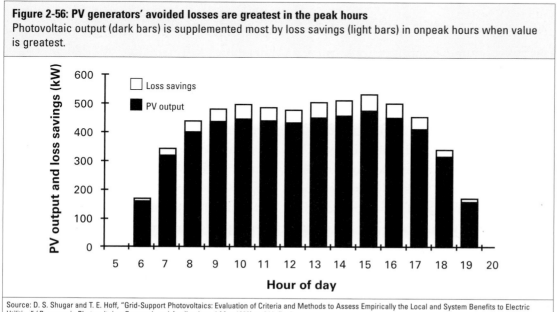

Figure 2-56: PV generators' avoided losses are greatest in the peak hours
Photovoltaic output (dark bars) is supplemented most by loss savings (light bars) in onpeak hours when value is greatest.

Source: D. S. Shugar and T. E. Hoff, "Grid-Support Photovoltaics: Evaluation of Criteria and Methods to Assess Empirically the Local and System Benefits to Electric Utilities" (*Progress in Photovoltaics: Research and Applications*, 4 May 1993), p. 244, fig. 6

the distributed resource connects). Connecting so far downstream multiplied the losses avoided upstream. How big are those losses? It depends on the system, time, weather, and load. For example, PG&E's Kerman study states (637) that "On average, [PG&E's] subtransmission and distribution losses average about 8%, while they could range from 3% to 13% in specific systems."

A widely used rule-of-thumb avoids these complexities by simply stating (217) that "In general, for every percent displacement of remote system supply by [distributed]...output, there will be roughly a 2-percent drop of losses associated with the load downstream from the interconnection point of the [distributed resource]....This simple relationship is an approximation that applies to all situations, including cases where the output of the facility exceeds the load of the feeder it is connected to." (In that case, however, where the excess locally generated power flows back through the distribution substation transformer to support loads on neighboring circuits, the avoided losses may be somewhat lower.)

This rule-of-thumb is intuitively appealing, but with more than fourfold variation between different utility systems or parts of them, it obviously has limited quantitative validity, especially where a distributed resource is carefully sited and its output timed to shave peaks from the most heavily loaded components. "Loss savings are especially significant for a [distributed resource] ...that is located within the [grid]...specifically to relieve local thermal overload, because relative power loss savings are greater with systems which are operating at higher currents ($P = I^2R$)." (605) Thus using an aggre-

gated rule-of-thumb obscures precisely the area- and time-specific benefits that distributed resources can most profitably exploit.

In general, "Computing the applicable credit involves two tasks: determination of the physical reduction in real and reactive power loss at different times, and determination of the value of the energy saved and support provided to the [grid]...system." (605) *Both* these factors often yield high area- and time-specific values for well-sited distributed resources, multiplying their value. With time-specific values as high as a dollar, or even several dollars, per kWh (§ 1.4.2), the number of avoided grid-loss kWh is worth counting very carefully.

Two cases must be distinguished when calculating a distributed resource's avoided grid losses (215). If a new distributed resource displaces generating capacity on the existing system, then the system energy displaced comes from no specifically identifiable location, so the loss reduction must be estimated systemwide. The most exact method is loadflow simulation that compares real and reactive system losses with and without the distributed resource. This is data-intensive, requiring "the system configuration, including generation and loads, and produces as output the voltages and currents throughout the system. The currents through various parts of the system are [then] converted into losses, given conductor and device specifications." [119] (605) Another method (215) is "to subtract from the overall system-wide loss factor the percentage of losses that would have taken place between the point of interconnection of the [distributed generator]...and the customers served by the feeder of interest in the absence of [that generator]...," starting

[119] Rather than performing the loadflow analysis for each hour of the year, it is conventional to compute only the loss reduction at the system peak, then convert it into an annual energy saving by multiplying by the square of the normalized load-duration curve, then applying the standard average avoided cost of produced energy, and then integrating over the year (assuming that voltage stays constant). (637)

with the design specifications of the primary and secondary feeder system and local load data. It appears that the latter method is less exact and may underestimate loss reductions.

Alternatively, if the distributed resource "displaces a new resource that would have been dedicated to meet load growth or capacity retirement," then the location of that resource, hence the avoided losses associated with it, can be determined from the resource plan or estimated from competitive bidding behavior. For example, in Los Angeles in 1992, such evaluations suggested that the marginal bulk power resource would be in the faraway Pacific Northwest or Canada. Long-distance DC Intertie losses could thus be avoided by in-city generation under capacity constraints.

2.3.2.3 Power factor and reactive power support

> **Benefits**
>
> **110** *Distributed resources can reduce reactive power consumption by shortening the electron haul length through lines and by not going through as many transformers—both major sources of inductive reactance.*
>
> **111** *Distributed resources can reduce current flows through inductive grid elements by meeting nearby loads directly rather than by bringing current through lines and transformers.*

So far we have discussed grid losses in terms only of real current (current in phase with voltage). But additionally, distributed resources can save reactive current (current out of phase with voltage). This has an engineering value, and therefore an economic value, that is hard to express intuitively but can be thought of as helping to maintain "the balance and functioning of the transmission and distribution system, rather than a commodity like real power. In fact, operationally, reactive power is more closely related to voltage levels than to real power." (606) Distributed resources close to loads decrease reactive power consumption by two separate methods. First, they shorten the length of lines and the number of transformers through which electrons must flow. Second, they reduce the current flowing through those grid elements; reactive losses vary as the square of current.

To review fundamentals already mentioned in Section 1.2.2 (p. 8) and elaborated in Tutorial 8: **power factor**—more precisely, **phase** or **displacement** power factor—is the ratio of actual power being used in a circuit (measured in watts) to apparent power drawn from the supply system (measured in volt-amperes). It therefore measures how much of the power drawn by the load is

Tutorial 8: Power Factor

Inductive loads like wires, motors, and transformers make current lag behind voltage, lowering power factor. Loosely speaking, power factor is the fraction of the power delivered that can do work. Power factor less than the ideal 1.00 consumes reactive power which the utility must provide but, save in grossly excessive cases, doesn't get paid for. The utility provides reactive power by installing special and rather costly devices, chiefly shunt capacitors, to create leading power factor and thus offset the lagging power factor of the inductors.[120] Otherwise, too low a power factor would cause excessive voltage drops in transmitting real power to customers, and would also increase grid losses and the capacity requirements for grid and generation.

The sensitivity of grid losses to power factor can be illustrated by a grid that delivers and bills for 1.0 kW of real power. At 100% power factor, apparent power is 1.0 kVA (kilovolt-amperes) and reactive power is zero. Let's call that a grid loss of 1.0 units. At 95% power factor, the same 1.0 kW is delivered, but so is 0.33 kVAR (kilovolt-amperes reactive) of reactive power, for an apparent power of 1.05 kVA and losses of 1.10 units (The relationship is: kVAR = $[(kVA)^2 - (kW)^2]^{-0.5}$.) At 90% power factor, with 0.49 kVAR of reactive power, apparent power reaches 1.11 kVA and the loss index hits 1.23 units. By the time power factor sinks to 80%, each 1 kW comes with 0.75 kVAR, apparent power is 1.25 kVAR, and losses soar to 1.56 units. At 70% power factor, losses reach 2.04 units; at 50%, 4.04 units. Thus as the 1 kW of billed power stays the same, the grid losses can quadruple and the utility's generating costs can double: twice the cost for no more revenue. In many developing countries, power-factor compensation could often cut electric demand by tens of percent for no more than a fifth the cost of new generating capacity, since resistive losses vary as the inverse square of power factor—actually slightly more because of avoided conductor heating.

Two decades ago, those increased losses were believed to account for as much as one-fifth of all U.S. grid losses. If that level still prevailed today (good estimates are hard to find), those losses would cost, at retail prices, about $3 billion a year, all upstream of customers' meters, not to mention further substantial losses between the meters and the customers' load terminals.[121] Moreover, since grid components must be sized in kVA, not kW, reactive power requires, say, a line or transformer serving a load at 80% onpeak power factor to be oversized 25% relative to the billable kW it will deliver. If national-average onpeak power factor were around, say, 0.97, then 3% of utilities' annual ~$15-billion grid investment, or about $450 million a year lately, would be paying for excess capacity required by low power factor. Moreover, reactive current heats grid components, increasing losses of real current and hence wasting capacity, plant life, fuel, maintenance, and pollution sinks to make power that cannot be sold. Low power factor also proportionately reduces how much real power a given conductor can carry: because resistive losses rise as the square of current, at 80% power factor a utility needs 56% more wire cross-section (25% more diameter) to serve the same billable load, so it doesn't take much shortfall in onpeak power factor to create grid bottlenecks.

---

[120] Motor-owners often use the same technique. Capacitive compensation of motors' power factor as close as possible to the load terminals is the most helpful method because it also reduces in-house distribution losses. However, since capacitors are nonlinear devices, care must be taken to avoid harmonic generation and resonance (§ 2.3.8.1), as well as the potential for self-starting (see note 122 below).

[121] S.F. Baldwin (51) stated that 1–2 percentage points out of the then-normal ~7–9 average U.S. grid loss "can probably be attributed to the reactive components of the load." It's not clear how these figures should be updated. On 3 April 1989, EPRI's Bob Iveson estimated in a personal communication that the national-average power factor at the peak hour is an impressive 0.95–0.98 lagging, with no annual-average figure available. However, despite some utilities' good efforts to approximate unity power factor, many are less assiduous. Inadequate compensation at the peak hour will imposes the greatest burden on the grid precisely when it is most loaded and hottest from transmitting real power that customers can be charged for.

Finally, low power factor worsens system voltage regulation: the voltage regulation of a transformer, for example, may degrade from 2% at 90% power factor to 4–5% at a 60% power factor (514). In extreme cases, locally heavy surges of demand for reactive power (*e.g.*, through loss of a transmission link importing reactive power into a region) can cause bus voltages in that area to drop abruptly, triggering protective relays to trip generators offline and potentially causing the whole interconnected grid to collapse. However, overcompensation of power factor (from lagging to leading) carries its own risks[122] and hence cannot be done to excess without being at least as bad as the disease it aims to cure.

[122] Leading power factor can interfere with the utility's or other customers' operations, especially where modern control equipment is lacking. It can also create dangerous overvoltages, especially in unusual conditions where excessive capacitance is in parallel with power-system inductances, creating the potential for resonances at a distance (§ 2.3.8.1). This is easily avoided by following standard codes such as the U.S. National Electrical Code's Article 460 or the National Electrical Manufacturers' Association standard MG2, and is normally prevented also by fuses built into most U.S.-built power-factor-correction capacitors. (Fuses are a backup protection; it is much better to ensure that system resonance frequencies do not coincide with common harmonic frequencies in the first place.) It is also worth remembering that a de-energized induction motor driven by an external torque can self-excite and self-start when equipped with a shunt capacitor (§ 2.3.2.10.4). Finally, special design rules apply to three-phase capacitor banks, DC motors, multispeed and reversing motors, electronic motor controls or starters, and several kinds of unusual situations.

real—in phase with voltage—and thus able to do work. Power factor is the cosine of the phase angle between current and voltage, so if inductance, say from a standard motor, causes current to lag behind voltage by, say, 18%, then the power factor is said to be 0.951, or 95.1%, lagging. This can be increased to or nearly to the ideal of 1.0 by adding capacitance, which causes current to lead ahead of voltage. Out-of-phase or **reactive** current is required to produce the essential magnetic flux from the coils in the induction motor, but does not represent a permanent transfer of energy from one place to another—rather, it is a sort of oscillation of energy—so it cannot be metered and sold, but it does require sufficient capacity to carry it throughout the grid, and it incurs its share of system losses. Thus shaftpower comes only from volts times *in-phase* amperes. Yet transformers, cables, transmission lines, etc. must be sized for volts times *total* amperes, since flowing current incurs I^2R losses whether it is in phase with voltage or not. In a more comprehensive sense, the term "power factor" can include not only the phase component just described but also a waveform or **shape** component (§ 2.3.3.8.1). In an attempt to avoid confusion, the measure of both together is sometimes called **true** power factor. Since these two effects are separate, a motor-drive inverter designed to improve phase power factor but allowed to reduce shape power factor can have the net effect of reducing true power factor. The remaining discussion in this section deals only with phase power factor; we return to shape power factor in Section 2.3.3.8.1 when discussing harmonics and power quality.

2.3.2.3.1 Distributed resources' reactive contribution

Distributed resources can provide reactive power (hence reduce reactive losses) in at least three ways:

Benefits	
112	*Some end-use-efficiency resources can provide reactive power as a free byproduct of their more efficient design.*
113	*Distributed generators that feed the grid through appropriately designed DC-to-AC inverters can provide the desired real-time mixture of real and reactive power to maximize value.*

- Some end-use efficiency measures directly improve the customer device's power factor as a free byproduct of their real energy savings: for example, most premium-efficiency motors are designed to achieve better power factor (compared to standard-efficiency motors' typical ~0.7–0.9 at full load, declining linearly with smaller loads) under all conditions, and especially at low loads, without needing to have so much corrective capacitance installed.

- Distributed resources of any kind decrease current in the [largely inductive] grid circuit they serve by displacing power flow from remote generators, and in round numbers, each 1% decrease in circuit current will decrease VAR loss by 2%.[123] (214)

- Certain kinds of distributed generators can directly generate reactive power and reinject it into the grid on demand. These are the generators that, like photovoltaics or fuel cells, use an inverter to convert direct to alternating current to send back into the grid. Modern inverters can instantly and continuously adjust the phase angle of their output current to lead or lag the voltage by any desired amount (*e.g.*, in a PG&E/Sandia-developed design, from 0.10 lagging to 0.10 leading [603]). This real-time adjustment can be in response either to grid voltage (a surrogate for power factor in the local grid) or to a command by radio, power-line carrier signal, etc.

This last and most flexible feature is essentially free—even though it performs the same function as a costly static VAR compensator.[124] Its only material costs are the potential but minor cost of signalling devices, and the foregone generation of real power, because producing reactive power correspondingly sacrifices production of real power: the same current is being produced, only out-of-phase. (Thus a given kW of generating capacity cannot be used to produce real and reactive power at the same time: their total must add up to the same output.) Most of the time, it will be much more lucrative to sell kVA than kVAR to the grid. However, when reactive power *is* worth more, it is quite valuable to have a virtually no-cost capability to produce it. This is because such an inverter can inject reactive current into the grid more cheaply than installing, operating, and maintaining switchable shunt capacitors. Whether the inverter *should* be so operated depends on the relative local economic values of real and reactive power at the time. But the option of doing so will make this type of inverter "the technology of choice at distribution-level voltage and power ratings." (153)

The benefit of adjustable-power-factor inverters in distributed generators is summarized thus by a PG&E team (633):

> ...[T]he use of reactive power by inverters has been the source of some concern over their effect on feeder voltage levels at high penetrations. Reactive power is always consumed by line-commutated[125] inverters,

---

[123] VAR losses, like I^2R resistive losses, are proportional to the square of circuit current. Assuming that fluctuations in voltage and in power factor are relatively minor, power is directly proportional to current. Thus adding a 2-MW distributed resource to a feeder under a 10-MW load will reduce VAR losses upstream of the interconnection point by approximately 40%. The exact value can be determined by loadflow analysis.

[124] This specialized utility device usually contains both capacitors and inductive reactors, rapidly switched in and out of the circuit by high-speed solid-state devices under computer control, to control system voltage by continuously matching system requirements for reactive power. New versions and analogous transmission-level devices developed through EPRI's FACTS program can inject reactive support or change power flows in a fraction of a cycle (328).

[125] "Commutation" is the rapid switching of power flow back and forth that lets the inverter convert direct into alternating current. Line-commutated inverters use the change of polarity of the AC power line to control switching, so they yield a harmonic-rich 60-Hz square-wave output (at single phase), which may be filtered into a more acceptable waveform at higher hardware cost, but still typically yield total harmonic distortion (THD) around 6–23% depending on filtration (667). Self-commutated inverters are controlled instead by an internal oscillator that typically runs at much higher frequency than the line, so it can digitally approximate an accurate sinusoidal output, and its high-frequency switching transients can be filtered out relatively cheaply, yielding a very low THD, often small fractions of 1%. Advances in electronics have made some self-commutated units cheaper than corresponding line-commutated units—previously considered the lowest-capital-cost option because of their simple circuitry.

since it serves to "drive" the real power output. Self-commutated inverters [like synchronous generators], however, can be made to perform at a wide range of power factors....At a leading power factor, the device is actually producing rather than consuming reactive power. The adjustability provides additional operating flexibility. Thus, as with active harmonic cancellation [see § 2.3.3.8.1], it turns out that the effect of inverters with regard to reactive power can actually be corrective rather than problematic for the distribution system.

2.3.2.3.2 Benefits

Excessive reactive current interferes with the transmission of real current in two ways.

> ### Example:
> ### A reactive-power-supplying inverter
>
> Capstone Turbines inverters now have the hardware capability to provide automatically either a fixed or a dynamically compensatory amount of leading or lagging reactive power in addition to the remainder as real power. The dynamic power-factor compensation mode, displacing the cost of capacitors or other compensators, is the reactive equivalent of the real-power load-following control already built into the unit. For example, a ~30-kW Model 330 microturbine at 480 VAC can supply 30 kW real plus 23.8 kVAR (91). Software supporting the easy exploitation of this valuable feature should ship in 2002.

The first and often the more important is to disturb normal system voltages. Improper power factor (consumption of reactive current) causes voltage to drop, while local injections of reactive power boost voltage. Distributed resources that reduce reactive losses or inject compensatory reactive current can thus increase effective grid capacity. Second, low power factor also proportionately reduces how much real power a given

> **Benefits**
>
> **114** *Reduced reactive current improves distribution voltage stability, thus improving end-use device reliability and lifetime, and enhancing customer satisfaction, at lower cost than for voltage-regulating equipment and its operation.*
>
> **115** *Reduced reactive current reduces conductor and transformer heating, improving grid components' lifetime.*
>
> **116** *Reduced reactive current, by cooling grid components, also makes them less likely to fail, improving the quality of customer service.*
>
> **117** *Reduced reactive current, by cooling grid components, also reduces conductor and transformer resistivity, thereby reducing real-power losses, hence reducing heating, hence further improving component lifetime and reliability.*
>
> **118** *Reduced reactive current increases available grid and generating capacity, adding to the capacity displacement achieved by distributed resources' supply of real current.*

conductor or transformer can carry. Out-of-phase current heats those grid components just as much as in-phase current; it simply can't do work, and customers can't be billed for it. Both of these first-order benefits are often counted, as they should be.

In conventional first-order-only assessments of the value of improved power factor, load-flow analysis is first used to calculate reactive currents. These are then multiplied, as an economic surrogate, by the avoided cost of installing shunt capacitors to correct the usually lagging power factor resulting from the inductance of the grid conductors and transformers. For example, in PG&E's Kerman analysis (597), that value per kVAR (kilovolt-ampere reactive) ranged from $16.9/kVAR for 300-kVAR capacitors to $5.30/kVAR for 1800-kVAR capacitors pole-mounted on a distribution feeder and including controls (1992 $). Transmission and subtransmission shunt capacitors ($40–60/kVAR) were analyzed because the installed capacitor cost rises with voltage depending on size, *e.g.*, $8.90/kVAR for a 12-kV distribution substation installation, $58/kVAR at the 70-kV substation level, and $59/kVAR for 230-kV transmission.[126] In this particular case, the 500-kWAC PV generator eight miles from the substation was estimated to save not only 58.5 MWh/y of real power losses (5% of plant output) but also 350 kVAR of reactive losses (735). Those saved reactive losses were allocated 17% to distribution, 42% to the substation transformer, and 41% to transmission. Multiplying by their respective costs of shunt capacitors, the present-valued 1992-$ reactive-power monetary savings of $27,224 were only 3% distribution, 48% transformer, and 49% transmission (619). These values were conservatively low because the inverter was assumed to have unity power factor, when in fact it had a real-time-adjustable power factor and hence could provide even more reactive power support when desired (618). Thus the total value calculated for the Kerman PV system's reactive power support—$9.60/kWy, or ~$0.0042/kWh[127]— would have been larger if the inverter had been credited for its adjustable-power-factor feature (§ 2.3.2.3.1).[128]

Voltage regulation also has a direct and avoidable cost. Devices to control voltage within the narrow ranges required for efficient distribution of real power and for reli-

able operation of customer devices include, but are not limited to, the same capacitors used to supply reactive power. Other kinds of voltage regulators may also be used. To the extent that they are needed and normally used for voltage support, they may represent an avoidable capital and operating cost beyond that of the shunt capacitors avoided by certain distributed resources (§ 2.3.2.3.1).

For example, the Kerman analysis (641) found that stretching the normal 5-year servicing of in-place substation and line voltage regulators to 7 years, because of lower line currents, had a present value of $26,145, or $9.88/kWy (1992 $), or $0.0043/kWh, two-thirds of it at the substation (§ 2.3.2.7) and the rest on the line. This is only the value of eliminating almost two service operations over the assumed 30-year life of the voltage regulators. In fact, "the PV generation would likely increase the life of the regulators [beyond 30 years], [but] no calculation was made of this value." (621)

Other, second-order benefits of injecting reactive power often go uncounted. For example, reduced conductor heating also reduces resistivity, amplifying the reduction in both real and reactive grid losses. Moreover, to the extent grid capacity is constrained by voltage stability, improved stability can achieve additional economic benefits just like those of reduced real-power losses. Those benefits are similarly the product of the size times the value of the reduced losses. They can be quite large at

---

[126] Although the effect is probably not important, in principle an inverter's nearly free option of reactive power support should be credited with a longer operating life and greater availability than switched capacitors, which are more subject to chemical deterioration. The inverter's routine maintenance for reliable production of real power will also encompass the reactive-power-support capability at no extra cost, rather than being dedicated to the reactive function as in the case of the proxy capacitors. And it should incur almost no maintenance costs.

[127] Converted, as described above, at the evaluation year's 25% capacity factor or 91% Performance Factor, when it produced more than 1,080 MWh and saved a further 58.5 MWh of losses (735).

[128] Fuldner (262) distinguishes transmission-level MVAR costs for adding an additional capacity step or providing a complete new installation.

the downstream end of the distribution system if the distributed resources are carefully deployed in space and time.

Improving voltage stability may increase reliability or add customer value. For example, voltage regulation, whether real or reactive, is added in discrete steps by switching capacitors in or out or by changing taps on a transformer. This has costs not just for utilities but also, less visibly, for customers. Even with good active control that prevents gross overvoltages, the control steps may cause problems with some customers' possibly voltage-sensitive special loads. Moreover, voltage control can have major hidden value because many common customer devices have an operating life that falls steeply with even modest overvoltages applied over long periods. Such values do not appear to be analyzed in the literature, but may be substantial.

Reactive power support can be provided at no extra cost (except signaling and software) as a coproduct of real power by modern inverters (§ 2.3.2.3.1). Such inverters are conventionally used by DC-output distributed generators, but may also be part of demand-side distributed resources, notably adjustable-speed (variable-frequency) electronic drives on customers' motors. This too does not appear to be reflected in the literature on the economics of end-use efficiency for motor systems—an important demand-side distributed resource, since most electricity goes to motors; many of those are suitable for inverter drive; and the inverters can easily be designed for adjustable power factor.

2.3.2.4 Avoided voltage drop

Real as well as reactive power causes voltage to drop slightly on the way from genera-

tor to customers. "Conceptually, one might think of this voltage drop as providing the incentive for current to flow outward." (631) The bigger the load, the bigger the voltage drop in proportion. On a heavily loaded line, voltage drop may become excessive, requiring the installation of voltage-controlling equipment with high capital and operating costs, such as regulators, boosters, and capacitors. This is often particularly required at the end of long feeders, such as in rural areas, whose low load density makes such equipment even harder to amortize from revenue.

Local generating or demand-side resources can reduce or avoid such installation by reducing line current. The lower current reduces the voltage drop. Specifically, the distributed resource yields a voltage rise equal to the difference of currents at the two ends (because of losses in between) divided by the resistance of the line. If the distributed resource is well correlated with the loadshape, like end-use efficiency or like PV generation in areas with big air-conditioning loads on hot afternoons, then its economic value in avoiding voltage support increases, because high loads require more voltage support. Naturally, since the voltage drop along a conductor cannot be influenced from a distance, the distributed resource must be sited along the conductor in order to have this desirable effect. This is exactly where such distributed generators as household photovoltaics are likely to show their greatest popularity and benefit. Even the near-substation Kerman PV array provided 3V of support on a 120-V base (735).

2.3.2.5 Ampacity savings from daytime-correlated resources

Conductor ratings are expressed in the max-

Benefit

119
Distributed resources, by reducing line current, can help avoid voltage drop and associated costs by reducing the need for installing equipment to provide equivalent voltage support or step-up.

Benefits

120   *Distributed resources that operate in the daytime, when sunlight heats conductors or transformers, help to avoid costly increases in circuit voltage, reconductoring (replacing a conductor with one of higher ampacity), adding extra circuits, or, if available, transferring load to other circuits with spare ampacity.*

121   *Substation-sited photovoltaics can shade transformers, thereby improving their efficiency, capacity, lifetime, and reliability.*

122   *Distributed resources most readily replace distribution transformers at the smaller transformer sizes that have higher unit costs.*

imum number of amperes of current that can safely be continuously carried without making the conductor so hot that it anneals, loses tensile strength, and sags. Ratings depend in real time on the present and recent (because of thermal lags)[129] levels of

loading relative to windspeed,[130] wind direction, and ambient air and sol-air temperature (the latter measures how hot an object becomes when exposed to both air and direct sunlight).[131] Sophisticated new control systems measure these parameters in real time and feed the resulting real-time line capacity into grid control software so that operators can maximize system security and economy by squeezing the maximal safely available capacity from the transmission lines (328). However, the simpler and more traditional method was for utility managers to use current flow as the main, and operationally the most useful, predictor of ampacity—often using round numbers rated for summer and winter and for normal or emergency operation (646). For common sizes of aluminum transmission conductors, for example, the summer static ampacity ratings are several percent lower in the daytime, when the sun is heating the conductor, than at night (Figure 2-57).

This means that distributed resources that produce power in the daytime, like PV generation,[132] or in concert with daytime loads, like end-use efficiency or fuel-cell cogeneration, have a special economic advantage (604):

> Consider a PV facility (or other peaking distributed [resource]...) that provides enough current to make up the difference between daytime and nighttime rating. If this PV facility is reliably available during daylight hours, the [higher] nighttime rating of the conductor can be used [instead of the constant conductor rating, which assumes daytime ampacity]. This amounts to a 15% difference in the normal rating of 75.5 kcmil [thousand circular milli-inch[133]]

Figure 2-57: Daytime supply's line support *is* worth more
Conductor ampacity sags in the daytime as the sun heats the metal.

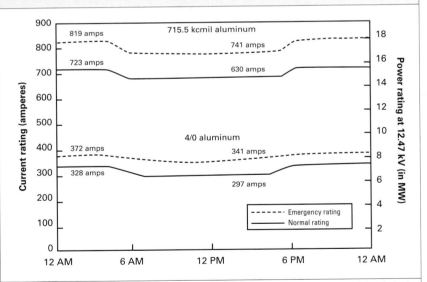

Source: D. Shugar *et al.*, "Benefits of Distributed Generation in PG&E's Transmission and Distribution System: A Case Study of Photovoltaics Serving Kerman Substation" (PG&E, November 1992), p. 3–7

[129] This lag makes it possible to overload lines briefly—perhaps in alternation—then allow them to cool before lasting damage is done. Sophisticated control practices therefore use dynamic, not just static, ampacity ratings, typically extracting a further ~10–20% of effective capacity. There is no general convention on which approach should be used as the baseline in calculating distributed resources' ampacity-expanding benefits, but dynamic ratings do provide an upper bound.

[130] Ampacity is so sensitive to air cooling that halving a 1.22-m/s windspeed can cut line capacity by one-fifth.

[131] The calculation of ampacity also takes account, as it should, of the conductor's emissivity—how well it radiates away infrared energy. This might be somewhat adjusted by suitable surface treatments.

Aluminum (Al) and a 0% difference in the normal rating of 4/0 Al. [The former size]...has a greater difference between its rating with and without sunlight effects than 4/0 Al because it has a larger diameter, and hence more surface area to be heated by the sun. This conductor benefit is unique to distributed solar generation technologies because they naturally become [more] available as conductor capacity decreases [and hence becomes more valuable].

This can avoid costly increases in circuit voltage, reconductoring (replacing a conductor with one of higher ampacity), adding extra circuits,[134] or, if available, transferring load to other circuits with spare ampacity. (In the Kerman case, unlike many others [622], "Unfortunately, there [were]...no obvious upgrades of conductor capability that could be eliminated, since some of the 12 kV circuitry was previously reconductored.") In principle, similar considerations apply to transformers that are also exposed to heating by ambient sun and air, since avoiding continuous or transient (dynamic) overload and hence overheating has a comparable value in reduced losses, extended life, improved reliability, and avoided upgrades or expansions of capacity. Such shading can be provided by photovoltaic panels installed at the substation, with due care not to interfere with conductors. While deferred or avoided transformer expansion is commonly counted as a distributed benefit,

the literature does not appear to treat this daytime-load-correlation benefit[135] for transformers—potentially a nontrivial one because they are rather expensive.

Nor does the literature seem to note an additional benefit—that the diseconomies of scale in relatively small transformers, shown in Figure 2-58, can be offset because those smaller units are most readily displaced by distributed resources.

2.3.2.6 Capacity expansion

An almost universal benefit of distributed

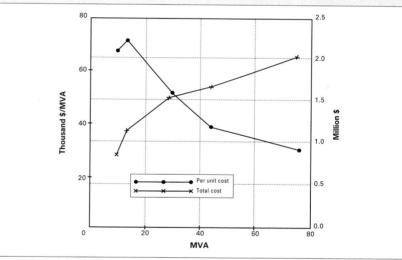

Figure 2-58: Distributed resources can offset distribution transformers' unfavorable scale economies
Typical transformer cost (1992 $ per MVA of installed continuously-rated capacity) is greatest per MVA for the smaller units most readily displaceable by distributed resources.

Source: D. Shugar *et al.*, "Benefits of Distributed Generation in PG&E's Transmission and Distribution System: A Case Study of Photovoltaics Serving Kerman Substation" (PG&E, November 1992), p. 3–7

[132] However, for PVs this value is partly offset because crystalline silicon arrays become less efficient when hot. For example, typical Siemens monocrystalline material yields ~13% lower output in an uncooled mount on a typical Sacramento house roof on hot, sunny days, when the silicon warms to ~52°C or more (*vs.* ~47°C if ground-mounted). (739) However, amorphous collectors generally become *more* efficient at high temperatures.

[133] One "circular mil" (cmil) is the cross-sectional area of a conductor 0.001 inch in diameter.

[134] Adding more conductors or increasing the diameter of existing ones adds weight and may therefore require structural strengthening of the towers or their footers to deal with the dead load, ice loading, and large lateral loads from windforce. Even with these changes, however, the cost is usually less than that of building a new line with the same marginal capacity (262).

[135] However, it should be readily calculable with such tools as PG&E's Transformer Capacity Analysis Program (TCAP), which "simulates transformer winding hot spot temperature as a function of loading, physical parameters, and ambient temperature." (623) Such a tool, augmented if necessary for basic solar and infrared physics, could also be used to test the hypothesis that there is significant capacity and reliability value to applying to transformers—normally painted dark gray—the sorts of high-emittance (up to ~0.97), low-solar-absorptance (down to ~0.07) paints developed for keeping buildings' roofs cool. Perhaps the leading authority on such materials is Dr. Hashem Akbari of Lawrence Berkeley National Laboratory, Berkeley CA 94720.

Benefits

123 *Distributed resources defer or avoid adding grid capacity.*

124 *Distributed resources, by reducing the current on transmission and distribution lines, free up grid capacity to provide service to other customers.*

125 *Distributed resources help "decongest" the grid so that existing but encumbered capacity can be freed up for other economic transactions.*

126 *Distributed resources avoid the siting problems that can occur when building new transmission lines.*

127 *These siting problems tend to be correlated with the presence of people, but people tend to correlate with both loads and opportunities for distributed resources.*

resources, and among the best-known and most obvious of all distributed benefits, is the ability to defer or avoid adding grid capacity (553). Although the potential for this lucrative investment avoidance is highly system-specific, a "first glimpse at utility-wide applications" applied to an unnamed utility with several million customers on 3,000 feeders, using many simplifying assumptions that cut both ways, found "potential impacts ranging up to 10% of total distribution capacity in 10 years in high growth scenarios, and approaching 100% of *new* distribution capacity at lower growth rates." (553)

Depending on where the resource is installed, distributed resources may displace grid capacity at all levels from the local tap or feeder all the way upstream to the power-plant switchgear and stepup transformer. As noted above in Section 2.3.2.2, the further downstream the distributed resource is sited, the greater are the avoided compounding grid losses and hence the more capacity is displaced.

Obviously, a distributed resource displaces the capacity it sends out (or saves if it is a demand-side resource). Transmission is

[136] The final evaluation (§ 2.3.2.3.2) (735) found 498 and 27 kW respectively for real power, plus 350 kVAR of reduced reactive power losses compared with the originally predicted 545. Transmission system firm capacity was increased by a total of 450 kW onpeak—90.4% of the PV capacity added (§ 2.2.8.4).

quite expensive—depending on voltage, capacity, and pole choice, about $74,000 (60 kV) to $340,000 (230 kV) per km, excluding right-of-way (265). However, a distributed resource *also* displaces the capacity it frees up by reducing line losses. In the Kerman case, this effect, counting both real and reactive power, increased the 500-kWAC PV array's nominal saving in transmission capacity to 534 kW (264).[136]

A standard text (768) offers this example of the value of "demand cost of losses," *i.e.*, "the total cost of the capacity to provide the losses and move them to their points of consumption":

> Consider a typical 12.47 kV, three-phase, overhead feeder, with 15 MW capacity…, serving a load of 10 MW at peak with 4.5% primary-level losses at peak (450 kW losses at peak), and having a load factor of 64% annually. Given a levelized capacity cost of power delivered to the low side bus of a substation of $10/kW, the demand cost of these losses is $4,500 a year. Annual energy cost, at 3.5¢/kWh, can be estimated as:
>
> 450 kW losses at peak × 8760 hours × (64% load factor) × 3.5¢ = $56,500

Thus, the losses' costs (demand plus energy costs) for this feeder are nearly $60,000 annually. At a present worth discount factor of around 11%, this means losses have an estimated present worth of about $500,000. This feeder, in its entirety, might include four miles of primary trunk (at $150,000 a mile) and thirty miles of laterals (at $50,000 a mile), for a total capital cost of about $2,100,000. Thus, *losses' cost can be a significant portion of total cost, in this case about 20%.* Similar loss-capital relations exist for all other levels of the T&D system, with the ratio of losses' costs/capital cost increasing as one nears the customer level (lower voltage requirement has higher losses/kW).

Saving transmission capacity is especially valuable because siting problems make it very difficult to build new transmission lines in most of the United States. Of the 10,127 line-miles of transmission additions originally planned for North America during 1995–2004, many "may be delayed for many years or may never be constructed." (768) The National Energy Plan proposed in 2001 posited a crisis in U.S. transmission capacity and called for a crash program of federally facilitated (even preëmpted) construction of new powerlines to augment a system now "strained to capacity." Distributed and demand-side alternatives were not considered in the plan, nor proposed in its implementing legislation, as alternatives to be compared with proposed transmission capacity. Fortunately, however, except in rare instances of encroachment on wild or sacred lands (such as the OLE powerline project in New Mexico, rejected for both reasons), objections to line siting typically come from local residents. But where there are the most people, making siting most difficult, the presence of people will also bring electrical loads and hence distributed resource opportunities.

Different elements of the grid may be added, expanded (such as adding a new bank of transformers), upgraded (such as reconductoring a line, or increasing the capacity of an existing transformer bank by replacing it with higher-capacity or lower-loss transformers), or reconfigured (such as switching loads between circuits). This may occur for a variety of reasons, including:

- meeting load growth,

- replacing aging equipment,

- strengthening the capability, topology, or flexibility of existing equipment to ensure reliable supply under various contingencies,

- improving system efficiency,

- facilitating load shifting between circuits or components to improve capacity utilization, improve reliability, or run equipment cooler to extend its life, or

- improving interconnection support with another system.[137]

Most utility planners prefer to evaluate displaced grid capacity as a deferral rather than an outright avoidance, since they are used to dealing with steady load growth that sooner or later outruns the previous "lump" of grid capacity installed. The deferral value is then the difference in present value between the normal installation schedule and the deferred one. (If the analysis uses a fixed time horizon, then the extra value of buying the capacity later and hence possibly having it last beyond that horizon must be taken into account.) Then (642):

> The value of the deferral is driven by the difference between the utility's cost of capital and the inflation rate. For example, suppose a utility delays a $100,000 investment for one year. The cost of the investment in year 2 would be higher for inflation, but the utility would have avoided having to raise the capital for the investment in year 1.

---

[137] This may be an option for primary distribution circuits on the periphery of a system, but may require special equipment such as a phase-shifting transformer, which is quite expensive and therefore valuable to avoid.

The deferred investment can be expressed by the following equation:

$$\text{Deferral Value} = \text{Investment} \times [\,1 - [(1+i)^n / (1+c)^n\,] - \text{Extra Life}]$$

where i is the interest rate, c is the cost of capital, n is the number of years of deferral, and the extra life is given by:

$$\text{Extra Life} = \text{\%Life Past Study} \quad [(1+i)/(1+c)]^n$$

For example (601), if

- a new distributed resource were located in the right place and at the right time on a radial distribution system where a 10.5-MVA distribution transformer is approaching its maximum capacity,

- the preferred alternative were an upgrade to a 16-MVA transformer with an installed cost of ~$1.15 million (1992 $),

- the old transformer had negligible salvage value because it was fairly old, and

- a 0.5-MW PV resource contributed power on the transformer's low-voltage side, highly available at peak loads, then

- that modest resource might enable the transformer to operate "within its load limit throughout the year, deferring the need for a larger transformer. Given load growth forecasts and the amount of distributed generation available, one can estimate the number of years for which the installation of the larger...transformer can be deferred"

and hence one can estimate the economic value of that deferral—in this case, $115/kWy for a five-year deferral (614). Reconductoring distribution lines through the same PV resource's ~25-A onpeak reduction in an aboveground, non-urban standard 12-kV line would save on the order of $27,000–$46,000/km (1992 $), depending on whether an old line were reconductored or a new line constructed (639). Transmission capacity directly deferred by the distributed resource's modest output would be relatively less important because a half-MVA is such a small part of a typical transmission line's capacity, but more "significant for the transmission system are loss savings and the transmission system capacity value associated with reduced load, which apply regardless of the reduction's magnitude." (602)

That is (637), "In addition to providing power loss savings, the reduction of current on transmission and distribution lines attributable to loss savings frees transmission capacity for service to other customers." The loss savings can be determined from loadflow simulations and the system average marginal transmission capacity cost ($282/kW in PG&E's 1990 General Rate Case filing), unless, preferably, a more site-specific and time-specific cost is known. For the Kerman PV installation, the avoided grid losses were expected to be worth only $21/kWy and the reactive power $8/kWy, but the transmission capacity was worth $44/kWy, ranking third (among grid-related benefits) behind the distribution transformer deferral at $115/kWy and the initially estimated[138] reliability benefits of $205/kWy (§ 3.3.5.5). Deferral values can be even larger in some circumstances. For example (583), Boston Edison Company's recent deferral value averaged $64/kWy, but reached $137/kWy in Hopkinton and $824/kWy for the North End area.

---

[138] The reliability benefit, the largest distributed benefit in this instance, was later reduced on reevaluation for site-specific reasons (§ 2.3.3.8.2). However (627), it could be larger with automated distribution than with the manual circuit assessed at Kerman, depending partly on localized value-of-service data.

This "decongestant" property is of course most valuable at the times and places where the grid is most congested, and should attract compensation based on the "congestion rent" that a market-priced common-carrier grid does or should charge (431). Naturally, evaluating the value of transmission capacity is very complex, depending not only on marginal cost of new capacity but also on the time- and space-varying capacity/demand balance, projected power-wheeling economics, and supply-side options and locations. However, this task is rapidly shifting from theoretical analysts to market actors whose real-time behavior will offer an increasingly available and convenient guide to economic value and who have profit motives to seek out the most lucrative hot-spots in the system (§ 1.4.2).

One intriguing hint that avoided grid costs and losses can tip scale decisions comes from a 1992 PG&E analysis (645) comparing fifty 1-MW distributed photovoltaic plants at the substation level with a single 50-MW central PV plant at Carissa Plains, PG&E's best PV site. Like the Los Alamos coal-plant comparison (§ 2.3.1.1), this comparison is valuable because it compares a single technology applied at two different scales—and in this case the difference is 50-fold, not 3-fold. The study found the distributed 50×1-MW version could yield a net cost of \$29/kWy—more favorable than the \$110/kWy for the centralized version. This difference was mainly due not to generation-related scale effects, as in the Los Alamos coal study (§ 2.3.1.1), but rather to grid-related benefits that were about one-third larger than the foregone economies-of-scale in maintenance and in the balance-of system investments.[139]

That is, *the grid advantages of the smaller PV plants more than offset their generation disadvantages.* The distributed configuration would save \$38/kWy less on energy output, minimum-load and QF savings, and pollution, and would cost \$71/kWy more for balance of system and maintenance, but would capture \$190/kWy of distributed benefits. This conclusion, though inexact, emphasizes the importance of grid-support benefits, especially capacity deferral or avoidance.

2.3.2.7 Life extension

Sections 2.3.2.2 and 2.3.2.5 above noted the many benefits of unloading conductors and transformers so as to reduce their operating

Benefits
128   *Distributed resources' unloading, hence cooling, of grid components can disproportionately increase their operating life because most of the life-shortening effects are caused by the highest temperatures, which occur only during a small number of hours.*
129   *More reliable operation of distribution equipment can also decrease periodic maintenance costs and outage costs.*
130   *Distributed resources' reactive current, by improving voltage stability, can reduce tapchanger operation on transformers, increasing their lifetime.*

temperature. This benefit has unexpected dimensions that emerge only on fuller examination.

If an aerial conductor becomes too hot, it may soften, irreversibly sag from its own weight or from wind loading, get too close to the ground for safety, and ultimately even melt.[140] If an insulated (*e.g.,* underground)

---

[139] These grid benefits were borrowed from the illustrative Kerman 0.5-MW PV plant—a case in which no deferral of line reconductoring was available as a benefit because in that particular area it had already been done. This would probably not be true for all 50 of a suite of dispersed sites.

conductor overheats, its insulation will deteriorate and ultimately fail. If a transformer overheats, its cooling oil will chemically deteriorate and may ultimately become conductive, causing a destructive short-circuit and possibly a fire or explosion that can in turn damage or destroy other utility and neighboring assets. All these acute failure modes, however, are mirrored by far more gradual processes of chemical and metallurgical change that over long periods may cause conductors, insulators, switches, transformers, capacitors, and other devices to deteriorate and become unreliable. Chemical reactions typically double their rate with each ten Celsius degrees by which temperature rises, or conversely halve their rate with each ten Celsius degrees by which temperature falls. Heat is therefore the prime enemy, and cool running is the friend, of longevity in all kinds of electrical equipment.

This physical reality, and the political reality that customers dislike prolonged outages, lead well-run utilities to be reluctant to run costly, long-lead-time, and mission-critical equipment near its thermal rating for long. PG&E's Kerman study summarizes (600):

> The reduction of current due to distributed generation [or demand-side resources] is

particularly significant for conductors and transformers. Even a small reduction of the peak current these devices experience [especially on the hottest days when they are already hot from ambient air and sun] can have a significant effect on their operating temperature and any overloading they were experiencing. Although the device may approach overload only during a few hours of the year, it would ordinarily be replaced with a new, larger device. With an eye toward future load increases and because of technical constraints [including "the intrinsic cost and nuisance of replacement, and the limited range of sizes available of transformers or conductors"], utilities would typically choose a much greater capacity for the new device than what was immediately required.

Running cooler because of lower current can therefore defer costly and lumpy (Figure 2-4) upgrades and expansions by extending the life of existing equipment, which is often approaching the end of its book life and hence has little or no salvage value. Periodic maintenance costs and outage costs may also be concomitantly reduced. In the Kerman case, the 0.5-MW substation-level PV installation looks like a modest load reduction for a 10.5-MVA transformer, as shown in Figure 2-59: Yet this seemingly minor peak-shaving would cool the transformer by ~4–7 C° on peak days, extending its life by ~5 y. That

[140] The stretching happens gradually, so lines have both a normal (continuous and indefinite) ampacity rating and an emergency rating (for a specific period such as several hours).

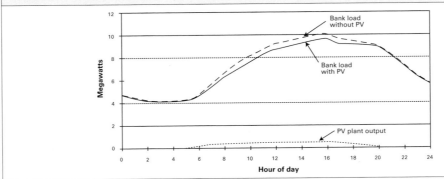

Figure 2-59: A little PV capacity goes a long way in relieving substation load
Photovoltaics with only 4.8% as much capacity as the transformer bank they support can have a disproportionate onpeak benefit. This is even true if PV output precedes system peak (see Figure 2-38).

Source: D. S. Shugar and T. E. Hoff, "Grid-Support Photovoltaics: Evaluation of Criteria and Methods to Assess Empirically the Local and System Benefits to Electric Utilities" (*Progress in Photovoltaics: Research and Applications*, 4 May 1993), p. 243, fig. 5

was worth an impressive $89 per kWy in deferral value in the original Kerman evaluation (621), or $0.039 per PV output-plus-avoided-losses kWh.[141]

Although normal sensors measure only transformer oil temperatures, the tools for simulating transformers' winding temperatures are readily available. For example, an appendix to the Kerman study (621) used a detailed heat-transfer dynamic model to simulate that under a 13.9-MVA peak load (32% overload) on an August afternoon rising to 42°C ambient, the nominal 10.5-MVA transformer bank, with rated losses of 14 kW at no load and 75 load loss at full load, would gradually heat to 154°C at the hottest spot in its windings. Being above the maximum allowable level of 125°C, this excursion would be expected to shorten the transformer's life by half a percent. Sustained operation under these conditions would be expected to burn it out in fewer than seven months. Both life-shortening and life extension as a function of loads and temperatures follow physical and chemical processes sufficiently well known that their economic value can be accurately estimated with such tools.

To put it another way, since deterioration is twice as fast for each ~10 C° of heating, the last few bins in Figure 2-60's histogram of top oil temperature represent most of the life-shortening. Therefore shaving just the rare peak loads that those hottest few hours represent—exactly as distributed resources (notably photovoltaics) can do—will capture most of the valuable life-extension benefit. Another kind of grid-equipment life extension comes from voltage support (from, say, a PV installation with a smart inverter),

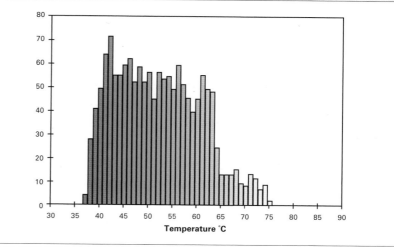

Figure 2-60: 1991 substation transformer bank 2 top oil temperature histogram
PG&E's measurements show how just the few hottest bins of temperature distribution cause most of the life-shortening.

Source: D. Shugar, "Grid-Support Applications for Photovoltaics in the Electric Utility System: A Test Case at Kerman Substation" (Distributed Utility—Is This the Future? EPRI, PG&E, and NREL conference; December 1992)

which, as described in Section 2.3.2.4 for real power and 2.3.2.3 for reactive power, reduces voltage fluctuations. Transformer tapchangers to change output voltage therefore need to be activated less frequently, reducing wear and tear. For example, the Kerman study (640) conservatively assumed a unity-power-factor rather than the actual adjustable-power-factor inverter. Yet it still found that the 500-kWAC PV generator on the feeder could extend the normal rated life of the substation transformer's top-mounted tapchanger (whose 32 taps can adjust secondary voltage by ±10%) from 20,000 tapchanges over 5 years to 20,000 tapchanges over *20* years, while increasing the interval between service calls from 5 to 7 years (617, 624). More broadly (604), "Each year, 5,000 substation voltage-regulating device operations are typically expected, with service required after 25,000 operations or once every five years....A typical cost for 12 kV [sub]station regulator service is about

---

[141] The final evaluation (735) shows a range of $16 to $88/kWy for the combined value of the transformer and tapchanger life extensions, compared with the original evaluation's total of $99/kWy. The difference is apparently due to the later finding that "it is relatively easy to switch load in the Kerman area"—a site-specific detail rather than a general truth.

$20,000. For a PV system which reduced variation of peak load current by about 25 Amperes, it is estimated that 1000 tap changing operations would be saved. This would result in a delayed service cost of about $5,000...." In the Kerman case, the present value of that deferred maintenance was worth $10 per kWy of substation-level PV output (620). That's a significant increase in value just from counting one kind of deferred substation maintenance. It compares, for example, with $47/kWy for avoided transmission capacity (in a case with no reconductoring deferral opportunities). There are probably other kinds of deferred maintenance awaiting similar analysis.

2.3.2.8 Repair, rerouting, and outage duration

Resources nearer loads inherently boost reliability: distributed resources reduce electrons' average haul length and hence

---

Benefits

131   *Since distributed resources are nearer to the load, they increase reliability by reducing the length the power must travel and the number of components it must traverse.*

132   *Carefully sited distributed resources can substantially increase the distribution system operator's flexibility in rerouting power to isolate and bypass distribution faults and to maintain service to more customers during repairs.*

133   *That increased delivery flexibility reduces both the number of interrupted customers and the duration of their outage.*

---

their exposure to mishaps in the grid. However, that is often not their biggest reliability benefit. A more important, though subtle, benefit is that once a fault does

occur in the grid, appropriately sited distributed resources can substantially increase the distribution system operator's flexibility in rerouting power to isolate and bypass distribution faults and to maintain service to more customers while repairing those faults. This rerouting reduces the product of the number of customers affected and the duration of outage they experience. However, the existence, size, and economic value of this benefit are all highly specific to the topology, circuit capacities, and outage characteristics of the grid, especially the part in which the distributed resource is located, so no generic valuation is possible.

In the case of the nominal 500-kWAC PV array supporting PG&E's Kerman substation (638), its close-to-the-customers support may enable customers normally served by a certain feeder be served instead by the distributed resource while repairs are being made. The outage will then last only long enough to locate the fault and open and close a few switches—perhaps a few seconds—rather than the full duration of the repair. Such "salvageable" outages are valuable in avoiding customer inconvenience and dissatisfaction. The final Kerman evaluation (735) stated: "Testing proves customer outage time can be reduced."

An analogous study for a 2-MW fuel cell in Los Angeles (219) reached qualitatively similar conclusions. It also identified six limitations on that outage-mitigating potential, discussed in Technical Note 2-5. Those limitations, especially the first, must be taken into account in evaluating the distributed benefit of rerouting in a particular situation. However, they do not, in general, contradict the basic idea that this benefit can be important, especially when we recall two key facts (at least for the United States):

- most outages (by some estimates as high as 99%) arise in the grid, and

- around 90%[142] (328) to more than 95% (110) of those stem from distribution failures, chiefly weather-related.

Almost all distribution failures, in turn, come from overhead lines and cables rather than from fixed equipment, although on the rare occasions when a substation transformer does fail, its repair time *averages* four days (560). This requires alternative power routing to avoid very unhappy customers. Yet with some exceptions noted below (§ 2.3.2.10.1), normal U.S. distribution design radiates feeders from a single, nonredundant substation. Avoiding transformer failure in the first place is thus a very high priority for any utility sensitive to its customers' needs, and carries a correspondingly high economic value.

2.3.2.9 Summary: Prospecting for grid-support distributed resource opportunities

Technical Note 2-5: Limits on distributed resources' ability to mitigate distribution outages

Paraphrasing the Los Angeles fuel-cell study's findings (102) *with some commentary added:*

- *Most customers whose loads can be transferred to a temporary-support circuit will lose power only until the switchover, not for the full duration of the repair. The difference between these two durations is the outage length avoided. However, while automated distribution systems can switch over in seconds (or less) to a few minutes, manually operated distribution systems may require one-half to one hour for an operator to transfer the loads. In such systems, outages shorter than that transfer time cannot be mitigated, with or without distributed generation. (However, if the utility's operating procedures permit continued operation of the distributed resource within an "island" isolated from the fault and from the entire rest of the grid, then customers near the resource may avoid shorter outages too.)*

- *Loads can be temporarily shifted to other circuits without having or needing distributed resources. Of course, if those resources are considered for siting where prolonged outages have lately occurred—revealing a weakness in rerouting capability—then this limitation isn't relevant.*

- *Distributed generation may be less necessary to support load transfers in offpeak hours, when the system may offer more flexibility in switching among lightly loaded circuits. However, this is true only insofar as the outage can be repaired before peak hours arrive.*

- *Depending on capacities and loads at the time of the outage and during the repair period, conductor ampacity may be too small to transfer loads, making some peripheral loads unsalvageable. This is obviously an issue specific to the characteristics of the primary and secondary feeders involved.*

- *Loads can't be transferred or salvaged beyond the generating capacity available for transfer to the temporary-support circuit. This limitation may not be important in practice, because peak demands pass quickly, outages usually don't last more than a few hours, and some utility resources, such as fuel cells, have large overload capacity at need.*

- *Owners of standby generators won't see a significantly shorter outage time if distributed resources are present (though in some cases it may let them ride through the outage without starting their generator at all). However, faster restoration of utility service does save standby generators some operating hours. Distributed resources may also make service sufficiently more reliable, in reality or perception or both, to dissuade some customers from buying standby generators in the first place.*

Capturing the benefits of rerouting supported by distributed resources is not automatic. It requires that those resources' siting be optimized by detailed consideration of where outages tend to occur and what switching options are available or could be added or moved. If siting choices are limited by land or other issues, one might, if possible, consider "moving switches in order to permit or enhance reliability gain....The cost of moving a recloser in the distribution system is approximately $10,000, which was small in comparison to the value of added reliability" calculated in this example (598). The Kerman study team also coordinated "prospecting" for the greatest potential benefits from reliability and from reducing real and reactive power losses. Happily, the same site was able to maximize all three benefits (622). This may be rather widely true, although there is not yet good evidence either way.

[142] EPRI's proprietary 1995 *TAG™, Distributed Resources* (64), gives a somewhat lower figure, but the difference does not seem important.

What SMUD Solar calls "Rules-of-Thumb for High-Value Locations" for distributed (specifically, solar) generation to support the grid may be paraphrased thus (525):

- transformers or feeders near capacity and requiring upgrade or expansion soon

- substations or feeders with high solar coincidence

- distribution planning areas with low or moderate growth (because if demand-side resources were not employed, rapid growth, forecast with high confidence, may call for large enough increments of supply that distributed resources would have less advantage from modularity and long lead time)

- urban areas with high grid construction costs and right-of-way problems

- high concentrations of low-rise commercial customers (partly because of their solar-coincident air-conditioning loads)

- summer-peak distribution planning areas

- long, skinny feeders (~25+ km) with voltage or power-quality problems

To these opportunities should be added one more, increasingly prominent in older cities: distribution planning areas with equipment that has reasonable capacity margins but is becoming old or unreliable (often both). Its looming replacements, and their inconvenient costs, could then be valuably postponed by supporting the equipment during peak periods to extend its life.

2.3.2.10 "Negaloads"

vs. engineering realities

An authoritative PG&E analysis in 1992 concluded for PVs, as it could well have done for virtually any other distributed generating resource, that (635):

> Over the past decade, since the technical and economic feasibility of photovoltaic technology has warranted the discussion of a variety of grid-connected applications, many analysts have raised concerns about the impact of dispersed PV generation on safety, protection, and power quality, especially in the distribution system....We... argue that given the technology progress[,] particularly during the past several years, the impact of dispersed PV need no longer be worrisome to the distribution engineer. Nonetheless, we emphasize that this conclusion assumes prudent planning and operating strategies.

A detailed National Renewable Energy

Laboratory review in 1993 concurred (723):

> Contrary to common perceptions that interconnections of intermittent renewable energy technologies and utility systems are problematic and costly and that major issues regarding harmonics, protection, and safety remain,...hardware and system design advances have eliminated most of the concerns about interface. Furthermore, the cost of hardware is going down.

To understand what those prudent strategies are, what advances have eliminated what concerns (especially a decade later), and how these developments help to create valuable distributed benefits, we must review certain technical aspects of grid structure and function not previously described.

At first it may seem obvious that just as a

customer's end-use efficiency provides the same or better services with less electricity, so distributed generation too can be interpreted as a "negative load"—a direct reduction in the net flow of electrons that must be generated and delivered. While physically correct, however, this mental model overlooks some important practical details of how electric grids actually work. These differences include how the many elements of the grid are laid out and connected, how they are switched and monitored, and what kinds of power normally or abnormally flow within them.

2.3.2.10.1 Grid topologies: radial *vs.* web

Electric grids have traditionally been designed to carry energy in one direction only—from the large power plant to the many relatively dispersed customers. More precisely,

- the transmission grid is usually like a network because it must be able to transmit large blocks of power through two or more alternative routes from generating to load centers—otherwise the failure of a single transmission corridor could shut off the supply, but

- the distribution system typically has a radial architecture, a "tree" that branches from the highest-voltage to successively smaller subtransmission lines, through substations (where subtransmission voltage is transformed down to distribution voltage), via feeders to local "taps," and thence to local transformers and retail customers.

There are important exceptions to this com-

mon pattern, chiefly in some metropolitan areas where the primary distribution system[143] is highly interconnected and looks more like a network than a tree. (Manhattan is the extreme example, with many intersecting nodes between crisscrossing distribution lines so as to provide rich interconnection and rerouting potential.)[144] Broadly speaking, however, the central-plant/transmission-web/distribution-tree model prevails. The web topology of the transmission system, combined with its large power flows, makes it more prone to stability problems but also more redundant. In contrast, the radial distribution system is more stable but usually less redundant, so failures in it are more likely to cause outages to customers—unless they have distributed generating resources that can stand alone at need.

2.3.2.10.2 Bi/omnidirectional flow

Utilities normally design transmission lines to accept power flow in both directions.[145] In contrast, most radial distribution systems traditionally accept real power flow only one way, from the central station to the customers, because they were designed for one purpose and one purpose only—delivering electrons from central stations to passive customers. (Reactive power can flow in either direction.) However, this limitation is not fundamental, nor is it technically necessary. The distribution wires and transformers can of course carry electricity equally well in either direction; the unidirectional bias comes rather from the way their protective devices and voltage regulators are customarily set up. Those devices can equally well be designed to accommodate bidirectional flows. If this is done as part of upgrades for distribution automation, adding bidirectional capability often incurs no extra cost but brings extra benefits.

The practicality of this approach is illustrated by the successful and routine operation of the Gardner feeder (§ 2.2.10.1) that often reverse-flowed during the daytime because up to 53% of its load was fed by photovoltaics.[146] (85) Similar reverse flows may be occurring at certain times and places in the Sacramento Municipal Utility District, which has over 10 MW of distributed PVs (the largest concentration in the U.S.) and has experienced no stability problems of any kind. However, such experiments are so far quite limited and not widely understood, so the electrical engineering profession may understandably be wary of such a radical departure from long-established practice.

Most utility engineers have not thought much about how the grid would work if it shifted toward high use of intermittent distributed generators on many feeders—though conceptually this is not very different from the existing ubiquitous presence of intermittent distributed *loads*. But ultimately, unfamiliar though the prospect is, there is no technical or, apparently, economic reason why the passive, radially organized distribution "tree" should not gradually evolve into a highly automated, intelligently active, omnidirectionally capable distribution

"web" that handles power flows in any direction with equal ease. It could keep its present radial form or could become, over time, more richly interconnected like denser and more web-like urban grids.

Either way, what will most matter to the "smart" grid's omnidirectionality is not so much its topology as its design intention and control intelligence. Like networked telephone or financial systems, or the lean- or no-inventory retail logistics systems used by such firms as Wal-Mart, or Federal Express's national kanban distribution system for IBM parts and products, the distributed intelligence will be "decentralized, collaborative, and adaptive." (378, 517) In general, adopting this architecture for an electrical, as for a logistical or financial, web can increase flexibility. Stability may become more difficult if highly centralized and hierarchical control is maintained, not relaxed. It appears that more distributed control may make stability easier to achieve by shortening communication loops and lags and by making microdecisions locally, as close as possible to the information that drives them. However, this will require significant new data-gathering—because so little has been measured about exactly how the present distribution system works—and probably some advances in applied control theory.

Other than (in part) widespread understanding among distribution planners, who are at a relatively early stage of evolving new tools for this task (73, 475), the means are at hand. Distributed microchips can do the arithmetic if more than Kirchoff's Law is

---

[146] It would be somewhat safer to wire a bidirectional web for three-phase power distribution (or collection) in delta rather than wye configuration—two different ways of hooking up the same wires—because then the voltages would be 42% lower for the same power flow. This change would require reconfiguring existing distribution wiring, hence an extra cost, although in new distribution systems it would cost the same. The choice also involves a potentially countervailing safety issue: faults in a wye system can send current into the neutral conductor and thence to ground, whereas faults in a delta system can create awkward imbalances between the other phases, possibly causing a hazard. These and related issues are discussed by Berning, *et al.* (65) See also Technical Note 2-6, bullet 5.

needed (the self-executing principle that says, loosely speaking, that electricity follows the line of least resistance). Distributed sensors and telecommunications are relatively cheap and are highly desirable for other reasons anyhow as part of distribution automation. The protective and safety devices, as we shall see presently, can adapt to distributed generation and to power backflows using sensible adaptations of established principles, operating practices, and hardware. And both reliability and service quality should be greatly improved.

2.3.2.10.3 Synchronization and dynamic stability

In general, "Synchronizing a distributed generator with the utility waveform is a basic requirement and has not resulted in major integration issues. Hardware for all generator types and capacities is commercially available and economical. Integration with solid-state inverter functions may further reduce costs for future designs" (708)—and has indeed done so since that was written.

Besides the stability and control issues discussed in Section 2.2.10.1 for PVs affected by cloud edges, wind turbines present special issues of dynamic and transient stability. First, a windfarm attached to a weak transmission line can upset the voltage profile or introduce unacceptable voltage flicker (as initially occurred in SCE's Tehachapi Pass 66-kV network). This can be fixed by changing the turbines' control systems, using more turbines to even out their fluctuations, strengthening the grid, and using smarter variable-power-factor inverters to provide

reactive support to the grid instead of sucking reactive power out of it (712). Voltage flicker caused by siting induction-backfed small wind machines on secondary distribution circuits can be resolved by changing to synchronous (*e.g.*, permanent-magnet) or self-commutated generators (712).[147]

More generally, wind turbines' integration in significant numbers into utility systems was at first thought to be problematic because the large, relatively slow rotors, geared up at high ratios to the synchronous generator speed, have quite different torsional properties than turbogenerator rotors. However, the mechanically soft coupling (295) (chiefly through "play" in the geartrain) means that the large wind-turbine inertia and the generator inertia are essentially decoupled. Therefore "short electrical transients tend to impact only the generator inertia whereas similar mechanical transients such as wind gusts primary affect the turbine inertia." (711) The same analysis continues (725):

> This unique behavior of a large wind turbine has the following implications:
>
> • Fault clearing time[148] and the duration of short-term load contingencies are not so critical as with the conventional rotating generators.
>
> • Synchronism with an electrical system under gusty wind conditions is not a problem.
>
> • Synchronization of a wind turbine with the electric power system can be achieved with speed errors of several percent and phase angle mismatches of 30–40 degrees.

Thus wind turbines and large turbogenerator systems, though quite different in their rotational characteristics, turn out to be quite compatible (294). Even with a large amount

---

[147] For this and other applications of distributed generation in parallel with utility distribution systems, ANSI/IEEE Standard 1001-1988 is a helpful introduction and information source.

[148] This is how long it takes for a protective device to interrupt a circuit after the device is actuated by a fault signal. This delay may range from one cycle (perhaps less with solid-state interrupters) to tens of cycles.

of wind generation, "under certain contingency conditions, wind turbines within a cluster could become unstable, but the system would remain stable." (296) These forgiving properties are a refreshing contrast to the rigorously demanding synchronization requirements of large turboalternators.

2.3.2.10.4 Self-excitation

A peculiar feature of induction (asynchronous) generators, or of induction machines intended to be used as motors, is that when equipped with shunt capacitors to correct their power factor, they can self-excite from the capacitors' reactive power and can then self-start, burning out the motor or capacitor or both, and possibly supplying voltages to circuits thought not to be hot, thereby endangering personnel. For de-energized induction motors that may come under an external mechanical load, like fans in a breeze or pumps under a gravity head, decoupling switches may be necessary to prevent such self-starting. This is a normal electrical engineering requirement, is readily provided, and does not apply to modern PV inverters and similar non-rotary distributed generators.

Concern has been expressed that arrays of grid-excited inverters may be able to excite each other without grid power, causing a supposedly "dead" circuit to be energized unexpectedly. It is extremely doubtful that modern inverters could behave in this way, and such behavior has not been observed where it might most be expected, such as the densely sited SMUD and Gardner (648) experiments. But if desired, extra layers of protection, such as harmonic sensors to ensure that AC signals received are actually from the grid and not from some nearby free-running inverter masquerading as the

grid, can be programmed into the inverter's control microelectronics. If necessary, inverters' output could be equipped with a special signal that would cause other inverters nearby to turn off and isolate rather than responding to them if the grid is down.

2.3.2.10.5 Fault protection

Protecting grid components and service personnel from excessive or unexpected power flows under fault conditions and during repairs—whether the fault originates in the distributed resource or elsewhere in the utility system—is a normal, if somewhat complex, part of utility procedures. While the utility industry, even within North America, has no uniform definition of acceptable power quality or reliability, typical operation seeks to maintain supply voltages within a fairly narrow range (95–106% of nominal) and to maintain grid frequency within margins as tight as 0.002 Hz. Deviations from these conditions are rapidly corrected and isolated in order to minimize damage to utility and customer equipment and potential hazards to people; keep the disturbance from propagating; minimize the area it affects; and if the disturbance is temporary, restore normal operation as soon as possible.

These requirements are maintained, and people and equipment (both utility and customer) are protected, by sets of standardized and highly reliable protective devices, such as relays that trip on over- and under-current, -voltage, -frequency, and sometimes -temperature, all set tight enough to maintain safety and standards but loose enough to minimize "nuisance trips." (In general, more precise, tighter-tolerance relays cost more, though the marginal benefits of the tighter performance standards may be less clear.) Protective equipment also often includes filters or traps to control harmonics and electromagnetic interference, and surge arresters to constrain overvoltages from lightning[149] or other conditions.[150]

The entire set of protective equipment must be designed to protect utility assets, loads, distributed generators, customers, and the general public under a wide range of conditions, both normal and abnormal: lightning, earthquake, storm, equipment failure, deliberate disruption, ground faults (short-circuits to ground), phase faults (short-circuits between the phases), faulty customer equipment, or whatever. Due attention should be, and is increasingly, given to preventing malicious interference with the software, sensors, or communications links.

Flowing from this demanding, engineering-intensive and properly conservative tradition, many utilities have requirements, some prudent and some seemingly superfluous, for the technical equipment required to interconnect distributed generators to their grids in order to ensure safety, voltage stability, etc. Automatic and rapid disconnection from the grid if grid power fails is nearly always required. However, some distributed resources raise novel technical issues (see Technical Note 2-6) about how to provide protection.[151]

More sophisticated controls and procedures

Technical Note 2-6: Special Interconnection Issues

- *PV or fuel-cell generators, having no rotors, cannot provide significantly more current than their inverters are rated for (i.e., they have quite limited, normally not over 120%, "short-circuit current capability"). This makes the normal method of protecting against ground or phase faults on the distribution system (overcurrent relays) difficult to use and perhaps undependable. Clearing such faults may therefore require voltage relays or transfer trip arrangements instead.*

- *Similarly, overcurrent protection is designed on the assumption that currents further from the substation will be smaller. Protective devices are therefore normally arranged in a sequence so that the overcurrent relay closest to the fault will clear it, thus losing as few customers' loads as possible. However, distributed generators could overturn that assumption by contributing current back upstream toward the fault. Current may then not flow properly through all of the protective devices whose functions are supposed to be coordinated (breakers, reclosers, fuses, etc.). This requires careful analysis and perhaps reprogramming.*

- *Multiple distributed generators may also increase the complexity of fault current flows from potentially multiple sources, requiring more sophisticated hardware and software to keep protective relays sufficiently sensitive (114). This may raise non-technical issues of who pays for such upgrades.*

- *In some circumstances, it is conceivable that failed distributed inverters might inject into the AC system some direct current, which its sensors are not normally designed to detect.[152]*

- *Most distribution systems use four wires—three for phase current and one well-grounded neutral. The neutral wire ensures that if one or more of the phase conductors get shorted to ground, the others will not experience high voltages. However, if a distributed generator is connected to such a system through a standard delta-connected transformer, then it uses only three wires. If the distributed generator gets isolated from the utility system during a line-to-ground fault that trips fast overcurrent relays, then the distributed generator will not experience the overcurrent and will take longer to trip via its undervoltage or underfrequency relays. But until that trip eventually occurs, the isolated three-wire system can experience up to a 173% overvoltage that may cause metal-oxide-varistor (MOV) surge arrestors to fail. In some circumstances, the overvoltage may also reach customers and blow out their end-use devices.*

- *Some types of generators, chiefly synchronous rotating generators, may require special protective equipment.[153]*

- *Control systems require careful design. For example, in a windstorm, tree limbs may briefly brush against an overhead distribution line, causing a voltage sag. A downstream distributed generator's controls may interpret this as a fault requiring the generator to trip offline. But more commonly, some controls could interpret the voltage sag as a rapid increase in load, causing the generator to increase output. Since there was no increase in load, line recovery from the voltage sag could then cause an overvoltage trip, leaving the operator without the distributed resource at the time (a windstorm) when it is especially valuable (764).*

[149] Lightning strikes can induce kiloampere or kilovolt surges with rise times of a few microseconds. Even more difficult to stop can be local static-electricity sparks, which have low current but ~15-kV voltage and very short rise times, often fractions of a nanosecond, making them capacitively coupleable into electronics whose chips they can instantly burn out. Fortunately, extremely fast protective devices are available for both needs.

[150] For example, ferro-resonant overvoltages from temporarily islanded self-exciting induction generators.

[151] A more detailed and technical account by Vito Longo is section 3 of EPRI's proprietary *TAG*™ (62).

[152] This could magnetically saturate transformer cores, galvanically corrode intermetallic contacts, and cause other mischief. Solutions much cheaper than isolation transformers are available (700).

[153] Induction generators and line-commutated inverters cannot continue to produce fault current after the first cycle from disconnection (when they may "dump" current as big as their start-up inrush current) because they then lack line excitation. In contrast, synchronous generators can contribute transient fault currents up to about 3–8 times normal full-load current, for up to about five cycles, rapidly decaying to a steady-state contribution of about 1.0–2.5 times normal peak operating current for at least several more seconds. The near-unity level might not be detected by normal overcurrent relays, but could be tripped by overcurrent relays set to trip only if voltage is low (703).

can resolve these and similar issues, chiefly related to grounding, but they do require due attention and careful engineering. All the solutions are facilitated by the sophistication and programmability of modern distributed-generator solid-state inverters.

There are longstanding debates about whether even the smallest distributed generators, such as inverter-coupled home PV arrays, need the same sophisticated and highly reliable "utility-grade"[154] protective relays, switchgear, power and instrument transformers, etc. as high-capacity utility equipment traditionally meant for large rotating machines (710). Although the costs of utility-grade protection equipment have been gradually declining with higher production volumes, the cost can still be prohibitive for small generators. However, well-informed utilities now appreciate, as PG&E concluded in 1992 (636), that the solid-state, often adaptable, inverter trip devices that substitute for protective relays in grid-interactive inverters

> ...have been shown to perform well and reliably enough to meet the most stringent utility safety requirements. Since this conclusion was by no means warranted based on theoretical considerations or laboratory test data alone, several utilities have conducted studies of PV systems and inverters in the distribution system and examined their performance under "real-life" conditions. These experiments have indicated that introducing photovoltaic generation on distribution feeders need not pose any safety problems (143, 548).

For example, the Gardner experiment found (651) that the 1985 inverters tested "did not run on for more than 8 milliseconds

during feeder dropping tests," "provided only a limited fault current for no more than 8 milliseconds during faults" at a maximum fault current "no more than 150% of rated [inverter output]...current," and "did not create problems for either the utility or the customers during feeder experienced faults, induced lightning surges, capacitor switching, and large load changes."

As this becomes more widely accepted and the inverter and interconnection hardware industry matures, utilities' interconnection requirements are also becoming more standardized, reducing confusion, conflicting interpretation, and both soft and hard costs. Many utilities and some states, for example, have settled on UL 1741 as an interconnection standard for small customer generators. Some encouraging approaches to standard interconnection guidelines use functional objectives rather than prescribing specific kinds of equipment, encouraging technical innovation and simplification. However, some degree of confusion and of circular requirements,[155] common in the 1980s, persists in certain regions even in 2002. And while researchers "have found solutions that can be integrated into existing protection devices and generator hardware," utility protection designs, equipments, and procedures vary widely, so a more complete survey of potential issues may be warranted; some "identified solutions may be quite expensive, and gaining utility acceptance of lower-cost, alternative hardware may require substantial effort"; and above all, "Effective technology transfer and successful demonstration of new

---

[154] At least for protective relays, however, this term is not well defined, nor is its supposedly higher reliability necessarily a valid assumption (538).

[155] For example, some utilities required that for an interconnection to be approved, the equipment must have undergone prior safety inspection. Some safety inspectors refused to approve the installation without prior utility approval. In at least one instance, a wind-generator control panel required UL approval for utility acceptance—an impossibility because UL tests only components, not assembled panels (653). At this writing, a 120-W PV system is being rejected by Calgary inspectors as a hazard to the grid, despite 60 similar systems in Toronto and more than 10,000 in Holland.

technologies in the utility environment will reduce concerns." (707)

2.3.2.10.6 Normally interconnected, optionally isolated operation

"Islanding"—the ability of a distributed generator to run nearby loads even when both are isolated from the wider grid—is normally considered undesirable,[156] except when deliberately designed as standby generation or as a special operating procedure resorted to in certain kinds of grid collapse.[157] However, that tradition, which stems originally from the laudable desire to keep lineworkers safe (§ 2.3.2.10.7) and to protect equipment from possibly unsynchronized or otherwise interfering power flows, may no longer be optimal (724).

> In the future, utilities may conceivably explore modified protection and coordination schemes so as to allow maximal benefit from distributed generation. It can be argued, for example, that "islanding" after transmission faults should be permitted, since service for some customers near the distributed generation facility could thus remain uninterrupted. In this case, distributed generation may come to be viewed as essential to local service reliability. Such an operating policy would require that switches and circuit breakers be placed in different configurations, coordination and fault-clearing procedures be modified, and personnel take precautions and treat all lines as "hot." Although the technology required by this kind of strategy exists, it is not current operating practice of any utility in the

U.S. to permit islanding. We do not intend to suggest that these practices should presently be changed, but do wish to point out that the notion of a "proper" operating policy for transmission and distribution systems is a relative rather than an absolute one, and that shifting viewpoints regarding the role of dispersed generation in T&D systems are imaginable.

That is, if distributed generators "can be designed to operate properly when islanded, [*and* if lineworkers know where the islands are so they can use proper precautions for the still-energized lines,] then the value of these...devices to the distribution system is greatly increased because such capability gives local distribution systems the ability to ride out major or widespread outages. System requirements necessary to allow distributed generation to serve load while a section of the distribution system is islanded from the main circuit and substation need identification and analysis." (106)

The cost of such analysis and of required safety and stability arrangements appears relatively minor (though necessary), while the potential value of prudently operated islanding capacity appears large, as reflected by observed customer expenditures for standby generators. However, cost, value, and design cannot be determined without rather detailed load and other data that few utilities have yet gathered for their distribu-

---

Benefit

134
Distributed generators can be designed to operate properly when islanded, giving local distribution systems and customers the ability to ride out major or widespread outages.

---

[156] For example, when a distributed generator continues to energize a grid section isolated by line-switching for routine maintenance work. In this circumstance, the generator's sensors may not realize it is supposed to turn off, so it may continue undesired and uncontrolled operation in an otherwise de-energized system. Normally this is rare, because for it to occur, the real and reactive power flow between the isolated load and generator must be closely matched at the time of line-switching ("no-fault utility disconnect"); otherwise the islanded system's frequency and voltage will drift out of bounds and cause shutdown, just as would occur if the disconnection were caused by a fault. "A [1989]...investigation (561) concluded that islanding [in this undesirable sense] is an intrinsic possibility with all power conditioning systems, but the practical possibility of distribution system islanding for extended periods is limited by normal variations in load and intermittent renewable generator output." (706) These variations are now often enhanced by "internal destabilizing circuits and internal trip mechanisms" which make undesired islanding even less likely, though still not impossible (724). Several authors have suggested that undesired islanding could be detected and stopped by making the inverter trip on sudden changes in harmonic impedance, which changes dramatically if the utility power source is disconnected (707). Easiest of all, Kansai Electric Company found that islanding of PV systems can be prevented from the utility side by simply inserting capacitors (381).

[157] Under some system contingencies in which system frequency keeps declining even after underfrequency relays have shed loads, islanding has traditionally been a way to protect generators, and ease system restoration, by breaking the system into isolated "islands," each containing enough local loads to keep local generators in operation. This operating procedure is sound and desirable in these circumstances. Here the term is applied to the different context and scale of enabling isolated distributed generators to serve their isolated local loads during distribution outages, often without the utility's knowledge or intervention.

Benefit

135
Distributed resources require less equipment and fewer procedures to repair and maintain the generators.

tion systems. The industry's shift toward distribution automation will gradually correct this problem by permeating the grid with sensors and telecommunications. It will also gradually give operators a better intuitive feel for the improved stability that can come from a more distributed architecture of grid intelligence and control.

2.3.2.10.7 Safety

The electricity industry has extensive and meticulous procedures for dealing safely with both live or "hot" (energized) and dead (de-energized) lines and equipment. U.S. Occupational Safety and Health Administration regulations, for example, require manual disconnect switches at every possible source of power in an otherwise isolated line to be opened before work begins, and the open switches locked open and tagged so that nobody will reclose them by mistake before the work is finished. The workers must also check with the system operator the location of every known distributed generator to make sure it is isolated first. Moreover, de-energized lines are to some degree treated as if they were energized, much as every gun must for safety be treated as if it were loaded: the six-step standard safety drill comprises notification, certification, switching, tagging, testing, and temporary grounding. Lines *known* to be live entail extra precautions such as special insulating equipment (tools, platforms, stools, mats, gloves, etc.), without which, under OSHA regulations, utility personnel may not approach or touch a conductive object.

These redundant regulations and practices, despite occasional breaches, offer sufficient defense-in-depth to have proven highly effective. Utilities' concern is that self-excit-

ing distributed generators, such as those with free-running inverters, may render live a line thought to be dead, especially by adding power downstream in a radial network whose upstream supplies have been cut off. Alternatively, distributed generators normally requiring grid excitation in order to produce any power may, under abnormal conditions, self-excite (*e.g.*, through interaction with nearby compensating capacitors or other reactances in the system) and thus produce power when they're not supposed to (§ 2.3.2.10.4). The latter was indeed possible with some early inverters, although most models since the late 1980s have had internal sensors and switches that provide extremely rapid and reliable fault detection and shutoff if the grid fails. Such features are now widely available and can readily be designed to be fail-safe.

Unscheduled distributed generation can occur without directly creating a safety problem—*i.e.*, without energizing a line that workers expect to find de-energized—if:

1. all utility staff always use live-line maintenance practices (which are somewhat slower and costlier than dead-line procedures) on any system that contains distributed generators, *or*

2. a highly reliable automatic isolation relay or manual disconnect (the latter, in general, mounted outdoors so that utility staff can reach and operate it in order to provide absolute assurance of disconnection), or an automated but verifiable equivalent equally accessible to utility staff, is installed between every distributed generator and the grid, *or*

3. both.[159]

This is analogous to saying that all guns are safe from accidental firing if people invariably handle them using proper treat-as-

[158] Some other approaches may also be feasible, and isolation will be simplified if sectionalizing switches are added in more places so that each area containing distributed generators requires fewer disconnections.

loaded safety procedures, *or* if they are assuredly unloaded, *or* preferably both. Distributed generation advocates and utilities have debated for many years whether it is necessary to use safety approaches #1 *and* #2, or whether either alone should suffice. There is no definitive answer; #3 is the ideal but is slower and less convenient. But in general, approach #2 or #3 is desirable in any event because disconnection not only protects lineworkers directly from shock, but also prevents a downed or short-circuited conductor from sending a "fault current" flowing into the ground, "interfering with the coordination of fault clearing operations and potentially threatening people or equipment at the fault location. It is therefore important that the connection between dispersed generation and the grid be interrupted in the event of a disturbance." (635) This suggests that approach #1 may at times be inadequate, but for broader reasons than just protecting lineworkers. If so, then #2 may be preferable, and if #2 alone is considered adequate, then #3 would simply offer added assurance.

However (632), while interruption is desirable in any prolonged fault, automatic and *immediate* interruption is not always the right response. When it is not, modern distributed generators, especially those with smart inverters, are at least as likely as traditional protective relays to achieve the right answer, and quite possibly more so:

> For example, a quick voltage spike might occur at the instant that load from a neighboring feeder is switched over to (or away from) the PV host feeder. Since this is a benign, one-time event, one would wish the inverter output to remain unchanged [and the inverter connected to the feeder]. The proper diagnosis of disturbances is therefore essential. As in traditional power equipment, this is accomplished by examining the magnitude and duration of an excursion of

voltage and current from the expected waveform, or the magnitude of a frequency or phase deviation. If the measured values exceed certain limits, the inverter will trip off line. The accuracy and response-time (on the order of tenths of milliseconds) achievable by inverters continue to improve with technological development and market incentive, but experimental results indicate that the performance even of today's [1992] state-of-the-art inverters is perfectly adequate by utility standards.

In 1992, PG&E proposed new inverter standards (633): for systems below 1 MW, overfrequency trip at 61.0 Hz for 15 cycles, underfrequency trip at 58.5 Hz for 2.0 seconds or at 55.0 Hz for 30 cycles, overvoltage trip instantaneously at 120% of nominal, and undervoltage trip at 90% nominal for 3–5 seconds. All these attributes are particularly easy to arrange with modern off-the-shelf PV inverters that are digitally controlled, rapidly switched, and completely programmable. Thoughtful safety arrangements are even present where they will be most needed—in the "vernacular" micro-inverters used to make AC-out PV panels into a plug-in "solar appliance" that the utility is unlikely to know has been installed. In general, inverters became substantially better through the 1990s, and most—all rated UL 1741—are now "ready for prime time."

In principle, the equipment and procedures needed to repair and maintain resources that operate at lower voltage, current, and power should be less costly, elaborate, scarce, awkward, and long-lead-time than those needed for traditional large-scale resources with the same aggregate capacity (corrected for relative failure statistics). It should, for example, be easier, cheaper, and faster to sustain a given level of firm output with large numbers of standardized, off-the-shelf, "vernacular" modular PV inverters than with a very small number of highly specialized, special-order, several-year-lead-time Extra High Voltage switchgear or transformers. Whether this is true, and if so, how much it is worth, is not yet clear from the literature, though the technical logic is compelling.

2.3.2.10.8 Reclosing

After protective relays trip circuits open, and after the fault is diagnosed and corrected, the open switching devices must be reclosed in a certain sequence to restore service without causing hazards or damaging equipment. Standard utility practice is to try automatically reclosing a breaker soon after it opens, in the hope that the initiating fault was temporary, such as a lightning-induced transient. (In overhead distribution lines, many faults clear themselves as soon as the line-frequency current is temporarily removed and the insulation has had time to restore itself.) But adding distributed generators requires coordination to ensure that a fast recloser, in trying to clear a temporary fault on the distribution line and reestablish utility power flow, does not do so before a distributed generator's own protective relay has had a chance to open. (If the distributed generator had meanwhile drifted out of synchronization, this could severely damage it.) This issue is readily dealt with technically, even for synchronous generators, but it requires design attention and coordination by both utilities and distributed-generator equipment designers.

The cultural context of this seemingly straightforward coordination requirement is unrelated but revealing. Power engineers are particular wary of unexpected problems that might emerge from large-scale use of new technologies because of several unpleasant

recent experiences. For example, subsynchronous mechanical resonances seriously damaged large turbogenerator shafts due to series capacitors inserted into extremely-high-voltage transmission lines for reactive compensation, and turbogenerator windings suffered electrical damage due to high-speed reclosing of nearby breakers on transmission lines—a feature introduced to improve transient stability (105). While these unhappy experiences have no direct link to distributed generation, they do make some utility engineers doubt facile assurances that even apparently simple new technologies can be integrated into the grid with no surprises. This degree of technical conservatism is not unwarranted, and clearly, the kinds of evolution in grid architecture, equipment, and operation discussed here will require careful, step-by-step testing and validation to guard against unwelcome surprises.

2.3.2.11 Avoided grid connection (standalone operation)

To this point we have considered how distributed resources *connected to the grid* can provide such economic benefits as capacity deferral, reduced losses, reactive power support, and improved reliability. Grid-connected photovoltaics, for example, may support the lines or substations, or may be installed on customer premises as peak-shavers (like a demand-side resource). However, an entirely different and increasingly important category of benefits arises from the option of *not connecting to the grid at all*, but serving a customer directly—or serving remote facilities of the utility itself, such as for cathodic protection or sectionalizing switches.

When this is done, instead of deferring an

Benefits

136 *Stand-alone distributed resources not connected to the grid avoid the cost (and potential ugliness) of extending and connecting a line to a customer's site.*

137 *Distributed resources can improve utility system reliability by powering vital protective functions of the grid even if its own power supply fails.*

138 *The modularity of many distributed resources enables them to scale down advantageously to small loads that would be uneconomic to serve with grid power because its fixed connection costs could not be amortized from electricity revenues.*

139 *Many distributed resources, notably photovoltaics, have costs that scale far more closely to their loads than do the costs of distribution systems.*

expansion of line or transformer capacity, the *entire* cost of connecting to the grid, and perhaps of extending a line to the site, is avoided. Instead of reducing real and reactive power losses in the grid, they are avoided for the customer. Instead of improving reliability by expanding opportunities for grid rerouting and by easing the wear cycles of equipment such as tapchangers, reliability depends entirely on the characteristics of the distributed resource itself, because the customer can no longer draw on the grid for backup—but the grid, where almost all outages now originate, would no longer be used. Instead of relying on complex and diversified equipment maintained and serviced by utility staff on a large scale, the customer would rely on simple and usually undiversified (but perhaps extremely reliable) equipment maintained by oneself or by a contractor on a very local scale. Both the engineering and the economics of standalone operation are therefore completely different.

In principle, conditions that favor grid connection include being near the grid; using a relatively large amount of electricity (especially in relation to distance from the grid); being relatively near the upstream end of a feeder where power quality and voltage stability are better; having reliable generating and transmission capacity available and cheaply connectable to; having only ordinary power-quality requirements; and having limited local generation potential. Conditions favoring standalone operation include being far from the grid or using a small amount of electricity or (especially) both; being at the downstream end of a feeder where power quality and voltage stability are typically poorer; having reliable local generating options or stringent reliability/power-quality requirements or (especially) both; having attractive local co/generation potential; or having costly interconnection options.

This standard checklist is important because in relatively remote areas, it is now becoming common, especially in the western U.S. and in many parts of Australia, for utilities to offer a photovoltaics lease package in lieu of costlier line extension to rural homes.[159] Some U.S. jurisdictions require that customers be offered this choice whenever it might be cheaper. (That would be true in far more cases if the utilities bundled the photovoltaics with a suite of highly efficient

[159] Analogous industrial opportunities exist (488), as when running a new 10-km power line to a 3-MW customer costs ~$365–1,100/kW, while a modular gas-fueled generator could cost less than the high end of that range, even counting the fuel.

end-use devices, but even today, few do.)

Such standalone options are attractive to rural utilities because of the great cost of just maintaining, let alone extending, lines to remote customers. One southern U.S. utility was recently reported in an Internet user-group to have paid $70,000 to upgrade and maintain lines supplying a single customer who paid ~$400 a year for electricity. Not only do western utilities spend about three-fourths of their marginal investment expanding and upgrading the grid (630), often to serve newly developed areas, but those assets are often very poorly utilized: PG&E's typical distribution feeder runs at 50% of its capacity less than 40% of the time (§ 1.4.2, Figs. 1-35–1-37). (104) Maintenance, often over long distances and in rugged or remote areas, and meter-reading add further ongoing costs to sustain service to customers who may be providing little revenue. And the long lines have inherently high vulnerability, inductive reactance, and voltage drops, so they require costly boosters, capacitors, and maintenance effort to achieve acceptable power quality, voltage stability, and reliability to the customer on the other end of the line.

U.S. line-extension costs for distribution are typically on the order of $11–22/m for overhead or (more variably) $21–40+/m for underground lines, plus fixed per-installation costs on the order of $1,000.[160] Just a distribution transformer added to serve an extra 120-VAC load can have an installed cost approaching $2,000 (158). This sort of cost, compared with likely revenues, makes many line extensions uneconomic for the

utility or the customer or both.

This is especially important for many rural electric cooperatives, which as part of their largely accomplished New Deal mission to extend electric service to rural areas eventually refund line-extension capital costs to the customers who paid them. U.S. coops deliver ~8% of U.S. electricity, at roughly the national-average price (~$0.07/kWh), through nearly half the nation's total length of distribution lines. This combination of long lines and low load densities, hence low revenues per unit of line, makes long line extensions to customers with low revenue into a certain money-loser (224). An often aging rural population further increases the financial risk (320). Moreover, of those coops' two million miles of such lines (80 times the circumference of the Earth), half are at least 40 years old and are now or soon in need of renewal. The opportunity to install distributed PV-hybrid power systems instead has been estimated to be worth as much as $1–2.5 billion in net present value. A preliminary survey found some line-replacement projects that would cost the coop $0.50–0.60—even up to $1.50—per kWh for power that sells for only $0.10/kWh. In general, rural lines delivering fewer than 10 MWh/mile-y would be good candidates for microgrids (§ 2.3.2.12), typically using photovoltaic-engine hybrid generators. Replacing 7–16% of the coop lines would represent a photovoltaic market of 0.5-0.95 GW, yet would be cost-effective at an installed cost of $3,000/kWAC (reasonable in such a volume). (322)

In a significant market distortion, many cus-

[160] Such costs depend strongly on terrain, capacity, and other variables. They are also often broken into a fixed and a variable component. For example, Idaho Power's 1993-approved tariff for Line Extension Average Unit Costs, for single-phase overhead lines to single-family and duplex houses, ran $740 base cost plus $15.58/m for primary or mixed primary/secondary extensions, or $625 + $13.94/m for purely secondary extensions. The corresponding underground tariff was $1,550 + $24.60/m ($700 + $24.60/m for secondary only), plus substantial extra charges for surface restoration, going through rock, or other unusual conditions. The utility extended a free allowance of up to $1,500 of connection costs ($2,000 if using electric heat) that was ratebased (socialized to all customers); any further extension costs fell on the customer alone. As mentioned below, the tariff was discontinued after~40 successful installations.

Examples: Special remote applications

Large numbers of emergency telephones, signs, and advertising billboards along highways in North America, Europe, and Japan routinely use PV power to provide services that would otherwise be impossible, difficult, or excessively costly. Such systems are cost-effective partly because they typically operate at 12 or 24 VDC with direct battery connections, DC-powered electronic lighting ballasts (electronic ballasts rectify AC into DC anyhow before inverting it to the high-frequency output current to run the discharge lamps), and no inverter. Southern California Edison Company, the leading U.S. user of high-pressure sodium lamps, has developed a 50-W PV-powered package with integrated controls.[162] (524)

Another typical remote application is utility sectionalizing switches, which control transmission lines and must operate when faults de-energize the lines; otherwise power cannot be properly rerouted or restored. The loads served are typically 48-V switch-operating motors and low-voltage-DC telecommunications systems; the several voltages are easily provided at no extra cost, since both the PV arrays and the batteries come in standard voltage modules that can be wired in any desired configuration. Using ~50–400 W of peak PV power, these systems typically proved cheaper in capital cost (usually by ~2–5-fold) than line extension or stepdown transformer options. If properly engineered, they are also highly reliable. They are usually the method of choice for ≥22-kV switches (668).

Still other examples of successful remote applications include cathodic protection for pipelines, buried fuel tanks, metal transmission-line towers, bridges, wharves, docks, marinas, and other metal structures subject to corrosion. Again, as Florida Power Corporation found for transmission lines, stepping down the transmission voltage to run this minor load (~36 W/tower) would have a higher capital cost than the PV system (524). PV systems also find favor with remote water-pumping, with portable livestock fences, and with bubblers to keep ranchers' stock tanks from freezing over.

---

tomers may find extension and operating costs of rural lines socialized to all customers (especially members of rural cooperatives) and charged over decades, whereas the costs of a self-financed distributed-resource alternative are private and all up-front. Alternatively, customers who have to pay up front for line extensions, as is some utilities' policy, may install a noisy, smelly engine-generator set with less than ideal cost, maintenance, and reliability characteristics, or may even forego electricity altogether. However, utility installation, leasing, and maintenance of PV systems overcomes both these obstacles to better customer service at lower cost. It can therefore both reduce uneconomic line extensions and create new customer relationships that couldn't otherwise be served.

Typical lease fees, including perpetual main-tenance and performance guarantees, for AC-out, ~4-day-battery-backup PV systems that avoid rural line extension are about $125-350/month for PG&E photovoltaic leases, and a lease-purchase option was explored in 1993. Such equipment is designed for easy removal and relocation. (Naturally, there is no energy charge and no exposure to changing fuel prices.) Idaho Power's Solar Energy Service, available under the 1993 Schedule 60 tariff in or (by special arrangement) outside but near its service area, provided five-year initial leases with purchase option, assumable by the new owner if the house is sold; renewed those leases automatically every year thereafter unless canceled; was installed in a few hours within six weeks of order (sooner if needed); could include an optional backup engine-generator set in addition to the normal battery bank; and charged an initial fee of 5% of

[161] This effective charge rate of 18.2%/y—not counting the initial 5% fee, which may be presumed to cover the load monitoring and subsequent system engineering and procurement overheads—covered all fixed and variable costs. It looks rather lucrative for the utility, especially after the first five years, by which time the system's installed cost was 96.2% amortized on an undiscounted basis.

[162] Marketed by SCE subsidiary Energy Services, Inc. (James Clopton), 7300 Fenwick Lane, Westminster CA 92683, 714/895-0556.

estimated installed cost plus a monthly charge of 1.6% of the balance.[161] The successful pilot program was dropped in 1997 because the prospect of deregulation (which ultimately didn't occur) deterred the utility from holding long-term leases, and tempted it to serve bigger developing-country solar markets rather than its own customers.

Special remote applications (see box above), the residential uses discussed in a moment, and others involving remote sites are a significant portion of the photovoltaics market (Figure 2-46): in 1990 they reportedly accounted for about 97%, and in 1999 for 68%, of the uses to which photovoltaics manufactured in the United States in that year were put (746). However, the cost-effective scope for such applications appears to have been greatly underestimated: standalone can be better than grid hookup in surprisingly many situations quite unrelated to remote siting or line extension.

The conventional wisdom holds that the breakeven distance to the distribution grid, beyond which photovoltaics and end-use efficiency are a cheaper option than hookup, is on the order of 400–1,000 m, depending on climate, load, topography, preference for AC or DC end-use supply, etc. On this principle, tens of thousands of standalone PV-powered houses have been built in the United States alone. Of the first nearly 2 MW of peak PV capacity installed in California standalone applications, over half the capacity was in houses (communications and billboards brought that share to 83%). (733) In the capital of standalone household installations, rural southern Humboldt County, California, by some estimates 80% of houses are PV-powered and off the grid—partly, it is said, because of the unusual economic circumstance that some of the homeowners grow special crops and therefore don't want meter-readers visiting.

A typical representation of this traditional view of PV cost-effectiveness is a 1991 EPRI graph (Figure 2-61), showing that even the tiniest loads aren't worth doing with PVs if the grid is within ~61 m:
Idaho Power's "Solar Photovoltaic

Figure 2-61: EPRI's 1991 view of standalone-PV economics
The conventional wisdom simply compares PV cost with line-extension plus energy costs for remote grid hookups.

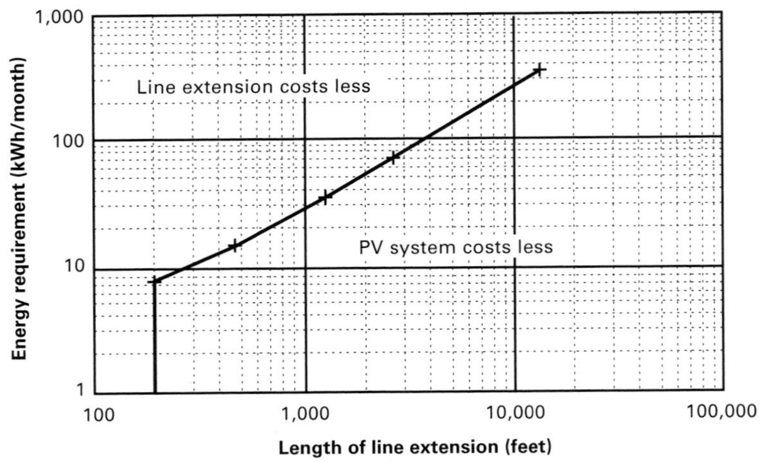

Source: D. E. Osborn, "Implementation of Utility PV: A Tutorial" (Solar Energy International, March 1995), part III, p. 21

Feasibility Guidelines" (341) are even more restrictive, recommending photovoltaic evaluation only for distances over a half-mile (805 m) even if the load is only 1 kWh/day or 42 average W, the lowest value considered. Even the Regulatory Assistance Project's "Economics of PV vs. Line Extension" chart (Figure 2-62), though it scales down to 0.1 continuous watt of load, still considers line extension potentially competitive with PV at that level (if the line extension is one foot long):

Figure 2-62: A more sophisticated but still incorrect view of standalone-PV economics
Line extension cannot actually compete for small loads as shown, because its fixed costs of connection are not justified.

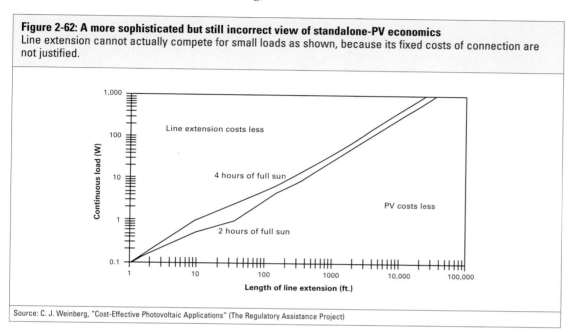

Source: C. J. Weinberg, "Cost-Effective Photovoltaic Applications" (The Regulatory Assistance Project)

But that is clearly not correct if the line extension entails the capital and maintenance costs of metering, utility-owned disconnect switch, customer-owned service entrance (the line from the utility's last pole into the building), or other code- or utility-required interconnection equipment, let alone the recurring cost of meter-reading and billing. Just the installed capital cost of a U.S. single-phase typical residential electric meter, wired by a licensed electrician in a new house but not counting other components of the service entrance,[163] is approximately $300, with a typical range of ~$275–$325 (336). Reading the meter in a typical rural area, sending a monthly bill, and processing payments adds roughly $2.30/month—a stream of costs that, discounted 5%/y over 30 y, has a present value of about $424. To this $700-odd cost of

installing and using the meter must be added the often larger capital cost of the service entrance and any interconnection equipment. Thus the whole *shape* of the standard breakeven-distance graphs is incorrect, because it ignores the different fixed costs of a standalone PV installation *vs.* a utility hookup.

To pick a somewhat extreme example, a continuous one-watt low-voltage DC load costs in principle less than $40 to serve indefinitely with photovoltaics in a normally sunny area where capacity factor is about 0.25. (The practical cost might be somewhat higher depending on mounting details and the higher cost of such a small storage battery.) But even though a continuous 1-WAC load (ignoring for simplicity the DC/AC distinction) costs only $0.61/y at a utility

[163] Including a typical household service panel would raise the cost to about $750 for a new or $1,400 for an old house.

Examples: Rethinking breakeven distance

- A few hundred meters from a 1,505-MW power plant, the Lower Colorado River Authority's engineers found it was cheaper to use photovoltaic power for six warning beacons (to keep boaters on the plant's cooling lake away from a baffle dam) than to supply power from the grid. Similarly, PG&E saved $200,000 by PV-powering warning strobe lights atop four 58-m-high transmission towers rather than rebuilding an aging wooden-pole distribution line across the mudflats (526). In both cases, extremely cheap and abundant power was available close by, but it was cheaper to use solar power instead. In fact, PG&E is among scores of utilities that have found cost-effective PV applications throughout its system, for such uses as water-level and -temperature sensors, automated gas meters and gas-grid controls, lights, cloud-seeders, weather towers, microwave repeaters, warning sirens, aircraft warning beacons, gas samplers—cathodic protection, rupture control valves, automatic gate openers, backup genset starters, etc.

- PV-powered outdoor walkway lights for house entrances are now routinely sold and widely used because for such a small load, photovoltaics and built-in overnight battery storage are cheaper than burying and connecting a cable to the house for even a very short distance. The light source is usually a low-voltage tungsten-halogen or infrared halogen miniature lamp with relatively high luminous efficacy. The latest units use even higher-efficiency LEDs, and some Japanese ones use cold-tolerant ultracapacitors.

- Analogously but on a larger scale, the Bent Tree Community Association in the West Miami, Florida suburb of West Kendall found that even in 1991, a $52,000 street-lighting system (26 lights run by 92 W of PVs with 48 Ah of 24-VDC batteries providing four days' reserve) had a capital cost $2,000 lower than that of utility power, because that would have required trenching the street. Of course, the PV system also had no operating cost, and kept operating after Hurricane Andrew when the utility power was down for 33 hours (526).

- The Sacramento Municipal Utility District has found that to light alleys (narrow little back-streets) even in downtown Sacramento, it is typically cheaper in capital cost ($2,500 instead of $3,000 per typical installation) to use PV power than to connect to the wires that are already in the alley, including required trenching and conduit (527). This comparison counts just the cost of installation and connection. Adding the solar system's avoided energy cost, maintenance for both systems, and the cost of reading the meter, sending bills, and processing payments—all of which the PV system makes unnecessary, since the "God utility" sends out no bills—would strengthen the solar advantage, probably to something on the order of $3,000 vs. $5,000 in present value.

- In Eindhoven, Dutch physicist/engineer C.C.H.T. Daey Ouwens built an unusually efficient house, with an average load of only ~50 WDC (slightly under half the load of the larger household area at the RMI headquarters building, which uses ~120 WAC, mainly because Dr. Ouwens used a gas- rather than electric-powered refrigerator/freezer). He then did a standalone PV installation rather than connecting to the grid a few meters away. The cloudy, high-latitude Dutch climate required extra battery storage. Yet the avoided capital cost of that grid connection—trenching, service entrance, meter—plus the avoided marginal cost for the common utility assets and operations (but not fuel) saved the utility a present value of about $5,000—close to the capital and maintenance cost of the PV system in 1991 and substantially below its likely cost today.[164] Ouwens estimated, on a true-marginal-cost basis, that Dutch installations like his can now or soon repay their PV-and-efficiency investments in about 10–12 years (540), equivalent to about a 9–11% pretax ROI—much better than money in the bank. He also believes such systems should be attractive even "in a densely populated area" at average loads up to about 90 average WDC—readily achieved for a household even using good 1988 appliance technologies and nonelectric space-conditioning and water heating (541), at marginal appliance costs below typical short-run marginal generating costs. (An often-cited rule-of-thumb states that each dollar spent on superefficient appliances will save about $3 worth of PV capacity.) The case is of course stronger with today's end-use technologies. It's also stronger at latitudes of up to 40°, where even at 1991 prices, the ~6-m² PV system with ample storage (at 10% array efficiency and 2,000 kWh/m²y insolation) has "investment costs ($5,310)...lower than for a grid connection ($6,000) if efficient appliances are utilized. Even though the per-kilowatt-hour costs in this case are high [$0.56/kWh], the annual costs for the [690 kWh/y of] electricity produced are very reasonable [$390/y]." This argument has led to large and successful PV installations in Dutch, Swiss, and other European housing in the past few years, and was confirmed in the world's then-largest residential solar development at the Sydney Olympic Village (§ 2.2.10.1).

Rethinking breakeven distance (cont.)

- In essence, Dr. Ouwens's conceptual argument, which appears to be correct, is that the marginal cost of household photovoltaic systems—especially the DC-out systems he prefers—decreases almost proportionally to load because the arrays and batteries are modular, while the marginal cost of household connections to the electric grid is more dominated by fixed costs and does not decrease much with load.[165] Therefore adopting very efficient appliances to achieve roughly fourfold load reductions, he suggests, can make the capital-plus-capitalized maintenance cost slightly lower for the PV standalone system than for the grid connection—using the avoided utility fuel cost to offset (or more) the extra cost of the efficient appliances. While the exact numbers depend strongly on local conditions, the qualitative conclusion seems plausible. It is worth emphasizing its premise: that the modularity of at least this kind of distributed resource is advantageous not only in scaling up (the risk-reducing thesis of Section 2.2.2) but also in scaling down.

- Building on his Dutch experience, Dr. Ouwens helped an Indonesian village, near a transmission line, to install standalone PVs and efficient end-use devices in each house, rather than the conventional interconnection (stepdown, switchgear, distribution wiring and meters). The whole installation was financed at the utility's normal discount rate (no subsidy) with ten-year amortization—severalfold faster capital recovery than the normal ~30-year straight-line depreciation of utility assets. Yet even from the beginning, the villagers had a positive cashflow—because servicing the debt for their independent, standalone, house-based energy systems cost less than they had already been paying for radio batteries and lighting kerosene! Similar results have been achieved by others in both Indonesia and the Philippines. If that works as well as it appears to work for people right next to a transmission line, it must be true for billions of other people too—and not only for the two billion in the South (i.e., about 70% of all people in developing countries) who currently have no electricity at all. As *The Economist* remarked, those people, who've never seen a pole with wires on it, now probably never will, because they'll get their power from photovoltaics and their telecommunications from wireless. The Solar Electric Light Fund, SunLight Power International, and similar organizations are starting to follow this business logic worldwide, often providing local revolving-fund cooperative financing for PV microsystems.[166]

These examples suggest that the conventional wisdom about breakeven distances needs serious reexamination in light of avoidable connection and metering costs.

---

[164] For the dense grid of the Province of North Holland (excluding Amsterdam), with 20-year amortization and 4%/y real interest, Ouwens reports (541) that marginal total-grid costs per average house are $2,650 (1989 US$ = NEF 2). Of this, $380 is for the house connection and meter and $1,345 for the distribution grid. (The rest is for transmission, $475, and other utility fixed assets. House wiring isn't included because it will be required in a standalone system too.) Generation and transmission have a present-valued corresponding marginal cost of $1,500/house; operation and maintenance, $1,950/house. The total marginal cost per average North Holland house is thus about $6,100, *excluding* the cost of generator fuel, which in Ouwens's calculation is used to pay the extra cost of the superefficient appliances, many of which are low-volume-production or special-order items. Using those efficient appliances, however, to achieve an 80-average-WDC load reduces the estimated grid cost only slightly, to about $2,450/house, and the system marginal cost to about $4,950, even though the marginal transmission capacity requirement is assumed to be halved. In contrast, his 1991 installed cost of a nominal standalone DC-out PV system (575 peak W, 5.75 kWh of battery storage lasting 10 y with no salvage value, controller, wiring, and maintenance at $4.3/m²y, but no inverter) was ~$5,320 ($9.25/peak W). Current costs would be far lower.

[165] This is because many of the costs of connection and distribution are fixed costs (Ouwens reports that distribution cable is only a tenth of distribution grid investment cost), and because the grid must still have a sizable peak capacity to deal with such appliances as washing machines, hair dryers, etc.

[166] This generally works best if PV systems are integrated with culturally appropriate packages of superefficient end-use devices. When South African authorities tried leapfrogging rural grid extension with photovoltaics but didn't provide end-use systems to match the PVs' capacities, customers plugged in such high-load devices as electric cooking elements and water-heaters. This not only made the PV systems fail; it also convinced those customers that solar power isn't "strong" enough and that they really needed the "better" grid power—which will take a great deal of time and money to reach them. Such episodes can set back rural development.

tariff of \$0.07/kWh, or under \$10 present-valued over 30 years, that doesn't make it cheaper: just a code-compliant connection to draw that watt from the grid at zero distance, without even needing a meter, would probably cost more than \$30-odd worth of parts and labor. Adding an installed meter cost and its present-valued usage cost, together exceeding \$700, would clearly make the utility hookup permanently uncompetitive, since it's hard to imagine that the PV system for such a small load could ever cost as much as just installing and using the utility meter. Indeed, \$700-odd (let alone the omitted additional costs of service entrance, etc.) can buy a simple do-it-yourself PV system on the order of 50–150 peak watts—about enough, with very efficient end-use, to provide the necessities of a decent life in a small home.

For this reason, for small loads the PV-*vs.*-hookup breakeven distance *can actually approach zero*—and not only for the PVs that power most of the world's pocket calculators. Graphs like EPRI's or RAP's look at the variable cost of energy, but apparently overlook the *fixed* costs (and perhaps some of the variable costs such as meter-reading) of *connecting* to the grid. The error of their conventional breakeven-distance approach becomes obvious when one considers examples like those in the box "Rethinking Breakeven Distance."

2.3.2.12 The intermediate case: micro-grids

A seminal 1997 paper (326), elaborated in 2000 (318) and subsequently (773), suggests a new market opportunity for distributed resources, especially renewables, that can emerge in the

competitive environment. As defined by pioneering analyst Tom Hoff *et al.*, "A micro-grid[167] is an electrically isolated set of generators that supply all of the demand of a group of customers. Micro-grids are not burdened with the [embedded] costs of the existing system (which can result in a cost savings) but they must reliably supply all of the demand without the benefits of a diverse set of loads and generation technologies (which can result in a cost increase)." (326–7) Hoff *et al.* explore conceptually how to estimate the technical and economic feasibility of a particular micro-grid, and find that micro-grids can indeed make sense. Their advantages are not confined to the kinds of special circumstances that otherwise require major distribution investment, as discussed above for rural electric cooperatives (§ 2.3.2.11).

The essence of the micro-grid approach is that:

- because of load diversity even among a modest number of customers, peak demand does not rise as steeply as the number of customers, but rather starts to flatten out;

- as more generating units are added, especially if they are small units, their collective reliability in supplying firm power rises rapidly (§ 2.2.9.2); and

- even if smaller units cost more per kW (including their installation and connection cost), the optimal unit size can still be relatively small.

A closed-form analytic solution (326–7) suggests that a notional 100-kW constant load in an isolated micro-grid can be delivered with a one-day-in-ten-years loss-of-load probability at a levelized cost of \$0.071/kWh on the following assumptions: 20-year system life, \$0.04/kWh O&M cost, 5% unit forced outage

[167] Robert W. Shaw, Jr. of Aretê Corporation, a coauthor of the original 1997 paper, reports that he originated this term (personal communication, January 2001).

rate, and capital costs ranging from $2,000/kW for a 1-kW unit to $1,000/kW for a 1-MW unit. The optimal unit size is then found to be 2.5 kW, and 50 such units (a total of 125 kW) are required to meet the 100-kW load with the requisite reliability. On consistent assumptions, installing a single 100-kW distributed generating unit and using the grid for backup would have a levelized cost of $0.059/kWh, so the micro-grid would be cheaper if grid backup cost more than the optimistically low value of $0.012/kWh.

Hoff and his colleagues also examine hybrid systems that may have variable loads and more than one kind of generator. This can be quite attractive if some generators match the loadshape well (§ 2.2.8). For example, if tracking photovoltaics provided the same loadshape match as the one-axis tracker at Kerman did during the eight peak load days in 1994, then the same level of reliability could be provided not with 50 2.5-kW fuel cells but with 25 2.5-kW fuel cells plus 25 2.5-kW PVs having a 1% forced outage rate. Interestingly, the total system cost would also be about the same as the 50-fuel-cell system, even if the PV, assumed to have only a $0.01/kWh O&M cost, had a higher capital cost ($2,500/kW). Thus PV generation "could be profitably included in the generation mix under the right conditions." A newer analysis suggests that, partly because residential housing developments can also avoid connection costs, the photovoltaic market is much larger and tolerates much higher PV prices than the market for grid-connected net-metered or clustered PVs, at least for PV system costs above about $2,000/kW (Figure 2-63).

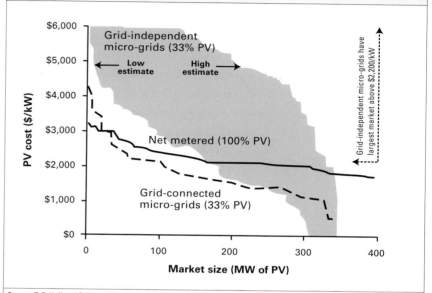

Figure 2-63: Residential micro-grids represent a huge PV market
U.S. market for single-family houses in housing developments without economic incentives. Most of the U.S. residential photovoltaic market at near-time prices is in standalone micro-grids, not in clustered or net-metered grid-connected applications. Even at $6,000/kW, comparable to 2002 prices, the annual grid-independent U.S. micro-grid residential market is estimated at 5–120 MW/y, depending on the avoidable cost of utility interconnection. For comparison, global shipments of PVs totaled 288 MW in 2000 (481).

Source: T. E. Hoff and C. Herig, "The Market for Photovoltaics in New Homes Using Micro-Grids" (National Renewable Energy Laboratory, 27 Jan. 2000), www.clean-power.com/research/microgrids/

This may help to explain why in late 2001, as noted earlier in Section 1.2.12.1, some of the biggest U.S. merchant homebuilders, such as Beazer, D.R. Horton, Shea, Morrison Homes, and U.S. Home, announced plans for hundreds of complete grid-connected PV systems in new subdivisions (531). If designed for islanding, and especially if equipped with optional storage, such "ultra-reliable power services" also offer a marketing edge if wired to a particular circuit with different-colored outlets that occupants can use for their most critical equipment (§ 2.3.3.8.2). A Beazer Homes survey in 2001 promoting its "Powerhouse" equipped with 3.3 kW of PV as standard equipment reported that "Over 95% of respondents to a... web survey expressed interest in purchasing a solar electric equipped home." (660)

Aggregating loads at the scale of a micro-grid requires careful attention to the "spikiness" of individual customers' loads. If there are too few customers, distributed generators "cannot be sized by comparing...to the kVA capacity of service transformers that can serve a site," because transformers can withstand momentary overloading by needle-peak loads that could trip protective gear on local generators or inverters (766). This is nicely illustrated by measurements of daily load curves for groups of 2, 5, 20, and 100 homes in a large suburban area (Figure 2-64). Not only does load per customer decrease with aggregation (note the shift of vertical scale), but needle peaks are smeared out:

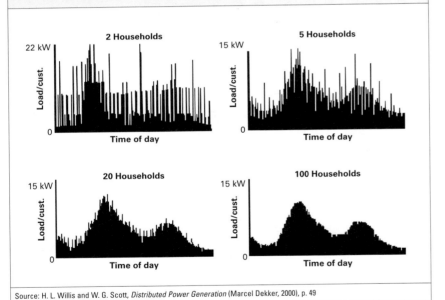

Figure 2-64: Peaky loadshapes smooth and decrease with aggregation
Decrease of load per customer and smearing of needle-peak loads with successively greater aggregation of residential customers. Note the modest number of customers needed to diversify the load—even more modest if they are of diverse kinds.

Source: H. L. Willis and W. G. Scott, *Distributed Power Generation* (Marcel Dekker, 2000), p. 49

In practice, it should be especially valuable to avoid capacity in a micro-grid by careful attention to motor soft-start devices, space-conditioning peak-load management devices, thermally efficient building envelopes that reduce and smear out peak space-conditioning loads, and other cost-effective ways of avoiding needle-peak loads in the first place.

The micro-grid concept—in essence, what traditional grid operators would call a deliberately created island (§ 2.3.2.10.6)—is being explored in depth to see when it might be preferable to either customer-level standalone operation or full grid interconnection. This will depend on the characteristics of the assumed generators, loads, and demand-side resources and on exogenous uncertainties. The preliminary screening analysis, however, does suggest that micro-grids could be a useful and perhaps quite an important new market for distributed resources, taking direct advantage of their modularity, speed, and flexibility. The more stranded-asset costs are loaded onto wires charges, the greater the incentive for customers to leave the grid altogether and set up their own micro-grid—an intermediate scale, big enough to be affordable and reliable, but independent so it needn't be burdened by the sunk costs of the old system.

2.3.3 Non-grid operational benefits

2.3.3.1 Energy generation

Distributed generators obviously provide energy (and demand-side resources save energy) that would otherwise have to be generated by the marginal plant at that moment, backed up by its spinning reserve, and delivered through grid losses to the same location. Traditionally, regulated utilities estimated avoided energy value much as they estimated avoided generating-capac-

Benefit

140 *Distributed generators provide electric energy that would otherwise have to be generated by a centralized plant, backed up by its spinning reserve, and delivered through grid losses to the same location.*

ity value (§ 2.2.9), namely by a simple- or combined-cycle gas-turbine surrogate, or sometimes by a baseload or intermediate-load-factor steam-plant surrogate with similar dispatch functionality.

Where purchased energy is the norm or the marginal resource, it can be priced from market observations, with due adjustment for the point of delivery. For example, an approximate lower bound on the value of non-firm onpeak (0600–2200) bulk energy at California's north and east borders could be set by late-1990s prices, when California was awash in cheap natural gas, virtually eliminating the normal summer premium. Those historically low prices were around $0.015–0.020/kWh, comparable to PG&E's avoided costs in the 1990s, which ranged from about $0.0184 to $0.0296/kWh;[168] as noted in Section 2.2.9, making the energy firm adds about $0.004/kWh to its price—namely, the capacity cost of a combustion turbine. A reasonable upper bound for onpeak energy price in California before the 2000–01 power crisis would have been the common 1980s expectation of around $0.06/kWh (187). Many analysts would even have argued for a lower value, since new combined-cycle gas plants were approaching

$0.025/kWh including their capital cost (§ 1.2.12), and because of their high thermal efficiency, they are relatively insensitive to the price of natural gas. While these ranges may sound like a bygone era while Western regional markets are still re-equilibrating after the California shock, the fundamentals of generating cost have not changed and should re-emerge in time. Meanwhile, energy costs may attract a market premium.

The avoided cost actually being incurred, *e.g.* running and repairing a nuclear plant already in operation but subject to backing down (or, ultimately, shutdown) if displaced, may not be properly counted if the motivation is to keep it "used and useful" so that it remains in ratebase. In a more competitive environment where it is tempting to use price markers discovered in the real-time wholesale market, therefore, any such differences between theoretical and actual operating behavior must be borne in mind. That is, every resource should be the marginal resource. A utility that is dispatching costlier-to-run capacity (usually its own) than it could buy in the wholesale market should not thereby burden other marginal competitors, such as proposed marginal resources, with an unfair comparison simply because it is not observing proper merit order in dispatching its entire portfolio of resources.

Energy purchases, *e.g.* in the wholesale market, may not have the same financial risk profile, *e.g.* from price volatility, as the distributed resource. The two cost streams must then each use the appropriate risk-adjusted discount rate as described in Section 2.2.3.

---

[168] The final evaluation of the Kerman plant (735) assigned an energy value of $143–157/kWy ($0.0626–0.0687/kWh)—lower than the originally assigned $194/kWy (595), but higher than the general daily price because the PV output was largely onpeak and highly correlated with load. However, the regional capacity surplus reduced the expected capacity value of $65/kWy to a final range of $12–53/kWy. That surplus proved temporary, showing the importance of a long view for managing system risks.

2.3.3.2 Reduced keep-warm (minimum-load) operation

To meet daytime capacity needs reliably, most classical utilities use many small and midsized (~100–200-MWe) fossil-fueled units that are difficult or costly to keep turning on and off. Those units must therefore be run at minimal output on weekends and at night so they will be warm and ready for peak-hours service. However, the minimum-load operation yields considerably lower thermal efficiency than operation at the normal design load, much as a car engine is less efficient at idling speed than under full power. For this reason, minimum-load power costs more to make than the cheapest baseload generation dispatched in offpeak hours. Yet it is operationally essential.

Distributed resources available onpeak (§ 2.2.8) can reduce the need for these costlier-to-keep-warm units at the system peak, and hence also for running them at minimum load. The saved fuel cost is simply the capacity of such plants, times 8,766 h/y, times capacity factor at part load, times the difference in heat rate between minimum-load and intermediate-load operation, times the fuel price (599). For the 500-kW Kerman PV station on the PG&E system, this turned out to be worth $28/kWy (1992 $), (613) or about $0.0122 per PV output-plus-avoided-losses kWh.

2.3.3.3 Reduced spinning-reserve operational cost

The example cited in Section 2.3.1.2 implied that the avoidable operating cost of spinning-reserve combustion turbines might

have a present value on the order of ~$1,457/kW (1995 $) less the operating costs of the fuel cell. Those should be lower because of its higher thermal efficiency (even net of reformer costs and losses). However, this concept does not appear to be valid, because in the spinning-reserve role, the fuel consumed by the surrogate combustion turbine is only enough to maintain the rotor in synchrony against the angular deceleration of friction and windage (air resistance). Once the unit actually comes under load, it thereby shifts from a reserve role to a generator of electricity for sale, so the extra fuel needed to meet the load becomes an ordinary operating cost of generation. To first order, then, it appears that the true marginal operating cost of spinning reserve is a negligibly small amount of fuel plus a very small amount of variable O&M (*e.g.*, faster exhaustion of rated hours' operation between bearing renewals)—perhaps on the order of $0.001/kWh or less. This conclusion may not hold if the *actual* spinning reserve is provided by a steam rather than a combustion turbine, although care must be taken not to double-count keep-warm fuel (§ 2.3.3.2).

2.3.3.4 Reduced startup cycles

Power plants, like any equipment, work more efficiently and reliably under steady loads than under the wear and tear of stop-

and-go or variable operation. This is especially true of thermal stations because of the
thermomechanical stresses, and perhaps
also corrosion-related stresses, of cycling the
boiler. Part of the complexity of deciding
whether to keep a plant warm overnight
(§ 2.3.3.2) is that it is difficult to compute
the true cost of cycling generating units—
turning them off and on again.

A very preliminary estimate by Ontario
Hydro for one of its thermal stations,
including incremental routine maintenance
costs and a "rough guess of major component replacement needs" but no fuel costs
nor reliability impacts, suggested that each
cycle might cost around C$5,000–15,000.
However, more complete assessments by
other consultants, including estimated
future system impacts such as higher fuel
costs and lost revenues, are in the range of
C$30,000–100,000 (477). Clearly these values
depend strongly on the plant, system, and
methodology, and are highly variable.
Equally clearly, where they can be properly
evaluated, they seem large enough to affect
the kinds of unit commitment decisions
described next—decisions that some distributed resources can make valuably
unnecessary.

2.3.3.5 Fast ramping

As explained in Tutorial 1, changes of supply or demand must match the "ramp[ing]
rates" (rate of change in output) provided
by system resources in order to keep supply and demand in balance and thus maintain frequency, phase, and voltage stability.
The most difficult ramps to deal with are
usually not the gradual fluctuations and
trends of load, but rather the "square-

> ### Benefits
>
> **144**   *Inverter-driven distributed resources can provide extremely fast ramping to follow sudden increases or decreases in load, improving system stability and component lifetimes.*
>
> **145**   *By combining fast ramping with flexible location, often in the distribution system, distributed resources may provide special benefits in correcting transients locally before they propagate upstream to affect more widespread transmission and generating resources.*

wave" shocks caused by the instantaneous
connection or disconnection of large loads
(such as electrometallurgical plants) or by
faults that trip offline whole blocks of generation or transmission capacity. Subject to
complex stability limits, the faster such
abrupt changes in supply/demand balance
can be dealt with, the less the likelihood of
awkward system stresses and outcomes, up
to and including system collapse; the less
the wear and tear on rotating machinery;
and the more gracefully the transients can
be smoothed to accommodate normal
changes in steam flows, angular momenta
of rotors, and the other ordinary tools of
fine-tuning the system.

Ordinarily, the fastest ramp rates available
to the utility dispatcher come from inter-
rupting load: this can be done instantaneously, but inconveniences customers (and
even those who pay interruptible tariffs
appreciate due notice to help in the orderly
planning of their affairs). Also quite fast is
the option of loading spinning reserve
capacity (Tutorial 1) maintained for this
purpose, since it is already synchronized
with the grid. However, this is not a true
square-wave response, not only because of
alternator reactance, but also because the
suddenly loaded rotor will lose momentum

if the prime mover does not keep pace by rapidly flowing more steam, fuel, etc. Ramp rates that require the *acceleration* of large unsynchronized rotating masses, such as steam or hydroelectric generators, obviously take even longer, and stationary rotors that must be accelerated to synchronous speed take longest of all.

In this context, distributed resources offer intriguing new ramp-rate options not valued in the literature. For example, some extremely reliable low-temperature fuel cells currently in advanced development for automotive uses, but equally applicable to buildings, can go from zero to full load[169] in a few *milliseconds*[170]—a small fraction of a cycle (687). Similarly, being purely electronic with no angular momentum ("inertialess"), thermal inertia, or massflow, solar-cell arrays can be connected to—or tripped off of—the grid instantaneously to achieve step-function changes of output in either direction. With suitable inverter and controller design, this can be done as rapidly and as often as the switchgear permits,[171] with no damage to the equipment. Only electrons are inconvenienced.

In principle, conventional power stations can also be connected and disconnected at will; they are then being treated as "unloadable generation." (726) However,

- this is more complicated operationally;

- it must be done carefully in order not to damage valuable equipment;

- it tends to decrease equipment life to some degree, however small; and

- once tripped offline, a thermal plant cannot be quickly brought back online, so such a tripping decision is irreversible in the short term and may therefore require the operator to schedule other, costlier unloadable generation instead.

In contrast, trippable PV resources are completely reversible (assuming the array is still illuminated), providing fast *and instantly reversible* downramping. Because of the short timescale involved, this attribute may even be considered more dispatchable than the PV's energy output.

The operational value of this new bidirectional fast-ramp capability is unknown. It is akin to, though faster than, the six-second load demand-side interruptibility being brokered in emerging "stability markets" for New Zealand's Transpower and soon for others (§ 2.3.1.2). The stability value discovered in those markets can presumably be used as a surrogate for the value of certain distributed supply-side resources' fast ramp rates. However, there are also significant differences. For example, fast-ramp distributed resources, both supply- and demand-side, can be activated not just in one place but in certain parts of the grid or in many parts at once. Properly deployed and dis-

---

[169] For a car, typically ~25–50 kWe, but fully modular at any desired scale.

[170] This is a far cry from a standard 2-MW molten-carbonate cell's one minute from hot standby to full load—let alone its rated 6-*hour* cold-start time (213). The cold-start time to full load for a typical proton-exchange-fuel-cell stack (*e.g.*, a 13-kW Ballard stack) can be several minutes if a liquid-fuel reformer must be warmed up to produce the hydrogen, but for a stack fed neat hydrogen, the optimal approach (440), the startup time is only a few seconds—much less if the auxiliaries are already running and the electronic startup sequence partly performed. For example, Ford tested in 2000 a P2000 direct-hydrogen car that could go from zero to full throttle in 0.2 seconds even though that vehicle's peculiar packaging required the hydrogen to travel from the tanks at one end of the car to the stack at the other. For any PEMFC stack, the cold-start-to-full-power time can be reduced to just milliseconds, especially in low- or ambient-pressure and passively humidified designs, simply by designing the gas-flow channels ("flow field") to deliver an adequate massflow of hydrogen and oxygen that quickly. Once the gases are delivered to the catalytic membrane, their conversion into electricity and hot water is instantaneous.

[171] A suggestive micro-example is that in the Gardner experiment (651), "Cold load pick-up [when restarting the grid after it has collapsed] can result in the simultaneous switching-on of all [line-excited] photovoltaic generation on the circuit soon after the distribution voltage and frequency [are]...stabilized within normal bounds. While the transient event may cause a momentary voltage surge, once interconnected[,], the generation will help boost the voltage which normally drops during cold load pick-up periods" because all the previously dropped loads are simultaneously coming back online, many of them motor-driven and therefore drawing large startup current surges.

patched, they may prove especially valuable in protecting the distribution system, at and downstream of the substations, from the transients caused by major generation and transmission faults. In principle, fast timing might enable such resources to damp the "ringing" of the grid very effectively, especially after large disturbances.[172] Such grid stability issues are very complex and far beyond our scope here, but we hope that experts in this field will consider whether resources that are both fast-ramp *and* distributed can derive special economic value from those novel attributes, singly or (especially) in combination.

2.3.3.6 Net-metering advantages

Many utilities suppose that distributed resources selling back to the grid require elaborate metering and accountancy. However, such pioneers as SMUD Solar took the plausible view in the 1990s that for small- to medium-sized customers generating power with distributed resources, "net metering" should be the common practice. By early 2002, it had been adopted in at least 34 of the United States for small (and in some cases not-so-small) power producers, and was being considered in most of the rest.[173]

Distributed-generator metering traditionally uses two back-to-back meters ratcheted against reverse flow. Net metering's innovation is that the utility bills or refunds only for the difference between energy bought and energy sold, counting both at the *same* price. Net metering pays the customer more for PV power, since the utility's average tariff is typically several times its short-run avoided cost. This can permit a Sacramento 4-kW PV system to cut a typical household's net electricity bill by about 75% (742). Net metering also reduces metering cost by permitting the use of a single meter that spins forward or backward.

Electronic meters can do the same with two different registers, for forward and backward flows of power, or can net them out internally, and can more easily measure power factor, time-of-day, peak power flow, harmonic content, etc. For example, Metricom meters can resolve roughly five watts and one millisecond on any phase(s), can record over 200 different data streams every few seconds, and can be accessed by four different means including bidirectional packet-switching radio.

This practice of counting flows in either direction as equally valuable can be advantageous to the utility. Not only is it the cheapest metering and accounting method, but it also provides the utility with valuable peak power from photovoltaics and similar resources, in a flat-price trade for what is

> **Benefit**
>
> *146*
> *Distributed resources allow for net metering, which in general is economically beneficial to the distribution utility (albeit at the expense of the incumbent generator).*

---

[172] Large disturbances are those "for which the nonlinear equations describing the dynamics of the power system cannot be validly linearized for purpose of analysis." (103) For example, during the 10 August 1996 collapse of the western U.S. grid, "unexplained grid power system oscillations began in which voltage and power transfers fluctuated wildly. As two parts of the system fought each other, power transfers fluctuated by ±1000 MW and ±60 kV. Within minutes, several more lines tripped...and both the Pacific AC and DC Interties opened (no longer carried power)." (328) Distributed resources would have to be very large or numerous to correct such massive power swings, but might not have to be nearly so large in order to offset or damp the initial local disturbance that ultimately swelled to such a disastrous size.

[173] For example, from 1 January 1996, all California utilities were required to provide net metering for residential PV systems of up to 10 kW, up to a total PV capacity equal to 0.1% of that utility's 1996 peak demand. Statewide, this was equivalent to 50-odd MW—an insignificant fraction of capacity, but a substantial increase in the installed PV capacity (four times U.S.-made PV installations during 1995 for power generation [185]) and hence reduction in the technology's marginal cost. Two California investor-owned utilities initially proposed standby charges, but were denied permission to implement them because they would defeat the law's intent. Even in 2002, some regulatory wrinkles remained to be ironed out, *e.g.* over who should pay for capacity upgrades for customers wishing to sell larger amounts of power back to the grid than their local feeder could accommodate. But the California power crisis of 2000–01 caused the 10-kW threshold to be increased to 1 MW for wind and solar additions through 2002. The maximum size was driven not by engineering but by politics, which often depend on whether the utility thinks it makes or loses money on net metering; ignoring distributed benefits can make it look like a loser.

often inexpensive offpeak power. Given many utilities' actual onpeak generating and grid costs, this can often save the utility enough money that it is more than reasonable to use net billing that pays customers the retail price for their generated electricity. Of course, if the retail price is not time-differentiated, then customers may be losing potential value, depending on whether the retail price properly reflects both short- and long-term marginal costs. Customers charged time-differentiated prices may very reasonably expect a symmetrical payment at the same prices for production they sell back. But either way, utility customers collectively will often benefit from net metering and should not suffer. Even in a worst case, net metering's potential impact on general electricity prices was found by SMUD to be at most 0.0009% (742). Irrationally, some state laws let the utility pay nothing for any net excess kWh produced over a year or even a month (§ 3.3.3.1.1, note 9).

> **Benefit**
>
> **147**
> Distributed resources may reduce utilities' avoided marginal cost and hence enable them to pay lower buyback prices to Qualifying Facilities.

2.3.3.7 Lower payments to QFs/IPPs

Ever since Section 210 of the Public Utilities Regulatory Policy Act (1978) was upheld by the Supreme Court in 1984,[174] U.S. utilities have had to buy back Qualifying Facilities' (QFs') electricity at whatever their state regulators decided was the appropriate avoided marginal cost. Reducing that marginal cost therefore, with some regulatory lag, reduces the payments. The utility benefits of this effect (at the expense of a given QF's owner) vary, depending on the capacity marginal

cost of the utility before and after the installation of the new local power plant. It is solely a regulatory effect, and has nothing to do with technical improvements. In the case of the Kerman substation, these benefits amounted in 1992 to $46/kWy. Presumably as avoided costs fall and competition increases, this type of benefit will gradually disappear, and even today it is often smaller than the 1992 calculation. Of course, for any utility that could successfully argue that demand-side investments were its marginal resource—a position readily defended by investing in that way—this would be a moot point because QFs would be paid nothing anyway, but in the confused strategic and regulatory climate of 2002, few investor-owned utilities were so aligned.

2.3.3.8 Unbundled service quality: harmonics, power quality, and reliability

Hodge and Shephard's penetrating 1997 analysis of "The Distributed Utility" (303) suggests that in the emerging competitive market for electricity and electrical services, with distinct markets for energy, delivery, and stability,

> The dynamics of the delivery market will probably be the primary trigger input to the dispatch function for distributed generators, as it will likely be more volatile than the energy [commodity] price, and will signal clearly local [transmission and distribution] system constraints.

If that occurs, as seems plausible, then those

---

[174] During the 1980s and 1990s, however, the Act was largely vitiated through redefinition by the FERC during the Reagan/Bush era to qualify many nonrenewables as quasi-renewables (e.g. waste coal counts as waste, not coal) and to gut provisions meant to favor smaller and more fuel-efficient generators. In the mid-1990s it also came under attack by the conservative Congress, where calls to repeal it were often heard; but in truth there is not a great deal left to repeal, especially after recent state-level interpretations, contrary to the statutory language, often seem to treat the avoided-cost level of buyback price as a ceiling rather than as the intended floor. Other sections, such as the seemingly clear intervener funding provisions, have also been gutted by bad caselaw. Formal repeal was a recurrent theme in the current Congress in spring 2002. There do appear to be material public benefits still captured under PURPA in some jurisdictions, but they're hard to quantify.

> **Benefit**
>
> **148** Distributed resources' ability to provide power of the desired level of quality and reliability to particular customers—rather than just a homogeneous commodity via the grid—permits providers to match their offers with customers' diverse needs and to be paid for that close fit.

volatile "dynamics of the delivery market" will provide strong signals also for the *installation* (or relocation) of distributed resources to the places and times where their dispatch will be most valuable in reducing costs and risks.

Meanwhile, a similar unbundling is already clearly emerging in the way electrical services are offered for sale. Traditionally, most electricity customers pay for a relatively high level of reliability and a moderate level of power quality, but those attributes are fairly uniform and immutable. But alert utilities and other service providers are already starting to unbundle offerings of both higher and lower reliability (UPS-based and interruptible services) and of higher power quality (premium, filter- and UPS-based data-quality power services). Rather than plain-vanilla, one-size-fits all commodity kilowatt-hours, many customers are starting to get wider choices in what levels of service they actually want and are willing to pay for. The rapid expansion of these and similar unbundlings of valued customer attributes is described elsewhere (157). The more it occurs, the more easily the discovered value of unbundled attributes can be counted as benefits of the distributed resources that can provide those attributes to the specific cus-

tomers they serve.

A natural counterpart of this unbundling is a greater symmetry between the values sold to and bought from customers. For example, many utilities penalize low (normally lagging) power factor, but few if any reward (buy back) high or leading power factor. As customers' distributed-generator inverters become able to generate reactive power at will, albeit at a concomitant sacrifice of real power output, there will be times at which the grid should be willing to pay a good price for reactive power that's produced by customers more cheaply than the grid could otherwise obtain it by installing capacitive compensation. The same could be true of real-power generation that provides similar voltage support, or of customer-provided improvements in reliability, flexibility, or other valued attributes. We next examine two obvious opportunities for such buy-backs of customer-generated value: harmonic reduction and improved reliability. They are listed here, rather than in Section 2.4.1, because they provide operational benefit to the utility as well as value to the customer.

2.3.3.8.1 Power quality, harmonics, and active harmonic compensation

Power systems do not deliver a perfectly sinusoidal waveform. Such ugly realities such as switching transients, lightning pulses, high-frequency noise, and harmonics intrude. Of these, the last two are the most commonly caused by customer devices, and the last, harmonic distortion, is the most relevant to grid-connected distributed generators.

A review of distributed utility valuation (107) comments:

Benefits

149   *Distributed resources can avoid harmonic distortion in the locations where it is both more prevalent (e.g., at the end of long rural feeders) and more costly to correct.*

150   *Certain distributed resources can actively cancel harmonic distortion in real time, at or near the customer level.*

151   *Whether provided passively or actively, reduced harmonics means lower grid losses, equipment heating (which reduces life and reliability), interference with end-user and grid-control equipment, and cost of special harmonic-control equipment.*

152   *Appropriately designed distributed inverters can actively cancel or mitigate transients in real time at or near the customer level, improving grid stability.*

Until recently, electricity supply throughout the U.S. was characterized by a (more or less) pure sinusoidal signal that could be relied upon for precise control and measurement. Commonly encountered measuring devices (such as the residential kilowatt-hour meter) were designed, and their indications were accepted with confidence, on the basis that a sinusoidal signal was available. Generating sources (such as synchronous rotating machines) were specifically designed at great expense to minimize harmonic output; thus energy losses due to harmonics or other signal distortions did not have to be considered in system models or calculations. When only a few devices causing distortion were being connected to the system[,] they were tolerated because a local "fix" could be applied to manage the adverse effects on the purity-of-signal attribute; system-wide impacts were not considered. With large number[s] of distributed generating devices that introduce high harmonics into the electric system, system-wide impacts are likely. Local generation is a high impedance source and will compound latent and future harmonic problems.

This is all true in principle. However, in practice it is a somewhat overly sanguine view of the current situation, for reasons having nothing to do with distributed generation and everything to do with nonlinear loads—those whose current does not vary smoothly with voltage, like capacitors, diodes, and power-switching devices. For example,

- The often inefficient switching power supplies in modern computers, other office equipment, televisions, and other consumer electronics are rich sources of harmonics, especially third harmonic. It is not unusual for the third-harmonic current in the neutral conductor of an office building to be half again as large as the fundamental (60-Hz) current in the

Electricity is generally supplied as alternating current at a standard system frequency of 60 back-and-forth cycles per second, or Hertz (Hz), in North America, 50 Hz in most of the rest of the world. Worldwide, alternating current has a nominally sine-wave pattern of alternation. However, interaction with any nonlinear device or impedance mismatch will create **harmonics**—"ringing" at integer multiples (or sub-multiples) of the 60- or 50-Hz fundamental frequency—that add to the fundamental frequency to produce a complex waveform. The nonsinusoidal part of the waveform performs no useful work and is a significant nuisance. It is measured in aggregate by Total Harmonic Distortion (THD) and by percentage content of individual harmonics, using special meters and power-quality engineering skills. Depending on where harmonics are injected into the grid, and the specific technical characteristics of the grid, the harmonics may travel back upstream and cause heating (hence inefficiency, lost capacity, and shorter lifetime) in wires, transformers, and generators. They may also interfere in other ways with utilities' or customers' equipment.

In general, even-numbered harmonics cancel out because their pulses are of both polarities. In three-phase systems, any harmonic number divisible by three (third, ninth, twelfth,... harmonics) will also be canceled out within each phase, but those "triplens" will add together across all three phases and end up heating the neutral ("return") wire. An added National Electrical Code section (NEC 110-4, 1993) requires fatter neutral wire to prevent overheating. A good tutorial on practical harmonics engineering in an industrial context is at pp. 282–297 of the E SOURCE *Drivepower Technology Atlas* (1996).

Impedance is an AC circuit's resistance to power flow, both real and reactive. It combines the effects of resistance and reactance, not by addition but by square-root-of-the-sum-of-the-squares.

phase conductors. Several office buildings have even burned down as a result of harmonically overloaded neutrals.

- Some adjustable-speed inverters for motor drive—especially early, small-scale, or low-quality designs—propagate harmonics, especially fifth and seventh. (Modern, high-quality units carefully suppress harmonics, and that attribute is sought by intelligent procurement practice.)

- The early and lower-quality kinds of electronic lighting ballasts were often rich in mainly third harmonic, although modern units have good harmonic traps and usually emit harmonics comparable to or less than those normally found in grid power. However, the older magnetic (core-coil) fluorescent and other discharge-lamp ballasts that are still very widely used are rather strong harmonic sources (654), typically producing 20–30% THD for nominal $2 \times F40$ ballasts (20), or 60% THD for a nominal 70-W metal-halide ballast (19).

- A few kinds of nondischarge lighting devices, such as certain halogen capsule lamps used in retail display and certain disks made for insertion into incandescent lamp sockets to stretch lamp life, incorporate halfwave rectifiers whose diodes are rich sources of harmonics (though these are generally stopped by the service transformer).

- Low-quality lighting dimmers, like those used in the widespread residential halogen torchieres, yield ~96% THD when dimmed to one-third of full output (89). (Fire hazards of these lamps caused their UL approval to be withdrawn, but many remain in service in homes, college dormitories, etc.)

- Such intensive loads as arc-welders, arc-furnaces, and diathermy machines propagate strong high-frequency "hash" and other electronic noise that flow into the grid and heat every conductor they encounter.

- Even the capacitors and transformers installed by the utility itself distort the waveform.

Harmonic voltages are worth minimizing because they cause heating, insulation breakdown, and other irreversible harm to utility and customer equipment. This harm may increase nonlinearly with THD.[175] Thermal aging at THD levels actually encountered in the grid can shorten appliance and motor life by as much as tens of percent (261). Harmonics can also disrupt data-processing equipment, sometimes in ways that are hard to identify.

Harmonics' effects on the electrical network depend largely on the *network's* electrical characteristics rather than on where the harmonics are injected, so the effects often concentrate far from that location, complicating diagnosis. This is especially true of resonances that may occur when harmonics are injected into a grid containing both capacitors (to compensate power factor) and normal line, transformer, and motor inductances. Uncontrolled resonances of this kind can quickly create destructive overvoltages, safety hazards, and equipment failures. (Fortunately, several effective control methods are available.)[176] Such issues are especially common at the end of long feeders, largely because of all the inductance along their length; "rural feeders may have almost three times the impedance of urban feeders, resulting in three times the harmonic voltage for the same harmonic current levels." (704) This may place a power-quality premium on distributed generation in exactly the same rural locations where it also has the greatest reliability, voltage- and power-factor-support,

---

[175] However, THD is a somewhat crude and aggregated measure, because eddy-current heating, one of the effects of harmonic currents, heats to a degree proportional also to the square of the harmonic frequency. Some particular pieces of equipment may also be especially sensitive to even low levels of a specific harmonic.

[176] These include relocation of the power-factor-correcting capacitors, using variable programmable capacitors, or installing active harmonic cancellation (704).

and grid- and loss-displacement benefits.

Harmonics are ubiquitous and indeed are necessary to the functioning of some devices. However, they can be greatly reduced by good design; readily controlled by traps, filters, or active devices[177] that detect and cancel them; and, importantly, confined by design to locations where they can do little harm. Harmonics exist widely and persistently; the problem is where they go and what they do. Since harmonics, like reactive power, are not measured by standard wattmeters, not well understood by most customers (and even by some utility personnel), and not charged for by most utilities, they tend to be an invisible cause of heating and hence of greater real-power losses and shortened equipment life. Without careful measurements, these problems tend to persist undetected. But with knowledgeable measurement and modern computer models (which closely fit the measurements), "simple mitigation measures" can provide practical solutions "for those cases where harmonic distortion levels become unacceptable, regardless of their source." (654)

Newly designed distributed generators can and usually do use modern techniques to limit harmonic generation and control its destination and effects. However, many utilities that unhesitatingly sell electricity to the worst contaminators of the grid's waveform, such as arc devices and switching power supplies, still impose far more stringent harmonic limits on small-scale generators than they do on ubiquitous, and sometimes larger-scale, customer loads.[178] In theory, utilities[179] typically limit THD to less than 5% of the current and 2% of the voltage signal, with no single harmonic (typically the third) contributing more than 3% of current or 1% of voltage—standards that continue to evolve.[180] In practice, such requirements are honored mainly in the breach—except for requirements placed on distributed generators.

However, a more modern view is rapidly emerging as better controls, software, and switching devices enable distributed generators' inverters to turn into part of the solution. The harmonic problem of early modified-square-wave PV inverters has "been completely remedied in more recent [as of 1992] high-quality designs, such as models tested by PG&E and other utilities [that]... can [produce] virtually pure sinusoidal outputs." Such self-commutated residential inverters' third-harmonic distortion (normally the harmonic of greatest interest for such units) is "much less than [for] most of the household loads, while [harmonic] distortion from a line-commutated inverter was comparable to the distortion from a window air conditioner." (697)

Some early studies assumed that large numbers of low-quality line-commutated inverters would be rapidly deployed, all run at peak load simultaneously, and produce har-

---

[177] With modern fast-switching thyristors, a device that cancels essentially all the amplitude of harmonics up to 13th can fit into a box of roughly a cubic meter for a continuous rating of several MVA at transmission voltages (518).

[178] There is often a similar apparent bias against motor inverters, which are often more prone to be disrupted by poor quality coming from the grid than vice versa.

[179] However, dedicated power-conversion devices are required by IEEE Standard 519-1992 to have THD—measured at the point of utility interconnection, and compared with the maximum load on the distribution system—less than 10% of the fundamental for that device and 5% for the whole facility.

[180] The commonly cited original ANSI-IEEE Standard 519-1981 listed only total line limits rather than specifying how much THD a given device may inject into the system. The 1992 revision was better but still vague, so IEEE recently launched a revision project (519A).

monics all perfectly in phase with each other. Combining these unrealistic assumptions led to estimates that harmonics might limit inverter deployment to the equivalent of as little as 13% of available line capacity. But with modern self-commutated inverters, this constraint has essentially disappeared (705). For example, the Gardner experiment (§ 2.2.8.1) found, as is now fairly common, that its residential PV inverters (using the best *1985* technology [650]) were adding less THD to the grid than they were receiving *from* the grid (85). On a feeder with up to 53% PV saturation, THD was therefore generally less from the PV inverters than from the ambient grid power. Moreover, even if unsophisticated, high-harmonic-output PV inverters had been used, their harmonic injection per kVA would have been only about one-third that of a typical variable-speed industrial motor drive of the same era (652).

Better yet, modern sinewave-output inverters can "even...reduce the amount of high-frequency noise from other devices in the grid and thus correct the utility waveform," (633) leading to a striking new distributed benefit (622):

> Another potential benefit of distributed generation may be the ability to improve the utility waveform in distribution systems. In general, the harmonic content of distribution systems has been rising with the introduction of nonlinear loads such as fluorescent lights and variable speed [motor] drives, while sensitivity to such harmonics has also increased with the prevalence of computers. New inverter designs could potentially alleviate distribution harmonics through active harmonic cancellation...[whereby] distortions of the utility waveform are cancelled by equal but opposite (out-of-phase) distortions controlled by power electronics.

These utility analysts recall that self-commutated inverters can provide reactive power at will (§ 2.3.2.3.2) rather than burdening the grid by consuming reactive power, so "as with active harmonic cancellation, it turns out that the effect of inverters with regard to reactive power can actually be corrective rather than problematic for the distribution system." (634) The PG&E team concludes (634):

> Given these developments in inverter performance, we believe that distributed PV systems and inverters will come to be valued for their beneficial effect on power quality. It is quite conceivable that power quality benefits such as harmonic cancellation would be included as an additional category of distributed benefits in future studies. While criticism and concern about safety and power quality was certainly in order during the early days of grid-connected PV systems and inverters, the time has now come to consider this technology an asset rather than a burden to the [transmission and distribution]...system.

With careful design, and within their operating and geographic limits, such fast-responding smart inverters on distributed generators could also create additional value by providing some degree of real-time cancellation of switching, lightning, and other grid transients and of voltage sags—thus addressing all power-quality issues, not just harmonics. Some aspects of this opportunity are being addressed in Utility Photovoltaic Group TEAM-UP ventures, such as the 1995 100-kWAC project by Niagara Mohawk and the 40-kWAC project by UtiliCorp United and Nevada Power, both testing photovoltaics as a means of power-quality correction.

Such active harmonic correction is now starting to enter the market. Jeff Petter[181] correctly points out that:

> Inverters on the outputs of disturbed generation devices can be designed to cancel the current harmonics locally. This will reduce

[181] Senior R & D Engineer, Northern Power Systems, 182 Mad River Park, P.O. Box 999, Waitsfield VT 05673, 802-496-2955 x257, FAX 802-496-2953, www.northernpower.com, jpetter@northernpower.com (personal communication, 15 November 2001).

system losses, increase the life of transformers[,] and reduce the need for the harmonic filters used to absorb these currents. It costs very little to add this active harmonic filtering capability to a modern inverter. Mostly the additional cost is only in the design and engineering of the control firmware. In addition our inverters are designed to help regulate the grid voltage and stabilize any resonances in the utility grid or loads. These inverters are designed to do their share of improving the power quality of the local grid in proportion to their size. I like to call them socially responsible inverters.

Northern Power Systems is currently developing inverter controls specifically to take advantage of this and many other potential benefits of having a fleet of socially responsible power electronic inverters distributed in a utility system.

The size of these benefits depends on the sensitivity of loads, especially customers' digital equipment, to the amounts and types of harmonics present, but the benefits can be important, and can be both local and sys-

temic. The analytic approach is analogous to but even broader than that used for power-factor improvements (§ 2.3.2.3.2), and the benefits can be larger because they also include avoided interference with customer equipment—a problem whose solution can otherwise, in some instances, be difficult and costly. Petter's suggestion of zero- or very-low-cost design improvements to make inverters into active harmonic compensators is an obvious opportunity for standards-setting organizations, *de facto* industry best practices, and grid operators. If incorporated into many inverters, it could lead to important and pervasive systemwide benefits as such distributed-resource inverters became widespread.

2.3.3.8.2 Premium reliability

Some uses or customers may be content with quite low reliability. A water heater doesn't notice interruptions, on a scale of at least minutes, that would be fatal to a person kept alive by an iron lung or the electronics in an operating theater. Other applications, like paper-making machines, chip-making steppers, mainframe computers, or air-traffic-control radars and radios, need much higher reliability than is now commonly delivered. For example, the Computer and Business Equipment Manufacturers' Association (CBEMA)[182] published in the early 1980s and updated in 2000 a standard design goal for most computing equipment (Figure 2-65).

Although most computers can tolerate 6% overvoltage or 13% undervoltage, or even a

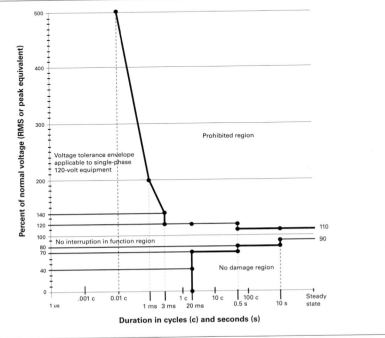

Figure 2-65: Typical design goals of power-conscious computer manufacturers
Voltage cannot spike or sag too much or for too long without losing or corrupting data.

Source: Technical Committee 3 of the Information Technology Industry Council, ITI (CBEMA) Curve (ITIC, 2000), www.itic.org/technical/iticurv.pdf

---

[182] Now the ITI Council: www.itic.org/technical/iticurv.pdf. This widely used curve is now in IEEE Standard 446-1987. For information systems managers and associated electrical engineers, CBEMA authors Dugan, McGraghanan, and Beaty have created the helpful and comprehensive power-quality guide *Electric Power Systems Quality*, November 1995, $55, McGraw-Hill (New York).

wider range for one or a few cycles of AC current, higher-voltage spikes lasting a fraction of a cycle risk voltage breakdown of power-supply or downstream components. Depending on the degree of magnetic and capacitive energy storage in the particular equipment, undervoltage may also cause data loss or corruption.

Electric supply reliability isn't easy to measure, and the value placed on it by different customers is notoriously slippery. Customer surveys usually yield a bimodal distribution—many customers who don't care, plus an array of customers who care much more, or very much, somewhat like the following graphs:

Measuring the economic *value* of reliability for electric service in general is notoriously

difficult because customers and their economic preferences and circumstances are so diverse. Moreover, reliability can be improved in a variety of ways: a reevaluation reduced the Kerman reliability benefit from $205/kWy (595) to $4/kWy (735), not by using a different value-of-service assumption, but by noting that "a capacitor bank could be added to the Kerman circuit and provide the same operational benefits at much lower cost...."

However, there is a growing consensus that a significant and increasing fraction of economic activities require "digital power quality"—whatever that means. EPRI's Consortium for Electric Infrastructure to Support a Digital Economy (CEIDS) takes perhaps the most expansive view of this need. Its 2001 study (406) extrapolated from a statistical sample of 985 firms, in segments representing 40% of GDP and showing special sensitivity to power disturbances, that U.S. power outages and disturbances cost more than $119 billion annually to digital businesses, continuous process industries, and "fabrication and essential services, which includes all other manufacturing industries, plus non-electric utilities and transportation facilities." There is obvious latitude and much ambiguity in defining those industries for which brief power disturbances are actually important, but in a complex and interconnected economy, the distinction isn't easy. Of the estimated cost, 87% was due to outages, 13% to briefer power disturbances. Digital business, the sector most sensitive to both, tends to have the lowest outage costs because it has already invested the most in protective equipment. (A stated example—a Miami data center that spent $300/ft², or 53% of its total cost, on power-conditioning equipment—indeed makes one suspect that an isolated onsite

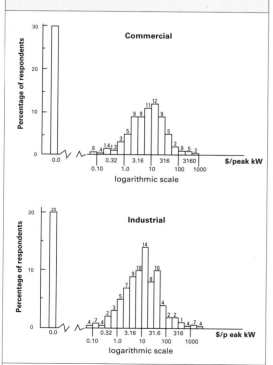

Figure 2-66: Different customers place very different values on electricity supply
A significant fraction hardly care, while some others value reliability highly.

Source: D. W. Caves *et al.*, "Customer Demand for Service Reliability in the Electric Power Industry: A Synthesis of the Outage Cost Literature" (*Bulletin of Economic Research*, May 1990), pp. 79–119

solution would have cost far less and yielded greater benefits.)

Widely cited views that the fraction of U.S. electricity that must be delivered with digital reliability will grow from, say, 10% to 60% in the coming decades should be viewed with skepticism: the mere presence of microchips, as in ubiquitous home appliances, does not necessarily imply a genuine need for digital reliability nor a specific definition of what "digital reliability" means. Equally dubious, when advocates seek to define the term, are many widespread statements about how many "9"s of power-supply reliability are required. (Extremely high power *quality* can readily be achieved by onsite power conditioning if the power supply is available, but that supply *reliability* is the hard part. The two are often confused, sometimes deliberately.) In particular, many observers claim that digital businesses, and such critical industries as chipmaking and pharmaceuticals, "require" five, six, or seven "9"s (six nines would be 99.9999% availability), or even ten "9"s at the chip (780). But as Dr. Joe Romm correctly notes, based on discussions with system reliability experts from many disciplines,

- "No moderately complex manmade system has ever delivered more than seven "9"s of reliability for an extended period of time, whether related to energy or otherwise. In fact, even six "9"s is hard to beat." (569) This is because of irreducible human error, common-mode and common-cause failures, natural disasters, and the like. It is meaningless to say that a given customer system "requires" 8–9 "9"s power reliability, because that level of reliability can't realistically be delivered. Ability to achieve even seven "9"s reliability consistently from any human system is doubtful, except on paper.
- Few if any customer processes are anywhere near that reliable anyway, for rea-

sons having nothing to do with power supply, and there's no point providing a power supply that's far more reliable than the system that uses it. The global challenge of sustaining six-sigma manufacturing quality standards, which seek error rates below two parts per million (*i.e.*, twofold short of six "9"s)—using sophisticated statistical process controls, highly trained specialists, and a complete change in manufacturing culture— illustrates how difficult this is.

Any user of the most common personal-computer operating system, for example, knows that its pre-Unix-kernel versions crash so often (at least when using complex multiple applications) that many users are lucky to get two, let alone three, "9"s of operating availability. This is due to bad code, not bad power quality; most PC users would be overjoyed if their computer software were anywhere near as reliable as today's commodity grid power. Even the improved version, Windows 2000 for servers, was advertised by Microsoft as "The mythical five nines. 99.999%. As close to perfect as you can get without breaking some law of nature."

In Internet-based digital businesses, most of which use even more robust operating systems, a 2001 survey by The Standish Group found that 37% of information systems' downtime was due to application software bugs, 12% to main-system hardware failure, 10% to database error, 8% to main-server system bugs, 8% each to network failure and operator error, 4% each to offsite servers' hardware failures or system bugs, 3% to environmental conditions, 3% to planned outage, and 3% to "other," presumably including power failures (544). Even if some of the hardware failures were actually induced by undetected power glitches, the dominant causes of failures clearly have nothing to do with the electricity system, and would persist even if power quality were perfect. That helps to explain why Dr. Romm remarks: "We've actually been talking with one of the world's largest telecommunications companies that ultimately decided that six '9's was more than it needed, and decided to go with something closer to between four and five '9's." (570)

For at least some customers who depend on computers or other loads vulnerable to even brief interruptions, the lower bound of the *perceived* value of reliability can be inferred—and only the lower bound, since it's also tempered by ability to pay—from what people actually spend on uninterruptible power supplies (UPSs)[183] and similar protective devices. Those market-driven expenditures are impressive (see box).

Grid power is cheap, but turning it into

Examples: Illustrations of the value of premium power quality and reliability

- In 1996, the U.S. market for power-quality protective equipment (500) was $3.5 billion uninstalled, of which about $2.75 billion was for UPSs and the rest for active filters and small generators.[184] In 2001, the global UPS market was estimated at $5.3 billion (148).

- The 1996 U.S. UPS market implies an installed cost plus present-valued maintenance cost (both together add typically around 25% to uninstalled cost) on the order of $4.4 billion in 1996 alone (500), and a power-protection growth rate since 1987 on the order of 12%/y (~9%/y since 1990) (670), marketed like insurance policies (500). Anecdotal reports indicate these purchases accelerated further after the California power crisis of 2000–01, and may also have been stimulated by the 9/11 attacks in 2001.

- The UPS expenditure is roughly 2% of national expenditures for electricity—an impressive figure, since actual UPS usage is spread over a modest fraction of customers.

- Those customers insist on paying an estimated ~15–20% premium for four "9"s protection from interruptions of grid power (500)—chiefly voltage sags, which account for over half of retail power disturbances and have been reported by one source to occur at a rate averaging ~0.7 per day (670).

- Although most users suffer relatively minor losses from brief (say, 5-second) outages, several of 29 high-tech manufacturing, software, and research customers surveyed by PG&E in 1994 reported losses in the vicinity of $30,000, and one as high as $1 million, per incident (563). Another firm reported that its interruption costs can exceed $8 million per event (564). Recovering from even a brief power glitch can take many orders of magnitude longer than the glitch itself (567).

- One estimate (564) pegs the aggregate cost of retail power disturbances at "more than $12 billion annually in the United States"; another, for the U.S. commercial and industrial sectors in 1989, at roughly $13–26 billion (129, 386). If so, customers are still underbuying (or cannot afford) protection, implying a market several times the level already discovered for UPSs.

- Soon the UPS market will expand massively as telecommunications firms switch from copper wires, which also carry the 48 VDC needed to run the telephones, to fiber optics, which can't. Since the phones must work even during a power outage in order to provide the emergency capabilities that customers expect as a fundamental telephonic function (and value even more highly in emergencies), telecom companies will ultimately need to invest in small UPSs or their equivalent (such as cellphone batteries) for each individual phone instrument or customer system (129, 386). Larger onsite backup power systems for cellphone switches and relays are of course already a major market.

- Being realists, nearly 40% of the able corporate energy managers responding to E SOURCE's 1996 survey considered power reliability and quality "the most important non-price features of energy that need improvement"; over 80% reported investing in protective equipment for critical loads; half said they would pay more for premium power services; and only 10% "expect high-quality power as part of the price they already pay." (399)

high-quality and -reliability grid power is not cheap. In the right circumstances, it may even be cheaper to start from scratch with clean-waveform, transient- and noise-free, highly reliable power from a PV array or wind turbine and a battery bank (and inverter in the unlikely event that an AC load must be served) than to use similar equipment (except the PVs) to back up and clean up the grid power. In such cases, the PV system's array and controller could cost less in present value than the sum of an equivalently reliable and high-power-quality UPS arrangement with the same battery bank but also

- a rectifier/battery-charger,

- control gear to switch between line and battery supply,

- possibly an inverter for AC output and a power supply for AC-to-DC reconversion for the circuitry, plus the losses incurred by both,

- heavy filters for line transients (because long supply lines are prone to lightning, switching, and other sources of spikes and sags),

- line-purchased kWh and their metering and meter-reading,

- line connection and perhaps extension costs, and

- probably either additional battery capacity or a standby generator plus its fuel and maintenance—necessities for critical missions, where user equipment must not fail even if the grid power supply failed and repairs were delayed beyond a normal UPS's battery life.

Compared with this array of avoidable costs, the ultrareliable PV array starts to look like an excellent buy. In this case, the avoided cost is an engineering-economic calculation. Increasingly, however, it will be inferred from market transactions for ultra-reliable electricity or its ultimate services.

This approach is being further explored by a Gainesville (Florida) Regional Utility 10-kWAC photovoltaic project—part of the Utility Photovoltaic Group's 1995 TEAM-UP project slate—specifically aimed at uninterruptible power supply. Indeed, Cypress Semiconductor announced in November 2001 that it would be partly powering its San Jose headquarters with a 335-kW photovoltaic system, the largest in Silicon Valley. CEO T.J. Rodgers noted that the "project was entirely justifiable on economic merits alone"—thanks to net metering, California's higher post-crisis electricity tariffs, but reportedly not crediting state subsidies or improved power quality and security of supply (579). That is, in California's current circumstances with PG&E's net-meterable tailblock tariff at a robust $0.24/kWh, one can often justify onsite power production, even from PVs, without counting power-quality and reliability benefits—but counting them makes the case even stronger.

A concept starting to enter residential and commercial building design is that providing built-in ultrareliable power circuits for critical loads can offer distinctive marketing benefits, as mentioned in Section 2.3.2.12 for merchant homebuilders' new PV-roof projects. A homebuilder, for example, could provide one islandable PV-powered circuit with specially colored outlets conveniently located for such loads as refrigerator, radio, clock, a few basic lights, and telephone. Homebuyers would then gain the peace of mind of knowing that even in a prolonged power failure, these loads could still be served, avoiding most of the cost, inconvenience, and potential danger normally expected from the outage. In a commercial building, such a circuit could power life-safety and security systems, emergency lights (perhaps avoiding local battery packs), computer servers, and telecommunications. Conventional at-the-desk plug-load wiring

could also be designed for easy conversion to PV supplies that might be installed later, so as to reserve this premium-reliability and -quality power for the light-duty office equipment that most needs it.

2.4 OTHER SOURCES OF VALUE

2.4.1 Customer value and marketing considerations

Distributed resources may be able to add significant economic value in several dozen categories not considered above. These forms of value are surveyed next in as logical a sequence as their great diversity permits. Most of them can be expected to become more important in a competitive environment as customers get to express a wider range of purchasing choices.

2.4.1.1 Green sourcing

Benefit
153 *Many distributed resources are renewable, and many customers are willing to pay a premium for electricity produced from a non-polluting generator.*

As a result of extensive research over recent years, many utilities are finding that substantial fractions of customers—from 50% to 95%—express a willingness to pay a modest premium (typically on the order of 10%) on their normal electricity price in order to get electricity notionally or actually sourced from renewable or environmentally benign

generators. More than 90 utilities and almost every retail choice market offers it, and some kind of green power choice was available to more than one-third of U.S customers in 2001—a fraction that may soon approach half. Actual participation in these novel markets is relatively low. Most green power markets have experienced penetration rates of one to two percent in initial years (771). However, the experience in some markets has been much more positive.

For example, when the Sacramento Municipal Utility District in 1994 established the PV Pioneers program, more than 700 homeowners volunteered to "host" the first 100 available installations of a 37-m², nominal 3.45-kWAC PV array on their roofs. This meant paying a 10% premium (~$4/month) for 10 years, even though they would not own the systems and could use the power only to cut their bills to the degree it matched their loadshape, not to sell back to the grid. They do, however, receive a form of price guarantee, in that their premium is reduced *pro rata* for any increase in general residential electricity prices.[185] As of March 2002, the Los Angeles Department of Water and Power (LADWP) topped the nation in customer participation with more than 87,000 customers. The LADWP premium averages $3 per month and includes a gift of two compact fluores-

[185] SMUD's next program expansion, approved 17 July 1997, offered to pay half the cost of the householders' purchase of PV systems that can resell surplus power to the grid with net metering. This effort was expected to increase the number of PV houses in Sacramento County from fewer than 500 to more than 2,500 by 2002. But the California power crisis of 2000–01 unexpectedly boosted demand. In 2001, the customer-owned PV Pioneer II program was swamped by more than 1,500 signed letters of intent. By the end of 2001, SMUD had over 10 MW of PV online in over 1,000 systems—more than half the total U.S. grid-connected PV capacity. With a backlog of more than 2,000 new orders in 2001, ten times the previous year's demand, the program was planning to install another 2.2 MW in 2002, using a newly tripled staff and contractor force. The local CalSolar PV factory even planned to add a third shift (660), but shut down with cashflow problems in 2002. Time will tell whether the reliability and technical consistency of the thin-film PVs meet long-term goals, and hence if their apparently low costs are valid.

cent lamps at signup (392). In Austin, Texas, customers of Austin Energy's GreenChoice program number more than 6,600 residential, 125 small business, and 30 large business customers (21). Interestingly, the Austin Energy program includes a 10-year commitment to flat price premia in lieu of the traditional and highly variable fuel-cost-adjustment charges. During natural-gas price spikes, this removal of fuel-price volatility has even made the green power program *less* expensive than conventional power (§ 2.2.3.1), giving customers both monetary savings and extra bragging rights.

In total, green power markets have supported a considerable expansion in renewable energy generation in the US in recent years. By January 2002, nearly 650 MW of new renewable energy generating capacity has been built to serve green power customers, with an additional 440 MW under construction or announced (69).

In general, green pricing of renewable resources—which as a practical matter are usually distributed—financially benefits the utility and can be ascribed to the distributed resource that attracts such premium payments. For example (740), SMUD calculated that the 1996-$ present value of the green pricing benefit for PV Pioneer resources ranges from $100/kW if residential rates stay constant to $25/kW if they increase at an average rate of 5%/y. The projected rate of increase was about 2.3%/y for the next ten years, yielding a green-pricing present-valued benefit of $44/kW. In addition, the installations supported by this benefit to the District are producing a public good by steadily reducing the turnkey bid costs of PV equipment. For the whole system (array plus balance of system, but excluding

~$1–1.50 of non-hardware program costs per peak watt), these were ~$5.50/W in 1996, or ~$0.165–0.18/kWh, and were projected from recent experience and component-specific expectations to fall to only $2.98/W (~$0.08/kWh) in 2002 (661)—comparable to their value to the system (737). The actual 2002 level, $3.18/W, and continuing decreases in manufacturing cost as PV markets boom worldwide, means the original 2002 target—already entering an economically attractive range without counting any distributed benefits—may be met by 2003.

Some green power programs raise questions about whether old resources are simply being rebundled and relabeled or new resources are being acquired with the marginal revenue. Even the former result is helpful because unbundling the green attribute expresses this market preference explicitly and may therefore elicit more green supply. However, new resources directly linked to the green pricing option are even more attractive. In order to prevent concerns about "double charging" (recovering costs for renewable energy facilities through ratebasing while also selling green or other attributes at a premium to green power customers), nearly every utility green pricing program relies on newly constructed renewable energy facilities. Certification standards, such as the voluntary Green-e Certification Program administered by the Center for Resource Solutions in San Francisco, also require an increasing commitment to new renewable energy generation as a condition of certification (277).

A relatively recent market innovation, the selling of green attributes separately from the electricity via sellable and tradable certificates (sometimes called "tags" or "credits"), offers an opportunity for sellers of distributed renewable resources to seek both the best price for their electrical output *and*, in a potentially independent market, the best value for the green attributes. This increases liquidity for renewable-based energy service markets, and may in time lead to an increased volume of renewable construction. It may also give developers more-reliable revenue streams from more-diverse sources. The Center for Resource Solutions adopted a certification standard in 2002 for tradable renewable credits, and certified the first such credits soon thereafter (101). In principle, other attributes such as the constant-price attribute—a valuable but often neglected attribute of most renewable energy (§ 2.2.3.1)—could also be unbundled and traded separately from the electricity.

2.4.1.2 Community sourcing and local control

> **Benefit**
>
> **154** Distributed resources allow for local control of generation, providing both economic-development and political benefits.

For both economic and political reasons, increasing numbers of communities are following Sacramento's lead by preferring energy resources that are locally chosen, made, and controlled. The economic benefits can be real, substantial, and beneficial to electricity providers themselves (§§ 2.4.10.3–2.4.10.4). The political benefits of such responsiveness can be even greater. Many distributed resources lend themselves to local manufacturing using relatively widespread skills, and

nearly all can support local operation and maintenance activities, often integrated with traditional building trades.

2.4.1.3 Amenity, comfort, productivity, and customer value

> **Benefit**
>
> **155** Certain distributed nonelectric supply-side resources such as daylighting and passive ventilation can valuably improve non-energy attributes (such as thermal, visual, and acoustic comfort), hence human and market performance.

With careful design integration, modern end-use efficiency plus some distributed nonelectric supply-side resources can create new forms of customer value that are *an order of magnitude more important than the entire energy bill*. These nonelectric resources notably include daylighting and passive ventilation (which is usually driven by solar-induced heat differences). A prominent example is the 6–16% increase in labor productivity observed in many well-designed energy-efficient buildings (571) because of their improved thermal, visual, and acoustic comfort. That is, if officeworkers can see better what they're doing, hear themselves think, and feel more comfortable, they tend to do about 6–16% more and better work. But a typical office-based business pays about 100 times as much for people as for energy, so a 1% gain in labor productivity would have the same bottom-line benefit as eliminating the entire energy bill. Comparably valuable gains have been observed in manufacturing throughput and

Benefits

156   *Bundling distributed supply- with demand-side resources increases many of distributed generation's distributed benefits per kW, e.g., by improving match to loadshape, contribution to system reliability, or flexibility of dispatching real and reactive power.*

157   *Bundling distributed supply- with demand-side resources means less supply, improving the marketability of both by providing more benefits (such as security of supply) per unit of cost.*

158   *Bundling distributed supply- with demand-side resources increases the provider's profit or price flexibility by melding lower supply-side with higher demand-side margins.*

quality, in ~40% higher retail sales pressure in pleasantly daylit stores, and in 20–26% faster learning in well daylit schools. These very large benefits reflect comfort and amenity considerably better than Class A offices normally provide, yet come with much lower operating costs and often lower capital costs too (769).

These results are available *only* from distributed resources; they bear no relation to any centralized energy supply or delivery choices, but must be delivered by design at the customer's space. In addition, such design—typically passive-solar and daylit—often fits well with other design choices, such as underfloor displacement air distribution in large buildings, that create additional customer and marketing value by making the use of the space completely flexible. Making changes in space-use very quick and inexpensive is important in modern offices, where the average person can easily move into a new physical location or configuration more often than once a year, at a conventional "churn cost" several times the total energy bill. This cost can be reduced by many-fold through efficiency- and amenity-enhancing integrated design.

2.4.2 DSM integration

The compelling history of competitive restructuring in other industries (515) suggests, and we argue elsewhere (433), that most customers care about an intricate bundle of service attributes, not just commodity price. But in a competitive world where all vendors of electrons buy at the same competitively leveled prices, the only important way to distinguish one's offering from those of other providers is to bundle the electrons with their far more productive use, so as to give the customers better service and far lower bills. This means that skillful delivery of demand-side resources, far from being an unaffordable or irrelevant frill, is the sharpest weapon in the competitive armory.

Typically the combination of supply- with demand-side resources is far more effective than either in combination, both in meeting this requirement and in achieving overall cost-effectiveness. This results from not only better matching the loadshape to the renewable output, but also, even more importantly, reducing the required size of both. Naturally, a very efficient building lends itself to onsite power generation that will cost less but do more to meet the reduced electricity (and, if cogenerating, thermal) demand. Examples were given in Section 1.6.4. Such bundling will also increase many of distributed generation's benefits per kW by increasing, for example, flexibility of dispatch, loadshape matching, and contribution to system reliability. Bundling makes the offer more marketable than either resource alone, because it gives a wider range of benefits at lower total cost. And it is more profitable (or, if market share is the objective, more price-flexible) for the provider, because of the higher margin normally available on the demand side. This strategy is now emerging in PV markets.

2.4.3 Local fuels

Benefit
159   *Certain distributed resources can valuably burn local fuels that would otherwise be discarded, often at a financial and environmental cost.*

Many industrial and agricultural processes produce modest amounts of combustible byproducts, from coffee-grounds to peach-pits, sawdust to rice-hulls, scrap-wood to refinery offgas, that cannot be economically transported for long distances. Some of these wastes are seasonal, some are more regular; some are uniform and others variable in their composition and fuel value; but most are costly disposal problems awaiting conversion into value. Often the magnitudes are very large. For example (465),

- the U.S. pulp-and-paper industry profitably gets upwards of half its total energy from its own wastes such as black liquor and hog fuel;

- Sacramento built a gasifier to consume 20,000 t/y of tree, lawn, and garden

trimmings (one-third of the early-1980s volume going to landfill) to heat and cool the Capitol Complex;

- Diamond/Sunsweet got a three-year payback by cogenerating 4.5 MW from its 100 t/d of walnut shells; and

- Boeing's Everett complex built a cogeneration plant powered by cartons and other factory wastes.

However, many other large opportunities remain to be tapped: to illustrate their diversity,

- Los Angeles County sends to landfill each day 4,000–8,000 short tons of pure, separated tree material (on the order of 1 thermal GW), not counting mixed truckloads;

- the cotton-gin trash burned or dumped in Texas is about enough to run every vehicle in Texas (if converted to liquid biofuels) at modestly increased vehicular efficiency;

- the distressed grain in an average year in Nebraska in the early 1980s was sufficient to fuel a tenth of the state's cars at 60 mpg (3.92 L/100 km);

- at that efficiency, the straw burned in the fields of France or Denmark could fuel every car in those countries.

From pecan shells to rice straw to peach pits to apple pomace (left from squeezing cider), there are so many locally significant waste streams that they add up to quite enough to fuel an entire U.S. transportation system run at cost-effective levels of technical efficiency. Alternatively, where wastes are used to generate electricity, often this new-fuel opportunity comes together with an even more important one—thermal integration.

Benefits

160   *Distributed resources provide a useful amount and temperature of waste heat conveniently close to the end-use.*

161   *Photovoltaic (or solar-thermal) panels on a building's roof can reduce the air conditioning load by shading the roof—thus avoiding air-conditioner and air-handling capacity, electricity, and the capacity to generate and deliver it, while extending roof life.*

2.4.4 Thermal integration

Central power stations generate ~1–3 GW of waste heat in one place—far more, but often at lower temperatures, than most applications can use or pay to transport to where it could be economically used. (Notable exceptions include the electricity-and-process steam cogenerators associated with the huge refinery, petrochemical, and commodity-chemical industries around areas like the Houston Ship Channel.) Distributed resources, however, typically provide a useful amount and temperature of waste heat conveniently close to the end-use. Thus the waste heat from a fuel cell or a Stirling or internal-combustion engine can heat or cool buildings or run industrial processes. This is done commercially at system efficiencies from 80 to ~95% using engine generators and heat-recovery devices, and will almost certainly be the key to the widespread early use of proton-exchange-membrane fuel cells (758). Even on a scale of a single rooftop, heating domestic water with low-temperature waste heat from photovoltaics in a simple nontracking concentrator (Winston collector, § 1.2.2) can displace fueled water-heating while boosting solar-cell efficiency, probably achieving cost-effectiveness overall in most situations.

Thermal integration also includes the *avoidance* of unwanted heat. Mounting photovoltaics on the roof of a California house typically reduces its cooling needs by up to 16%, according to a building simulation performed by DOE (595); SMUD Solar's standard figure is 20%. This saves not only energy but also air-conditioning capacity and the onpeak capacity of generators and grid to run the air-conditioner, and can thereby add on the order of one-tenth to the direct economic value of rooftop-mounted photovoltaics. Of course, this benefit is not available if the PVs are roof-integrated. PowerLight's free-lay foam-base flat-roof PVs add about R-20 of roof insulation; this plus the shading can cool the roof by over 40 C° and extend roof life by 10–15 years (731).

Example: Thermal integration of a microturbine array
Off-the-shelf heat-recovery equipment can achieve considerable value; some systems exceed 90% efficiency from input fuel to useful work.

Harbec Plastics, a power-quality-critical rapid prototyper near Rochester, New York, is successfully using distributed trigeneration. A Capstone microturbine array is almost entirely fed to micoGen™ heat recovery boilers (each taking waste heat from up to four turbines) to make ~180°F water at a system efficiency over 70%. The hot water runs radiant slab heating in winter and absorption cooling in summer. At a contract gas prices of $6.85/10³ ft³, the generating cost of ~$0.104/kWh is credited for ~$0.03/kWh of thermal value for a net electricity cost of $0.074/kWh, undercutting the grid price of $0.10/kWh. The redundancy of the turbines also valuably reduces lost production due to power disturbances. In addition, the factory remelts scrap thermoplastics into plastic lumber, and surplus microturbine output charges a fleet of battery vehicles (270).

2.4.5 Byproduct integration

> **Benefit**
>
> **162**   *Some distributed resources like microturbines produce carbon dioxide, which can be used as an input to greenhouses or aquaculture farms.*

Combustion of any hydrocarbon fuel, renewable or nonrenewable, produces carbon dioxide that by itself, or combined with low-temperature waste heat, can be a valuable input to operate greenhouses, aquaculture, etc. This in turn can offer still further opportunities for integration: in one Cape Cod greenhouse, for example, each freestanding water tank paid for itself annually by each of its two main functions (storing heat and growing fish), with the other output being free (462).

An example from dairy-farming illustrates some of the multiple levels of integration that distributed energy resources can make possible (462). An anaerobic digester converts manure to an improved fertilizer, which saves energy to make and apply it, plus methane. This homemade natural gas then runs a diesel generator, which powers the farm and produces an exportable surplus of electricity. The generator's waste heat makes hot water to wash the milking equipment, thus saving more fuel. Waste heat recovered from the washwater then preheats the cows' drinking water, boosting milk yields. Dried residues from the digester, whose heat kills germs, is used as bedding for the cows; this cleaner bedding leads to a reduction in mastitis, which by itself saves enough money to pay for the

digester within a few years. These functions are integrated with on-farm production of fuel alcohols from crop wastes, using waste heat for successive processes and sharing other infrastructure, then selling the alcohol or using it to run farm vehicles. The "stillage" residues from alcohol production have a high yeast and hence protein content that makes them a premium livestock feed. The carbon dioxide from fermentation, as also from the engine, can boost production in a greenhouse that yields crops and crop residues, which in turn can be used for tilth improvement or fed to the digester for C/N balancing. The digester can be heated in the winter with waste heat from the bulk milk chiller—boosting methane yield so much that one 40-cow dairy farm can often meet all its own energy needs plus those of five other farms. Still further levels of valuable biological integration are also possible, such as those demonstrated at a Namibian brewery by the Zero Emissions Research Initiative (www.zeri.org).

2.4.6 Structural integration

> **Benefit**
>
> **163**   *Some types of distributed resources like photovoltaic tiles integrated into a roof can displace elements of the building's structure and hence of its construction cost.*

Ground-mounted photovoltaics typically incur an extra mounting cost on the order of 20–30% for a building permit, foundations, structure, interconnections, fence, and site maintenance. In contrast, PV arrays integrated into the surfaces of a building receive an economic *credit* on the order of 15–25% of system costs—a 45% swing, significantly enhancing PVs' competitiveness (595). This approach not only takes advantage of existing distribution lines; it also uses a better-than-free mounting surface by displacing some normal structural and weatherproofing elements of the building skin.

As noted in Section 2.2.8.2, the arrays can face in a wide range of directions without significantly harming their economics, and some oblique orientations may even increase economic viability. And the more expensive the land, the larger, in general, will be the free roof or wall area economically available for PV integration.

Such building-integrated photovoltaics (BIPVs) are rapidly expanding in scope and attractiveness and are extensively promoted and applied in Europe (347, 513, 655, 686). Major expansion is also expected from multicolor thin-film modules, such as those sputtered in a continuous process onto stainless-steel strip by Energy Conversion Devices (Troy, Michigan); from solar shingles developed by Sanyo and others in Japan; and from solar windows and spandrel glass.

Solar shingles or equivalent roof- and wall-integrated structures are weatherproof, attractive, lightweight, and easily installed by ordinary builders. Some versions also use recycled materials. The electrical connection can be made by wires, by twist-locking modules into a simple metal frame, or in some Japanese systems by nailing through special on-shingle contact patches into a conductive metal strip beneath. And integrating PVs into the roof structure avoids the potential nuisance of having to demount, unwire, remount, and rewire sep-

arate, unintegrated roof-mounted PVs in order to renew the roof shingles.

Some manufacturers offer PV coatings on commercial glazing units; light transmission is adequate and not unduly tinted. A Barcelona office tower using such windows is a net producer of electricity. Alternatively, opaque glass-mounted PVs can be used as spandrel units (spanning between the view glass on successive stories of a curtainwall building), as was done, to notable commercial advantage, on the south and west elevations of the flagship Condé Nast Building at Four Times Square in New York. Emerging techniques should even permit stick-on PVs to be applied as a retrofit to the spandrel glass on existing buildings.

2.4.7 Infrastructural displacement

Benefit
164 *Distributed resources make possible homes and other buildings with no infrastructure in the ground— no pipes or wires coming out— thus saving costs for society and possibly for the developer.*

An idea starting to emerge among some technologically adventurous real-estate developers is the potential to build tract homes, or larger buildings, with no infrastructure in the ground—a concept Buckminister Fuller devised as early as 1930 and popularized in 1952 as the "Dymaxion® Autonomous House." (50) Now the technology to do this is rapidly maturing. Electricity can come from photovoltaics or other onsite renewables; gas can be bottled, replaced by biogas or spare-PV hydrogen, or displaced by solar water heating and efficient electric cooking; water can be obtained from wells, roof-collection/cistern systems, or advanced water recycling (some completely closed-system devices have been successfully developed); wastewater can be handled onsite, preferably through urine-separating toilets or onsite biological treatment like the Living Machine™; stormwater can be handled by landscaping; and telecommunications can be coupled wirelessly, now off-the-shelf at 11–54 Mbit/s with the 802.11b and 802.11g or "WiFi" series or with analogous spread-spectrum technology. At least one no-digging project on these lines, a cash-flow-constrained eco-village, is in conceptual design.

Traditionally, developers of tract homes, for example, count the considerable cost of trenching their sites for this infrastructure as inevitable. But now this is no longer obvious: with no wires or pipes, there need be no trenches. Many of the onsite systems offer such significant economic advantages to the developer—let alone to society, which typically pays to build the facilities at the other end of the pipes and wires—that it is well worth considering onsite systems as a package that collectively displaces trenching. Especially in hilly, rocky, or fragile sites, this can both save construction cost and help protect environmental values. The cost of trenching for buried infrastructure may be better spent elsewhere in the project, or taken as lower cost or higher profit. The savings on distribution and collection systems, and on the connected remote facilities, may be very large. This is especially important for the distressingly large number of places, from Kabul to East Timor, where conventional infrastructure has been shat-

tered by war and there is an opportunity to consider replacing it with cheaper, and comparable or better, distributed systems.

2.4.8 Land-use integration, land value, and shading

Benefits
165 Because it lacks electricity, undeveloped land may be discounted in market value by more than the cost of installing distributed renewable generation—making that power source better than free.
166 Since certain distributed resources don't pollute and are often silent and inconspicuous, they usually don't reduce, and may enhance, the value of surrounding land—contrary to the effects of central power plants.
167 Some distributed resources can be installed on parcels of land that are too small, steep, rocky, odd-shaped, or constrained to be valuable for real-estate development.
168 Some distributed resources can be double-decked over other uses, reducing or eliminating net land costs. (Double-decking over utility substations, etc., can also yield valuable shading benefits that reduce losses [#121] and extend equipment life.)
169 The shading achieved by double-decking PVs above parked cars or livestock can yield numerous private and public side-benefits.

Many distributed resources can share land with other uses, such as windfarms with grazing or farming. Photovoltaics are not, in general, constrained to remote (*e.g.*, desert) siting by their land intensity (607). In ordinary rural or agricultural sites, this is hardly a consideration, because 4 ha (10 acres) will accommodate a generously spaced 1-MWe PV array even assuming only 10% conversion efficiency. At $74,000/ha ($30,000/acre) and $8 per installed whole-system peak watt, land costs would total only $300,000, or 4% of total plant. Cheaper PV systems and cost-

Figure 2-67:
Windpower can enrich farmers and ranchers

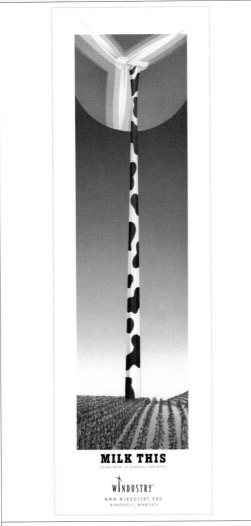

Source: www.windustry.org, reproduced by kind permission of Lisa Daniels

might be preferable" in order to achieve the distributed benefits of close support to dense and perhaps heavily loaded grids, plus other benefits mentioned next (607). It is now commonplace for farmers and ranchers in windy states to report more net income from farming wind than from growing crops or cattle—hence www.windustry.com's famous "Milk This" poster.

Moreover, "since PV systems are modular, they can be constructed on irregular land plots, such as L shaped plots," or on small, rocky, steep, roadless, or otherwise hard-to-build plots, that would be unsuited to normal development and therefore discounted by the market. Such plots are also often available around older substations or other utility facilities whose original site setbacks are loose. And PVs can be added *above* such facilities (with due allowance for any overhead conductor), simultaneously using no additional land (at some possible extra cost for supporting structures) and shading the equipment beneath (§§ 2.3.2.5, 2.3.2.7).

This concept is already widely used in sunny Western U.S. climates such as SMUD's. Installing PVs above parking lots, by shading the cars, improves their users' thermal comfort, extends interior finish materials' lifetimes, and saves fuel otherwise used for automotive air-conditioning (and the pollution caused by consuming that fuel—an important smog-former because photochemical reaction rates double with each 10 C° of increased air temperature).[186] It also makes the parking-lot paving material last longer, cause less radiant thermal discomfort to pedestrians, and contribute less to the urban "heat island," which can have a huge effect on the temperature of an entire conurbation and hence on

[186] This is especially important to air quality because three factors coincide: the biggest air-conditioning load, when the driver first climbs into a heat-soaked car, also occurs when the car's engine is least efficient and most polluting because it hasn't yet warmed up, and also when photochemical smog formation is vigorous because of the accumulated heat and pollutants of the day.

lier land will gradually increase this share, but there's a big safety margin in these figures: actual 1992 Northern California land prices, *e.g.*, for "much of rural Sacramento and the San Joaquin Valley," were typically 20 times smaller, only around $3,700/ha ($1,500/acre), if undeveloped and 3–6 times smaller, $12–22,000/ha ($5–9,000/acre) if in agriculture. Land costs "will therefore not limit the siting of a [grid-supporting PV plant]...except in expensive urban and suburban areas, where rooftop or parking lot siting

the electricity and electric capacity needed for air conditioning. A 1 C° increase in a Sunbelt city's temperature can add many hundreds of MW to the regional utility's peak load; Los Angeles, for example, incurs an extra ~1–1.5 GW of peak load and pays an extra ~$100 million a year for electricity to offset its heat-island effect. Yet considerably more cooling than that is available from making urban surfaces, such as parking lots, less solar-absorptive (393).

At least as advantageous for PVs is that under conventional real-estate appraisal practices, land without an electric utility connection is significantly discounted. Typically for a new homebuilder—especially if water, wastewater, and similar infrastructure can be locally provided without trenching for long pipes (§ 2.4.7)—PVs and efficient end-use devices can be provided at lower capital cost than the land discount. In effect, therefore, choosing the no-utilities land makes the permanent, no-operating-cost electrical supply better than free. This is because appraisers often use (explicitly or implicitly) the cost of owner-paid powerline extension as a surrogate for the value of the discount, even though PVs may cost far less (§ 2.3.2.11). This market failure may not last, but while it lasts, it's a good deal for savvy land buyers.

Other notable land-use advantages come from the potential for dual use. Many Western ranchers and farmers, for example, find that their royalties from a windfarm, which coexists nicely with their grazing livestock, roughly equal their previous net agricultural income. In Storm Lake, Iowa for instance, ranchers can receive up to $3, 000 per 750 kW turbine on their property (22). In some circumstances, renewables' shade in

otherwise sun-blasted areas may have value for livestock, improving animals' health and temperament while reducing their water consumption.

Finally, central thermal stations often have spillover effects that devalue nearby land, such as noise, air pollution, nuclear exclusion or evacuation zones, or freezing fog from cooling towers' emissions in the winter (often a contentious issue in cold climates where black ice can form on roads). In contrast, other than diverse personal reactions to the size and sometimes the noise of wind turbines (§ 2.4.10.5)—some people consider them nice kinetic sculptures, others an intrusive nuisance—renewables tend to have limited environmental impacts confined to their own sites.

2.4.9 Avoided subsidies

Technical Note 2-2 showed for broad classes of renewable generating technologies that distributed resources may receive different subsidies and tax treatment than centralized resources. If "different" means "smaller" or "less favorable," then choosing the distributed resources may reduce society's subsidy payments compared with the centralized resources if those would otherwise have been bought instead. The distributed resource may actually receive less subsidy if it is one of several kinds of renewables in some categories of ownership (22), and it will almost certainly get far less subsidy if it is a demand-side or a storage resource. Such comparisons can become quite complex, partly because most energy subsidies are still poorly documented, and most are subject to change every time the tax code of the particular jurisdiction is revised.[187]

> **Benefit**
>
> *170*
> *Distributed resources may reduce society's subsidy payments compared with centralized resources.*

[187] A current database of U.S. state, local, utility, and selected federal renewable energy subsidies is at www.dsireusa.org.

2.4.10 NEEDs

Some of the hardest benefits to quantify, yet some of the most politically potent and societally important, are those that an analyst once summarized as "Not Easily Expressed in Dollars," or "NEEDs." NEEDs are also frequently controversial because those who cause societal costs may have strong reasons to deny or minimize those costs, while those who suffer them may have equally strong reasons to demonstrate or maximize them. It is impossible to avoid some attempt to quantify NEEDs, because deciding not to do so is a decision to value them at zero (72). Quantification is always difficult, but its importance and methodological issues are clear (362).

Societal costs not internalized into prices—called by economists "external" costs and by the late Garrett Hardin "larcenous" costs—often become a political football between inflictors and inflictees. Thus when an authoritative Pace University survey of the value of air pollution by power stations (539) established that, for example, coal-fired power plants emit pollution whose societal cost is several times the initial capital cost of the power station (33), it provoked a storm of generally unconvincing attempts at rebuttal.

The Pace analysis encouraged many state regulatory bodies and utility executives to take account of the externalities that many distributed resources, especially demand-side and renewable resources, can largely avoid (729), and that all distributed resources help to avoid indirectly by reducing grid losses. While some regulators and utilities merely think about those avoided externalities, or reflect them in paper studies,[188] others embody them in actual resource acquisition decisions:

- some states, like the Wisconsin and Pacific Northwest regulators, have offered demand-side resources a 15% cost credit in comparisons with fossil-fueled resources as a proxy for avoided environmental costs;

- others provide "set-aside" quotas for fractions of new capacity that must be renewable in order to achieve the public good of their lower externalities;

- the national Clean Air Act Amendments provide explicit wind and solar credits for avoided sulfur emissions;[189] and

- numerous jurisdictions have adopted explicit "shadow prices" that did not internalize externalities into prices,[190] but at least allocated investments as if this had been done ("shadow pricing").

Representative recent values from the more detailed proceedings in eleven states[191] (333) appear in Table 2-5.

---

[188] For example, the Kerman PV study proposed 1993 emission adders extrapolated over the project's life on assumptions ranging from their falling to zero in seven years to their increasing at 5%/y for 30 years. The resulting value was a range of $22–$62/kWy or $0.008–$0.022/kWh (643). The final evaluation (735) used a narrower range of $31–$34/kWy, far avoiding 155 t/y CO_2 and 0.5 t NO_x/y.

[189] Title IV (section 404(f)(2)(F), 42 U.S.C. 7651c(f)(c)(F)).

[190] With some exceptions: Oregon, for example, recently developed externality guidelines that could actually be reflected in higher customer charges for renewable energy to offset their avoided externalities (56).

[191] By 2000–01, upwards of 30 states recognized externalities in some formal way, often as part of the integrated resource planning required by the Energy Policy Act of 1992. However, some of these recognitions were subsequently undone by utility restructuring that simply ignored all externalities, apparently in the mistaken belief that they would somehow be recognized in the market—precisely what markets do *not* do, by definition.

Table 2-5: Regulatory externality values for selected U.S. jurisdictions in the 1990s (various years' $/ton; see notes)

| | Pollutants | | | | | | Greenhouse Gases | | | Other (¢/kWh) | |
	SO$_2$	NO$_x$	VOCs	Particulates[o]	CO	Air toxics	CO$_2$	CH$_4$	N$_2$O	Water use	Land use
Bonneville Power Administration[n]											
West (WA, OR)	1,500	884		1,539							0–0.2
East (ID, MT)	1,500	69		167							0–0.2
California Energy Commission[a]											
South Coast[b]	7,425	14,488	406	47,620	3		7.64				
Bay Area[c]	3,482	7,345	90	24,398	1		7.64				
San Diego	2,676	5,559	98	14,228	1		7.64				
San Joaquin Valley	1,500	6,473	3,711	3,762	0		7.64				
Sacramento Valley	1,500	6,089	4,129	2,178	0		7.64				
North Coast	1,500	791	467	551	0		7.64				
N. Central Coast	1,500	1,959	803	2,867	0		7.64				
S. Central Coast	1,500	1,647	286	4,108	0		7.64				
Southeast Desert	1,500	439	157	715	0		7.64				
Out-of-state Northwest[d]	1,500	730	0	1,280			7.64				
Out-of-state Southwest	1,500	760	5	1,280			7.64				
California PUC[e]											
SCE and SDG&E	19,717	26,397	18,855	5,710			7.64				
PG&E	4,374	1,904	3,556	2,564			7.64				

[a] California Energy Commission *Electricity Report*, Tables 4-1 and 4-2, November 1992. 1989 dollars

[b] Includes Ventura County

[c] Values for resources located inside California

[d] Values for resources located outside California

[e] California PUC values from California Energy Commission Staff, "In-State Criteria Pollutant Emission Reduction Values" (Testimony), November 19, 1991, Table 2. 1989 dollars.

[n] Bonneville Power Administration, "Application of Environmental Cost Adjustments During Resource Cost Effectiveness Determinations," May 15, 1991. "Land and other" values vary from 0 for DSM to 0.2 ¢/kWh for coal and new hydro. SO$_2$ value is zero if offsets are purchased. 1990 dollars.

[o] Values for California and Minnesota are per ton of particulate matter smaller than 10 microns (PM10); all other values are per ton of total suspended particulates (TSP).

Table 2-5 (cont.): Regulatory externality values for selected U.S. jurisdictions in the 1990s (various years' $/ton; see notes)

	Pollutants						Greenhouse Gases			Other (¢/kWh)	
	SO_2	NO_x	VOCs	Particulates°	CO	Air toxics	CO_2	CH_4	N_2O	Water use	Land use
Massachusetts DPU[f]	1,700	7,200	5,900	4,400	960		24	240	4,400		
Minnesota PUC (interim)[g]											
Low		69		167			5.99				
High	300	1,640	1,200	2,380			13.6				
Nevada PSC[h]	1,560	6,800	1,180	4,180	920		22	220	4,140	SS[p]	SS
New Jersey BPU[i]	4,060	1,640	1,180	2,380			13.6				
New York PSC[j]	832	1,832		333			1.1			0.1	0.4
New York State Energy Office[k]	921	4,510	3,188	2,645	307	75,490	6.2				
Oregon PSC[l]											
Low		2,000		2,000			10				
High		5,000		4,000			40				
Wisconsin PSC[m]							15	150	2,700		

f Massachusetts DPU Decision in Docket 91-131, November 10, 1992. 1992 dollars.

g Minnesota PUC Decision in Docket No. E-999/CI-93-583, March 1, 1994. 1994 dollars. Values shown for NO_x and particulates have been rounded from $68.8 and $166.6 respectively.

h Nevada PSC Decision in Docket No. 89-752, January 22, 1991. NO_x, and VOC values are only for areas that comply with federal ambient ozone standards, as all areas in Nevada currently do. The Nevada PSC says its NO_x value for areas that do not comply is "equal to or greater than" those listed, and that its VOC value for such areas is $5,500/ton. 1990 dollars.

i Values adopted by the NJBPU were $0.02/kWh for electric utility DSM programs and $0.95/MMBtu for gas utility DSM (23 NJR 3383). These values are based on Ottinger, Richard, David Wooley, Nicholas Robinson, David Hodas, and Susan Babb, *Environmental Costs of Electricity*, New York: Oceana Publications, 1990; this source is the "1990 report prepared by the Pace University Center for Environmental Legal Studies" cited in 23 NJR 3383 response to Comment 134. 1991 dollars.

j NYPSC, "Consideration of Environmental Externalities in New York State Utilities Bidding Programs," 1989. Values: 0.25¢/kWh for SO_2, 0.55¢/kWh for NO_x, 0.1¢/kWh for CO_2, 0.005¢/kWh for TSP, 0.1¢/kWh for water discharge, and 0.4¢/kWh for land use impacts for a total of 1.405¢/kWh for a NSPS coal plant. Values are translated to dollars per ton by Sury Putta "Weighing Externalities in New York State," *The Electricity Journal*, July 1990. 1989 dollars.

k NYSEO, 1994 Draft New York State Energy Plan, Volume III: Supply Assessments, February 1994, p. 529. Values shown represent "mid-range" values. For utility planning, NYSEO estimated low as 50% of mid-range values and high values as 200% of mid-range values. 1992 dollars.

l Oregon PUC Order No. 93-695, May 17, 1993, p.5. 1993 dollars.

m Wisconsin PSC Order in Docket No. 05-EP-6, September 18, 1992, p. 95. 1992 dollars.

o Values for California and Minnesota are per ton of particulate matter smaller than 10 microns (PM10); all other values are per ton of total suspended particulates (TSP).

p Site-specific.

Such values cannot be exactly correct, but are obviously more sensible than the practice, found in some backward jurisdictions historically and in most "deregulated" electricity commodity markets prospectively, of ignoring externalities and thus implicitly valuing them at zero, which is definitely not the right number. A point often overlooked by those ideologically opposed to internalization is that counting fuzzy external costs at zero value presents risks not only to the public but also to the investor, because future regulations may impose stringent environmental controls that are costly to retrofit—perhaps so costly as to force abandonment of the asset—or tort law, which has not been repealed, may directly impose liability on those responsible for harming the public. This issue now looms large for many proprietors of old "grandfathered" U.S. coal-fired power stations.

The Pace study's conclusion that externalities are real and important was so unpalatable that its senior author recently had to issue a measured but devastating rebuttal (333) to three 1994–95 counterstudies commissioned (after changes of political leadership) by New York and U.S. utility representatives, the U.S. Department of Energy, and the European Commission—all apparently eager to show that the pollution wasn't important or wasn't worth much. Those counterstudies' conclusions that the total of *all* externalities of new coal-fired plants was worth only a small fraction of a cent per kWh was easily reached by changing many technical assumptions,[192] omitting most of the larger terms, and labeling the results "new and improved." Just as the original Pace study had been widely cited by those who found its conclusions plausible and congenial, the counterstudies—

typical of the products of a rather large segment of the consulting industry—were soon equally widely cited by those who preferred the altered conclusions. Unedifying public debates often turn on such my-study-*vs.*-your-study comparisons, especially when studies are being wielded as weapons rather than transparently explained in the context of their assumptions (362).

How can externalities be valued even in theory? Economists have a favorite method: Section 1.6.3 explained that undesired outcomes can be valued by asking their victims either how much compensation they require to be paid in order to accept them (the economically correct method [411]) or how much they are willing to *pay* in order not to have those outcomes imposed on them (the common but economically fraudulent method [2]). Both these approaches have their roots in economic theorists' quest for Pareto optimality, in which at least someone becomes better off while nobody becomes worse off (because any losers are compensated by winners). This approach quickly founders on even one individual's unwillingness to accept a given injury for any amount of compensation (2); on the theoretical and practical impossibility of comparing the way different people value different outcomes (367); and on the complexities—some of them not resolvable in principle—of reconciling actuarial risks or experts' opinions of risks with the risks that people perceive.

Most studies of externalities therefore value costs like air pollution by either what they cost to avoid ("control cost," like the extra cost of stack scrubbers or cleaner fuels) or what their consequences cost the victims ("damage costs," like the loss of forest value to acid rain, or the loss of human life to pol-

[192] Risk assessments, especially those meant to produce low estimates, often multiply a long chain of successive terms dealing, for example, with pollution source terms (how much of what is emitted), dispersion mechanisms, population exposures, dose/response curves, and economic cost of those responses. Modest changes in each of a large number of successively multiplied terms can of course yield almost any desired answer. This technique was long used (and still is) to show that such events as major technical disasters are extremely unlikely (411). Reality is often more convincing.

lution-caused illness and premature death). Usually these values are compared and the lesser one is taken to represent what avoiding the pollution is worth. If public policy is working properly, then abating the pollution should cost less than incurring it. But that abatement cost can still be large. The Electric Power Research Institute estimates (629) that over two-fifths of the capital cost of U.S. power stations is for environmental compliance—not surprising when we recall (§ 1.1) that the plants produce one-third of the nation's emissions of carbon and nitrogen oxides and two-thirds of the sulfur oxides.

Like cashflows, externalities are often discounted at inappropriate rates. Society is less impatient than individual investors are, has a broader and more diversified view of benefits than individuals do, and is not limited to individual lifetimes. Future generations, too, may value more highly than do present cohorts the environmental benefits from which they will largely benefit, and intergenerational equity requires that those future generations' preferences be taken into account in present decisions that affect them. For all these reasons, externalities should generally be discounted at the Social Rate of Time Preference rather than at a private-market rate appropriate to a particular level of systematic risk, as in consumer borrowing or business returns. Reasonable levels of STRP, calculated by adjusting market-based rates for personal and corporate tax effects, are on the order of 2.6 to 3.5%/y nominal or –0.4 to +0.5%/y real.[193] (30) Thus its (nominal) value "approximates the real,

long term growth rate in the economy... (about 3% [nominal]) for cost/benefit streams whose systematic risk equals that of a widely diversified financial portfolio."

2.4.10.1 Security of supply

> **Benefit**
>
> **171** *Distributed resources can significantly—and when deployed on a large scale can comprehensively and profoundly—improve the resilience of electricity supply, thus reducing many kinds of social costs, risks, and anxieties, including military costs and vulnerabilities.*

In the aftermath of the 9/11 terrorist attacks on the World Trade Center and the Pentagon, security against disruption of critical infrastructure, such as energy supplies, has become a hot topic. This attention to a serious and pervasive problem is long overdue. The definitive unclassified analysis showed in 1981–82 (442) that in the U.S. as in nearly all other countries, energy infrastructure is often fatally vulnerable to accidental or deliberate disruption—even more vulnerable in many cases than oil imported from the Persian Gulf. However, this problem is unnecessary and would be cheaper not to have. An invulnerable energy system is feasible, costs less, works better, and is favored in the market, though not by current energy policy in the U.S. or in most other countries.

[193] Contrary to an opinion widespread among those for whom a little knowledge of economics is a dangerous thing, a negative discount rate is wholly consistent with economic theory, as explained to one of us (ABL) by the late Nobel Prize-winning economist Professor Tjalling Koopmans. The choice of social and indeed private discount rates is a cultural construct, not a requirement of theory. Indeed, quite a few cultures that have taboos against lending money at interest (at least to their own tribe), or rituals governing how long-term and even intergenerational obligations are recompensed, appear to have practices operationally equivalent to implicit negative real discount rates. One could also infer that this result is implied by the Second Law of Thermodynamics. And the phenomenon is not merely a theoretical nicety but deeply rooted in the human emotions of love and hope. Any parent or grandparent who strives to pass on assets undiminished and indeed enhanced to future generations, anyone who educates the young or improves the world, and arguably anyone who chooses to have children could be said to care more about the long future than about short- or medium-term gratification.

Its foundation is efficient use of energy provided by diverse, dispersed, and often renewable sources—that is, distributed generation and its nonelectric analogues. In an increasingly dangerous world, as well as for natural-disaster insurance (549), this security imperative takes on such importance that it merits a fuller explanation here.

The architectural features mentioned in Section 1.2.9 make today's electrical (and for that matter fuel) infrastructure astonishingly prone to disruption. A few people could cut off the electrical supply to certain cities or any sizeable region—possibly for months or more if major equipment were damaged. This could be caused without physically attacking that equipment—as a result, for example, of certain kinds of interference with the grid's control and communication systems that could plausibly occur (128). Indeed, in the last half of 2001, computer attacks in the U.S. rose sharply, and "power and energy companies suffered an unusually high rate of attacks that appeared to originate in the Middle East...." (491)

This vulnerability in the existing energy systems' architecture flows from their complexity, stringent control and synchronization requirements, reliance on vulnerable telecommunications and information technologies, hazardous materials, inflexibility of fuels and equipment, interdependence, specialized equipment and labor needs, paucity of key spare parts, and difficulty of repair. These problems are not just theoretical. Seven successive accidental failures in four cables over 42 days, some during attempted repairs, caused the virtual evacuation of the downtown business core of Auckland, New Zealand, in 1998. Power could not be restored for five weeks (491).

Increasingly, electricity supply failures are linked to, and may be caused or exacerbated by, failures in other technical systems. In preparing for potential Y2k issues in San Diego, one heard many conversations in which the electricity provider said it would work fine as long as it had water, while the water provider said it would work fine as long as it had electricity. When other infrastructures, such as transport and telecommunications, are also involved, failures can quickly cascade out of control (442, 545).

After World War II, such Nazi leaders as Goering and Speer said that the war could have been shortened by two years if the Allies had bombed the highly centralized German electrical infrastructure early. In contrast, 78% of Japan's electric generating capacity in that era (like virtually all of Vietnam's later) came from dispersed small-hydroelectric plants that sustained only 0.3% of the bombing damage. In the past two decades, however, attacks on centralized electrical infrastructure came to be part of standard tactics for U.S., Soviet, and other armed forces. It would be foolish to assume that terrorists would not take a similar approach; indeed, they already do in many countries. Partly for this reason, the governments of China, Sweden, and Israel, among others, have long favored energy decentralization as an important element of their security policy.

Distributed electrical resources have a resilient architecture, especially if they are designed to work with or without the grid (that is, designed for isolation and islanding capability). The essential design elements conducive to resilience have been described (442) thus (see also § 2.2.9.8, Technical Note 2-3):

An inherently resilient system should include many relatively small, fine-grained elements, dispersed in space, each having a low cost of failure. These substitutible components should be richly interconnected by short, redundant links....Failed components or links should be promptly detected, isolated, and repaired. Components need to be so organized that each element can interconnect with the rest at will but stand alone at need, and that each successive level of function is little affected by failures or substitutions at a subordinate level. Systems should be designed so that any failures are slow and graceful. Components, finally, should be understandable, maintainable, reproducible at a variety of scales, capable of rapid evolution, and societally compatible.

These attributes, systematically designed in, can make local failures benign and widespread failures impossible. This in turn does not merely provide protection from disruption; it also removes the power system from terrorists' lists of attractive targets where a modest effort can produce a large and dramatic effect. This is true whether terrorists are of the old-fashioned variety seeking largely a theatrical effect (gaining attention to air grievances) or of the new variety seeking to do profound physical and psychological damage (as major blackouts surely would). The world, and especially the United States, seems to be entering a period of asymmetric warfare between disparate antagonists, one with elaborate means and the other with prodigious will. Making the power system no longer an attractive target is thus vital to national security.

The onsite and neighborhood-scale generation that some customers adopt for reasons of patriotism or convenience could in time come to yield incalculable security benefits on a much larger scale. And the marketing is starting to express a preference for onsite supply. For example, The Durst Organization won in the Manhattan real-estate market by equipping its flagship Four Times Square office tower (now the Condé Nast Building, § 2.4.6) with 400 kW of phosphoric-acid fuel cells and with photovoltaics integrated into the south- and west-facing spandrel (the opaque wall surfaces between floors). The extra cost of these onsite generators was offset by savings elsewhere, chiefly in the sizing of mechanical systems as better design choices elsewhere (glazings, lighting, daylighting, etc.) cut the building's energy use by 40%. The availability of the two most reliable known power sources right in the building helped the developer recruit premium tenants quickly at premium rents.

2.4.10.2 The megaproject syndrome

An important or even overriding concern to many communities and regions is the perception that a planned giant energy project imposes a long list of cascading social costs and social stresses grossly disproportionate to their local benefit. Often what is at stake is then not just the perceived impacts but also a sense of injustice because those who will get most or all of the benefits seem unwilling to bear the costs themselves, but all too willing to impose them on politically weaker groups at the other end of the transmission line (§ 2.4.10.9).

Major energy projects have classically brought, or at least were expected to bring, many social and environmental impacts that can seem, to a small host community, anywhere from major to unimaginable. Often such concerns are well justified. Many case-studies have documented serious environmental and social problems from the concentrated and fast-growing populations and activities in small communities where large

Benefits

172   *Technologies perceived as benign in their local impacts make siting approvals more likely, reducing the risk of project failure and lost investment and hence reducing the risk premium demanded by investors.*

173   *Technologies perceived as benign or de minimis in their local impacts can often also receive siting approvals faster, or can even be exempted from approvals processes, further shortening construction time and hence reducing financial cost and risk.*

174   *Technologies perceived as benign in their local impacts have wide flexibility in siting, making it possible to shop for lower-cost sites.*

175   *Technologies perceived as benign in their local impacts have wide flexibility in siting, making it easier to locate them in the positions that will maximize system benefits.*

176   *Siting flexibility is further increased where the technology, due to its small scale, cogeneration potential, and perhaps nonthermal nature, requires little or no heat sink.*

---

energy facilities are sited. Stress on land-use, labor and capital markets, infrastructure, social fabric, mental health, and quality of life can be severe (251). Distributed resources can avoid or mitigate many of these rate-of-change-driven impacts by spreading their impact in both time and space. Even comparing one 3-GW power station with six 500-MW plants—far larger than distributed resources—shows that the smaller units would have much less impact on separate, isolated towns than the giant plant has on one town, due to dynamic effects that tend to multiply and spread boomtown problems once they arise (251).

Low-impact microprojects' ability to avoid such concentrated impacts is often reflected in greater political receptivity and hence lower regulatory hurdles. This in turn creates tangible distributed benefits to developers, analogous to but different from those described in Section 2.4.10.7 for small scale. These benefits arise instead from low impacts (real or at least perceived), which are often linked to small scale but in this case come from the choice of technology, not just from small unit sizes. This is especially true where the technology permits flexible siting and needs little or no heat sink, thus reducing or avoiding land-use conflicts and allowing the choice of sites with lower cost but higher system value.

2.4.10.3 Keeping the money on Main Street

Projects that use local or renewable inputs produce greater local economic benefits than those that haul in fuel and other inputs from far away. While an economist may view employment for (and even induced by) a project as one of its costs, the host community and its political and business leaders are more likely to view the jobs at as a benefit and their salaries and wages as a source of local respending, stimulus, and prosperity. This perception can make the community not just willing but eager to accept the project and facilitate its siting and other approvals, thus reducing its costs and risks.

Benefit

177
Distributed resources' local siting and implementation tend to increase their local economic multiplier and thereby further enhance local acceptance.

<table>
<tr><td>

Benefit

178
Distributed resources can often be locally made, creating a concentration of new skills, industrial capabilities, and potential to exploit markets elsewhere.

</td></tr>
</table>

Input-output analyses that can capture the important respending, induced, and other multiplier effects are presented in studies cited by Wenger, Hoff, & Pepper (1997) (738), and performed by Laitner, Goldberg, & Sheehan (1995), Hoerner, Miller, and Muller (1995), Geller, DeCicco, and Laitner (1992), Clemmer (1994), and Roberts *et al.* (1995).

2.4.10.4 Support of local economies, employment, and trade balance

Acquiring distributed resources can support local industries that make, install, and maintain those resources, therefore adding and internalizing value and multipliers. For example, SMUD's five-year commitment to purchasing an average of 2 MW of PVs per year is sufficient to support a new PV factory, and in fact, Sacramento anticipates more than 300 new jobs from two recently announced factory setups there to supply PV equipment (661). Local manufacturing not only reduces some costs (such as shipping), shortens lines of communication, suits the product to local requirements, and spreads fixed program costs over larger production volumes, but also expands local business and job opportunities. Even with the very conservative approach of counting extra utility surplus (revenues minus costs) from the new PV factory—assumed to be a 10 MW/y, three-shift thin-film plant drawing a possibly low estimated load of 1.8 MW—the present-valued benefit to the District is estimated to range from $1.1 million (if SMUD is the only customer so the plant shuts down after five years) to $6.7 million (if there are other customers and it keeps running for 30 years). The upper value corresponds to a distributed benefit, for the PV-resource purchase commitment, of a sizeable $708/kW (741).

However, the uncounted indirect effects—net revenues from electricity bought by local input suppliers, workers, and business stimulated by respending of their respective earnings—could be comparable or larger: for example, another Fairfield PV manufacturing analysis found that indirect effects could be seven times as large as direct effects alone (151). Moreover, all such assessments omit "the possibility that the region can develop global market preeminence" in "emerging growth technologies," as Michael Porter has shown occurred with Italian tiles (33).

It is also important to avoid a common fallacy: using the myopic indicator of direct job creation rather than the broader measure of wealth creation (33), which depends also on how well the jobs are dispersed by location, income, trade or discipline, and other attributes, and on whether people's talents and efforts are being used most productively. Whether a given technology is more or less labor-intensive than another is far less important than how it creates a durable regional advantage and makes the local economy more efficient.

2.4.10.5 Noise and aesthetics

<table>
<tr><td>

Benefit

179   *Most well-designed distributed resources reduce acoustic and aesthetic impacts.*

</td></tr>
</table>

Most renewable resources, such as PV, as well as microturbines and fuel cells, do not have significant noise or aesthetic drawbacks. (A typical Capstone ~30-kW microturbine, for example, emits with its normal silencer option only 58 dBA of noise at 10 m.) Architectural photovoltaics can be artfully designed to look just like normal building elements such as walls or windows. Windfarms are among the few distributed renewable resources sometimes faulted on aesthetic and acoustic grounds. The noise of wind turbines was initially an issue with downwind designs that made noise as each blade passed through the tower wake, but modern designs largely eliminate this by placing the blades upwind of the tower. A modern turbine in the hundreds-of-kW range emits about 45 dBA at 250 m, and is typically inaudible above existing wind noise beyond 200 m. Noise does not bother 80% of the population around a wind farm in the Netherlands (although it was a major source (~30%) of accidents for golden eagles). (688) Early raptor-collision problems, chiefly in California's Altamont Pass, have been resolved by better siting and design, so the risk to birds is typically less from modern turbines than from the conventional power lines they supply (390).

Recent studies of windpower's aesthetics, acoustics, and raptor kills suggest that these effects can be very largely mitigated by modern design (543), although many small U.S. turbines have been slow to adopt proven solutions (272). In very round numbers, the number and size of wind turbines is often roughly comparable to the number and size of transmission-line towers otherwise needed to bring a similar amount of power from central stations. The difference is that the wind turbines are, in the opinion of some, a more interesting sort of kinetic sculpture. Not surprisingly, expansion of windpower is widely favored in the countries where it is most prevalent, with typically 70% or higher approval in such countries as Denmark, Germany, and Holland. Opposition in parts of Britain delayed development there for some years, chiefly in scenic and sensitive coastal or mountain areas, but now appears to be dissipating as the merits of modern turbines become more widely known. Interestingly, over 80% of Denmark's wind turbines are owned by individuals or cooperatives, with over 100,000 families owning shares in 6,000 machines (147).

Several European and U.S. developers are planning major offshore windpower installations, including an unopposed 520-MW project expected to make a tenth of Ireland's electricity, comprising 200 80-meter-tall turbines on a sandbank in the Irish Sea, as little as 7 km south of Dublin and visible on a clear day. Subject to normal issues of marine wildlife and navigation, this seems a sensible solution for crowded landscapes. If the marine engineering works as hoped, total costs might even be lower than onshore, because its cost and that of cables (or hydrogen pipelines) could be more than offset by free sites and stronger, steadier wind.

Another important dimension of aesthetics is visibility. Air pollution, such as sulfate aerosols and particulates from coal-fired plants, has seriously degraded visibility in the once-pristine American Southwest, so visitors often have trouble even seeing the Grand Canyon. A 1982 Los Alamos study cited by Ford, Roach, & Williams (258) found that just asking people what they would be willing to pay to preserve visibility in such parklands (the wrong test, as mentioned in Section 1.6.3 and Section 2.4.10) yielded a *visibility* value of abated SO_x of nearly $29,000 per metric ton.

> **Example: Macroeconomic benefits to a state economy**
>
> Influential 1994–98 analyses used a dynamic macroeconomic model to simulate renewable generation's net benefits to the Wisconsin economy. During 1995–2020, a 750-MW renewable mix would raise Gross State Product by $3.1 billion, and real disposable income by $1.6 billion or about 2¢/kWh (both in 1987 $), more than 775 MW of coal and gas plants with the same ~118 GWh/y output. That's mainly because the renewables, being more locally sourced, generated over three times more jobs and state economic benefits per GWh than the nonrenewables. Donald Aitken estimated for Union of Concerned Scientists a net benefit of over 5¢/kWh for investments in energy efficiency instead of nonrenewable generation. (3, 130–1)

> **Benefit**
>
> **180**
> Distributed resources can reduce irreversible resource commitments and their inflexibility.

> **Benefit**
>
> **181**
> Distributed resources facilitate local stakeholder engagements and increase the community's sense of accountability, reducing potential conflict.

2.4.10.6 Irretrievable commitments of resources

It is well established in public policy, and codified in the National Environmental Policy Act of 1969, that irreversibly committing public resources to a specific use forecloses other potential uses and is thus a serious decision calling for careful balancing of alternatives. In general, distributed resources involve smaller resource commitments, commit them less irreversibly, and commit them to uses that are often portable to other locations. It is difficult in the abstract to quantify any resulting benefits, but their common-sense obviousness may help to win approvals.

2.4.10.7 Conflict avoidance: stakeholders and trust

Centralized resources tend to be built by large, bureaucratic institutions that are relatively opaque, slow, and inflexible as seen by outsiders. The impression that such an organization is trying to impose its will on relatively powerless citizens can create a sense of injustice, reaction, and revolt, and this perception in turn can exacerbate resistance to local impacts perceived as relatively large. Distributed resources fit better with stakeholder engagement at a community scale, with flexible siting sensitive to local needs, and with the sense that the enterprise is of a comprehensible scale more likely to prove politically accountable. These attributes can reduce the potential for conflict, and hence can moderate cost, financial risk, and delays in approvals.

2.4.10.8 Health and safety issues: risk and perception

> **Benefits**
>
> **182** Distributed resources generally reduce and simplify public health and safety impacts, especially of the more opaque and lasting kinds.
>
> **183** Distributed resources are less liable to the regulatory "ratcheting" feedback that tends to raise unit costs as more plants are built and as they stimulate more public unease.

Any energy system has health and safety effects. There is a huge literature on them. They range from obvious to subtle, local to global, and immediate to long-delayed. In general—though no doubt exceptions can be found—electrical resources that are distributed and renewable tend to have lower, easier-to-understand, easier-to-measure, and more temporary health and safety impacts than those that are centralized and nonrenewable. This should have an economic value to the extent impacts are internalized, and a political value, which translates into reduced cost and risk, even if they are not internalized, so long as they are at least perceived.

It is also noteworthy that giant facilities tend to attract the sort of political and regulatory scrutiny and "ratcheting" feedback described in Section 1.2.3 (Figure 1-8) that can increase unit cost geometrically with the number of units built. Distributed facilities generally avoid this disadvantage.

2.4.10.9 Equity

> **Benefit**
>
> **184** *Distributed resources are fairer, and seen to be fairer, than centralized resources because their costs and benefits tend to go to the same people at the same time.*

Regardless of the type of actual or perceived impact considered, it is generally true of distributed resources that their impacts affect those who use their energy. In contrast, the impacts of centralized resources tend to affect most those nearest the facility, who are by definition remote from most of the users benefiting from the energy production. Distributed architecture and appropriate scale, in contrast, tend to deliver the costs and benefits to the same people at the same time, thus tending to reduce both actual and perceived inequity. Since perceived inequity is at the root of conflict, small can be especially profitable because it can be less contentious.

2.4.10.10 Accessibility

> **Benefit**
>
> **185** *Distributed resources have less demanding institutional requirements, and tend to offer the political transparency and attractiveness of the vernacular.*

Large, complex, arcane technologies require specialized institutions and skills. This makes them less accessible to ordinary people and less straightforward to form opinions about (§ 1.2.8). In contrast, small, simple technologies—at one extreme, an AC-out photovoltaic panel that one can buy at the lumber yard and plug into the wall socket—can engage ordinary people in both judging and applying them without technical intermediaries. This can speed implementation by involving a far wider range of actors: there is much historic evidence that in general, it is faster to do many small things than one big thing. It can also help to reduce political resistance and to avoid a feeling of alienation between citizens and the technologies proposed in their name.

2.4.10.11 Accountability and local control

> **Benefits**
>
> **186** *Distributed resources lend themselves to local decisions, enhancing public comprehension and legitimacy.*

For the same reason, technologies that are relatively easy to understand, due to their technical characteristics and their human scale, can enhance both the feeling and the reality of political choice at a sufficiently local level to provide reasonable accountability. Decisions about deploying a wind-farm or a PV array whose principles are easily grasped (even though it requires special skills to make) and whose scale is comprehensible are likely to be perceived as more legitimate, durable, and accountable decisions than those made by remote institutions on the basis of inaccessible knowledge held only by experts or elites. History suggests that this not only reduces conflict and hence cost; it also tends to lead to sounder decisions.

Benefits

187   *Distributed resources are more likely than centralized ones to respect and fit community and jurisdictional boundaries, simplifying communications and decision-making.*

188   *Distributed resources better fit the scale of communities' needs and ability to address them.*

2.4.10.12 Community and autonomy

Human affairs tend in most cultures to be organized, conducted, and conceived at the scale of the community. In general, technologies with a comparable scale are better suited to community action and acceptance than those whose scale spans diverse communities and crosses jurisdictional boundaries. Moreover, while rigid autarky is seldom a desirable goal, an appropriate degree of interdependence and independence may be better served by technologies whose scale fosters relative self-reliance than by those whose scale subsumes the needs of the community within a far larger, more fractious, and less cohesive area. "A region," said planner Paul Ylvasaker, "is an area safely larger than the one whose problems we most recently failed to solve." A community, in contrast, is one whose scale both requires and permits solutions.

2.4.10.13 Learning institutions, smaller mistakes

The modern era's rapid and accelerating change—technological, cultural, geopolitical—is among its most basic defining characteristics and greatest challenges. It requires that institutions learn at least as quickly as the world changes; otherwise they are always reacting to conditions that no longer exist, with results ranging from ineffectual to counterproductive. Learning organizations tend to be organized as networks, not hierarchies. Their technologies tend to be atomistic and pervasive, not monolithic and concentrated. They look and act less like classical organization charts and more like ecosystems. Technologies built at appropriate scale fit this model of social structure, and their ability to improve rapidly (§§ 1.6.5, 2.2.2.3) further increases their ability to suit and even lead, not retard, the pace of organizational learning and societal change.

Any technology deployment is bound to make mistakes, due to imperfect or late information if nothing else. Technologies deployed gradually in small modules are better able than big, slow, massive ones to keep up with the latest information, to send and receive information faster (being less encumbered by layers of bureaucracy), to

Benefits

189   *Distributed resources foster institutional structure that is more weblike, learns faster, and is more adaptive, making the inevitable mistakes less likely, consequential, and lasting.*

190   *Distributed resources' smaller, more agile, less bureaucratized institutional framework is more permeable and friendly to information flows inward and outward, further speeding learning.*

191   *Distributed resources' low cost and short lead time for experimental improvement encourages and rewards more of it and hence accelerates it.*

learn quickly from initial errors, and to reduce the total size and consequence of mistakes.

Moreover, the process of technical innovation is utterly different: distributed technologies can elicit and adopt innovation by individuals more broadly, deeply, and rapidly than can centralized technologies whose development and deployment are more bureaucratic. The cost of an experiment, especially a failed one, is low and its results will be known promptly, so there is more incentive and likelihood to try a variety of solutions. Distributed resources' openness to the just-do-it, keep-trying spirit further accelerates their improvement.

2.4.10.14 Public image

Benefit
192 *Distributed resources' size and technology (frequently well correlated) generally merit and enjoy a favorable public image that developers, in turn, are generally both eager and able to uphold and enhance, aligning their goals with the public's.*

An overwhelming majority—lately around 92%—of Americans, and similarly in most other countries, favor renewable and small-scale energy sources. The main reasons given typically include environmental, security, and societal benefits. While experts may quibble about details, this societal opinion seems basically sound. Most developers of distributed technologies are well aware that continued public acceptance depends on fulfilling these expecta-

tions: hence microturbine developers, for example, are eager to make their technology's noise, emissions, and other impacts more like those of fuel cells than of engine generators. While the attributes that merit and win public approbation are at least as much due to technology as to scale, and the two do not always go together (as in classical engine generators), their correlation is likely to increase as local aversion to nuisances gets expressed in regulatory and siting decisions.

2.4.10.15 Avoided air emissions

Air emissions, whether regulated or not, are a classic and principal externality of most electricity generating technologies. However, two additional scale-related remarks are needed. First, many comparisons of emissions are expressed per unit of electricity generated, not delivered. This doesn't properly credit distributed resources for avoided grid losses nor, where available, for coproduction of heating, cooling, or other emission-displacing services. Second, distributed resources' emissions, if any, are typically at or near ground level and are thus directly experienced by any surrounding population, rather than being put up a tall stack and spread over a regional or even continental area. This localization is not only more equitable, as noted above; it also tends to result, very understandably, in insistence on zero or very low emissions. This short and direct political feedback from those with siting authority or influence to the project developer is likely to prove more effective in reducing total emissions than indirect feedback through diverse, faceless, and heavily lobbied legislators or bureaucrats in a faraway national capital. This

Benefits

193   *With some notable exceptions such as dirty engine generators, distributed resources tend to reduce total air emissions per unit of energy services delivered.*

194   *Since distributed resources' air emissions are directly experienced by the neighbors with the greatest influence on local acceptance and siting, political feedback is short and quick, yielding strong pressure for clean operations and continuous improvement.*

195   *Due to scale, technology, and local accountability informed by direct perception, the rules governing distributed resources are less likely to be distorted by special-interest lobbying than those governing centralized resources.*

tends to yield better rules, more aligned with the public interest and less beholden to political and economic power.

For example, engine generators with ground-level emissions, running most or all of the time, are unlikely to be acceptable unless they use technology comparable perhaps to that of Ultra-Low Emission Vehicles.[194] Even relatively clean natural-gas microturbines could yield order-of-magnitude greater ground-level NO_x emissions per unit of delivered service (even counting waste-heat recapture and avoided grid losses) than a remote combined-cycle gas turbine. Of course, such unacceptable ground-level emissions may be imposed on people anyway, just as asthma-inducing fine-particle emissions from diesel trucks are today. Injustice always remains possible. The difference here is that those perpetrating the injustice are not vast and remote, but are local and must daily deal with their offended neighbors, so a just outcome is more likely and an unjust one more risky. This simply means that social feedback from ground-level exposures will tend to drive technology choice in the direction of clean, safe sources far more than it can for centralized units.

2.4.10.16 Land conservation

Centralized facilities typically require a large site (sometimes surrounded by a hazard exclusion zone), shipping/receiving facilities for fuel and discharges, and two large kinds of land areas omitted from most assessments: dedicated or shared transportation corridors, such as coal-hauling rail lines, and offsite areas where fuel is extracted or treated (and perhaps wastes disposed of). The total area can be very large, as was found, for example, by the massive 1981 study *Energy in a Finite World*, led by nuclear advocate Wolf Häfele at the International Institute for Applied Systems Analysis. Using renewable technologies 20 years inferior to those now on the market, IIASA confirmed that even California-style central solar-thermal power stations (about

Benefits

196   *Distributed utilities tend to require less, and often require no, land for fuel extraction, processing, and transportation.*

197   *Distributed resources' land-use tends to be temporary rather than permanent.*

---

[194] This is not a small issue: just San Jose, California, has more than 1 GW of emergency diesel generators.

the most materials-intensive renewable source) would use over their lifetimes about the same amount of land as a system producing the same electricity from strip-mined Western coal—but without the permanent land damage of coal or uranium mining. The land-intensive solar power plant would use 18 times or 1.6 times more land than a light-water reactor, depending on whether it used high- or very-low-grade uranium ore. Modern renewables such as PV and wind would typically do better (§ 2.4.8), and PVs, as mentioned in Section 2.4.6, would often be integrated into buildings rather than requiring additional land areas. The portability of most distributed resources and the nature of their technology also make their use of land usually temporary rather than permanent.

2.4.10.17 Fish and wildlife conservation

The interaction of fish and wildlife with energy facilities is a complex, site-specific subject. Nonetheless, distributed resources of the more benign kinds can be safely presumed to have advantages in this regard because their impacts can be minimal for some technologies, milder due to small unit scale for others, and easier to avoid by a small detour around them.

Fish and other aquatic and marine life could also benefit from reduced cooling requirements because distributed resources are more likely to be able to co- or trigenerate, greatly reducing the total waste heat discharged to the environment. It is possible that distributed resources may follow variable loads more than centralized resources serving more diversified loads would do; the resulting fluctuation in heat discharge

Benefits
198 Distributed resources tend to reduce harm to fish and wildlife by inherently lower impacts and more confined range of effects (so that organisms can more easily avoid or escape them).
199 Some distributed resources reduce and others altogether avoid harmful discharges of heat to the environment.
200 Some hydroelectric resources may be less harmful to fish at small than at large scale.
201 The greater operational flexibility of some distributed resources, and their ability to serve multiple roles or users, may create new opportunities for power exchange benefiting anadromous fish.

could be beneficial or harmful, depending on local ecological conditions.

Hydroelectric turbines' damage to fish is probably easier to avoid with microhydro than with giant hydro plants, because run-of-the-river (no-dam) options and fish-diversion structures are often more attractive and effective at small scale and because there may be greater opportunity for new kinds of turbines that are far less risky to fish.[195] Distributed resources' flexibility can also facilitate power swaps beneficial to anadromous fish such as salmon. For example, the Los Angeles Department of Water and Power has had the nice idea of using fuel cells to run city buses during the day, using them as a stationary generator when the buses are parked at night, and using the power to supplement the existing seasonal exchange that releases more Pacific Northwest water for salmon migration and displaces more Southern California gas-turbine generation for NO_x mitigation.

---

[195] These include the helical turbine developed by Prof. Alexander M. Gorlov (Director of the Hydro-Pneumatic Power Laboratory, MIME, Northeastern University, Boston, MA) and the experimental biomimetic centripetal turbine vortex-laminar-flow rotor invented by Australian naturalist Jayden Harman of PaxResearch (paxresearch@compuserve.com), described in (566).

Benefits

202   *Well-designed distributed resources are often less materials- and energy-intensive than their centralized counterparts, comparing whole systems for equal delivered production.*

203   *Distributed resources' often lower materials and energy intensity reduces their indirect or embodied pollution from materials production and manufacturing.*

2.4.10.18 **Less indirect pollution**

In addition to the general advantages described above under "Avoided air emissions (§ 2.4.10.15)," distributed resources often enjoy an advantage in indirect environmental emissions thanks to their reduced materials intensity. This is contrary to a sometimes cited but clearly erroneous belief that distributed resources have unusually large materials requirements. While it is probably true that, say, a 1-MW wind turbine in a good site uses less mass of total materials than ten 100-kW machines—consistent with the larger machine's economies of scale—it is also generally true that such renewables are no more, and often less, materials-intensive nowadays than equivalent central thermal plants. This is partly because the latter are dealing with high temperature, pressure, and mass requiring more robust structures; partly because of special (*e.g.,* nuclear) hazards requiring uniquely strong containment; and partly because of the greater capacity that centralized systems may require to offset grid losses and reduced cogeneration opportunities.

For example, competitive modern renewable energy equipment is so materials-frugal that the entire lifecycle's embodied energy of a wind turbine is repaid by its output

within months according to three careful European studies (269), and that of PVs within a few years (691).

Almost every materials-related industry, including PVs, poses some environmental risks, even if slight. But as a relative of the semiconductor/microelectronics industries, PV technology profits from their wide experience in minimizing environmental risks, which are becoming minute as thin-film solar cells dramatically shrink materials needs (382, 587). A kilogram of silicon in such solar cells can produce more electricity than a kilogram of uranium in a light-water reactor.

2.4.10.19 **Less depletion**

Benefits

204   *Many distributed resources' reduced materials intensity reduces their indirect consumption of depletable mineral resources.*

205   *The small scale, standardization, and simplicity of most distributed resources simplifies their repair and may improve the likelihood of their remanufacture or recycling, further conserving materials.*

Similar advantages apply when materials flows for the construction and operation of energy facilities are considered from the perspective of resource depletion rather than pollution—which is, after all, simply a resource out of place. Moreover, components of distributed resources are more likely to be small and standardized enough to be relatively easy to collect for remanufacture, for reuse, or to recycle, and more likely to be repairable during operation. There seems to be no *a priori* reason to expect a significant difference—in either direction—in lifetime or in manufacturing resource productivity between centralized and distributed resources.

2.4.10.20 Less water withdrawal and consumption

> **Benefit**
>
> **206**   Many distributed resources withdraw and consume little or no water.

In areas where water is scarce, unreliable, remote, or contaminated, it can be not just valuable but make-or-break that such distributed resources as windfarms and photovoltaics, not being heat engines, require no heat sink and need no water. At a minimum, this can save the considerable expense of dry cooling towers. But in desert areas, where competition for water is often fierce and worsening, the advantage of waterless technologies can even drive the entire business strategy of a utility.

2.4.10.21 Psychosocial benefits

People are complex bundles of needs, wishes, hopes, fears, myths, and beliefs, all conditioned by history and culture. People therefore exhibit an almost infinite range of reasons for their energy choices, just as for any other choices. As a trenchant antidote to excessively narrow economic reasoning, the Canadian engineer D. Gordon Howell, PE (344), building on a 1984 analysis (345), has described the valuation expressed by his client Hélène Narayana, for whom he is

designing a small household PV system in the face of astonishing institutional resistance. For that reason and because of immature local markets (Canada has only ~120 grid-connected home PV systems), the system is expensive. Yet the payback is under three months when Ms. Narayana's personal preferences are properly counted. She assigns subjective probability-weighted values, based on her willingness to pay for them, to 11 outcomes besides lower electricity bills: higher esteem from her daughters, fun, creating curiosity, educating people, helping save the environment, creating a personal green image, being the first on her block with green power, sprucing up the neighborhood, greening the neighborhood, leadership, and Kyoto compliance. On this basis, she is prepared to spend up to ten years' worth of benefits, or C$86,841. Some economists may scoff, but at their peril: this is precisely the point of consumer sovereignty in a free market and a free society. It is why people buy all sorts of things, such as inefficient SUVs, for which a conventional economic case is hard to discern. Such NEEDs make the world go round. They are not peripheral but central to individual and societal choice.

We have now surveyed more than 200 distributed benefits—a list that is impressively if not tediously long, and even now may not yet be complete. But how can those benefits actually sway investment decisions? What trends and driving forces are shaping the emerging energy industry that will apply distributed resources? How might markets develop in ways less or more receptive to distributed resources, and how can thoughtful public policy make distributed benefits a real source of value to market actors? We turn next to these questions in Part Three.

> **Benefit**
>
> **207**
> *Many distributed resources offer psychological or social benefits of almost infinite variety to users whose unique prerogative it is to value them however they choose.*

Part Three

A CALL TO ACTION: POLICY RECOMMENDATIONS
AND MARKET IMPLICATIONS
FOR DISTRIBUTED GENERATION

3.1 A FRAMEWORK FOR ACTION

The evolution of the energy sector is determined by the interaction of technology, policy, and markets. The regulatory and legislative policies adopted have a major influence on how the market environment for distributed generation will evolve, and therefore, on the behavior of private market players. The preceding chapters have set out the context for the emergence of distributed generation resources and the benefits of their greater use. This chapter seeks to explain the real-world policy issues and tradeoffs related to the rapid development of the distributed generation sector, and the implications for the major private-sector players.

Part 3 identifies the broader energy policy goals and discusses the key policy issues for distributed generation in light of the distributed benefits identified in Part 2. Within a U.S. context—but using an approach adaptable to other societies—it offers a portfolio of policy recommendations at the federal and state level that support the rapid development of distributed generation. Given the ongoing debate over further restructuring of the power sector, we provide separate recommendations both for states that have adopted or will adopt some degree of restructuring and for those that have decided to continue traditional utility regulation. As of 9 May 2002, seventeen of the United States had adopted or were implementing "retail choice," twenty-six had chosen not to, one (California) had abandoned it, and six had deferred action; these proportions are constantly changing, but U.S. restructuring seems at best stalled.

We have not made specific recommendations for the private sector, because each company has unique strategic objectives, market conditions, and organizational capabilities. Instead, we provide *implications* for the private sector: for investor-owned utilities, public power utilities, financial markets, commercial and industrial customers, and real estate developers. The implications provide insight into distributed generation's threats to and opportunities for current business models, and into the issues that arise as organizations attempt to respond, drawn from the practical experiences of early market adopters of distributed generation options. Finally, Part 3 addresses the question of relevance—why the outcome of the distributed generation debate matters to the customer.

Like distributed benefits themselves, market and policy issues are highly company-, geography-, and time-specific. While it would be impossible to capture a fully detailed understanding of these issues in every specific context, we have endeavored to identify the common issues facing most regulators, managers, developers, users, and supporters of distributed generation technologies. From this basis, we define exciting and rewarding opportunities to accelerate distributed generation (DG) and to capture its wider benefits to society.

3.2 POLICY GOALS AND OBJECTIVES

3.2.1 Overview

The formulation and implementation of policy is ultimately concerned with the proper degree to which collective values should be imposed on private individuals and firms. Because the operations of energy utilities are related to vital public services, they have historically been deemed "affected with the public interest"[1] and subjected to varying degrees of regulation. Even where substantial progress has been made in liberalizing or restructuring utility businesses, they remain subject to regulation; no jurisdiction has truly "deregulated" the electric or natural gas businesses in the pure laissez-faire sense of the word.

Policy is never formulated or implemented in a vacuum. More than a century of commercial, legal, and policy development has shaped the energy services industries we know today, and forms the foundation on which the future of distributed generation will be built. Decades of emphasis on the central station model of electrical supply, transmission, and distribution are reflected in laws and regulations governing construction approval, siting, cost allocation and recovery, and operations. These regulations in turn derive from a suite of policy decisions typically summarized as serving "the public interest." Laws and regulations advance policy objectives, and policy objectives are based on maximizing collective value according to current views of public demands, technological options, and economic, social, and (increasingly) environmental and security benefits.

For distributed generation, the policy questions are framed by understanding three questions: What are the relevant objectives? Which barriers must be overcome? And how can social tradeoffs be most efficiently managed?

3.2.2 U.S. energy policy goals and objectives

In theory, the regulation of electricity production, distribution, and consumption is designed to achieve overarching policy goals. Hence, we must arrive at a consensus on our energy policy goals before articulating a regulatory framework to achieve them. That consensus largely exists but has seldom been articulated. Despite decades of dispute over the goals for U.S. energy policy, the National Energy Policy Initiative in 2002 achieved a bipartisan consensus on the key goals of energy policy as seen by an impressive group of experts informed by very broadly based constituency interviews.[2] These goals that have very broad bipartisan support include, in paraphrase,

---

[1] "When private property is affected with a public interest, it ceases to be *juris privati* only." Britain's Lord Chief Justice Hale (1609–1676). In the words of Chief Justice Waite of the United States Supreme Court, in the case of Munn *v.* Illinois, 1877, "Property does come clothed with a public interest when used in a manner to make it of public consequence, and affect the community at large. When, therefore, one devotes his property to a use in which the public has an interest, he, in effect, grants to the public an interest in that use, and must submit to be controlled." Subsequent eminent jurists have specifically found that electricity is "peculiarly affected" with the public interest.

[2] The National Energy Policy Initiative was a bipartisan process to define the energy goals and policy options for the U.S. It interviewed 75 and convened a further 22 internationally recognized policymakers from the public, private, and nonprofit sectors. For a discussion of the policy goals, see National Energy Policy Initiative, Appendix B, Section II, Energy Policy Goals, pp. B4–B9, March 2002. This and all other papers are posted at www.nepinitiative.org.

1. **Improve domestic supply from diverse sources**. Reduce national dependence on foreign sources of supply and diversify national sources of supply.

2. **Increase efficiency of production and use**. Improve efficiency in energy production, transmission, distribution, and end-use applications.

3. **Promote stable, efficient markets and pricing**. Foster the development of truly competitive electricity and gas markets, with appropriate oversight to minimize the potential for abuse of market power.

4. **Enhance delivery infrastructure and systems**. Improve the physical infrastructure and systems for energy transmission and distribution to complement and enable the reform of the markets.

5. **Minimize health and environmental harm**. Apply appropriate and cost-effective regulation and innovation to reduce the health and environmental impacts of energy production and use, while maintaining affordability and reliability.

6. **Develop new technology**. Promote new technologies that enable achievement of national energy policy goals through public sector investment in energy technology research, development, and demonstration (RD&D).

Underlying these goals is the notion that America's energy policies should *simultaneously* provide energy security, economic stability, and environmental protection. Given the increasing volatility of the energy sector, particularly electricity, improving the energy system's ability to adapt and strengthening its governing regulatory institutions should be recognized as worthy goals in their own right.[3]

[3] For more on the importance of creating an energy system that is adaptable to external shocks, see *Brittle Power* (442).

From the sharp divergence between and within the Houses of Congress over 2001–02 energy legislation, there appears on the surface to be far less agreement on national energy priorities and how to achieve them. For example, the furious political debates over whether to allow oil drilling in the Arctic National Wildlife Refuge, and whether to raise light-vehicle efficiency standards, reflected divergent views of how best to advance national security, economic, and environmental goals, and about whether these goals are even consistent with each other. In contrast, the National Energy Policy Initiative was able to bridge these apparent gaps by focusing on existing areas of consensus, reframing the issues in an integrative vision-across-boundaries fashion that turned tradeoffs into synergies, and suggesting innovative win-win policy options. Its key hypothesis was that focusing on what most Americans agree about—such as efficiency, innovation, competition, and fairness—could make less necessary the things they don't agree about.

The result of testing that hypothesis was gratifying. The NEP Initiative's consensus could achieve security, prosperity, and environmental quality simultaneously and without compromise—achieving "an energy system that will not run out, cannot be cut off, supports a vibrant economy, and safeguards our health and environment." The NEP Initiative's vision, goals, and strategies have been endorsed by a politically diverse group of 33 distinguished experts—half current or recent senior executives in the energy industries, and the other half with such credentials as two Presidential Advisors, two Deputy Secretaries of Energy, five other Subcabinet members, a Director of Central Intelligence, two senior staff economists from the President's Council of Economic Advisors, chairs or members of two federal and three State energy regulatory commissions, and a House energy leader (505). Their wide political spectrum makes their message especially timely for a fractured Congress and for the electorate it serves. It is as if policy wildcatters had drilled through thick strata of partisan polarization and found beneath…an astonishing gusher of consensus.

Ideological polarization, perhaps less acute but clearly troublesome, also surrounds the narrower issues of distributed generation policy. Some utilities are concerned over revenue loss, stranded assets, and system performance. Distributed generation's advocates claim reduced environmental impact and seek increased market access. Everyone agrees that the current patchwork of regulations is undesirable. But what is the appropriate framework to resolve these disputes and to seek an effective consensus?

3.2.2.1 Policy portfolio framework

The answer may be to structure a portfolio of policies that can hedge against the risks and uncertainties that are inherent in today's energy system. A balanced portfolio of policies that hedges against risk will be diverse, robust, and adaptive (66). Diverse portfolios reduce risk and increase returns by attempting to reach the "efficient frontier" of diversification against risks. Robust portfolios tend to perform well against a variety of projected outcomes for the energy sector, and provide good hedges against downside risks. Adaptive portfolios evolve over time, operate with clear near-term goals, and have credible exit strategies.

As discussed in Part 2 of this book, the energy system faces several risks that distributed generation can protect against. The critical risks are:

- loss of system reliability in congested zones

- extreme price volatility

- utility financial distress

- environmental degradation and climate change

- unreliable customer service (relative to emerging needs)

Distributed generation can also create new risks if policies meant to promote it are developed inappropriately. The areas of greatest concern are:

- creation of market power within a congested zone

- increased environmental pollution

- instability of distribution systems

The challenge facing regulators is to craft a specific set of policies that can manage these risks, level the playing field for distributed generation, and allow society to capture the benefits fairly and expeditiously.

3.2.3 Key barriers and issues facing distributed generation

All serious observers of the electricity industry recognize that there are many barriers to rapid market capture by distributed generation. Not all are within the control of any one set of actors. For example, lack of information or understanding of distributed generation reduces expressed demand for these technologies and services. Similar lack of information on the part of regulators and utilities may preclude their considering distributed ways to meet the need for reliable, least-cost service. Business and individual customers typically apply implicit hurdle rates to investment decisions that do not necessarily reflect, and often exceed, common rate-of-return or return-on-investment indices, implying higher risk when in fact risk may be lower. And both regulators and utilities typically approach utility investment decisions from a perspective developed for evaluating central station facilities.[4] Some barriers are related to the immaturity of technologies and of supporting service and repair industries. And some of these barriers are firmly entrenched in legislative or regulatory provisions governing utility revenue collection.

3.2.3.1 Key barriers

Much has been written about the barriers facing deployment of distributed generation (5, 694). Seven major barriers stand out:

Public sector barriers

- *Interconnection standards.* Utility standards for interconnection and protective equipment to allow on-grid operation of distributed generation sources vary widely and can create potentially prohibitive costs. A utility that wants to prevent such sources from connecting can impose strict connection, protection, and insurance criteria. Because of the complexity, variation, and potential costs of interconnection requirements, uniform standards are under development that

will make interconnection requirements more predictable.

- *Siting, permitting, and environmental regulations.* Existing air quality regulations under the Clean Air Act (CAA) and its most recent amendments of 1990 are designed for large central generating stations. Conventional DG technologies installed for emergency standby power are exempt from this process. In most jurisdictions, however, existing standby generators will probably have to re-apply for permits or exemptions in order to operate in a dispatchable mode for peak shaving or grid support. While DG sources are generally too small to trigger New Source Review activity under the CAA, many potential DG applications will be in non-attainment areas for NO_x. In these areas, DG will receive increasing scrutiny with regard to air emissions. This is bad news for reciprocating engines (at least using current standard technologies) and probably for gas turbines, but it is good news for fuel cells and renewables.

- *Utility pricing practices.* Distributed generation can help distribution utilities by deferring investments in distribution capacity, providing voltage support and reactive power, and improving reliability. However, existing utility tariff structures do not generally recognize these benefits, and may not result in their proper allocation, recovery, and feedback to investment decisions.

- *Wholesale market access.* Distributed resources currently have limited access to the wholesale power and ancillary services markets due to current Independent System Operator (ISO) and Regional Transmission Organization (RTO) rules.

[4] Thus on 16 May 2002, the Tennessee Valley Authority, when voting $1.7–1.8 billion to revive a nuclear reactor mothballed for 17 years while increasing its design life 50% and its design output 30%, declared that it had considered "every option available"—all of which just happened to be nuclear or fossil-fueled central generation, as if it were still the 1960s (236). Oddly, the same board had two months earlier abandoned a $150-million investment in a $360-million gas-fired power plant on grounds of insufficient demand. That plant would have supplied half as much power but at a fivefold lower price. Two of TVA's three directors were appointed by President George W. Bush, and the TVA Board has no accountability to either markets or voters, so it is ideally suited to make investment decisions that no private-market actor could make.

- *Retail market access.* No states allow direct retail wheeling of distributed resources, which thus lack access to the retail markets as well. Most states' restructuring, even though launched in the late 1990s when distributed generation was already conquering many markets, was still designed as if the only competitors were and would remain central power stations. Moreover, distribution companies are often barred from owning distributed generators, thus splitting ownership from benefits.

Private sector barriers

- *Manufacturing scale.* Many distributed resources are currently expensive on a unit basis ($/kW of new capacity). In part, this is due to the recent emergence of the several of such new technologies as PEM fuel cells and microturbines, where manufacturing facilities are clearly subscale. Manufacturing experience in the turbine, wind, and solar industries suggest that the unit costs will drop by 30–50% or more from current prices once production attains minimum efficient manufacturing scale. In some cases, notably PEM fuel cells, long-run production costs at very large volumes could become significantly lower than for gas turbines.

- *Financing uncertainty and cost.* The costs of DG technologies are generally concentrated in relatively high capital costs that, like those of some energy efficiency measures, can be difficult to finance. Power generation projects are more complex and have significant transaction costs. Because of the relatively small scale of distributed generation projects, these costs make up a larger share of the total project cost than for larger conventional projects. These costs are fully at risk in the early stages of project development, so their contribution to financial risk is amplified.

3.2.3.2 Regulatory response

Regulatory responses to these barriers address isues that can be grouped into three major areas: technical interfaces, economic and financial, and environmental. The technical interface issues address which markets distributed generation will participate in, and at what cost. The economic and financial issues address what economic value will be realized and what costs will be borne among the stakeholders. The environmental issues address how the environmental impact of distributed generation will be managed compared with centralized generation. A 1999 Arthur D. Little, Inc. white paper asserted there are eight fundamental distributed generation issues (14). Updating this starting point to 2002, we would add three additional issues. How regulators respond to these eleven distributed generation issues, summarized in the box on p. 316, will ultimately determine whether these regulators have met the widely shared policy objective of creating a competitive environment for distributed generation.

Eleven policy issues for distributed generation (DG)

Technical interfaces

1. *System interfaces:* Should DG interface with grid operations and markets?

2. *Interconnection:* Should the interconnection's technical requirements, processes, and contracts be modified for DG?

Economic and financial

3. *Utility ratemaking (price formation):* Should utilities' primary financial incentive continue to be based on selling more kWh?

4. *Grid-side benefits:* Should grid-side benefits of customer DG be monetized and allocated among stakeholders?

5. *Energy pricing:* Should the price of energy fed into the grid reflect the incremental value, net of costs, to the system?

6. *Stranded costs:* Should utilities be compensated for stranded costs associated with DG installations?

7. *Fixed charges:* Should utilities be compensated for providing standby and reliability services?

8. *Disco participation:* Should distribution companies (Discos) participate in DG?

9. *Public support:* Should DG technologies be supported by financial incentives, subsidies, or public funding of RD&D?

Environmental

10. *Siting and permitting:* Should siting and permitting requirements be modified for DG?

11. *Technology differentiation:* Should environmentally friendly DG receive differential benefits?

These issues are interrelated—how one issue is addressed will affect the results from addressing another. Action taken to address any particular issue relating to distributed generation is informed by and influences a broad range of additional regulatory issues. As such, the preferred approach for policy makers seeking to capture any specific set of benefits from distributed generation is to undertake such action within a broader agenda of regulatory reform. Further, these issues must be resolved at either the federal or state level, or in some cases both (Table 3-1). (For sim-

plicity, this treatment omits other jurisdictions, notably Native Tribes. Yet those sovereign entities happen to hold about one-fifth of U.S. fossil fuel reserves and enormous renewable energy flows. Just Tribal land in the Dakotas, for example, has Class 4–6 windpower resources on the order of 250 GW—equivalent to one-third of total U.S. generating capacity! These lands' unique legal status may permit unusual kinds of commercial transactions.)

Table 3-1: Policy issues for distributed generation

Issue	Commercial importance to DG	Jurisdiction			
		Wholesale	Transmission	Distribution	Retail
Technical Interfaces					
1. System interface	High	FERC	FERC, RTO	State PUC	State PUC
2. Interconnection	High		FERC, RTO	State PUC	
Economic and Financial					
3. Utility ratemaking	High	—	—	State PUC	State PUC
4. Grid side benefits	Moderate	—	FERC, RTO	State PUC	—
5. Energy pricing	Moderate	FERC, ISO	—	—	State PUC
6. Stranded costs	Moderate	—	FERC, RTO	State PUC	—
7. Fixed charges	High	—	—	State PUC	—
8. Disco participation	Low	—	—	State PUC	State PUC
9. Public support	Low	DOE	DOE	State PUC	State PUC
Environmental					
10. Siting and permitting	Moderate	—	RTO, EPA	State*	—
11. Technology differentiation	Low	EPA	—	State PUC	State PUC

Multiple state agencies involved, including public utility commissions, land use councils, and environmental agencies

3.3 POLICY RECOMMENDATIONS

3.3.1 Overview

Today, there is broad recognition of the importance of distributed energy resource technologies and services, but there is only sporadic specific support. Policy makers, regulators, and industry players have some general sense that there are private and public benefits to be economically captured from increased use of distributed generation. As the benefits this book catalogs and explains are made more tangible through practical experience, pressure will increase to devise policy that accelerates the capture of these benefits. Already, forward-thinking legislators and regulators are implementing measures designed to speed the launch of distributed generation markets.

In spite of all the benefits of distributed generation, a smaller, right-sized energy infrastructure will not supplant existing systems overnight. Indeed, many of the benefits of distributed generation derive directly from their interaction with the existing system. Pre-peak photovoltaic generation, for example, is valuable in part precisely because without it, the distribution system heats and degrades under normal operation (§§ 2.2.8.4, 2.3.2.7). Similarly, the load-following benefits of microturbines and fuel cells help save the fuel and maintenance costs of large plants kept warm to provide spinning reserves (§ 2.3.3.2). Capturing system-related benefits beyond energy value can pay for, and in many cases exceed, any above-market premia inherent in the prices of technologies early in their commercialization life cycle. Price and cost reductions that come from manufacturing economies of scale for distributed generation can over time put distributed generation on a more competitive first-cost footing with thermal central stations. First cost then becomes the key discriminator for choice of technology and service to meet customer demand. But meanwhile, fair competition requires that real distributed benefits be recognized in the market or in public policy or, preferably, both.

Not surprisingly, then, a number of distributed generation advocates and industry experts have articulated a need for policy reform in order to create greater opportunities for use of these technologies and services. Regardless of the specific mechanisms chosen for implementation, however, policy makers will continue to rely upon and justify their proposals on the basis of a few basic concepts. These include the goals of economic efficiency, protecting customers from improper discriminatory treatment, preserving reasonable opportunities to earn returns on investments, preserving and enhancing safety and system reliability, and preserving such public goods as a healthy environment.

Obviously, the degree of emphasis on each of these values varies from jurisdiction to jurisdiction. A great benefit of the electric utility restructuring or liberalization debate has been a reinvigoration of the debate about the best means for accomplishing these goals. Broad underlying policy principles evolve slowly, however, and those debates are likely to continue. Advocates of distributed generation and of obtaining the benefits that distributed generation offers have, in recent years, begun to argue for adopting specific mechanisms that both promote increased opportunities for these services and technologies and serve broad underlying policy objectives. As this book has stressed, it is fair to argue that a shift to greater reliance on right-sized, smaller-scale energy resources is, in sum, better policy, according to even the most restrictive views of what public policy is for.

3.3.2 Getting there— crafting an effective policy agenda

The ultimate question, then, is which basket of policy initiatives is best suited to helping distributed resources contribute to the broader policy goals just enumerated for energy services. As with all public interest questions, there is no single answer, but rather a portfolio of policies that should be robust in achieving the enumerated energy goals. This section reviews the policy positions supporting distributed generation and recommends specific policy interventions for federal and state regulators.

3.3.2.1 Analysis of proposed policy reforms

From a historical perspective, distributed generation has been a part of the energy industry picture from the very start. But in recent years, and as a result of the forces and trends discussed in Part 1, the debate about distributed generation has grown exponentially.[5] A review of some of the leading authorities in electricity policy reveals a remarkable austerity in the converging policy debate about distributed generation.

Policy advocates and policy makers offer recommendations for advancing distributed generation in four general forms.

[5] A recent Google search of the World Wide Web immediately found some 28,200 references to "distributed generation."

- *Level the playing field.* Remove barriers to distributed generation by allowing greater access to markets on an equal basis with centralized generation.

- *Capture the benefits.* Design policy innovations to enable stake-holders to realize the benefits of distributed generation.

- *Advocate specific technologies.* Combine measures to remove barriers to greater use of the particular technology with measures to allow the benefits of the technology to be captured more effectively.

- *Advocate specific issues.* Custom-design solutions for a particular issue such as environmental protection.

The menu of measures designed to advance distributed generation utilization is already quite large, seems limited only by proponents' imagination, and will certainly grow over time. Still, most of these measures are captured within six general categories:

- *Financial assistance mechanisms*
  Buy-downs, tax credits, set-asides, portfolio standards, hook-up fees and feebates, etc.

- *Technical standards*
  Interconnection and safety standards, building codes, environmental standards, "plug and play" standards, etc.

- *Regulatory and tariff provisions*
  Net metering, tariff unbundling, avoided-cost determinations, portfolio management oversight, market structure reform, tariff structure reform, etc.

- *Market innovations*
  Tradable permits, biddable curtailment markets, tradable negawatt markets, green power markets, etc.

- *Technology development mechanisms*
  Publicly funded demonstration programs, government sponsored research and development, international aid and economic development programs, etc.

- *Public technology procurement (teknik upphandling)*
  Developed in Sweden, this innovation combines government incentives with guaranteed orders from organized and aggregated buying groups (such as apartment managers or public housing authorities) in a competitive solicitation for efficient, environmentally-friendly products that were not previously developed because of a perceived risk of an inadequate market (the "chicken-and-egg problem").

3.3.2.2 Emerging consensus on a policy agenda

A review of the rapidly growing field of distributed generation and resource policy discourse reveals the need for policy reform in two key areas.

First, the emergence of distributed energy resources and the difficulties faced in deploying them show a clear need for policy reform focused on *creating a level playing field* in which all technology and service options can compete fairly to meet the need for energy services. On reflection, it can be no surprise that an electric system built on the central station model is not scale-neutral. In order to advance broader public policy objectives of economic efficiency, environmental protection and enhancement, and competitive opportunity, it is incumbent on energy policy makers to take seriously the economic, financial, engineering, and environmental benefits from the distributed generation sector. In short, creating a level playing field is something policy makers should undertake regardless of whether they support or oppose distributed generation per se. So long as they believe that full and fair competition gives better answers than bureaucratic preferences, they need to ensure that such competition flourishes.

Second, there is established and growing evidence and that reform is needed to *enable the capture of distributed generation benefits.* While the many supporters of distributed generation have offered both specific and general recommendations in this regard, the many benefits of distributed generation described in Part 2 merits a broader policy agenda. This agenda should be both technology-neutral and scale-neutral—but based on a recognition that absent a meaningful portfolio of policy reforms, society will too long be denied access to distributed resources' benefits.

3.3.3 Recommendations to federal regulators

The federal government has a role in regulating the U.S. electricity system that derives from its powers under the interstate commerce clause of the U.S. Constitution and congressional legislation. The line between federal and state jurisdiction is not always clear; thus, both federal and state regulation will affect the market evolution of distributed generation. At the federal level, three agencies play a critical role in regulating distributed generation: the Federal Energy Regulatory Commission, the Department of Energy, and the Environmental Protection Agency.

3.3.3.1 Recommendations to the FERC

Within the electricity sector, the Federal Energy Regulatory Commission (FERC) regulates the transmission and wholesale sales of electricity and natural gas in interstate commerce.[6] The FERC has a powerful role in fostering competition by creating national standards for wholesale market access and operation. In the 1990s, it became clear that the existing patchwork of contracts and tariffs allowing third-party access to the transmission grid was hindering competition. Under Order 888, the FERC pried open the grid by requiring each jurisdictional transmission provider to file an open access tariff that met minimum national standards. Order 888 made access to decisions transparent by requiring all transmission business to be done on an Internet-based information system, OASIS. Order 888 also allowed vertically integrated utilities to continue to operate the grid, but required to them to unbundle the transmission operations functionally from the merchant kWh business. While Order 888 initiated competition in the wholesale markets, it proved insufficient due to fragmented transmission grid management across individual utilities.

The FERC therefore issued Order 2000 to establish Regional Transmission Organizations (RTOs) across the nation. RTOs have the potential to enhance competitive markets by separating operational control of the grid from private utilities' merchant operations, improving grid reliability through centralized responsibility for congestion management, system emergencies, and new transmission siting, as well as expanding the liquidity of wholesale mar-

---

[6] The FERC's legal authority comes from the Federal Power Act of 1935, the Natural Gas Act of 1938, the Natural Gas Policy Act of 1978, the Public Utility Regulatory Policies Act of 1978, and the Energy Policy Act of 1992. For a full description of the FERC's roles, see www.ferc.fed.us/about/about.htm.

kets through better pricing and broader geographic scope. By spring 2002, approximately 12 RTOs had filed for approval from the FERC.

The FERC will play a primary role in defining how distributed generation will be accommodated in the wholesale power markets. The FERC will ultimately be the arbiter for the following decisions:

- National interconnection standards for distributed generation

- Inclusion of distributed generation and demand-side bidding in wholesale power and ancillary services markets

- Transmission system planning and rate design

Our recommendations to the FERC are straightforward, and are aligned with the FERC's stated goal of increasing competition in the wholesale power and ancillary services markets.

3.3.3.1.1 Create uniform national interconnection standards for distributed generation

National standards for distributed generation interconnection are needed to enable distributed generation to enter the wholesale power markets on an equivalent basis with centralized generation. The FERC's *pro forma* tariff under Order 888 standardized transmission service *across* the transmission grid. The FERC needs to adopt a *"pro forma interconnection agreement"* to standardize access *to* the grid, and indeed, has initiated a docket to begin the rulemaking process.[7]

The creation of national interconnection standards for distributed generation enjoys broad support among state regulators, as well it should.[8] Net-metering laws in 34 states vary widely, reflecting political convenience rather than engineering necessity. For example, net metering is available up to only 10 kW in eight states, but up to 1 MW in California and without limit in Connecticut, Iowa, Ohio, and (for the moment) New Jersey (665). Some states have stringent limits on total net-metered installations, such as 0.1% of peak load, while others have no limit. Some states allow utilities to require costly and elaborate engineering studies and tests, liability insurance, and interface equipment, while others require only that basic national interface standards (UL, NEC, IEEE) be met. There is also extremely wide variation in financial terms.[9] Such a patchwork of inconsistencies cries out for a uniform federal standard based on the principles of sound engineering, simplicity, transparency, and fairness, so that mature markets can develop efficiently and rapidly.

---

[7] The FERC initiated an Advance Notice of Proposed Rulemaking (ANOPR) entitled "Standardizing Generator Interconnection Agreements and Procedures" (Docket No. RM02-1-000, issued 25 October 2001). The full NOPR was issued 24 April 2002, and differentiates large generators from "small" ones (<20 MW).

[8] NARUC supports the establishment by Congress of national interconnection and power quality standards. See "Resolution Endorsing Model Interconnection Agreement and Procedures," www.naruc.org/Resolutions/2002/winter/elec/model_interconnection.shtml.

[9] For example, only California currently allows (but may not consistently require) symmetrical bidirectional time-of-use metering. At least 11 states require monthly or annual generation in excess of customers' usage to be given to the utility without compensation—an especially utility-favoring provision for summer-peaking areas and for customers who combine photovoltaic generation with end-use efficiency or passive cooling. Most such provisions have their political roots in unsophisticated utility views that net metering is a net cost rather than a net benefit to them. Where utilities realize that net metering is actually a money-maker for them, especially for such load-correlated resources as PVs, they are more likely to support its expansion. Where the misperception is due not to an underappreciation of distributed benefits but rather to a rational concern over lost contribution to margin—because regulators reward the distribution utility for selling more kWh and penalize it for cutting customers' bills—then the appropriate remedy is to decouple profits from energy sales (§ 3.3.4.2.1; see www.rapmaine.org).

3.3.3.1.2 Integrate distributed resources into wholesale power markets

Wholesale markets for energy and capacity must be restructured to accommodate both supply- and demand-side distributed resources.[10] (749) Distributed generation and load reduction have an equivalent impact on the transmission grid and power markets, in that both reduce the demand for power at a particular node. Therefore, wholesale market reform must include both of these distributed resources in a nondiscriminatory manner compared with market provisions for supply-side resources. The Independent System Operators (ISOs) are responsible for managing the regional power markets and should take two actions.

1. Create markets for negawatts

Market rules must be developed to allow demand-side bidding. Demand-side bidding requires customers or their load-serving entities (LSEs) to place binding bids for reduced loads alongside supply-side bids. LSEs should have the ability to structure demand-side bids so that the quantity requested can vary with the price sought, thereby revealing the customer's aggregated demand elasticity.[11] (753) Demand-side resources must be dispatchable, *i.e.*, they must be able to respond to real-time signals to activate the amount of load reduction that has been bid.

Such resources include peak-load controls, compressor cycling, light dimming, and other load management measures. Their technology is well established, typically using radio or ripple control, and so are some sophisticated institutional arrangements. For example, the California Energy Coalition organized in California, New York, Massachusetts, Illinois, and Sweden a total of ~17 industrial/commercial voluntary load-management cooperatives that *collectively* dispatched pre-defined load reductions when demanded by the utility. The coops coordinated which members actually reduced their loads so as to minimize their cost and inconvenience, and shared monthly payments from the utility for this standby resource.[12]

Onsite distributed generation is one of the mechanisms that can serve to reduce loads. However, most distributed resources are too small individually to enter the power markets directly at reasonable cost. Hence, our expectation is that these resources would be aggregated by load serving entities or power marketers in order to create a larger market for "negawatts."[13] In order for negawatt markets to function, the ISO will need to create standardized metering techniques that allow for reliable *post hoc* assessments and ISO communication protocols that facilitate dispatch of aggregated distributed resources (754). All this is well within the current art—evaluation methods for demand-side resources became highly sophisticated in the late 1980s and early 1990s—but its systematic market application requires policy attention.

---

[10] This study contains an extensive description of the benefits of distributed resources in the wholesale power markets and recommendations for integrating these resources into the wholesale power markets.

[11] Also see Cowart (2001). (142) Cowart notes that distributed resource bidding would require use of interval metering to allocate peak and energy responsibilities among load-serving entities.

[12] See www.energycoalition.org/coop. The latest coops run for Southern California Edison Company provided up to 18 MW of dispatchable load management at a cost to the utility of $25/kWy, and most coops totaled around 5–10 MW (241).

[13] Roughly 20 ways to make markets in saved electricity were devised at Rocky Mountain Institute in the 1980s and 1990s. Many are summarized in *Factor Four* (473).

Demand-side dispatch opportunities will expand as other jurisdictions emulate the New Electricity Trading Arrangements, which since March 2001 in England and Wales have allowed load reductions—demand-side resources—to compete directly against generators' supply bids. So far, for reasons that are not clear, this is only permitted via the Balancing Account mechanism, starting 3.5 hours before real time, so only about 2% of all trades occur via that mechanism, and price volatility, though decreasing, was initially pronounced (476). But there is no obvious reason to limit demand-*vs.*-supply-side competition just to this spot market. Demand-side resources should be equally entitled to bid on all timescales. If traded by aggregators who use them to short the supply-side market and take money from extortionate suppliers (§ 3.4.2.2.1), demand-side resources can greatly increase the public and private benefits of other distributed resources.

2. Support development of multi-settlement power markets

In three regions of the United States, the power markets are designed as multi-settlement markets in which the markets are cleared more than once in order to accommodate adjustments to real-time conditions.[14] By contrast, in single, real-time settlement markets, settlement is determined after resources are dispatched, which prevents LSEs from planning to incorporate load reduction in order to manage real-time price spikes. In multi-settlement markets, distributed resources would be brought on line by load-serving entities in the "day-of" market, whenever price spikes occur as a result of imbalances in the "day-ahead" markets. In essence, the LSEs would use distributed resources to lower their load requirements and "sell back" excess power into the grid (750). The multi-settlement market structure greatly enhances the value of dispatch-capable distributed resources to LSEs or power marketers, since it allows them to profit from "day-of" price volatility. Allowing them to do this will of course reduce that volatility, and the resulting benefits will be shared between the distributed resource providers and other market actors, making everyone (or almost everyone) better off.

3.3.3.1.3 Integrate distributed generation into ancillary services markets

Ancillary services refer to the ability of the power system to *deliver* energy in a usable form after it is produced by power generators. Ancillary services were previously bundled in the energy and capacity prices, but are now separately purchased by the Independent System Operator in order to meet the reliability needs of the bulk energy system. As discussed in Part 2, certain distributed resources can provide particular ancillary services.[15] Ancillary service prices have tended to be extremely volatile, due to lack of enough market participants during periods of crisis. In fact, generators were able to earn greater revenues from the

---

[14] Multi-settlement markets are in operation in PJM and the New York ISO, and are under development by the New England ISO. Multi-settlement markets clear before physical generation and consumption activity, *vs.* single-settlement markets which use real-time markets to adjust for imbalances after resources have already been dispatched. The California power markets of 2000–01 were an extreme example of single-settlement markets with large system imbalances.

[15] Weston (752) notes that distributed resources are generally well suited for Network Stability and Contingency Reserves when connected to the grid, and providing they are dispatchable by the ISO. The characteristics of the distributed resources, in particular its response time, response duration, and ability to be dispatched, will determine its suitability in helping to maintain or restore the real-time balance between generators and loads (*e.g.*, Regulation, Load Following, Frequency Responsive Spinning Reserves, Supplemental Reserve, and Backup Supply.).

ancillary service markets than the power markets during the California electricity crisis of 2000–01 (150). Distributed resources can improve both the liquidity of these markets and overall system reliability. Currently, distributed resources are not included in ancillary services markets, which are limited to larger-scale generators. Ancillary service markets should be designed to allow distributed resources to participate by allowing any technology that is capable of providing the service to enter and to compete fairly.

3.3.3.1.4 Support locational marginal pricing for transmission resources

Wholesale power markets can be fully functioning and competitive only if the problem of network congestion is resolved. Transmission congestion makes it impossible to complete all the proposed transactions to move power from one location to another across the grid.[16] Transmission congestion increases overall costs to the system, since less efficient generation units are, by definition, required to meet the load. The pricing question is how to allocate these costs and send efficient price signals. The debate around whether to use locational pricing or broader measures such as zonal pricing has centered around the complexity of defining locational prices and concerns over potential market power abuses.[17] In order for locational pricing to move forward, these concerns must be resolved

within the broader context of how the RTO and ISO will manage regional power markets. Practical experience will be gained in the PJM (Pennsylvania–New Jersey–Maryland) power pool, which adopted locational pricing in 1998. In general, the more location-specific the price signal, the greater the incentives to manage loads or to site distributed generation in the constrained area. Hence locational pricing is an important enabler to allow distributed generation to capture the benefits it provides to the wholesale power system.

A further important innovation required in marginal transmission pricing, and an issue for both national and state authorities controlling that pricing, is whether pricing is symmetrical between losses incurred and losses avoided. In principle, a distributed resource (generation, grid-improvement, or demand-side) that *avoids* a transmission loss should be paid for doing so. Such a decongestant or "Dristan"[18] rent would appropriately reward the installation and dispatch of distributed resources that reduce losses and free up grid capacity for other transactions (430–1). This practice is surprisingly rare, but not unknown. For example, in 1999, the Alberta Transmission Administrator's tariffs stated that location-specific transmission loss charges would be charged *or paid* as appropriate. One wind-power operator was therefore paid an extra 12.93% onpeak and 11.93% offpeak for the transmission losses it avoided, calculated

---

[16] Transmission congestion can be caused by several physical factors, such as the thermal, voltage, or stability limits of particular transmission lines. Transmission planners avoid the actual overloading of lines by constraining generation dispatch based on contingency analysis. Transmission planners monitor the system and will perform contingency analyses to determine whether the system will fail because the line will overload *if* the contingency occurs.

[17] For in-depth discussion of this debate and potential vehicles to resolve it, see (332). In spring 2002, press reports suggested that some market actors in the California crisis had used sham transactions to create an appearance of grid congestion, which they were then paid to relieve—even though the congestion did not actually occur and was not relieved.

[18] This registered trademark of Whitehall Laboratories was used for a popular over-the-counter nasal decongestant, before it was withdrawn from the market due to safety concerns about one of its ingredients, phenylpropanolamine.

by a systemwide model and revised quarterly.[19] (159) Current FERC policy appears to be moving healthily in this direction.

3.3.3.1.5 Provide greater access to information on the transmission system and wholesale markets

Independent developers of distributed resources need access to information regarding flows of power and potential constraints across the transmission grid in order to determine which locations would have the greatest potential value. Similarly, owners of distributed generation need access to wholesale market information at reasonable cost in order to make informed decisions about whether and how to participate in these power markets.

Transmission system information was formerly provided to the public through OASIS sites. Since the 9/11 terrorist attacks, much of this information has been restricted due to fear that terrorists might use it to sabotage the power system. Restricting this information is counterproductive to improving the security of the system, since it withholds information from the very parties who could help make this system more resilient and immune from such potential assaults. Further, it is anticompetitive, since it provides incumbent utilities with an asymmetric information advantage for developing new distributed generation projects—or for preventing the development of such projects by others in order to advantage their existing generation projects. And restricting this information is unlikely to be effective, since terrorists will have little difficulty estimating the likely location of criti-

cal power flows by observing power-plant operations, weather, and obvious arrangements of major physical assets whose maps are widely available. In case of uncertainty, redundant attack is cheap, since any transmission line can be quickly knocked out by one person with a rifle or other readily available means (442).

3.3.3.2 Recommendations to DOE

The Department of Energy plays a key role in funding Research, Development and Deployment (RD&D) of energy technologies as well as administering numerous programs that subsidize centralized power generation. Indeed, given the inherently political nature of the process for allocating federal funds, eliminating subsidies to all energy technologies may be the best way to ensure a level playing field. We have published several leading studies that speak to the need to level the playing field between centralized power and alternative power sources, which we will not repeat here (291). Instead, the key question is what criteria DOE should use in determining the extent of public support for RD&D in distributed power.

3.3.3.2.1 Accelerate funding of RD&D for distributed generation

New technology development will be an important enabler for any energy strategy to achieve our national technology goals. Appropriate criteria for public RD&D support of distributed generation are whether the technology is in the early stage of commercialization (and therefore unable to gar-

---

[19] This provision remains in the Alberta tariff, though values change frequently as new generation or load come online. The 2002 method of calculating Loss Factors is at www.eal.ab.ca/ts/loss_factor_calculation_methodology_public_rev_1.pdf; the Table in that URL's section 3.1 illustrates positive loss factors (credits).

ner private-sector support), and its ability to produce significant public benefits (14). Public benefits are defined by the degree of progress against the energy policy goals discussed in Section 3.2.2. As discussed in Part 2, distributed generation has the promise to provide significant benefits to the overall energy system. Given the barriers described in Part 3, RD&D funding will also be needed to investigate and improve the communication, metering, and control technologies needed to integrate distributed resources with ISOs and RTOs. The recent establishment of a distributed-resources center at the National Renewable Energy Laboratory is an encouraging step, but in truth it only begins to rebuild capabilities destroyed in ill-conceived budget cuts years earlier. It may be hoped that the national-security imperatives revealed by the 9/11 attacks may help to strengthen distributed-resources RD&D and to organize its guidance within DOE in a more coherent fashion, much as is already occurring within many military organizations reluctant to depend on vulnerable centralized power systems.

3.3.3.3 Recommendations to EPA

The Environmental Protection Agency regulates pollution from major U.S. electricity generating facilities, primarily air pollutants under the Clean Air Act. Since EPA has focused its regulatory effort on major facilities, emissions standards do not exist for facilities smaller than 1 MW. The problem this creates is that the vast majority of the distributed generation installed in the United States is diesel generators providing backup power for critical loads. An estimated 60 GW of backup generators now exist, equivalent to over 7% of total installed U.S.

capacity. Diesel generators emit 5 to 10 times more criteria pollutants (SO_x, NO_x, PM-10) than typical coal- or gas-fired generators, which are required to have emissions controls. It is estimated that the country's annual NO_x emissions would increase almost 5% if 0.5% of U.S. demand for power were met by uncontrolled diesel engines (489). Worse, most of those emissions would be at ground level and in or near well-populated areas, rather than dispersed through tall stacks. Although diesel backup was called on extensively during the California crisis, significant issues were raised regarding air quality permits and the availability and cost of NO_x permits. Operators who tried to run their backup generators for much longer than the permitted hours were, quite properly, threatened with prosecution.

The issue of environmental standards for distributed generation points to the need for broader regulatory reform. Most jurisdictions have not adapted environmental regulations to address the overall pollution reduction benefits associated with generation cited close to the load. Indeed, few environmental regulatory schemes address small generating units at all. The lack of such regulatory structures continues to pose a barrier to rapid development of distributed generation markets because the rules are unknown and therefore a risk.

A second issue is the ownership of environmental credits or offsets associated with the use of clean generation. For criteria pollutants under the Clean Air Act, allowances are given to the polluting entity, which must demonstrate reduced emissions in order to generate credits for sale. In this case, valuable offsets are created if a polluting entity were to invest in clean distributed generation that directly offset more polluting fos-sil-fuel generators.

In the case of carbon offsets or allowances no rules have been set defining the alloca-tion of allowances or credits. The U.S. is a party to the U.N. Framework Convention on Climate Change (UNFCCC) and a signa-tory to the Kyoto Protocol. However, the Bush Administration has indicated that it will not accept the emission reduction target negotiated in the Kyoto Protocol, which would require a 7% reduction from 1990 emissions by the period 2008–12, equivalent to a reduction of almost 20% from 2000 emissions. This U-turn has not only dam-aged U.S. foreign-policy interests, but also seems certain to disadvantage U.S. firms against foreign competitors, which can both gain CO_2 credits tradable in other regions and also, through end-use or conversion efficiency, cut their fuel bills. It is also awk-ward for multinational firms that can take valuable credits for their carbon savings abroad but not at home. However, as of July 2002, the U.S. Administration seemed adamant, so the U.S. will have little or no influence in the continuing refinement and extension of the global climate-protection regime for many years to come.

Despite this policy-driven handicap, some U.S. private traders have begun making markets in carbon emissions reduced within the United States, for sale to U.S. or foreign parties. These traders, as traders do, are making up their own rules rather than wait-ing for the official rules. There are already strong arguments that renewable generation and demand-side management indirectly reduce carbon emissions by displacing

emissions somewhere other than the project site. The trouble here is that multiple par-ties, including owners of the emitting source (*e.g.*, utilities), vendors, and developers could each claim ownership of any official or unofficial carbon credits created.[20] (573) As a result of this ambiguity, developers of distributed resources have been generally unable to capture the value created. This is especially ironic at a time when European countries are enthusiastically promoting renewable generation in order to capture the resulting well-defined carbon credits as major tools for compliance with their Kyoto commitments—and are capturing the other environmental, economic, and national-competitiveness benefits as byproducts.

3.3.3.3.1 Create emission standards for distributed generation

Emission standards should be technology-neutral, scale-neutral, and fuel-neutral. Different emissions standards should be applied to all modes of distributed genera-tion (*e.g.*, emergency, peaking, and base-load), reflecting their relative potential to pollute. Exemptions for very small units clearly designed for emergency use are needed. However, the emission standards should not necessarily grandfather all exist-ing units; otherwise there will be no incen-tive for technology improvement or replace-ment. Emission standards should be designed in phases with predictably increas-ing stringency in order to encourage contin-uous improvement. Uniform emission stan-dards, if adopted nationwide by state juris-dictions, would alleviate barriers to siting and development of distributed generation. The Regulatory Assistance Project's working

---

[20] In addition, significant questions exist regarding the measurement of emission reductions, which depend on the mix of generation resources offset by the distributed resources, and how to measure, monitor, and certify the reductions.

group has developed emission standards for distributed generation that go a long way toward many of the these objectives (755).

3.3.3.3.2 Clarify ownership rights to pollution credits created by distributed resources

The fundamental governing principle is simple: the owners of distributed resources that directly or indirectly reduce pollutants from fossil fuels should be entitled to the pollution credits. Yet for pollutants such as SO_x, where the emission allowances have already been allocated to existing generation units, this principle cannot be legally applied. Where such rights have not yet been allocated, as in carbon credits, ownership rights to these credits can and must be extended to the owners of distributed generation in order to allow them to capture

the benefits that they have invested capital to create.

3.3.3.4 Summary: Actions needed to adopt the suite of federal recommendation

Overall, the suite of federal government recommendations is designed to further the goals of increased market competition in the wholesale and interstate transmission markets. The good news is that the federal government has promising initiatives underway on several of the recommendations that support distributed generation, most notably the development of national interconnection standards. However, progress on integration of distributed resources into the wholesale power and ancillary services market has generally been slow. Significant

---

Table 3-2: Federal recommendations summary

Recommendation	Action required			
	Responsible agency	Continue & accelerate	Take new action	Stop existing action
A. Create uniform interconnection standards	FERC	√		
B. Integrate DR into wholesale markets	FERC, ISO, RTO	√	√	
B1. Create negawatt markets			√	
B2. Support multi-settlement markets		√		
C. Integrate DG into ancillary services market	FERC, ISO, RTO		√	
D. Support locational marginal transmission pricing	FERC, ISO, RTO	√	√	
E. Provide greater access to information	FERC			√
F. Accelerate RD&D for DR	DOE	√		
G. Create emission standards for DG	EPA		√	
H. Clarify ownership rights to pollution credits	EPA		√	

acceleration is needed if distributed resources are to play a meaningful role in avoiding further power crises like those that occurred in 2001.

3.3.4 Recommendations to State regulators

State governments primarily permit and regulate intrastate generation, transmission, and distribution, and set the retail tariff structure and quantities, at least for privately owned utilities. (Publicly owned utilities, such as muncipal utilities and rural electric cooperatives, may experience little or no state regulation.) In general, the tariffs set by investor-owned utilities for customers, the charges to developers of distributed power, and the payments for services provided are traditionally subject to state regulation by elected or appointed agencies in nearly every state, typically known as the Public Utility Commission (PUC) or Public Service Commission (PSC). In most cases, multiple state agencies (such as environmental, planning, and land use) also regulate the construction and siting of new power plants and transmission or distribution lines.

The restructuring of the electric power industry is taking place on both the state and federal level. Federal government actions have restructured the wholesale markets and provided open access to the interstate transmission system, both of which are regulated exclusively at the federal level. Roughly one-third of the states have taken the initiative in restructuring electric service at the retail level; the rest have not yet done so, and many do not wish to. The retail restructuring experience has at best been mixed, with some notable failures, such as California, and some potential successes, such as Pennsylvania. While we tend to favor careful restructuring of the electric power industry, at least at the wholesale level,[21] under appropriate circumstances,[22] we are not looking to make those arguments here.[23] Instead, we must recognize that the decision to engage in restructuring is a political one.

As Section 3.2.3.2 discusses, many of the policy decisions for distributed generation will appropriately and necessarily be made at the state level. In several instances, the issues are in parallel with issues presented at the federal level, differing only in the geographic scope of the regulated activity. Further, in the absence of federal action, it is incumbent on the states to take action. Since the implications of distributed generation for electric utilities are distinctly different under traditional regulation than under the

---

[21] The benefit of more competitive bulk power generation is already obtained by wholesale competition, which has been federal law since 1992, and cannot be obtained twice. It is vital that this significant benefit be achieved without sacrificing the manyfold larger benefit of using electricity efficiently. Achieving that benefit depends largely on state policy—specifically, on forming electricity prices of regulated distributors in a way that rewards them for cutting customers' bills, not for selling more energy (§ 3.3.4.2.1). Under retail choice, so many complex precautionary mechanisms seem to be needed to prevent gaming and abuse of market power that the theoretical second-best solution—a well-regulated monopoly rewarded for meeting clear societal goals—may be similarly or more efficient in practice, especially for small jurisdictions. So far, many experiments with retail competition seem to have produced, as one of us (ABL) predicted in 1994, far more losers than winners.

[22] These do not obtain everywhere. For example, two of the authors (TF and KRR), long before the California fiasco, advised the states of Colorado and Alaska against retail choice because it didn't suit those states' particular conditions. The authors' cautious recommendations were adopted.

[23] For discussion of the benefits of restructuring, what mistakes to avoid, and suggested regulatory approaches, see (433, 439). Interestingly, California and many other jurisdictions continued to imitate Britain's initial auction procedure long after it had failed, and are only belatedly moving toward the harder-to-game New Electricity Trading Arrangements adopted in England and Wales on 27 March 2001 after three years' redesign. Had proper attention been paid to the British experience, the worst excesses of the California debacle could probably have been avoided (476). However, NETA complicates and discourages interconnection of and payment for distributed generation.

restructured environment, this section provides state regulators with a different set of recommendations for each environment. Our recommendations that do not depend on the status of restructuring are first presented in a separate subsection (§ 3.3.4.1) called "Universal State Recommendations." And to avoid repetition, we will not duplicate here the many recommendations made above that, while needed at the federal level, can also be usefully echoed, supported, and elaborated at the state level.

3.3.4.1 Universal state recommendations

3.3.4.1.1 Adopt "plug and play" interconnection standards for distributed generation

In the absence of national federal standards, state regulators must fill the void; indeed, only state regulators can tell the utilities in their jurisdiction how to behave in most intrastate matters, since the Federal Power Act gave the FERC the authority to regulate wholesale prices "and no others." Adopting sound technical, safety, and regulatory state-

level standards for interconnecting[24] distributed generation technologies will eliminate one of the most important barriers to their wide and competitive deployment.

Addressing each of a broad range of technical standards often adds significant costs to each installation of a distributed technology. "Plug and play" standards instead create categories of standards for installation and interconnection of distributed generation. The categories are often delineated according to output (*e.g.*, under 10 kW, between 10 and 100 kW, etc.). "Plug and play" standards provide that equipment meeting general safety standards can be installed using standardized, often expedited, and sometimes no, review and approval processes. In simple cases, inverters on an approved list, and readily available in the marketplace, can simply be plugged in without even informing the utility. Such standards reverse the traditional burden imposed on distributed technology installers to prove that their equipment is locally safe, is properly configured, and will not harm grid reliability, power quality, utility equipment, or lineworker safety. The Texas Public Utility Commission, for example, took a bold and commendable step in developing "Plug and Play" regulations described below.

Technical standards already exist or are under development for DG. For example, the Institute of Electrical and Electronic Engineers (IEEE) has adopted an interconnection standard for small-scale photovoltaic devices (352),[25] and in spring 2002 was nearing Board submission of a broader standard for interconnecting distributed generators.[26] Underwriters Laboratory (UL) develops safety standards applicable to a broad range of electric components and technologies, and

---

[24] The Interstate Renewable Energy Council (IREC) publishes "The Interconnection Newsletter" free every month: www.irecusa.org/connect/newslettersub.html. The EPRI/CEC/DDE Distributed Utility Integration Test should also yield valuable models.

[25] As described by the U.S. Department of Energy, "This recommended practice contains guidance regarding equipment and functions necessary to ensure compatible operation of photovoltaic (PV) systems that are connected in parallel with the electric utility. This includes factors relating to personnel safety, equipment protection, power quality, and utility system operation. This recommended practice also contains information regarding islanding of PV systems when the utility is not connected to control voltage and frequency, as well as techniques to avoid islanding of distributed resources." (693)

[26] As described by the U.S. Department of Energy, "In March 1999, the Institute of Electrical and Electronics Engineers (IEEE) Standards Association Board voted to undertake the development of uniform standards for interconnecting distributed resources with electric power systems. The IEEE Standards Coordinating Committee 21 (IEEE SCC21), the committee responsible for developing technical standards for distributed technologies, is now working to develop IEEE P1547, the Standard for Distributed Resources Interconnected with Electric Power Systems. The consensus standard will contain specific requirements related to performance, operation, testing, safety, and maintenance of interconnections between distributed resources and other electric power systems. The U.S. Department of Energy (DOE) is funding IEEE to develop the standard on an accelerated schedule of two to three years—about half of the time period usually required. Information on officers and members of the IEEE P1547 working group, upcoming meetings, and the status of the standards development process are available online at the P1547 web site." (351, 693)

has several current and pending standards for distributed generation systems.[27] And of course the National Electrical Code governs the safe installation of all kinds of electrical equipment. In general, revising these national codes and standards is an extremely slow and tedious consensus-based process that can be blocked by a small number of objectors, who may have a variety of motives. Those revisions and refinements that are achieved therefore merit close attention, wide adoption, and high praise for the industry leaders who patiently move them through the process.

Under the Texas Commission's "Interconnection of On-Site Distributed Generation" rule (559), one of the most comprehensive of its kind, distributed generation sellers operate in nearly a "plug and play" environment that greatly reduces barriers to distributed generation interconnection. The rule addresses, *inter alia*, disconnection and reconnection, demand charges, pre-interconnection studies, equipment pre-certification, time periods for utility response to interconnection applications, technical requirements, general interconnection and protection requirements, and other technical and economic issues. The rule was adopted under a legislative provision passed as part of the state's restructuring legislation, and provides that a customer in Texas is entitled to "to have access to providers of energy efficiency services, to onsite distributed generation, and to providers of energy generated by renewable energy resources." (695) Clearly one of the most progressive statutes and regulations enacted to date, the Texas approach to interconnection stands as a model of what states can and should do to level the playing field for distributed generation.

3.3.4.1.2 Create net-metering rules with buyback rates based on system value

Net metering is a reform that typically pays customers for every unit of energy they generate and sell back to the grid at the same price they are charged for every unit they buy. Net metering usually requires the installation or modification of meters to allow them to rotate backwards, in effect giving customers full retail value for self-generation. Some net metering is instead done by back-to-back meters, one for power flow in each direction. The size of qualifying generators should not be limited to be less than the size of the consuming load. By May 2002, net metering had been adopted in 34 states and was under consideration in most of the rest (353).

Buyback rates are typically calculated under the avoided-cost formulas originally created by the Public Utility Regulatory Policies Act of 1978 (PURPA).[28] PURPA required utilities to purchase power from non-utility generators when that power was offered at a price at or below the utility's avoided cost of generation. While in some jurisdictions the avoided cost was calculated to be basically equivalent to incremental fuel costs, some other jurisdictions, like California, created standard offers that included avoided capacity costs. Neither method is adequate for distributed generation, since these buyback rates capture only a fraction of the distributed benefits to society discussed in Part 2.

---

[27] The Underwriters Laboratory standards are available at http://ulstandardsinfonet.ul.com/. Note particularly UL 1741, which ensures (at least before installation) compliance with IEEE 929-2000. See also www.irecusa.org/pdf/guide.pdf.

[28] 6 U.S.C. 2601ff.; for a map of U.S.C. sections, see www.ferc.fed.us/informational/acts/purpa.htm.

Buyback rates for distributed generation should be calculated based on the full value they bring to the system. At a bare minimum, the value calculations should include the energy value, capacity value, and distribution system value to arrive at a loss-corrected "total facility avoided cost." These values can and should be determined by a revised least-cost planning process (see the discussion of ERIS below, Section 3.3.4.2.2). To reduce administrative complexity, this process could result in a set of "locational standard offers," which must be updated periodically for utilities that remain under regulation. For distributed generation, the timing is as important as the location, since distributed generation often defers, but may not replace, the need for distribution system upgrades (772). Obviously, many significant values described in Part 2 are not captured by this procedure.

An alternative proposed by the Regulatory Assistance Project's David Moskovitz is for two related regulatory concepts: de-averaged distribution credits and distributed resource development zones (498). In essence, these credits seek to define the distribution investment deferral value based on deferral or avoidance of distribution upgrades. Note that in order to achieve a significant deferral value in the distribution system, distributed resources—generation or demand-side measures—must offset a material amount of capacity in a given distribution area. The typical criterion, for simplicity, is that DG or demand-side management (DSM) must be able to displace its area's load growth for at least one year (679). Obviously it would be counterproductive if this quantity were too big to qualify for net-metering treatment under state law, as would be true in many states today. It may

also be necessary or desirable or both for the DG or DSM resources to be combined in various aggregations to meet this criterion.

Proposed as a pilot program, these options present a low-risk opportunity for regulators to begin to harvest the benefits of distributed resources while building the analytical foundation necessary to support broader regulatory reform. To reduce transaction costs, the locational standard offer should be in the form of a standardized contract. The more distributed benefits can be counted, consistent with transparency and simplicity, the better.

An encouraging model comes again from Alberta, whose Transmission Administrator has combined "postage-stamp" charges with location-based loss charges/credits. Moreover, the Administrator "may create incentives for new generators to set-up in areas beneficial to the transmission system [such as near the main load centers, especially Calgary]. This approach encourages suppliers to locate facilities for the maximum efficiency of the interconnected system,"[29] including avoidance of upgrades to the north-south transmission backbone (275).

3.3.4.1.3 Adopt emissions standards for DG

In the absence of federal regulations, states should adopt emissions standards for DG. Our recommendations for emissions standards are the same as proposed for federal regulation and presented in Section 3.3.3.3.1 above (755).

[29] The policy framework was established in May 2001 by the Alberta Energy Utilities Board (275).

3.3.4.1.4 Provide public support to distributed generation RD&D through wires charges

Many states already have wires charges to promote such public-goods programs as energy efficiency and low-income energy bill support. The magnitude of these charges is determined by each state according to its energy and demographic situation, as well as its political circumstances, so we can appropriately comment not on the amount of the wires charges but rather on how they are used. In our view, the wires charges should support the most effective distributed resources, whether demand-side mangement or distributed generation, that will best help to achieve the state's energy policy goals.

3.3.4.1.5 Update building codes and real estate development covenants to accommodate DG

Building codes and development covenants (such as those of homeowners' associations) are a common feature of local and community regulation. These requirements, which typically take the form of local ordinances or covenants that run with real estate titles, are intended to protect public safety and economic value. Most such codes and covenants were adopted at a time when local distributed generation was not common or even contemplated, so at best they are silent on the issue. Building permit writers and homeowners' associations often simply don't know whether, for example, rooftop solar systems or microturbines are allowed, much less what requirements to impose for safe and aesthetically pleasing installation. Some covenants and rules say explicitly, or are interpreted to mean, that distributed generation, such as roof-top PVs,

are forbidden. But reasonable requirements can generally be met at reasonable cost through modern design and construction practices. Building codes and state law need to be updated to accommodate distributed generation. Where necessary, solar access laws, which prevent the blocking of a neighbor's access to sunlight or which create tradable rights to solar access, may also be necessary, and have long been successfully used in some states, based in part on the English common-law doctrine of "ancient lights."

3.3.4.2 Recommendations for states with traditional utility regulation

3.3.4.2.1 Decouple utility revenue requirements from kWh sold, and create incentives to lower customers' bills, not price per kWh

The most profoundly important regulatory change to support distributed generation and efficient end-use is also the simplest: *decouple utility revenue requirements and profits from kWh sold.* This decoupling of revenues from sales, through revenue caps or balancing accounts, fundamentally changes the incentives and hence the culture of regulated utilities (497). Regulated utilities should be rewarded not for selling more kWh, but for helping customers get desired end-use services at least cost. Utility shareholders should share in the savings if overall revenue requirements are reduced. This can be done by a performance-based approach to providing utility incentives.[30] This regulatory approach was a precursor to industry restructuring and was successfully practiced in the early 1990s in order to foster least-cost investments and accelerate the adoption of energy efficiency measures.[31]

[30] For more on performance-based ratemaking for regulated utilities, refer to www.naruc.org. Note that what is needed is a revenue cap, not a price cap.

As we have argued in Part 2, wider adoption of distributed generation would dramatically lower the total costs of the existing energy system and improve reliability. Therefore, we believe that if this disincentive for utilities to encourage and adopt distributed generation and efficient end-use as key elements of their resource portfolios were removed, and incentives were created to increase profits by lowering customers' bills, utilities would become proponents of distributed generation. They would then have happier customers and investors, whose incentives are no longer opposite but fully aligned.

3.3.4.2.2 Require mandatory ERIS planning as the basis for prudent cost recovery

With revenues decoupled from sales, a regulated distribution utility has the incentive to identify and implement the least-cost options to serve incremental demand growth. The inclusion of distributed generation and targeted DSM, in addition to traditional distribution solutions, can help reduce system costs by significantly expanding the menu of available resources that must compete with each other, including:

- small-scale *DG facilities* located near the source of load growth,

- *differentiated tariffs* to encourage customers to limit demand during peak hours,

- targeted *energy efficiency and load management* for customers or uses that drive the peak demand, and

- central-grid power, incurring the costs of *new T&D capacity* to transport the power to customers with new and/or increasing loads.

In the 1990s, the comprehensive approach to least-cost distribution planning was often called Local Integrated Resource Planning (LIRP).[32] It designed the demand-side and distributed-generation portfolio to maximize distribution savings, reducing capital intensity by up to 90% (§ 1.4.1). In the restructured environment, its new applications are now labeled Energy Resource Investment Strategy (ERIS). One such application is being conducted for the City of San Francisco.

Mandatory ERIS for a regulated distribution utility would provide the basis for implementing the least-cost combination of distributed resources and T&D upgrades and allow the recovery of prudent costs by the utility. This measure alone could greatly expand the realization of distributed benefits and the market demand for distributed generation. It could also restore to compliance with the 1978 PURPA law the many states that now ignore its mandate for state-level least-cost integrated electric resource planning.

---

[31] Decoupling energy sales from profits, typically by using a simple balancing-account mechanism, was practiced in up to nine of the United States in the early 1990s, after the National Association of Regulatory Utility Commissioners, in November 1989, unanimously approved the principle that the least-cost investment for the customers should be the most rewarding for the utility, and vice versa. Many of these states also shared with utility investors the savings that their efforts achieved on customers' bills (472). All 8–9 states but Oregon, however, got distracted during the restructuring enthusiasm of the late 1990s, so by spring 2002, only Oregon and (subject to pending implementing regulations) California were rewarding what they wanted—lower customer bills—rather than the opposite, higher energy sales. This irrational distortion leads to immense misallocations of capital in nearly every jurisdiction worldwide.

[32] LIRP methods are presented in *Tools and Methods for Integrated Resource Planning: Improving Energy Efficiency and Protecting the Environment* (680). For LIRP case studies, see E SOURCE, *Local Integrated Resource Planning: A New Tool for a Competitive Era* (397).

3.3.4.2.3 Restructure distribution tariffs to reduce excessive fixed charges

Fundamentally, we recommend that the distribution tariff structure be progressively shifted toward a greater proportion of volumetric pricing (usage-based unit prices) rather than fixed pricing. The unit prices should aim to approach the long-run marginal costs of the system in order to send correct price signals and promote economic efficiency.[33] From a practical perspective, some degree of fixed charges will be necessary and desirable to achieve both these regulatory objectives, as well as to reflect the nature of costs imposed on the system.[34] (747) As a matter of principle, utility distribution tariffs, standby charges, and backup charges should be changed to reflect the *actual* costs imposed on the system by distributed generation and incorporate the benefits that distributed generation provides to the system (in capacity deferral, increased reliability, or other attributes). (503)

For example, standby tariffs should reflect the actual costs of performing the service, net of the savings that distributed generation provides to the distribution system. As discussed by Weston, the issue is the likelihood that the self-generator demanding intermittent service will contribute to an increase in distribution capacity requirements. Therefore, the tariffs are calculated on the *probability* that the self-generating customer will contribute to peak needs, thereby increasing total system costs, rather

than the *presumption* that each self-generating customer's peak draw on the system will be entirely peak-coincident—an extremely unlikely event.[35]

3.3.4.2.4 Adopt renewable portfolio standards (RPS) and tradable credits

States that continue with traditional regulation need some form of RPS in order to provide a systematic hedge on fossil-fuel prices and to enhance energy security. Renewable portfolio standards set minimum renewable generation requirements, expressed as a percentage of net electric generation or capacity within a particular jurisdiction. By April 2002 eleven states had adopted a renewable portfolio standard and a further three had adopted a renewable portfolio goal (155). Renewable portfolio standards should be coupled with renewable credit trading systems designed to reduce compliance costs for affected industries and customers alike. The range of qualifying technologies should include all forms of renewable energy, generally including fuel cells (perhaps subject to conditions on their hydrogen source), but should exclude such environmentally questionable facilities as large-scale hydropower. We do not advocate minimum set-asides for particular categories of renewable generation. We do propose programmatic cost ceilings or other provisions such as "just and reasonable" standards to prevent overall costs from exceeding predetermined limits. Where states wish to promote other resources, such as the recycling of previous-

---

[33] For a detailed analysis of distribution rate design and the economic fundamentals of ratemaking (the U.S. term for regulatory price formation), see "Charging for Distribution Utility Services: Issues in Rate Design." (747)

[34] Fixed charges support both regulatory goals of predictable and stable revenues for utilities and practical considerations (they're easy to administer).

[35] NEM notes that demand charges and backup standby charges should be in accordance with Section 210 of PURPA, 18 CFR Section 292.305(c)(1), which states that "The rates [*i.e.* fees charged] for sales of backup power or maintenance power: (1) Shall not be based on the assumption (unless supported by factual data) that forced outages or other reductions in electrical output by all qualifying facilities on an electric utility's distribution system will occur simultaneously, or during the system's peak, or both." This clear requirement of federal law is typically honored in the breach.

ly wasted power-plant heat, they should do so through separate standards, not by qualifying these resources under the RPS: since both goals are desirable, they should both be pursued, not traded off against each other.

3.3.4.3 Recommendations for states adopting restructuring

3.3.4.3.1 Decouple distribution companies' revenue requirements from kWh throughput

After restructuring, the retail business is supposed to be competitive and unregulated, but the distribution business, as a natural monopoly, remains under regulation. Under traditional ratemaking, distribution companies' (discos') tariffs are based on the throughput of kWh over the wires. Therefore, distributed generation would reduce revenues, and, in the worst case, could potentially strand distribution assets. It is not surprising that distribution companies would want to adopt distributed generation only in geographical locations that require new distribution assets but lack the underlying load growth to use those assets efficiently. Other distributed generation, particularly on existing distribution assets with no new requirements, would be contrary to such a firm's financial interests.

As with regulated vertically integrated utilities, the solution is to decouple distribution companies' revenues from kWh throughput and to institute incentives to have shareholders share in the savings from lowering total revenue requirements. This process should be linked to the recommendation mandating ERIS as the planning tool to determine system design and prudent addition of assets.

3.3.4.3.2 Restructure *and* unbundle distribution tariffs

In addition to the tariff reforms proposed in the prior section, state regulators should unbundle distribution tariffs to increase customer choice in a deregulated environment. Unbundling will allow customers to choose products with respect to time (time-of-use pricing), location (geographic pricing), power quality (premium service pricing based on power quality standards), and reliability (interruptible or curtailable pricing). These increased customer options will allow more informed decisions of what services to purchase from the distribution utility versus investing in onsite distributed resources. It will also bring greater clarity in setting the buyback prices for distributed resources. Although unbundling adds complexity to the tariff-setting process, several industry working groups are currently addressing the methodological issues (160).

3.3.4.3.3 Impose stranded costs only after production threshold is exceeded

Stranded costs were created by the alleged need to compensate utilities for generation plants not needed or no longer competitive due to the advent of restructuring. Stranded costs are generally imposed as some form of Competitive Transition Charge (CTC), which typically is charged on the wires and cannot be bypassed. These charges impose high exit costs on customers wishing to leave the system, and represent a barrier to market entry for distributed generation. Most states that have imposed such charges have experienced little shift in customer choice of suppliers because the charge removes most of the potential advantage to be gained. Critics of CTCs say they compensate utility investors for risks which they have already been compensated for bearing, and that they often leave the utility with fuller and faster amortization of its sunk costs than it would have achieved without restructuring. Advocates of CTCs claim a "regulatory compact" in which utilities were supposedly promised full recovery of their investments.

Our recommendation is that where CTCs exist, onsite generation should not need to pay stranded-cost "exit fees" for reduced customer purchases of power from the grid until (at least) the total kWh production for all new distributed generation exceeds a threshold equal to the expected revenue growth rate, plus asset amortization or depreciation rate, implied by the utility's current tariff-setting regulatory case. Distributed generation should not pay exit costs until the total reduction in kWh sold imposes real costs to the system that other customers must bear (*i.e.,* causes a real increase in electricity prices beyond the normally projected baseline)—if then. For example, in New Jersey, new onsite generation does not pay any exit fees until total kWh production reaches 7.5 percent of the 1999 kWh distributed by electric utilities; otherwise the utility would be compensated by its competitors for revenue losses which it is not actually suffering (490). The same logic could be extended to stranded distribution assets that might arise from new distributed power facilities. In many cases, distributed generation may cause a decrease in revenue requirements by deferring the need for planned generation and distribution assets, which could lower tariffs, thereby eliminating the need for stranded cost recovery. Where present, this condition too should be taken into account.

3.3.4.3.4 Allow discos to participate in DG only if all competitors enjoy equal access to system information

Distribution companies are often best positioned to capture the value from distributed resources, particularly grid-related deferrals and system reliability benefits. The regulatory conundrum is to prevent discos from having market power arising from their privileged knowledge of access to customers (15). Discos' unique access to information regarding customer loads, system requirements, and potential constraints has the potential to create conditions for undue local market power.

We recommend that distribution companies provide open access to information to all competitors regarding the distribution system requirements, performance, and constraints, in order to create a level playing field. Customer load information is proprietary to the customer, but distribution utilities must provide historical area load information to competitors if the customer requests it. If these conditions are met, then competition can occur on a more equal basis, mitigating concerns over market power. Under these conditions, distribution utilities should be allowed to participate in the distributed generation market and to own DG—from which they can often most directly gain the benefits of reduced grid costs and losses and improved grid operations.

In practical terms, distribution companies are often required to be the retail provider of last resort. Therefore, they are exposed to the volatility of the power market, and seek load control opportunities that can be directly dispatched in order to hedge their exposure. The ownership of distributed generation will enable them to perform the same hedging function, as well as to manage the distribution system at least cost.

3.3.4.3.5 Uniform and reasonable retail wheeling tariffs

Uniform and reasonable retail wheeling charges (for moving electricity, at least notionally, over the lines of another supplier) should be developed to enhance customer choice from distributed generation. Just as with wholesale power, fully competitive retail markets would provide open and non-discriminatory access to the distribution system in order to allow transactions between distributed generators and retail customers. Although resource planning should in principle reflect area-specific marginal costs, retail wheeling tariffs should probably be based on uniform "postage stamp" rates to avoid excessive complexity.

3.3.4.3.6 Provide public support for green markets

If consumers are provided with retail choice and transparent market prices, renewable portfolio standards are no longer needed. Instead, retail power marketers will sell green energy to consumers, allowing the public to make an informed choice. As with all new commodities, the initial market development cost will be prohibitive without some degree of public support. The simple reality from the most recent experiences with retail choice is that absent any public support, 10–15% of customers who

switched will choose green energy.[36] (149) When a price credit of approximately 15% is provided, enabling green power to be cheaper than conventional fuels, the proportion of customer choosing green can rise to 40%.[37] (149) We do not believe that support should be continued indefinitely, but it should be applied for the first three years of retail choice to prime the market. A more attractive alternative may be to offer renewably generated power as a constant-price resource, as discussed in Section 2.2.3, so that customers who value this attribute can, perhaps at a modest cost premium, avoid the price volatility that plagues everyone else. Regulators who do not allow the constant-price attribute to express its market value, whether by this means or otherwise (perhaps by unbundled sale), are in effect confiscating one of the renewable generator's most valuable products and socializing it to competitors.

3.3.4.4 Summary: Actions needed to adopt the suite of state recommendations

As of spring 2002, only three states—California, New York, and Texas—had largely completed regulatory treatment of distributed generation, while 14 others had initiated some degree of action.[38] California and Nevada had experimented with decoupling of sales and profits in the early 1990s, but these efforts were folded into broader industry restructuring. Several states have ongoing initiatives to address interconnection

---

[36] The experience in Pennsylvania is that 15% of all customers who switched suppliers in 2000 switched to green power. This total was raised to 20% only after Green Mountain Energy purchased large blocks of PECO customers as part of the Basic Generation Service auction.

[37] In California, the 1 cent per kWh green credit resulted in over 40% acceptance by both business and residential customers. The program was ultimately discontinued when California's retail restructuring ended.

[38] For an updated review of state regulatory actions regarding distributed generation, see www.eren.doe.gov/distributedpower under "State Activities."

issues. In general, however, most of these recommendations will require new initiatives for regulatory action. Given the great diversity and rapid change of state laws and rules, there is little point summarizing the current status of which states have done what, but it is important to recognize that unlike the federal government, state governments and regulatory bodies are often a vibrant source of action and a valuable locus of parallel policy experiments.

3.3.5 Summary: A balanced portfolio

We offer this set of 23 recommendations (§§ 3.3.3–3.3.4) as a balanced portfolio to help achieve the overall national energy policy goals. In our view, six questions should be posed in evaluating the efficacy of these policy recommendations:

- Do they meet the energy policy goals?

- Do they hedge against the major risks in the electrical power system?

- Do they address the barriers to distributed generation?

- Do they resolve all the key regulatory issues?

- Does the cost/benefit ratio of proposed actions justify their adoption?

- Who wins and who loses if these recommendations are adopted?

The following section examines the proposed set of recommendations in light of these six questions.

3.3.5.1 Achieving the energy policy goals

Section 3.2.2 suggested that the national energy policy goals, often assumed to have inherent tradeoffs, can be achieved together by integrative policies. Therefore, while individual recommendations may further some goals and hinder others, the whole portfolio of recommendations helps to achieve the full suite of goals developed by the National Energy Policy Initiative (§ 3.2.2, note 2). For distributed generation, two broad policy goals have been to level the playing field and capture the wide range of benefits discussed in Part 2.

The primary objective of the federal recommendations is to improve market efficiency and competition by integrating distributed generation into the wholesale power and transmission markets. Including distributed generation in these markets creates value in five major ways:

- *Market liquidity.* The wholesale power and transmission congestion markets have suffered from a lack of market liquidity during periods of high energy use. This has led to extraordinary price spikes—one to two orders of magnitude above the average—as well as reliability concerns. Distributed generation can provide for rapid deployment of new generating capacity within constrained market zones, thus augmenting market liquidity and damping volatility. The increased competition for provision of power and ancillary services will also decrease prices.

- *Economic efficiency.* Price signals and system information in each of the wholesale power and ancillary services markets, which incorporate the value of services to the system in terms of time *and* location, will improve economic efficiency by providing incentives to add new resources to the system when and where needed.

- *System reliability.* Distributed generation will augment the reliability of the whole-sale power markets by providing power and ancillary services where they are most needed, and can do so faster than adding conventional generation and transmission.

- *System security.* A dispersed, networked energy supply system is inherently more flexible and resilient than a centralized, radial one. The overall security of the energy system would be enhanced by widespread adoption of distributed gen-eration, combined with state-of-the-art control, communication, and power elec-tronics technology. These technologies are needed to coordinate the grid, inte-grate both central and distributed gener-ation sources, and provide intelligent, real-time control to respond to external signals ranging from electrical distur-bances to market prices.

- *Environmental impact.* By adopting environmental standards and clarifying the ownership of environmental credits, environmentally friendly distributed generation resources will be favored over more polluting ones. This will avoid the unintended consequence of greater environmental impact from increased use of distributed resources.

The states clearly desire more competitive markets in wholesale and retail power, so in this aspect, their objectives are similar to the federal government's. The primary objectives of the states differ from federal objectives in four important ways. First, the states desire to create the incentive for utilities to provide power and delivery services at least cost to customers. Second, the states want to remove the major barriers to more widespread adop-tion of least-cost resources. Third, the states must ensure that the utilities themselves are

financially solvent and hedged against major shocks to the broader energy system. Finally, the states are concerned with equity among classes of customers.

Our recommendations collectively work to achieve these objectives both for states retaining traditional ratemaking regulation and for states that have chosen to restruc-ture. The suite of universal recommenda-tions seeks to remove the barriers to distrib-uted generation and take the first steps in capturing the value by sending the right price signals within net metering. The suite of recommendations for states with tradi-tional regulation is intended to remove the disincentives for regulated utilities to pursue distributed generation and then provide positive incentives to allow these utilities to capture the value from distributed resources.

The combination of revenue price caps and performance-based incentives enables utili-ties to set prices to consumers on a usage basis. These price signals then promote more economically efficient consumption and energy investment decisions by con-sumers (751). The renewable portfolio stan-dards are one mechanism to hedge against price and security concerns related to fossil fuels, particularly natural gas.

The suite of recommendation for states that have chosen to restructure is designed both to achieve the same regulatory outcomes for the distribution companies and to support the development of a more competitive retail market. Efficient retail markets will be enhanced by better price signals, more com-petitive service providers, equal access to market information, and a wider spectrum of customer options. Unbundling of retail prices will provide both better price signals and a wider spectrum for customer choice. Retail wheeling and lowering the stranded cost charges will increase the ability of distributed generation to enter the retail market, augment the number of competitive service providers and increase customer choice. Fair competition requires that all participants have equal access to information. The initial pump-priming public support for green markets is proposed to increase the choices for customers, but is not intended as a permanent subsidy.

3.3.5.2 Creating a policy portfolio to hedge risks

Fundamentally, an efficient policy portfolio hedges against the critical risks that can affect the electric power system. As recent events have shown, there are five critical risks that should be of concern to federal and state regulators:

- *Electricity price volatility.* Even in a fully competitive market, power price volatility can be caused by inelastic demand, the underlying volatility of the natural gas market and of weather, and constraints in the power and gas transmission grids. At the extreme, price volatility can cause grave economic dislocation, as in California 2000–01.

- *System reliability.* The reliability of the power system can be threatened by localized shortages of supply due to unforeseen load, actual or artificial supply shortages, and transmission or other system constraints.

- *System security.* The existing electricity system is extremely brittle and vulnerable to deliberate or natural disruption.[39]

- *Financial stability.* The loss of utility financial stability and the ensuing credit risk can cause economic dislocation

throughout the energy system, triggering both reliability and price concerns.

- *Climate change.* The potential impact of climate change on both the environment and energy markets could create major discontinuities in the current planning and operation of the energy grid.

As discussed in Part 2, distributed generation has many benefits that create a hedge against price volatility and against disruption of system reliability and security. The portfolio of recommendations will accelerate distributed generation by leveling the playing field and creating the conditions necessary for participants to capture the benefits. Concerns over the financial stability of utilities are addressed by revenue caps, which ensure that utilities are able to cover the revenue requirements for their regulated entities. In the restructured environment, allowing distribution companies, which are often providers of last resort for energy supply, to own distributed generation gives these companies the option of creating a physical hedge against power-price spikes. Finally, the call for renewable portfolio standards implicitly addresses concerns over climate change, diversifies supplies in a way that supports both short- and long-term security concerns, and creates a hedge against fossil-fuel price volatility.

3.3.5.3 Addressing barriers to distributed generation

The proposed recommendations address all the public-sector barriers presented in Section 3.2.3 (Table 3-3). The recommendations designed to address interconnection and standards are shown on both the federal and state level, because if the federal government does not act on national standards, it is incumbent on the states to do so.

[39] See *Brittle Power* (442) for a broader description of these issues.

Table 3-3: Recommendations to address public sector barriers

Policy recommendations	Public sector barriers				
	Interconnection standards	Environmental regulations	Utility pricing	Wholesale market access	Retail market access
Federal					
A. Create uniform interconnection standards	√				
B. Integrate DR into wholesale markets				√	
C. Integrate DG into ancillary services markets				√	
D. Support locational marginal pricing			√	√	
E. Provide greater access to information				√	
F. Accelerate RD&D for DG	√	√			
G. Create emissions standards for DG		√	√		
H. Clarify ownership rights to pollution credits					
Universal state recommendations					
A. Adopt "plug and play" interconnection standards	√				
B. Create net-metering rules with buyback rates based on system value			√		
C. Adopt emission standards for DG		√			
D. Provide public support for DG RD&D through wire changes	√	√	√		
E. Update building codes and development covenants to accommodate DG	√				√

Table 3-3: Recommendations to address public sector barriers (cont.)

Policy recommendations	Public sector barriers				
	Interconnection standards	Environmental regulations	Utility pricing	Wholesale market access	Retail market access
States with traditional regulation					
F. Decouple utility revenue requirements for kWh sold and create incentives to lower bills, not kWh prices			√		
G. Require mandatory ERIS planning as the basis for prudent cost recovery			√		
H. Restructure distribution tariffs to reduce excessive fixed charges			√		
I. Adopt renewable portfolio standard and tradeable credits				√	
States adopting restructuring					
J. Decouple Disco revenue requirements from kWh throughput			√		√
K. Restructure and unbundle distribution tariffs	√		√		√
L. Only impose stranded costs after production threshold is exceeded			√		√
M. Allow Discos to participate in DG only if equal access to system information is given to all competitors					√
N. Uniform and reasonable retail wheeling tariffs			√		√
O. Provide public support for green markets		√		√	√

3.3.5.4 Resolving key regulatory issues

The proposed recommendations address all the key regulatory issues presented in Section 3.2.3 (Table 3.4). The recommendations designed to address technical interfaces and siting and permitting are shown on both the federal and state level, because if the federal government does not act on national standards, it is incumbent on the states to do so.

Table 3-4: Resolution of key regulatory issues

Regulatory issue	Outcome	Recommendation reference	
		Federal	State
Technical interfaces			
1. *System interface:* Should DG interface with grid operations and markets?	Yes	B, C, D, E	
2. *Interconnection:* Should the interconnection technical requirements, processes, and contracts be modified for DG?	Yes	A	A
Economic and financial			
3. *Utility ratemaking:* Should utilities continue to have a primary financial incentive based on selling more kWh?	No		F, J
4. *Grid-side benefits:* Should grid-side benefits of customer DG be monetized and allocated among stakeholders?	Yes	E	G, H, K
5. *Energy pricing:* Should the price of energy fed into the grid reflect the incremental value, net of costs, to the system?	Yes	D	B, H, K
6. *Stranded costs:* Should utilities be compensated for stranded costs associated with DG installations?	No*		L
7. *Fixed charges:* Should utilities be compensated for providing standby and reliability services?	Yes		B, H, K, N
8. *Disco participation:* Should discos participate in DG?	Yes		M
9. *Public support:* Should DG technologies be supported by financial incentives, subsidies, or public funding of RD&D?	Yes	F	D
Environmental			
10. *Siting and permitting:* Should siting and permitting requirements be modified for DG?	Yes	G	C, E
11. *Technology differentiation:* Should environmentally friendly DG receive differential benefits?	Yes	D	D, I, O

* Unless threshold is exceeded so that stranded costs are actually incurred.

3.3.5.5 The cost-benefit question

Little research has been done, and even less of it is publicly available, on the cost-benefit relationships justifying policy proposals on distributed generation. As shown in Part 2, the benefits of distributed generation are generally quite large. Yet we acknowledge that there may be diminishing returns to distributed generation benefits. In other words, the value of distributed generation will initially rise with each incremental MW as additional resilience is created across the grid, but may in time pass an inflection point beyond which the additional incremental MW may have decreasing marginal value. On the other hand, this speculation may also prove unsound: one can imagine circumstances in which distributed generation might yield expanding returns due to synergies not captured in the current power system.

Many of the recommendations require administrative changes to markets and standards, and thus have low costs when compared with societal value created, although the cost to the implementing agency *vs.* its current budget must also be considered. On the other hand, significant administrative costs now required to support centralized facilities—whose administrative requirements tend to be more complex—may also be saved. In general, it seems plausible that any net increase in the "soft" costs of regulation and administration should be modest compared with the "hard"-cost benefits, as the boxed examples illustate.

Example:
Utility case studies of fuel-cell and PV distributed benefits

Although it is difficult to generalize the economic benefits of DG, their magnitude appears to be significant. For example, a fuel-cell DG system would ordinarily provide, at a minimum, the following major benefits, neglecting many others described in Part 2:

- **Electric energy value (§ 2.3.3.1).** Regardless of DG benefits, the energy produced by a fuel-cell system would be worth about $100–150/kWy, assuming the system is sized to provide baseload power and operate almost continuously.

- **Thermal energy value (§ 2.4.4).** Especially in a well-designed commercial application, the waste heat recovered from the fuel cell can provide fuel savings of about $100–150/kWy.

- **Option value (§ 2.2.2.1).** In an area with fast but uneven growth, the added cost of overbuilding generation that could be avoided by widespread use of DG is about $50–200/kWy.

- **Deferral value (§ 2.3.2.6).** In a high-cost area, with distribution capacity constraints and moderate growth, the deferral value would be about $50–200/kWy, assuming that these areas are targeted with sufficient DG capacity to defer capacity expansion.

- **Engineering cost savings (§ 2.3.2).** In a "problem" distribution area, properly cited DG can avoid the cost to reconductor feeders, add capacitor banks, and install voltage regulators, worth about $50–150/kWy. Reductions in losses are worth about another $25/kWy.

- **Customer reliability value (§ 2.3.3.8).** In a commercial application with a high value of service, a highly reliable DG system that reduces outage risk for critical loads provides a reliability value of $25–250/kWy, depending on the customer's circumstances.

- **Environmental value (§ 2.4.10).** The environmental benefit of fuel cells' low emission rate is unlikely to be realized directly, but it makes fuel cells easier to site than other DG.

In addition to the electrical and thermal energy values, any one of the other DG values would raise the total DG value to about $400/kWy or higher. In an area where all these benefits are realized, the total DG value could reach about $800/kWy or higher. Assuming conventional commercial financing, these values translate into an allowable system capital cost of $2,000–4,000/kW. The $2,000/kW value is considered achievable in the near future by fuel-cell manufacturers (and approximates the DOE-subsidized net price of many phosphoric-acid units installed in the past few years), while the $4,000/kW value is commercially achievable today, or nearly so. This means that with proper design and siting, fuel-cell DG systems can be

(cont.)

Example (cont.)

cost-effective today, based on the value of their distributed benefits.

How about a reality check from the costliest distributed resource—photovoltaics? The nominal 500-kW PV plant near PG&E's Kerman substation (§ 2.2.8.4) was installed in 1992 at a turnkey cost of $8,900/kW, plus $1,000/kW of PG&E-paid sitework, and was initially evaluated to have a benefit/cost ratio of only 0.73. However, the benefit/cost ratio depended sensitively on the perspective considered; the calculated benefits were more than twice as great from a feeder perspective ($788/kWy in 1992 $) as from a planning-area perspective (Figure 3-1).

Of the $788/kWy of estimated total societal benefits, $521/kWy would accrue to PG&E and a further $267/kWy to its customers (615). Depending on perspective and on the availability of tax credits, the breakeven installed cost in 1992 was estimated at $2,600–7,400/kW (Figure 3-2).

This threefold range in breakeven value is 1.3–3.8× the $1,950/kW breakeven turnkey cost for PV plants counting only traditional generating-capacity and energy benefits but no distributed benefits (628).

In Figure 3-2, the "planning area" perspective counts traditional energy (§ 2.3.3.1) and generation (§ 2.3.1) capacity value, transmission capacity value (§ 2.3.2.6), and loss savings (§ 2.3.2.2).

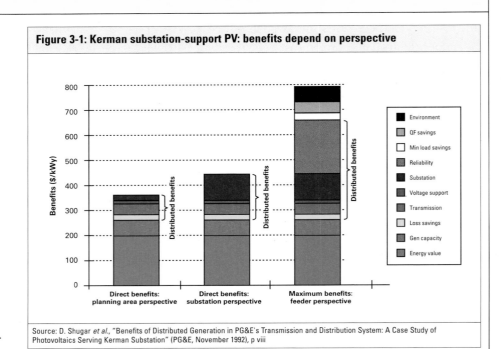

Figure 3-1: Kerman substation-support PV: benefits depend on perspective

Source: D. Shugar *et al.*, "Benefits of Distributed Generation in PG&E's Transmission and Distribution System: A Case Study of Photovoltaics Serving Kerman Substation" (PG&E, November 1992), p viii

The "subplanning area" perspective adds substation/distribution value (§§ 2.3.2.6–7) and reactive power value (§ 2.3.2.3), but excluding the option for the flexible inverter to inject additional reactive power on demand, (§ 2.3.2.3.1). The "feeder" perspective, with or without tax credits, adds minimum-load sav-

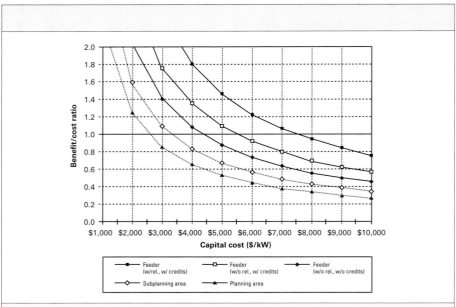

Source: Source: D. Shugar *et al.*, "Benefits of Distributed Generation in PG&E's Transmission and Distribution System: A Case Study of Photovoltaics Serving Kerman Substation" (PG&E, November 1992), p. ix

Example (cont.)

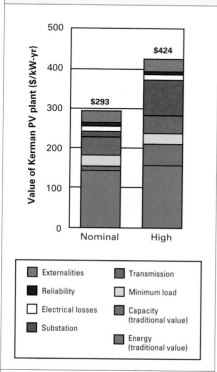

Figure 3-3: The value of the Kerman plant to PG&E (1995 $)
Post-operational validation of baseline and sensitivity-case value of Kerman PVs to PG&E, assuming 30-y project life, 9%/y cost of capital, and 3.5%/y monetary inflation rate.

Legend:
- Externalities
- Reliability
- Electrical losses
- Substation
- Transmission
- Minimum load
- Capacity (traditional value)
- Energy (traditional value)

Source: H. Wenger *et al.*, *Photovoltaic Economics and Markets: The Sacramento Municipal Utility District as a Case Study* (SMUD, CEC, and USDOE PV Compact Program via NCSC; 1996), p. 3, fig. 4. www.energy.ca.gov/development/solar/SMUD.pdf

ings (§ 2.3.3.2), Qualifying Facility savings (§ 2.3.3.7), and environmental benefits (§ 2.4.10), and in a separate increment, reliability benefits to the distribution service area (§ 2.2.9.4).

A 1994 re-evaluation revised the Kerman PV plant's distributed benefits from about $788/kWy (1992 $) to $293–$424/kWy (1995 $), equivalent to

$274–$397/kWy in 1992 $—a decrease of about half in the comparable higher case (Figure 3-3).

In the Nominal (lower) case, this decrease was due to lower natural-gas prices; a 1992 regulatory assumption that PG&E would need no generating capacity for the foreseeable future; lower reliability enhancement value because adding a capacitor bank on the Kerman circuit would achieve that benefit more cheaply than had been assumed in the original cost-of-service study; and reduced substation transformer value because in the Kerman area it happens to be relatively easy to switch load. The High case found a 45% higher value by assuming an earlier need for bulk generating capacity, using the "substation perspective" used in the original Kerman study, and assuming full operating shakedown to achieve the nominal annual output. (The plant's output fell 9% short of its nominal target in its startup year, due largely to inverter teething troubles during three months. Nonetheless, validation in that startup year [§ 2.2.8.4] confirmed that the Kerman PVs achieved 77% effective load-carrying capability—ability to displace conventional generation capacity—and an even more impressive 90% ELCC with respect to the transmission system.)

In short, the 1994 re-evaluation found that rather than roughly tripling value,

distributed benefits would only about double it, increasing equivalent levelized value of the PV resource from ~$0.07–0.10/kWh to ~$0.14–0.20/kWh. This implies a breakeven PV price of $2,700–$3,800 in 1995 $ if owned by PG&E, or $3,400 to >$5,000 if owned by an independent power producer.[40] Since the average price of PV arrays shipped in the U.S. fell in real terms by 48% during 1992–99 (205) and continues to fall, just these basic distributed benefits, under the Kerman site's relatively unfavorable conditions, are sufficient today to bring such installations into an economically interesting range. *Yet the Kerman evaluation omitted many of the distributed benefits evaluated in Part 2, including all the financial-economic benefits—typically the most valuable category.* Even the most basic consideration of such financial-economic benefits as lower investment risk, reduced exposure to fuel-cost or purchased-power price volatility, and portability would therefore have made this PV installation economically worthwhile—despite, again, the relatively unfavorable conditions of the particular Kerman site.

Moreover, the PG&E evaluators found the non-traditional distributed benefits to be "measurable, predictable, and significant" for grid-support PV, and their analytic methods "repeatable and generally applicable to other forms of distributed resources and applications."

[40] In 1994, the PG&E evaluators believed that IPPs would have financial advantages over investor-owned utilities, including cheaper capital, greater access to tax credits and accelerated depreciation, and more flexible financing ratios, enabling IPPs "to afford a 30 percent more expensive plant than an IOU, while maintaining profitability." In 2002 this looks overly optimistic for IPPs, though perhaps realistic for publicly owned utilities.

3.3.5.6 Who wins and who loses

The central question of this section is the scope and character of government policy necessary to help vibrant, fair, and orderly markets develop, and by extension, the success of traditional distribution utilities in managing disruptive technological change. In restructuring electricity markets, policy makers need to recognize that the capabilities, cultures, and practices appropriate to the growth and development of the industry as it exists today are valuable only under certain conditions. The pace of progress can easily be accelerated or retarded by the regulatory and statutory mechanisms described above. Policy makers must also recognize that distributed resources, which may not appear useful in yesterday's regulatory context, may squarely address the needs of tomorrow's customers by providing them the best energy services at the lowest cost and by dealing gracefully with new forms of risk. Because one cannot expect customers to lead the way to products and markets that they do not yet understand, this places a premium on policy makers to create market conditions appropriate to emerging opportunities, and faithfully to pursue open and competitive market structures.

Managing the restructuring process to produce the greatest public benefits is similar to the integrated resource allocation process developed by leading state regulatory agencies over the past 15 years. Surprising opportunities for providing energy services at lower financial, social, and environmental costs emerged when barriers were removed and nontraditional options were measured against traditional solutions on a level playing field. Managing resource allocation decisions in today's world, where multiple customers (and many new market players) are competing for limited resources, requires utility managers and their government regulators to focus on and invest in distributed resources even in advance of mature and profitable markets, because there is a strong advantage to early adopters and fast learners.

If the policy recommendations proposed above were adopted, there would be many winners. Society at large would prosper because electric service could be provided at lower cost with higher reliability. Regulators would achieve their objective of fair and competitive electricity markets at the wholesale and retail level, since distributed generation would add more competition and liquidity. Further, grid reliability and energy system resilience (hence security) would be enhanced. Business customers would benefit from a wider spectrum of options to manage their energy needs, greater grid reliability, and the ability to reap commercial profits from advantageous sites. Progressive utilities would benefit by sharing in the savings from the lower revenue requirement—in effect, earning a higher return on assets. Clearly, distributed generation manufacturers and energy service companies would become high-growth industries, attracting capital and creating jobs. The environment will benefit from lower air pollution than with centralized generation. The benefits of distributed generation are widespread, and accrue across the value chain.

The losers are those parties that resist technological change and do not adapt to it. Incumbent utilities would lose significant revenues and have a new class of stranded costs ("stranded wires") if they clung to traditional regulation. Similarly, under traditional regulation, costs could be shifted to customer classes that do not or cannot adopt distributed generation. Regulatory agencies will need to develop increasingly sophisticated capabilities to set prices, terms, and conditions in a distributed environment, and must be able to monitor multiple markets to protect against localized market power. It is the fear of these losses that creates resistance from the incumbent players to widespread adoption of distributed power.

Despite these understandable fears, to the extent distributed generation moves the energy system towards a more efficient frontier, everyone wins. Now is the time to capture the prize. Here are some of its dimensions.

3.4 IMPLICATIONS FOR THE PRIVATE SECTOR

This section explores DG's implications for the private sector, more specifically investor-owned utilities, public power utilities (viewed here as firms, not government agencies), financial markets, commercial and industrial customers, and real estate developers. The implications provide insight into distributed generation's threats to and opportunities for current business models, and issues that arise as organizations react, drawn from the practical experiences of early market adopters of distributed generation.

From the outset, the most fundamental implication for all private sector stakeholders is that *distributed generation is a disruptive technological change*. What do we mean by that? Disruptive technological changes occur when a new technology outstrips the current boundaries of cost, performance, and value of the incumbent process for delivering a product or service. This irreversibly changes the value chain's economics. For distributed generation, an analogous example of the impact disruptive technology can have on a value chain is the introduction of the personal computer into the mainframe computing environment. The personal computer created a new distrib-

uted environment whose cost and performance characteristics made mainframe computing obsolete for all but the largest-scale applications. Companies, such as Control Data Corporation, that were unable to adapt went bankrupt, and others, such as IBM, struggled with the changes but ultimately found new opportunities to profit. What explains the differences in response between companies that successfully adapt versus those that perish?

Clayton M. Christensen, in his best-selling book *The Innovator's Dilemma* (127), describes the failure of companies to stay atop their industries when they confront certain types of market and technological change. The dilemma of innovation is not that it is rare; in fact, all firms ultimately face it. The dilemma is rather that extremely well run companies usually fail to understand the implications of new technologies and the markets they engender, because of the threat to their existing markets. As a result, the incumbent companies lose market share and are eventually replaced by new competitors. It is only by ignoring current customers and disobeying seemingly sound management practices that drastic innovation can be harnessed. Traditional electric utilities face this exact dilemma with distributed generation. So how should private sector companies adapt to the disruptive technology represented by distributed generation?

The first step in successful adaptation is understanding the implications of distributed generation for the current business model. Therefore, we have organized the implications discussion below to answer four basic questions:

- What are the threats posed by distributed generation to the current business model?

- What new business opportunities are created by distributed generation?

- What are the organizational impacts implied by these new opportunities?

- What are the overall financial implications of action or inaction?

This section (§ 3.4) primarily explores the implications of DG for electric utilities, which collectively do more than $200 billion worth of business annually in the United States. For investor-owned utilities, the implications of distributed generation are substantially different depending on whether they are under traditional regulation (and typically organized as vertically integrated utilities) or under some degree of both regulatory and organizational restructuring. Therefore, we present implications for investor-owned utilities in two different sections. We also separately address the implications for public power entities, since they are often outside the state regulatory regime and have an entirely different profit dynamic than investor-owned utilities. We next address the financial community, since we believe that the underlying valuation assumptions for traditional business models

made by Wall Street are brought into question by distributed generation. Lastly, we discuss the implications for two important classes of end-user communities—commercial and industrial (C&I) customers and real estate developers.

We reiterate that this section does *not* contain recommendations to the private sector, since each company's strategic direction, competitive circumstance, and organizational capabilities are different. Further, we have not addressed the vendors of distributed generation, because of the vast differences in technology and ultimate market applicability across vendors. Finally, we remind the readers that the implications presented will be broadly applicable to each of the business classes discussed. While your individual company's position may differ from the rest of the industry's, your collective fate and choice are shared.

3.4.1 Implications for electric utilities under traditional regulation

Despite all the media attention to restructuring, the vast majority of the nation's ~110 electric utilities remain under traditional utility rate-of-return regulation. Of these, over 90% are vertically integrated utilities with generation and distribution bundled together within the same company. To understand the threats to and opportunities for electric utilities, a brief digression is warranted on how utilities profit within the regulated environment.

Such utilities expect to recover their revenue requirements, defined as the sum of a fair and reasonable return on and of "used and useful" capital plus reasonable operating expenses. Tariffs are set by allocating the revenue requirements across customer classes, then dividing by the usage of each customer class to get the price in cents/kWh. Tariffs are set periodically in quasi-adjudicatory "rate cases," which typically occur every 3–5 years, but have tended to run for longer periods of time as utilities learned to become more efficient. Fuel costs, including purchased power agreements, are usually passed through directly to customers. [41]

In this environment, if the utility's total sales and costs were exactly as predicted in the last rate case, then the utility's profit would be limited to its allowed return on equity from the assets within its "rate base" on which the authorized rate of return may be earned. If the new assets are deemed to be prudent, used, and useful, they are generally allowed in rate base. If sales are higher than predicted, and the cost of realizing those sales is lower than the resulting revenues, then the utility earns additional profits. Utilities can also earn more profits if expenses are lower than predicted. Utilities use their profits primarily to pay dividends to their shareholders, who tend to expect a constantly growing stream of dividends with relatively low risk.

Unpleasant surprises can be disastrous in this business model. If new assets are partially or entirely disallowed because the costs are imprudent or the investments are not needed by the time they are finally built, shareholders must sustain the losses. If sales fall faster than operating-cost efficiency improvements, then the utility will either suffer losses or be forced to enter into a new rate case in order to raise its prices. If capital costs rise faster than projected in the last rate case, and this increase cannot be

made up by operating-cost efficiency improvements, then the utility will again suffer losses or be forced to enter into a new rate case in order to raise its prices. In general, utilities would rather avoid new rate cases until absolutely necessary, because they reset the baseline, negating all the additional profits previously captured by load growth or efficiency improvements. Also, most utilities try in their rate cases to underestimate future sales of electricity so that they can sell more than forecast and thus increase their profits—a perverse effect that rewards utilities for increasing customers' consumption and bills. Conversely, traditional rate-of-return regulation penalizes utilities for helping customers use electricity more efficiently. And while this regulatory system rewards gaming the demand forecast (wasting a lot of time of lawyers and expert witnesses), it exposes the utility unnecessarily to financial risks from external conditions, such as weather and business cycles, that are not under its control. All these problems are eliminated by decoupling utilities' profits from their sales volumes and then giving them a share of any savings they achieve for their customers; but as of May 2002, only one or two of the United States had done this (§ 3.3.4.3.1).

3.4.1.1 Threats to existing business models

Distributed generation poses four primary threats to the existing vertically integrated business model. First, distributed generation results in the loss of revenue under traditional tariff structures, because the customer simply purchases fewer kWh or fewer distribution services. Second, more substantial market capture by distributed generation can cre-

[41] For a broader understanding of ratemaking see, *Tools and Methods for Integrated Resource Planning: Improving Energy Efficiency and Protecting the Environment* (680).

ate a new class of stranded asset within the distribution system—grid capacity no longer needed. Third, the ability of distributed generation to enter more rapidly than centralized generation or transmission upgrades can partially strand new capacity additions. Fourth, the combination of the first three threats can create a "death spiral" in which the higher prices to remaining customers induce more of them to leave this system, creating a self-reinforcing cycle of ever-increasing unit prices. Let us consider each of these threats in more detail.

Even modest revenue losses have substantial impacts on profit. The problem for utilities is that their gross margins (revenues less cost of goods sold, *e.g.*, fuel) are quite high. In the short term, their operating and depreciation costs are relatively fixed versus sales. Hence, a reduction in sales volume reduces gross margins, tending to reduce profits. A highly simplified example can illustrate this point.

A typical vertically integrated utility with two million customers may have sales of 40,000 GWh/y, revenues of $5.5 billion/y, but net income of only about 10% or $500 million/y (which translates to about $0.013/kWh). If total sales revenue is on the order of $0.10–0.11/kWh, and fuel and purchased power costs are around $0.04/kWh, then gross margins are high—about $0.06–0.07/kWh. Direct costs are around $0.03/kWh, indirect costs (overheads) about $0.015/kWh, depreciation and income taxes are in the vicinity of $0.01/kWh, so total costs are around $0.05–0.06/kWh, leaving only $0.01/kWh in profit. From the utility's perspective, each kWh of lost sales results in lost gross margin, but does not change the operating costs or depreciation in the short

run. Therefore, even a modest penetration of distributed generation—on the order of 5%—would reduce profits by $160 million or almost 30%.

Clearly, a sustained drop of this magnitude would be difficult to make up in operating expense improvement, and therefore would require an increase in prices. Absent any improvement in operating costs, a price increase of 4% would be needed to make up for the lost revenues. Each increase in price makes the economics of leaving the system more attractive for large business customers—hence the concern among traditional utilities that distributed generation only be added to the system when it reduces the need for new capital investments. As will be seen in Section 3.4.1.2, this concern is often overblown in practice, but in theory, it is real.

In addition to the direct revenue losses, the regulatory question of what assets are stranded due to distributed generation is sure to arise in the ensuing rate case. If several large industrial customers in a localized area were to exit the system, the reduction in utilization to the distribution feeders and substations could be substantial. In such a case, the remaining industrial users could argue that these distribution assets were effectively stranded, and at least in principle, might demand that they be removed from the utility's rate base. If successful, such action would depress profits further, since prices would drop, but costs (which are largely sunk and therefore expressed in depreciation) would not. This concern underlies the move by utilities to change the tariff structure to both shift more of the tariff to fixed costs and charge high backup or exit fees.

Finally, as noted in Section 2.2.2.2, the long lead times of conventional generation and transmission present a risk to utilities. The advent of distributed generation by third parties increases the risk of investing in large lumpy assets to serve a load that may not materialize. Utilities experienced this problem during the nuclear cost overruns of the 1970s and 1980s. The concern over potential loss of load from restructuring led many utilities to build both fewer and smaller plants in the 1990s and transmission upgrades virtually ceased. Distributed generation would have a similar chilling effect on major, long-lead-time investment decisions.

In the minds of many utility executives, a high degree of market capture by distributed generation represents one of the worst outcomes for existing utilities. In our view, this is true *only* if they resist change, persist with traditional regulatory strategies, and do not adapt their planning and operating practices to profit from the new opportunities.

3.4.1.2 Opportunities for regulated utilities

If the threats appear so grave, how is it possible for distributed generation to present profitable opportunities for regulated utilities? The answer lies in embracing the benefits of distributed generation, reforming the perverse regulatory incentives, and understanding the difference between accounting profits and value. Even for regulated utilities, distributed generation can offer significant growth opportunities and lower risks.

Regulation does not necessarily have to mean traditional rate-of-return regulation. As we argued in Section 3.3.4.2.1, state regula-

tors should adopt revenue caps or balancing accounts to eliminate the disincentives discussed in Section 3.4.1.1. What impact would this have for the regulated utility? Using the prior example as a guide, the revenue losses would no longer harm the bottom line directly. Under a revenue cap, the utility receives the same revenues regardless of the sales volume, so profits would remain the same even if fewer kWh were sold. However, for this to occur, prices might have to rise for the remaining customers (though practical considerations suggest this would in fact rarely occur, and in effect, the effect is tiny).[42] How can the utility avoid the "death spiral" of supposedly continuous price increases?

The answer lies in understanding how distributed generation can lower total revenue requirements, thereby ultimately lowering bills to all customer classes. Vertically integrated utilities can lower revenue requirements through three strategies:

- leveraging distributed generation's option value

- de-capitalizing the wires

- lowering operating expenses

3.4.1.2.1 Leveraging distributed generation's option value

Distributed generation creates two types of option value. First, as discussed in Section 2.2.2, its inherently modular nature and rapid deployment can make distributed generation a lower-cost method of expand-

---

[42] Higher prices would be required to protect profit levels only if the competing technology cut contribution to margin faster than the sum of several compensating effects: depreciation (as old fixed costs get paid off), growth in kWh sales to other existing or new customers (§ 3.3.4.3.3), and the potential to use saved operating costs (because generating less electricity means buying less fuel and variable O&M) to prepay the costliest debt, thus avoiding interest accrual. This is similar to the obvious, though often overlooked, argument that investing in end-use efficiency is unlikely in practice to raise electricity prices, and if it did, *bills* would still go down, because consumption would fall more than price rose (423). Empirically, California's Energy Revenue Adjustment Mechanism balancing account had an average price effect far below 1%.

ing generation capacity than building new power plants, even though much distributed generation has a higher capital cost per kW and a lower thermal efficiency (unless cogenerating) than centralized power stations. Further, the smaller distributed units place a lower reserve margin requirement on the system (§ 2.3.1.1). The appropriate amount of distributed versus centralized generation required in each system can be determined using the LIRP/ERIS process described in Section 1.4.1. In short, any deferrals of planned centralized generation, transmission, or distribution capacity or upgrades due to distributed generation would *lower* future revenue requirements. However, that does not solve the short-term problem of direct revenue loss; realigned regulatory incentives do (§ 3.3.4.2.1).

Second, distributed resources can be designed so that they are capable of being centrally dispatched. Indeed, under the 5% penetration rate assumption in the prior example, a significant portion of distributed generation would need to be dispatch-capable in order for the system operator to maintain the voltage regulation. These distributed resources therefore represent a type of call option on additional power—more capacity and energy can be brought on at a particular set of strike prices. Alternatively, the generation resources of the vertically integrated utility displaced by must-run distributed generation can also be call options. These capacity options have value in the wholesale power markets, which the utility's wholesale power operations should optimize. In a regulated context, the value created from selling surplus capacity or power into the adjacent wholesale power markets would reduce the revenue requirements.

The magnitude of this benefit should not be underestimated. During the 2000–01 California power crisis, savvy utilities in the Pacific Northwest, such as Avista, made millions of dollars finding distributed resources at $0.06–0.10/kWh and reselling the power to the wholesale California markets at $0.15–0.50/kWh. Similar profit opportunities were available in the Midwest during 1999 and in the New York/New England power markets during 2000–01. In fact, virtually all wholesale power markets with low reserve margins will exhibit extremely high peak prices, as further discussed in Section 3.4.2.2.1.

3.4.1.2.2 De-capitalize the wires business

In the distribution business, most U.S. utilities require significant cash investment just to maintain and continually upgrade the distribution system (Figure 2-50). The annual capital cost often equals or exceeds depreciation, leading to a growing asset base. The net asset base sets revenue requirements and hence prices during the rate case. Once the rate case is completed, to the extent that the utility is able to maintain system performance with lower ongoing capital costs, the additional cash is available to pay the shareholders additional dividends, which greatly increase the value of the utility's shares. Ratepayers do not receive the benefits of reduced capital cost until the next rate case, when, *ceteris paribus*, the net rate base would be lower due to the lower capital investment. At this point, the lower revenue requirements translate to lower prices (though not necessarily lower profits), and the cycle begins anew.

As discussed in Part 2, once distributed generation achieves critical mass in a given area, it will defer transmission and distribution system upgrades while preserving and often enhancing system reliability. These deferrals will be particularly valuable to the utility when they avoid T&D system expansion projects that are considered high-risk. High-risk projects are characterized by having large new capacity requirements (lumpy), with uncertain or slow load growth.[43] These high-risk projects are likely to have very low utilization, and could potentially become stranded assets. Hence what matters to the distribution planner is location, location, location (§ 1.4).

Correctly sited, the distributed generation resource will provide financial benefits to both the utility and its customers. Although third parties are using their capital to invest in distributed generation, the utility enjoys lower capital investments, and ultimately the customers will enjoy lower revenue requirements. It is important to recall that since the capital investment is deferred but not eliminated, the economic value is equal to the discounted present value of shifting the capital cost by the number of years deferred, as opposed to the entire capital cost itself (see Section 2.2.2.5 for a broader discussion of deferral economics). In our view, the distributed generator should receive some of this economic benefit, which should be factored into the tariff paid for backup services. The remaining economic benefits will indeed improve utility cash flow, lower revenue requirements, improve delivery system utilization, and mitigate the risk of stranded system expansion projects.

3.4.1.2.3 Lower operating expenses

Distributed generation could lower operating costs of vertically integrated utilities in several ways. To the extent that distributed resources can be dispatched, as used by the grid operator for ancillary services, several forms of savings could occur, as discussed in Section 2.3. For example, distributed generation could provide virtual spinning reserve at lower cost than centralized power generation stations or could provide voltage or reactive power regulation. To the extent that distributed resources reduce the overloading and losses of distribution elements, operating costs would further decrease (§ 2.3.2.2).

Many utilities have the perspective that distributed generation could potentially add to system costs if unmanaged by the utility. Distributed generation can increase system costs for operation and control, particularly for must-run distributed resources. For example, in smaller systems with larger penetration of distributed resources, the utility must adjust its own production to manage the system voltage, often running plants at lower loading levels with associated heat rate penalties—which is not always the least-cost solution.

The reality is that the impact of distributed generation on operating expenses will be system-specific, and will depend strongly on the resource's location, operational protocols, technology choice, contractual arrangements, and configuration with respect to the existing system. Since significant penetration of distributed power has not yet occurred in most of the U.S. the empirical evidence of distributed generation's impact on total system operating costs is lacking. However, where it has, as with

---

[43] From a distribution planner's perspective, distributed generation in high load growth areas conventionally has only a modest deferral effect, since the load growth will outgrow the distributed resources and new system expansion projects will ultimately be needed. This might not be true if all kinds of distributed resources are systematically and comprehensively deployed; after all, they should in principle be faster to add than equivalent centralized resources, and they may very well not add up to a very large total—just as loads can do.

Danish windpower (§ 2.2.10.1), operational impacts appear to be slight and generally favorable (§ 2.3.2.10.2, note 145).

3.4.1.2.4 Organizational implications for vertically integrated utilities

Utilities will clearly need to enhance their organizational capabilities to manage distributed generation. This will affect several functions, particularly planning, distribution and transmission system operations, systems management, and accounting. The planning process should incorporate Local Integrated Resource Planning/Energy Resource Investment Strategy (§ 1.4.1). This implies a more detailed analysis of the distribution system and a detailed time- and area-specific understanding of incremental cost-to-serve.[44] Obviously, T&D system planning will change dramatically as distributed generation resources are routinely considered and compared with T&D upgrades. T&D system operations will have to manage new operational challenges related to unit performance and control, as well as system protection and coordination.[45] New operational safety procedures and protocols will have to be developed and implemented. In terms of contractual arrangements, the Customer Information System (CIS) will certainly require upgrading. Most utilities that have not undergone restructuring have legacy mainframe CIS systems that were not designed to accommodate net-metering billing or payments for distributed generating resources.

These changes will not come for free. For a large or mid-sized utility, CIS system upgrades required to manage distributed generation accounting and billing typically cost around $10–30 million. Distributed generation planning will require significant commitments of internal staff resources and external consultants. A considerable amount of training is needed for the distribution system maintenance staff in order to ensure safe operations. Regulated utilities will want and expect recovery of these costs. However, nearly all the same investments are desirable or necessary anyhow to modernize any traditional utility, especially under restructuring.

In vertically integrated utilities, the responsibility for the distributed generation business is straightforward. The corporate planning department is responsible for Local Integrated Resource Planning. The majority of distributed generation planning, operations, commercial contracting, and management will fall to the distribution and transmission organization. Both supply- and demand-side distributed resources will be treated as extensions of distribution planning, planning from the customer back upstream, rather than as part of traditional generation planning (§ 1.4.1).

3.4.1.2.5 Summary for regulated utilities

Distributed generation represents a grave threat to vertically integrated utilities under traditional rate-of-return regulation. The potential financial impact to balance sheets dwarfs industry concerns over environmental issues. Utilities that continue business-as-usual could ultimately suffer the same fate as Control Data Corporation—a continuing spiral of financial distress that ends in bankruptcy. It is not surprising that utilities have tended to resist this change by charging high fixed fees for backup, demand charges, or exit fees, and have been slow to interconnect distributed generators. In short, for distributed generation to be successful, the utility business and regulatory model must change.

[44] Software tools for this activity already exist; for example, ABB's Q2 has the ability to provide utility planners with local area costs on a dynamic basis, accounting for changes in load and dispatch.

[45] Most modern utilities have control area software capable of automatically managing the grid. Most can certainly accommodate distributed resources.

The good news is that there are solutions that allow utilities and their regulators to embrace distributed generation without industry restructuring. If the regulatory incentives are changed with revenue caps or balancing accounts, a Local Integrated Resource Planning/Energy Resource Investment Strategy process is employed to define the appropriate degree of distributed generation, and the net system benefits created by distributed resources are shared among the stakeholders, then all parties can prosper without conflict, tradeoff, or compromise. Indeed, each can be considerably better off than if nothing had changed—making more money with better service and lower risk.

If these reforms are adopted, we expect that in the period between rate cases, the utility will gain additional cash flow benefits. Although other customer classes will not enjoy the lower revenue requirements until the next rate case accounts for the reduction

in capital costs, this is an artifact of the tariff-setting process that already occurs today, and cannot be entirely avoided without a completely different method of price formation that does not look periodically to a "test year."

3.4.2 Implications for restructured electric utilities in states with restructuring

The implications for electric utilities in a restructured environment are quite different and far more complex. By April 2002, 17 of the United States had opted for retail customer choice, allowing their customers to choose their power providers.[46] To begin with, many states required legal separation of the unregulated and regulated utility business, and functional separation of generation and transmission. In several states, such as California and Massachusetts, the utility was required to divest itself of all gen-

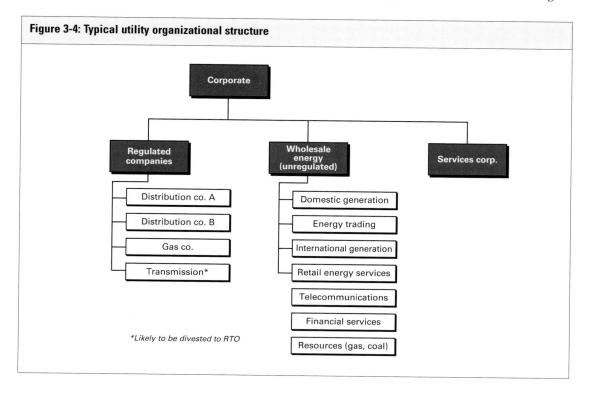

Figure 3-4: Typical utility organizational structure

Corporate
- Regulated companies
  - Distribution co. A
  - Distribution co. B
  - Gas co.
  - Transmission*
- Wholesale energy (unregulated)
  - Domestic generation
  - Energy trading
  - International generation
  - Retail energy services
  - Telecommunications
  - Financial services
  - Resources (gas, coal)
- Services corp.

*Likely to be divested to RTO

[46] See www.naruc.org for a listing of the regulatory status of all the states. Even in states that allow customer choice, the states still set the retail price structures based on the standard offer, or price to beat. Therefore, retail service providers can compete on only a limited set of service offerings such as onsite services or green power. Further, since the delivered retail prices have been very close to the standard offer in most states, only a small proportion of customers have actually switched.

erating units. Further, the FERC has ordered utilities to functionally separate the transmission business from the wholesale energy business and to group these transmission businesses into Regional Transmission Organizations (RTOs). Therefore, in the restructured world, a once vertically integrated utility will have many new businesses, and might typically have the organization structure shown in Figure 3-4.

The unregulated businesses include retail energy services and often wholesale power (which includes the generation and trading groups). The regulated businesses include the distribution company and the transmission company (the latter probably integrated into the RTO). The state PUC retains jurisdiction over the distribution company. The FERC regulates the transmission company rates and its operations, and retains market-power oversight over the wholesale power transactions of the generation company. The regulated businesses are prohibited from sharing non-public information or resources with the unregulated companies.

Since each of these companies has an entirely different competitive situation, we consider DG's implications for each of them separately.

3.4.2.1 Implications for the distribution company

Although the distribution company will remain regulated, senior management will expect the wires company to demonstrate profitability on a standalone basis. Indeed, due to separation of generation assets, there are several utilities that are solely wires companies. These companies will be managed to increase shareholder value.

Shareholder value is determined by free cash flow, not by simply by accounting profit (483). The two concepts are related, since free cash flow starts with net income and adds back non-cash items (such as depreciation, deferred taxes, equity AFUDC), then subtracts capital expense and cash taxes (290). Free cash flow is particularly important for correctly valuing capital-intensive business such as utilities, which have high depreciation that lowers accounting profits, but tend to have strong cash flow, which is primarily used to pay dividends. Indeed, the valuation of distribution businesses is almost perfectly predicted by the constant dividend growth model, indicating that investors view these stocks essentially as bonds with a growing annuity coupon. Further, the utility's bond ratings are defined by ratios that measure the cash coverage of interest payments.

Therefore, the corporate imperative is to manage the regulatory process not only to ensure recovery of costs, but also to increase free cash flow and dividends continually. This can be accomplished by underlying load growth (not a public policy goal!) plus efficiency improvements in operating and capital costs. In fact, the most profitable distribution utilities are those that are able to reduce operating and capital costs while maintaining service quality and acceptable levels of load growth. These distribution companies can avoid rate cases for seven years or more, and meanwhile, the shareholders receive the surplus cash from the capital or operating cost reductions.

There is a growing trend toward holding distribution utilities accountable for service and quality. The Performance Based Ratemaking (PBR) regulatory concepts provide financial incentives and penalties based on the utility's ability to achieve performance targets, which include both reliability measures (SAIDI and SAIFI, for example) and efficiency target benchmarks which are continually ratcheted down according to a set index (*e.g.*, CPI-X). Under PBR, utilities must manage both the timing and the trajectory of efficiency improvements to remain under the benchmark costs while maintaining performance levels. Given these circumstances, distributed generation will be a critical tool to manage both cost *and* service levels on the distribution grid.

3.4.2.1.1 Threats to the standalone distribution company

The same threats to revenues and profit margins that apply to the regulated company also apply to the standalone distribution company. However, the relative importance of the threat is far greater, because the standalone companies does not have other profit components. Using our early simplified example of the two-million-customer utility, the standalone distribution company selling 40,000 GWh/y would have realized revenues of $1.6 billion/y and net profits of $160 million/y. The cost structure is different, in that all purchased power is passed through and the distribution company only receives realized revenues equal to the distribution tariff, which would be approximately $0.04/kWh. Operating, maintenance, overhead costs, and depreciation will typically run $0.035/kWh, leaving approximately $0.005/kWh in profit. Therefore, the same instantaneous 5% penetration of distributed generation would lead—other things being equal—to lost revenues of $80 million, or a whopping 40% reduction in profits.

Again, these severe losses occur if distribution companies resist distributed power and persist with traditional regulatory practices, leaving distributed generation in an unmanaged state. But what happens if the distribution company were instead to embrace distributed generation?

3.4.2.1.2 Opportunities for the standalone distribution company

The same opportunities to defer system expansion investments that apply to the regulated utility will continue to apply here. What is different is that the relative value to the shareholder is magnified, due to the importance of cash flow to the standalone distribution company. In addition, the peculiar nature of U.S. restructuring creates two additional opportunities for distribution companies.

Distribution utilities do not necessarily lose their obligation to supply retail customers with power, despite the advent of customer choice. The distribution companies are the suppliers of last resort, obligated to serve those customers who do not switch. Since restructuring, very few customers have in fact switched, except in those few cases where the entire load has been auctioned off through a Basic Generation Service (BGS) auction.[47] Absent a successful BGS auction, the distribution utility will remain exposed to the power market's volatility and must still procure and deliver power. Typically, the distribution utility will purchase most of its obligation in long-term power supply agreements with generation companies (including the generation company belonging to the same holding corporation as the distribution company) and will be largely hedged. Nevertheless, distri-

[47] Basic Generation Service auctions are auctions held by the distribution utility for competitive suppliers to provide power to large blocks of customers, typically 50,000–100,000 or more. These auctions have been held in Pennsylvania, Ohio, and New England, and are planned in New Jersey. In general, less than 10% of the retail load has switched service providers outside of these auctions.

bution companies will typically have to procure some power from the spot market, particularly to meet unforeseen loads. While peak power prices tend to run about $40–60/MWh during most of the year, the prices have soared in almost every power market to $300–800/MWh or more during the highest peak periods—which is typically when distribution utilities are forced to go to the spot market to meet their load obligations. Distributed resources are an effective, predictable mechanism for utilities to acquire power (or shed load) rather than go to the spot-power markets. From the power-supply perspective, distributed resources are another form of hedge on the power markets, in the form of a call option, and are more under the company's own control, with minor and highly diversified technical risk of malfunction and with no counterparty underwriting risk (a real concern after Enron's collapse).

The magnitude of the spot-market exposure is great: a distribution utility can lose its entire retail profits for the year in a matter of days. Even if the state regulators allow the distribution company to pass through the purchased power costs (as in New York and New Jersey), the practical reality is that high costs will trigger a political reaction that subjects the shareholders to some degree of earnings exposure. Wall Street understands this reality. Utility bond ratings are penalized when their potential exposure to the spot market seems large, regardless of whether they are allowed to pass through these costs to customers (*e.g.*, GPU and ConEd). (228) And the risk of not being allowed to recover power-purchase costs is not just theoretical. In spring 2002, Sierra Pacific suspended its dividend and was threatened with insolvency when the Nevada PUC denied for impru-

dence the recovery of $437 million used to purchase unhedged spot-market wholesale power during the California crisis.[48]

Another opportunity for the distribution company is to use distributed generation to meet or exceed the performance targets set out in PBR at lower cost. As discussed in Section 2.3, distributed generation can improve system reliability by lowering the probability of system outages and helping to provide various ancillary services. Depending on the PBR targets, distributed generation may prove to be a highly profitable vehicle for realizing the additional revenues for meeting these targets.

In most states with restructuring, distribution companies are not permitted to own distributed generation. If regulators do not allow ownership, then the locational pricing tariffs—the distribution analogue of Alberta's locational pricing for transmission (§ 3.3.3.1.4)—can provide the price signals for third parties to site distributed generation where it is most needed. However, we believe that distribution companies should urge regulators to adopt our recommendation to allow such ownership, provided that the distribution companies create fair and open access to all competitors wishing to use the distribution grid (§ 3.3.4.3.4).

The economic benefit of distributed resources is far greater to the customers in restructured markets. By using distributed resources to reduce peak demands, distribution utilities *will change the market price for power during these peak periods* (§ 3.4.2.2.1). In essence, a reduction in peak demand lowers the price for all power purchased in the spot market in that hour. Therefore, the more MWh the utility needs to purchase after operating or buy-

[48] The FERC sets wholesale prices, but state commissions can still reduce or deny recovery if they find that the purchase was imprudent. Two conflicting legal theories (*Delaware* and *Pike County*) in state caselaw leave the extent of this state authority unsettled.

ing from distributed resources, the more money it will save by lowering the very high peak charges in the market. In essence, the greater the exposure, the higher the savings from embracing distributed resources. Since purchased-power savings accrue directly to customers, it is the customers that collectively gain from the ability of a small amount of distributed resources to leverage a substantial drop in price.

Distributed generation should be a strategic priority for distribution companies as one of the most effective ways to defer capital costs between rate cases, preserve system reliability, and hedge power market risks.

3.4.2.2 Implications for the generation company

In order to understand the implications of distributed generation for generation companies, we must first understand, at least in outline, how generation companies are valued and how the U.S. power markets work. Generation companies (gencos) own and operate electric generating plants, and are usually integrated with trading operations into a wholesale power division within a utility. Gencos' share prices depend on the market expectation of earnings growth, which is a combination of growth in both absolute MWh and the underlying margins associated with each MWh.[49] As market prices rise and fall, so do gencos' share values.

Shares in generation companies rose by an average of 215% during 2000–01 during the height of the California power crisis when power and gas prices across much of the U.S. were rising.[50] Indeed, many investment banks were making the case that the generation companies of vertically integrated utilities were worth more than the entire utility as an integrated company.[51] Therefore, they pressured utilities into spinning out their generation units into separate companies, as Southern and Reliant did, creating Mirant and Reliant Energy. The concept seemed justified at the time. Policy makers at the state and federal level called for hundreds of new power plants to be built in response to the alleged capacity shortage, with the National Energy Policy calling for between 1,300 and 1,900 new power plants—about one new power plant per week. California alone would supposedly require at least 5,000 MW of new capacity (504).

The sobering reality is that California was not short of physical generating capacity in the first place (§ 1.2.12.2), and that in any event, generation is a cyclical industry, prone to periods of overcapacity and undercapacity. When power market prices fall, the highly leveraged generation shares collapse, as they did in late 2001, falling over 50% from their prior peaks six month earlier[52] and much further in the first half of 2002.[53]

---

[49] For example, see research analyst reports on Calpine, Mirant, AES, and Reliant Energy from Morgan Stanley, Merrill Lynch Capital Markets, and Crédit Lyonnais Securities for optimistic projections of growth and the high resulting multiples on earnings. By April 2001, the price-earnings multiple for Calpine and Mirant had reached 28 and 22 respectively. To achieve this level of earnings growth to justify the valuations of the merchant energy sector, earnings would have to grow by 25–45% per year during 2001–04. Given the company projections of new plants, the implied generation margins were generation margins of $14–$18/MWh—far in excess of typical generation margins of $4–5/MWh.

[50] Based on stock close prices, adjusted for splits and dividends by AES, Calpine, Mirant, NRG, Reliant between 1 May 2000 and 1 June 2001.

[51] See the Morgan Stanley Dean Witter "sum of the parts" analysis by Kit Konlege for several U.S. utilities during 2000–2001, which routinely claimed that the generation and trading groups of integrated utilities should command at least 15–18× price/earnings multiples—far greater than the 10–11× multiples of typical vertically integrated utilities. www. morganstanley.com

[52] By 2002, the overcapacity in the generation market, combined with slowing economic growth, depressed power market prices, resulting in collapsing share prices and long term market expectations of growth. The price/earnings multiples for these same shares collapsed to 7–11× earnings—about the same as or worse than traditionally regulated companies (338, 542).

Naturally, the earnings of merchant generating units are very sensitive to peak power prices. In most power markets, 20–30% of the projected margins for new combined-cycle plants sited within a particular power pool are derived from the margins earned during the peak power period. Further, generators can increase earnings by 25–35% by strategically bidding between the power markets and the less liquid ancillary services markets (150). Therefore, the behaviors of the power and ancillary markets are absolutely critical to valuation of gencos. So how do we expect these markets to behave in the future?

The California crisis revealed a fundamental flaw with deregulated spot-power markets: sellers (generators or traders) have an unusual potential to increase market prices by withholding supply, because 1) buyers are short-run price-inelastic, since the social value of electricity is on the order of 100 times the marginal production cost, and 2) electricity cannot be cheaply stored in bulk (300, 368, 438). No collusion between sellers was necessary, as Joskow observes, since individual sellers had the economic incentive to withhold part of their generation portfolio up to the point where the marginal revenues from rising prices equaled the marginal costs from withheld production (365–6). Industry estimates are that California experienced over $9 billion in higher electricity costs during May 2000–June 2001, creating the largest interstate transfer of wealth in the shortest period of time in U.S. history (§ 1.2.12.2). (82, 233, 366, 558)

True, California's market rules exacerbated generators' market power by forcing the major buyers (utilities) to purchase their power on a spot basis from the state-created Power Exchange. Nonetheless, while politicians and regulators in other jurisdictions may believe that their market designs preclude generator market power, no market is immune. Most U.S. markets have load pockets created by congestion in the transmission grid—Boston, Dallas, Chicago, New York City, Wilmington (226, 509, 520, 551). Virtually all power markets have real-time balancing or ancillary services markets that essentially operate as spot markets (225, 510, 521, 550). A recent study by the consulting firm Mc-Kinsey suggests that virtually all the major power markets globally suffer from a similar vulnerability (70). They observe that when total capacity utilization within a given market approaches 85% of the installed capacity, peak power prices "fly up," reaching levels ten times the norm predicted by simple dispatch economics. So when power is in short supply, prices rise dramatically for both power and ancillary services.

The expectation is that power markets are cyclical. McKinsey observes that net reserve margin (the difference between the peak demand and the maximum available capacity) tends to decline in virtually all liberalized power markets after the advent of restructuring. The reason is straightforward. At the start of restructuring, there is excess capacity, so prices decline, typically by 20–40%, based on experience from Sweden, the UK, and even California. Utilities and independent power producers tend not to

---

[53] As Dynegy, Reliant Resources, and CMS Energy admitted in May 2002 that they had inflated reported revenues by sham transactions, and some had used Enron-style special-purpose vehicles to take debt off their balance sheets, their shares reached 52-week lows. Aquila, which denied the allegations, hit an 11-year low. Overleveraged Calpine, which had planned to add 70 GW of capacity (nearly 10% of U.S. total capacity today), cancelled 35 combined-cycle gas turbine orders from GE in March 2002 to save $3 billion over the next two years. GE reported plans to build 150 such units for U.S. power suppliers in 2003, down from 284 in 2001, and expected further shrinkage in 2004 (53).

build power plants when prices are low. Reserve margin, too, is a public good to which no individual firm has an incentive to contribute—quite the contrary. Thus, as load grows, net margins shrink, so prices rise. As prices rise, more plants are ordered and ultimately brought online. Since these are large, long-lived assets, the industry tends to create overcapacity, and prices fall. In fact, this exact cycle has occurred in California during 1998–2002 (§ 1.2.12.2), as it had done earlier in 1983–85 (§ 1.2.4).

Power markets that have a high dependence on hydroelectric or nuclear capacity will have greater volatility, because these large baseload units are vulnerable to drought or prolonged technical outages respectively. Power markets with mostly thermal capacity (coal, gas, or oil) will be comparatively more stable, but vulnerable to the underlying volatility of the fossil-fuel markets, particularly gas. For example, the U.S. Western power markets (WSCC) are largely dependent on hydro and backed up by gas. Hence these markets suffered far greater volatility for the same level of reserve margin than the Eastern and Southern U.S. markets, which have higher coal and nuclear shares.

Given these dynamics, genco strategy recommended by consultants and other industry pundits is to time both new capacity investments and asset sales to match the cycles of the market (71). The power markets, they say, may be down in 2002 due to overcapacity, but they will return to high prices in the future. (Of course, market timing is always difficult, especially given herd behavior). So how will distributed generation change this conventional wisdom?

3.4.2.2.1 Threats to existing generation business models

A revolutionary change is occurring within the U.S. power markets as customers are beginning to realize the power of managing their loads and harnessing distributed generation. The summer of 2001 was extraordinary because of the lack of blackouts or lofty peaks in power prices, even in such tight markets as California or New York City. As customers responded to higher prices and poor reliability, their own end-use efficiency, load management, and distributed generation added 50% more available power in both these markets than new central generation capacity added during 2000–01 (87, 435). Yet this is just a harbinger of the greater revolution that is now brewing.

That revolution comes from the *buyer's ability to change the market price* by harnessing the underlying option embedded in distributed resources and dispatching it into the market. The option inherent in end-user load is not simple interruption, since this can be used only infrequently (fewer than four times per month using 4-hour windows). Rather, it is the ability of commercial and industrial (C&I) customers to flex their net demand using distributed resources—both demand- and supply-side—in response to price signals or payments, coupled with their willingness to allow a third party to dispatch their negawatts or distributed kilowatts. The impact of changing the market price can be dramatic due to the very high degree of supplier price elasticity in the bidding process. Our research shows that if an additional 500 MW of dispatchable distributed resources had been available to California's default buyers in 2000, consumers would have saved *$1 billion*.[54]

[54] This analysis is drawn from an unpublished manuscript by Kyle Datta, Dan Gabaldon, and Isabelle Gecils written during their tenure at Booz, Allen & Hamilton, and is based solely on publicly available data from the California Power Exchange.

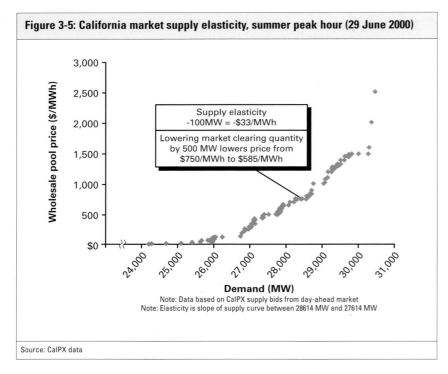

Figure 3-5: California market supply elasticity, summer peak hour (29 June 2000)

Supply elasticity
-100MW = -$33/MWh

Lowering market clearing quantity by 500 MW lowers price from $750/MWh to $585/MWh

Note: Data based on CalPX supply bids from day-ahead market
Note: Elasticity is slope of supply curve between 28614 MW and 27614 MW

Source: CalPX data

In essence, the supplier price elasticity (% change in price per 1% change in quantity) is extremely high at the end of the bidding stack—averaging 10–12 for the first 500 MW of reduced load (*i.e.*, a 10–12% lower price for each 1% of total quantity bid). The supplier elasticity flattens out sharply after the first 1,000 MW of reduced load is shed, dropping to a still-high but no longer astronomical value of about 4. The implication is that reducing load has a sharply decreasing marginal value. While shedding the first 500 MW can lower market price by $165/MWh, dropping the next 500 MW will lower price by only $85/MWh (Figure 3-5).

Our analysis of the California markets in 2000 shows that if utilities or the California's Department of Water Resources (which became the sole buyer after California investor-owned utilities' credit rating col-

lapsed in late 2000) had aggregated their disparate programs for distributed resources (roughly 500 MW) into 100-MW negawatt and/or kilowatt blocks capable of dispatch, they would have used these blocks almost 50 times over the course of the year, saving more than $1 billion. The savings would have occurred on the 45 specific days when prices soared, with the maximum daily savings reaching $80 million. Control technology and software that harness both the Internet and cellular telecommunications already exist to make this dispatch of capable distributed resources a reality.[55] While these would not necessarily have avoided all the blackout conditions on days when load was actually interrupted, it would have sharply reduced the ability of generators to impose high prices by strategic bidding and withholding supply (Figure 3-6).

While the California market experience represents an extreme in terms of both price spikes and value at risk, the same implications apply in every market. Even PJM, which maintains ample reserve margins (well over 15%), is not immune to price spikes; its peak prices rose to $800/MWh or more several times during 2000.[56]

The penetration of distributed generation will have a major impact on most power pool markets. In effect, the aggregated distributed generators act as "virtual peakers." Production cost modeling of several U.S. power markets tends to show that *just a 4% or greater penetration of distributed generation would effectively clip the peak*, eliminating price "fly-ups."[57] As a result, the average revenues earned from a new com-

---

[55] Refer to Silicon Energy's website at www.siliconenergy.com for a discussion of central dispatch of distributed technologies.

[56] PJM maintains comfortable reserve margins by requiring all retailers to procure adequate capacity and reserve margin to meet their loads, and penalizing retailers a capacity charge equal to roughly twice the carrying cost of a new combustion turbine. Despite these high reserve margins of 19%, peak prices ranged from $300 to $900/MWh during the summer of 2000.

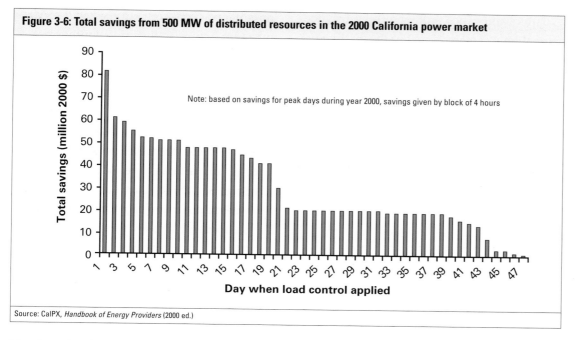

Figure 3-6: Total savings from 500 MW of distributed resources in the 2000 California power market

Note: based on savings for peak days during year 2000, savings given by block of 4 hours

Total savings (million 2000 $)

Day when load control applied

Source: CalPX, *Handbook of Energy Providers* (2000 ed.)

bined-cycle plant would drop by 10–15%, or roughly $6–9/MWh.[58] The number of hours run would fall by 15% as well. The combined effect would lower the total net free cash flow by 30–40%. Thus, the distributed generation penetration reduces the profits of new combined-cycle plants so much that they would simply not earn an adequate return on investment. Further, the profitability of utility gencos, which typically have a mix of coal, nuclear, and gas thermal plants, would fall by a stunning 15–25%.[59] This remarkable competitive leverage gives early adopters of distributed resources an important source of advantage.

Morever, the advent of distributed generation makes timing the market nearly impossible. A new thermal plant requires three to five years to site and permit. Even if the

plant developer invests the money and effort to gain approvals, and then waits to build until the market is right, there is still a two-year time lag for construction and plant shakedown. By then, the coveted market peak has probably passed. By contrast, stationary distributed generators can enter the market within 12–18 months; mobile distributed generators, within six months; dispatchable load management, probably in weeks to months. Since distributed generation and distributed demand-side resources act as a virtual peaker, developing combustion-turbine peakers is a far riskier investment decision—meriting a less favorable risk-adjusted discount rate (§§ 2.2.2–3).

Clearly, distributed generation would have a major long-term impact on power market

[57] These insights were developed by Kyle Datta from multiple analyses of the U.S. generation market during his tenure with Booz, Allen & Hamilton.

[58] The impact of distributed generation on power pool prices can be determined by applying production dispatch modeling to estimate the impact on energy prices, and capacity balance modeling to define the impact on capacity prices. Distributed generation lowers energy prices by lowering peak load and providing additional resources to sell into the peak power market. The lower peak load then creates higher reserve margins, which tend to lower the capacity prices. Thus, generator revenues suffer from both effects, which we are expressing in equivalent $/MWh.

[59] Profitability here is measured by the free cash flow earned from a mixed portfolio of nuclear, coal, gas combined-cycle, and gas combustion turbines, approximating the mix within the typical power pools outside of WSCC.

dynamics and genco profitability. Distributed generation is a major threat to the existing generation model. The conventional wisdom regarding centralized generation profitability and strategy is seriously obsolete in a distributed-resource world.

3.4.2.2.2 Opportunities for generation companies

Distributed generation is not, however, necessarily a complete replacement for centralized generation. Depending on what meanwhile happened on the demand side, there could still be a case for baseload and midmerit plants, and each choice of unit scale would need to be judged on its system-, time-, and site-specific merits. Distributed generation is a very effective technology for bringing peaking power to the market and for providing onsite high-quality power to C&I customers. Since onsite power is generally the business domain of the utility's

unregulated retail company, we will concern ourselves here with the opportunities presented for that business unit.

The ability to bring peak power to the market rapidly is very valuable in volatile power markets. As this section has demonstrated, distributed power, when used by load-serving entities to reduce their consumption at critical times, has the net effect of lowering peak power prices dramatically. If a generation company were to own distributed power in a power market where it has little or no generation, then the generator would use it as a peaker—essentially creaming off the value from the volatility as shown in Figure 3-7.

The economic question is whether distributed generation can be competitive in this virtual peaker role. Subject to wide variations, the capital cost for conventional distributed generation and microturbines typically falls within the range of $600–1,500/kW. Fuel cells are still far more expensive at $3,000/kW or more, though this is expected to decrease rapidly as production volume rises. Assuming a 20-year life for the equipment, the capital charge for these technologies is approximately 15%/y, so conventional technologies must earn $90–225/kWy, and fuel cells must earn ≥$450/kWy, in order to pay for the annual capital costs. By comparison, the carrying charges on a conventional combustion turbine are $45/kWy. (Of course, these comparisons are not risk-adjusted, and they should be.) Further, a 100-MW combustion turbine may be more fuel-efficient and have lower O&M costs than the smaller-scale distributed resources (though a midsized turbine, such as the 40-MW unit shown in Figure 1-23, may be more efficient still). So, in the years when generation supplies are

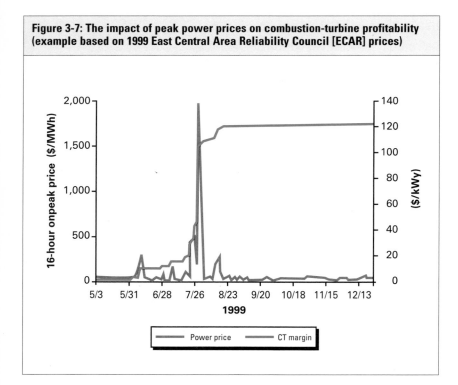

Figure 3-7: The impact of peak power prices on combustion-turbine profitability (example based on 1999 East Central Area Reliability Council [ECAR] prices)

short, the energy and/or capacity values available in the wholesale power market-place are likely to cover capital carrying charges on the order of $120–150/kWy. However, as noted earlier, large penetration of distributed resources will depress the market price, even in short years. Further, the 20-year lifespan of the peak resources will include many years when the power markets will have overcapacity and peaking units will be fortunate to earn $15/kWy. So, absent significant distribution value, the wholesale power markets will not be adequate to elicit confident genco investment in distributed generation as a stationary resource based on commodity kWh value alone.

The opportunity lies partly in recognizing that distributed generation can be a mobile resource. In this business model, the generator can either own the mobile distributed power or lease it to others. Caterpillar's Cat Rental Power division is an excellent example of the leasing model, and has been very profitable. Williams Distributed Power Services is another example of a new generation business based on placing rapidly deployable mobile generation units in high-value, short-term peaking markets.[60] The business concept is to provide reliable, just-in-time power supply to wholesale traders and large industrial users (760). This "Flex Peaking" concept serves both the wholesale and the retail markets. Williams believes this has several advantages over traditional merchant plants, including (337):

- higher capacity utilization

- no stranded investment

- higher reliability and dispatchability from multiple generator trains

- ability to take advantage of short term arbitrage across broad geography

Williams believes that WDPS will provide a major growth opportunity as a standalone business, and has already captured customers across the U.S. In essence, WDPS is capturing, on a somewhat larger unit scale, much of the portability benefit already described in Section 2.2.2.8.

3.4.2.3 Implications for trading

Wholesale energy trading can be a highly profitable enterprise. Despite the Enron debacle and the subsequent withering scrutiny of many energy traders, energy trading has proven to be the most profitable (in terms of margin) of all the utilities' new lines of business—though one might reasonably expect those margins to be increasingly arbitraged out as markets mature and competition spreads. Distributed generation will not change the way that energy traders make money.[61] Nor does distributed generation create much of a threat to the trading business, since most traders attempt to create a balanced book of business, so as to stay within the corporate value-at-risk limits. Of course, for the trader who goes long on generation supply in the power markets, we hope that the preceding discussion (§ 3.4.2.2.1) on implications of distributed generation to the power markets has been a cautionary tale. But rather than a threat, we believe that distributed generation creates a host of new opportunities for the trading business.

3.4.2.3.1 Opportunities for wholesale energy traders

What is lacking in wholesale power trading is the recognition that distributed generation can create a *fundamentally new business model based on lowering the market price: in essence,*

[60] In perhaps the ultimate kind of portability, Sierra Railroad's PowerTrain USA is offering 100 MW from 48 biodiesel-fueled 2.1-MWe locomotives under a 5-y California Power Authority "green power" contract. ABB inverters in commuter coaches match the voltage locally re-quired wherever the engines are deployed. It's like New York City's generator barges, but on rails (49).

[61] For a good overview of energy trading across all forms of energy, see *Energy Futures* (689), and the recently updated *Energy Futures: Past and Present* (690).

selling short. Distributed resources, particularly mobile distributed power, provide the trader with an excellent means of lowering the power price by reducing demand, or capturing the power price spike on the way down using distributed generation. The trader need not directly own the resources. The contractual relationships can be structured in many ways, for example, as call options with an upfront capacity payment. There are many forms of arbitrage that are possible with distributed resources. The arbitrage opportunities multiply when distributed generation is mobile and can be moved to where the trader believes there will be transmission constraints or generation capacity shortfalls leading to higher power prices. The scope of the arbitrage widens when renewable resources are brought into play, since these resources are hedges against volatile fossil-fuel prices (§ 2.2.3). For example, the trading value of distributed windpower can be determined by understanding how the deliverability of the windpower across the power transmission grid compares to the deliverability of gas across the gas transmission grid, and how both correlate with weather. As we have previously noted, the technology exists to both aggregate and dispatch large amounts of distributed generation and load management across multiple companies and geographies.

Needless to say, the business opportunities presented by widespread distributed generation are limited only by the creativity of the trader and the availability of resources in the market. The practical limitation has previously been the availability of dispatch-capable distributed resources in the market. Indeed, Williams Distributed Power is as much a trading play as it is a standalone new business model for generation. And the California

Energy Coalition model (§ 3.3.3.1.2) suggests a very large untapped opportunity for traders to encourage and reward the aggregation of dispatchable load management as virtual peakers.

3.4.2.4 Implications for retail electricity supply

Traditional business models for the unregulated retail business consist of energy commodity supply and services on either a regional or national scale. These are typically in separate business units with an integrated sales force effort around key accounts. Despite the great expectations for retail power and energy service companies, these business models are inherently low-value, for three reasons.

First, the energy commodity business is a 3–5% net margin business that has considerable risk in several jurisdictions due to the exposure to market structure of the deregulated power markets. Virtually all the retail service providers in California either went bankrupt or exited the market as power prices soared. Enron Energy Services even attempted to abrogate its long-term contracts, as it was unable to hedge its exposure despite a prodigious trading operation. Retailers face a difficult supply conundrum. If retailers are short of power, they are exposed to the volatile spot market for power and ancillary services. If they choose to purchase long-term contracts (generators need 15-year contracts to finance new plants), their customers typically sign up for supply agreements no longer than five years, so the retailer bears duration risk, much like the developer of a building who knows the tenancies will roll over several times before the investment is amortized. Alternatively, the retailer could purchase options on the spot market to hedge its exposure, but these options tend to be expensive, because the risk cannot be avoided. Hence, retailers face a margin squeeze.

Second, customer choice has turned out to be the big retail bust as the customers generally have not switched. In most jurisdictions, the regulators set a standard offer or price to beat as the benchmark. The delivered power price, however, has been so close to the price to beat that retailers are unable to provide their customers with significant discounts. Booz, Allen & Hamilton's research demonstrates that customers are generally unwilling to switch service providers unless a 10–15% discount is provided (149). As a result, switching rates have been low (less than 20%) in every state but Pennsylvania and New Jersey, which resorted to auctioning off large blocks of customers.[62] (149) Even green retailing has not proven successful. Although Green Mountain Energy has captured 20% of all customers that did switch, it remained an unprofitable business through 2001.[63]

Third, the energy services (esco) business has proven much more difficult to manage than expected. Industry players recognized the margin-squeeze problem in commodity energy retailing, and conventional wisdom was that the esco business was the solution. Although the services business can be an 11–15% net margin business, most players are only earning 6%, due to the fragmented buy-ing behavior of C&I customers.[64] Many players in this industry lost money because they attempted full service and/or national business models which have yielded higher fixed cost and lower staff utilization than projected.

Distributed generation is a strategic business opportunity for the retail energy business. Unlike the distribution business, retailers do not have default service obligations, hence distributed generation does not threaten them with lost revenues. On the contrary, it is the retail energy service companies that have been at the forefront of distributed energy and stand the most to profit from this trend. Where there is a mismatch of contractual periods between upstream and downstream cash flows, retail providers can lease portable resources with a suitably staggered portfolio.

3.4.2.4.1 Opportunities for retail energy businesses

Distributed generation presents three major opportunities for retail energy businesses. First, escos can offer premium power quality and reliability services to business customers using distributed resources. Second, distributed resources can perform the risk-hedging function for retail energy supply. Third, distributed power can form the bridge between wholesale and retail, creating an entirely new business model. This section discusses each of the business opportunities in turn.

Escos fully understand the value of distributed energy in providing premium reliability

---

[62] In a survey of over 1,000 business customers, the C&I switching rates were: Pennsylvania 48%, New Jersey 30%, Illinois 22%, California 13%, Massachusetts 13%, and Connecticut 8%.

[63] See S-1 filed by Green Mountain Energy in 2001. S-1 reports can be obtained at www.sec.gov/cgi-bin/srch-edgar.

[64] Refer to the annual reports of the publicly traded pure-play energy services companies Quanta and IES. In both cases, the retail energy business of providing demand-side management and onsite energy services was a 6% net margin business. These companies survived by expanding into telecom services.

services to customers. The electricity crisis of 2000–01 created a great deal of customer interest, particularly among customers with critical loads. Critical loads are defined by their high cost of business interruption, resulting in the perceived need for typically six "9"s reliability (but compare Section 2.3.3.8.2) and a high level of power quality. This level of service cannot be provided by the distribution grid, which provides at best four "9"s (99.99%) and in industrial countries, based on annual outage duration, averages about three "9"s (780). The value of the reliability premium to customers varies greatly (§ 2.3.3.8.2), but is largest for continuous process industries (such as refining, papermaking, or microchip manufacture) or data-intensive businesses (call centers, e-business). A few examples of business interruption cost illustrate this (160, 396, 407):[65]

- process industries (per episode >30 minutes): HP fabrication $30 million, Mobil Oil $10 million

- financial services (per hour): brokerage $6 million, credit card $2.5 million, banking $1.6 million

- call centers (per hour): airline reservations $0.9 million

These losses must be translated into opportunity cost per kW of peak backup by multiplying the expected opportunity cost of power outages in $/kWh by the expected outage frequency and duration per year. For example, ABB Energy Services performs these calculations to arrive at per a peak-kW opportunity cost of $240/kWy for petroleum refining (77).

The commercial problem is that providing 6 "9"s reliability is very expensive because of the generating unit redundancy and UPS power conditioning equipment needed. If the individual distributed generation unit is 99% reliable, then, in round numbers, three backup units are needed to achieve 99.9999% reliability (e.g., for every kW of peak load, 3 kW of distributed generation is needed configured as separate units).[66] Anecdotal information from escos suggests that the installed cost for "6-nines" distributed generation tends to run around $2,000/kW.[67] Although the carrying costs of $200–300/kWy are close to the reliability value boundary, the sheer upfront costs of these installations have retarded customer acceptance. Moreover, although what is actually needed (§ 2.3.3.8.2) is often not so much reliability as power quality, which can be much cheaper using onsite power conditioning, many customers do not seem to understand the difference, and equipment vendors may not want them to.

Once customers are obtained, the most difficult task facing retailers is managing the supply portfolio. Retailers are then in the same strategic situation as distribution companies. Thus, as described in the sections above, distributed generation has a critical role to play in taming the power markets and providing an effective hedge against high power price spikes during peak demand periods. At a minimum, retailers have the opportunity to aggregate distributed resources across their customer base and use this to manage their supply portfo-

---

[65] For further insight into reliability costs to industry, refer to the *Journal of Contingency Planning and Management* (223).

[66] Improving distributed generation unit reliability to 99.9% lowers the redundancy requirement to 2.5 kW/kW peak demand. The reality of distributed generation is that the units themselves are at best 99% available (e.g., a 1% forced outage rate) when used for peaking purposes, and lower when used as baseload. For example, the availability of microturbines drops to ~98% when used as baseload. Fuel cells can have a lower availability at 97%, unless specially designed. See Section 2.2.9.2.

[67] Industry interviews with esco service providers.

lios. The implication is that retailers should be advocates for the regulatory reforms in the bulk power markets that would allow for trading of negawatts and integration of distributed generation into the ancillary services markets.

Distributed generation may spur the development of more sustainable retail energy business models consisting of energy commodity supply, services, and trading linked together. This model is developed to make each of the business units profitable, and derives the dominant share of its revenue from the trading operations. Within the U.S., Williams Distributed Energy Services represents this new kind of thinking (Figure 3-8).

This business model has several implications. A successful trading model requires scale—in virtually all commodity trading operations the top three players within the market are profitable, while the remainder are marginal at best. Trading margins are highest on medium-sized customers. Thus the C&I retail operations must be designed to provide the trading group with enough scale to be profitable (when combined with the other residential retail positions). Thus C&I retail sales operations require either regional dominance or national presence to build scale, and then the right customer mix to balance volume and profit. Moreover, it will be the trading organization that ultimately capitalizes on the significant business value offered by aggregated distributed resources. Finally, the traditional esco business model of providing reliability and avoiding customer energy costs may not be a strong enough value proposition to support rapid penetration of distributed technologies. In essence, escos may have been

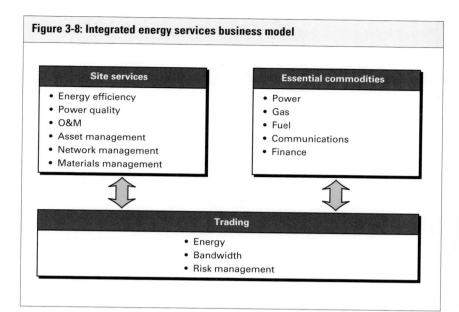

Figure 3-8: Integrated energy services business model

selling the right product to the right customers, but in the wrong way.

3.4.2.5 Organizational challenge: who should own distributed generation?

Restructured utilities wrestle with the organizational challenge of which business unit should own DG. The problem is that utilities must harness a wide spectrum of capabilities to prosecute a sustainable retail business model (Figure 3-9). Further, utilities must integrate the sales effort to avoid duplication and customer confusion. Today, these capabilities are resident in different business units, each with its own profit motive. Further, the unregulated units must be separated from the regulated distribution company. This raises an organizational challenge that has been faced by many utilities, and to which there is no single "right" answer.

The organizational solution will be formed by a combination of structure, process, alignment, and leadership. The structural solutions have proven to be the least important

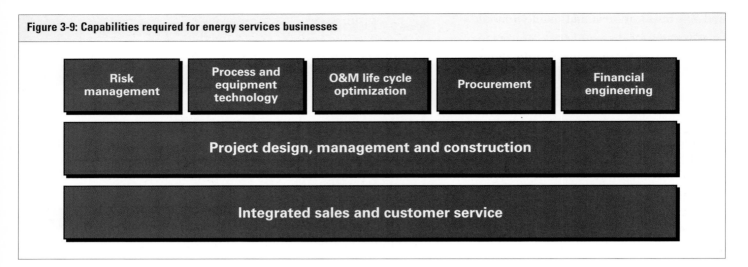

Figure 3-9: Capabilities required for energy services businesses

| Risk management | Process and equipment technology | O&M life cycle optimization | Procurement | Financial engineering |

Project design, management and construction

Integrated sales and customer service

of the three. The business process defines how the capabilities described in Figure 3-9 will be executed to achieve customer value. The alignment issues that must be addressed to harness these capabilities on an integrated basis include the sales force integration, transfer pricing, management incentives, and the matrix of profit/loss accounting. To a large degree, the answer will depend on the corporation's business focus. To the extent that the corporation is focused on its regulated business, distributed generation should reside in the regulated distribution company to capture the grid-side and generation supply benefits. To the extent that the corporate strategy is to grow the unregulated businesses rapidly, distributed generation should be integrated between trading and retail energy services.

3.4.2.6 Summary for restructured utilities

It should be abundantly clear by now that distributed generation is anything but business as usual. Forget the investment bankers, consultants, and other purveyors of conventional wisdom at high prices.

Distributed generation represents a game-changing threat to the conventional business models—and opportunity to outcompete them—in both the regulated and restructured environment. Distributed generation is indeed a disruptive technology that will restructure the entire electricity value chain.

There are several "key take-away" implications for utilities:

- Distributed generation is a major threat to traditional business models for distribution and generation companies. Utilities ignore this threat at their peril, as the lost revenues and potentially stranded assets could cripple the enterprise with surprising speed and thoroughness.

- Distributed generation represents an opportunity for some distribution companies to increase their valuations significantly by realizing higher than expected cash flows, if the regulatory disincentives are removed. Therefore, distribution companies should understand the full potential value of distributed generation to their system and change their regulatory strategy to advocate for the reforms proposed—most importantly, decoupling profits from sales volumes.

- Distributed generation linked with wholesale trading may be a breakout business play. There will be considerable first mover advantage, since the most strategic sites for distributed generation are limited.

- Organizational capabilities needed to understand, manage, and capture the value from this opportunity will need to be upgraded across the board. Since distributed technologies have a considerable learning curve, sooner is better. "Fast follower" responses will come too late.

3.4.3 Implications for capital markets

The implications for capital markets are profound. Distributed generation has the potential to change fundamentally the risk associated with the revenue streams for each business within the electrical power industry. Therefore, the valuation that Wall Street currently places on these companies could be wrong in both directions. Several key implications emerge from the prior discussion of threats and opportunities:

- The cash flow from regulated distribution companies may not be low-risk, as is currently believed. Even a 5% penetration of distributed power can create significant revenue losses. Further, distribution companies in constrained markets that have default obligations are exposed to the power markets. In the event of high power price episodes, regulators will question the prudence of utility actions, regardless of passthrough mechanisms. (Of course, distributed *generation* only heightens and makes more obvious the revenue risks already long present from the enormous overhang of unbought demand-side resources [468].)

- Distribution companies that embrace distributed power, and successfully remove the regulatory disincentives for

doing so, are likely to be increase their cash flows and valuation substantially relative to their peers. Distributed generation should be a strategic priority as one of the most effective means to defer capital costs between rate cases, preserve system reliability, and hedge power market risks.

- Distributed generation presents a major risk to generation company margins in every power market. Aggregated distributed resources collectively act as a "virtual peaker," reducing peak power prices and mitigating the potential for highly profitable price "fly ups." This further depresses long-run valuation on generation companies, which are already depressed because of 2002 oversupply in the power markets.

- Distributed generation can increase the profit potential of both the trading and retail energy services business, particularly when they are linked.

- There is a first mover advantage to commercializing distributed technologies, particularly in the wholesale power markets, as strategic sites are limited. Investors should reward the first movers accordingly and be suspicious of "fast follower" claims for equivalent growth.

- The valuation of distributed technology vendors should reflect the changes in the regulatory playing field and the rate of utility adoption of distributed generation. Like most new technologies, there are too many companies in the business space, so not all will succeed. As with e-commerce, hype is ultimately unhelpful for investors in this sector. Investors should value the companies with sound business models, and discount the rest.

- Distributed generation will create a new class of contractual relationships between the developer, the distribution utility, and the wholesale power markets. Financing is critical to the expansion of distributed technology, so lenders need to understand and correctly value the credit risk associated with these new financial instruments.

The regulatory recommendations provided earlier in this chapter should be viewed by Wall Street as milestones. As each group is adopted, a business window opens for the rapid expansion of distributed resources, especially in jurisdictions that correct the perverse incentive of rewarding energy sales rather than lower customer bills.

3.4.4 **Implications for public power**

Publicly owned utilities could be some of the most direct beneficiaries of the economic advantages of distributed generation and targeted demand-side management (DSM) programs. Public power involves a range of different types of entities, but the two most important are municipal utilities and rural cooperatives. In the U.S., there are roughly 2,000 municipal power utilities and power districts (many small) and nearly 1,000 rural cooperatives. Both offer promising opportunities for capturing distributed benefits.

Most communities have municipal utilities to provide water and wastewater services. Many also have municipal electric power utilities. Some of the largest include the Los Angeles Department of Water and Power and the Sacramento Municipal Utility District, both leaders (especially SMUD in photovoltaics) in creative and aggressive deployment of distributed resources. These and other large public power suppliers, such as the Salt River Project in Arizona, own generating capacity and produce a significant share of the power they sell to customers. Smaller municipal power utilities

buy most or all of their power from other sources and distribute it to customers. Some munis sell power only to other municipal departments.

Munis are among the types of utilities that are still subject to public policy oversight and regulation. Depending on the policy and regulatory framework, they are not necessarily forced to compete on price to keep customers, nor to maximize sales of kWh to earn a profit. Munis that do not own generation can be indifferent to the amount of power they buy and distribute, as it is simply a cost that they pass through to customers. For munis, therefore, it is relatively simple to design a regulatory framework in which they are rewarded for minimizing their customers' total cost of service, including power generation (or purchase), transmission and distribution, as well as environmental costs and other externalities.[68]

Munis that recognize the distributed costs and benefits enumerated in this book are bound to find distributed generation and targeted DSM to be attractive investments in many instances—especially where it can avoid unsightly and disruptive construction or upgrading of grid facilities in heavily populated areas. There is also an impressive tradition of technical and policy innovation among many munis. Some of the most successful DSM efforts, and those with the clearest and most remarkable benefits for local economic development, have been undertaken by small munis, such as that of Osage, Iowa, or larger ones, such as in Austin or Seattle—further supporting customer identity and loyalty. Munis may also be in a better political position than investor-owned utilities to coordinate their

[68] The comprehensive approach to least-cost distribution planning called LIRP or ERIS is a promising application of the ATS costing method (§ 2.1.4). For case studies, see *Local Integrated Resource Planning: A New Tool for a Competitive Era* (397).

distributed generation needs with reforms of public policy, such as in building codes.

Rural electric cooperatives (§ 2.3.2.11) proliferated after the creation of the Rural Electrification Administration in 1935 and the enactment of the Rural Electrification Act in 1936, with the goal of bringing electricity to millions of unelectrified farms and rural communities. Most rural electric cooperatives are owned by their residential and agricultural customers, to whom they provide distribution services. Also, many such distribution coops are themselves members and owners of generation and transmission cooperatives. Distribution coops are often still regulated entities, and most do not own generation facilities. Like munis, they are likely to benefit from a regulatory framework that takes distributed costs and benefits into account. More than any other class of electricity providers, most coops (except where suburban growth has caught up with them) have long distribution lines, very low load and revenue densities, aging populations, and other precursors of serious economic stress—a natural fit with distributed resources. Coops also have a special opportunity to avoid or defer relatively high costs of transmission and distribution line extension and the renovation of their aging lines and substations. For example (§ 2.3.2.11), one analysis showed that the cost of refurbishing distribution lines could more than double the cost of service for about 25% of the rural electric cooperatives in the U.S., giving these coops and their customers ample incentive to adopt DG promptly (321).

3.4.5 Implications for commercial and industrial customers

Commercial and industrial customers have an unparalleled opportunity to capture the benefits of distributed generation. The core of this book has provided insight into the types of value created by distributed generation, and its importance to the ability of utilities to keep prices and risks low, and for all stakeholders to tame the power markets. It is the C&I customers that should be lobbying hardest for the regulatory reforms presented, as they stand to benefit the most.

C&I customers face several challenges in adopting distributed generation. The initial costs of distributed generation may seem high, since C&I customers are, in effect, buying premium services and insurance. Current utility pricing can make the distributed generation decision uneconomic for energy savings alone, due to high backup charges, exit fees, and the like, coupled with relatively low buyback rates. Although the value of reliability may be high for some businesses, the need for reliability insurance is often cyclical, reflecting the cycles of the overall power market and distribution system investment. Therefore, the distributed generation assets purchased may have relatively low asset utilization over its lifetime, and correspondingly limited return on assets. While the investment decision will depend on site-specific economics, there are several implications for C&I customers as a class:

- *Understand and capture your site value.* The value of distributed generation depends on both location and time to market. Owners of strategic sites should recognize their value.

- *Seek aggregated business models.* Aggregation of distributed resources is far more valuable than standalone projects. The opportunity cost of segregating individual sites rather than managing them as a larger portfolio is high.

- *Inaction can increase exposure.* C&I customers are ultimately harmed by the inability to tame the power markets, and by higher distribution costs. Under traditional regulation and current business models, power prices will remain high. Under restructuring, volatility increases. Active management of energy costs and regulatory strategy should be a priority.

- *Active involvement can significantly increase profits.* Negawatts are valuable, as aluminum smelters recognized when they shut down Pacific Northwest smelters in 2001 and resold hundreds of MW of power into the soaring wholesale markets, making more profit in a few months than they would have made in a year of production.[69] Even when energy costs are a relatively modest proportion of the cost structure, their potential profitability can be far higher when harnessed as dispatch-capable distributed resources.

- *Getting the pricing right matters.* The current tariff structure based on system average prices may feel safe and stable, but it masks the true underlying economics and leads to poor investment decisions that business customers ultimately pay for through higher prices. Pricing that varies by location and time will provide the right price signals for sound investment decisions by both the utility and the C&I customer.

- *Reliability is your responsibility.* The distribution grid will be able to provide perhaps four "9"s levels of reliability, but the distribution company probably cannot sensibly upgrade to provide higher service levels. Therefore, higher levels of reliability will ultimately be the C&I customer's investment decision, and distributed resources typically the method of choice.

Ultimately, C&I acceptance of distributed generation will matter more than that of any other stakeholder. There is tremendous business value locked up in the current power system, and distributed generation can unlock it. The time to act on this opportunity is now, rather than waiting for the next energy crisis to remind us of the importance of managing this resource prudently.

3.4.6 Implications for real estate developers

Distributed generation must create value at every phase of the real estate development-ownership cycle: design/entitlement, finance, construction, marketing, operations, and disposition. Value begins with underwriting new acquisitions and entitling new development. It persists through securing critical debt and equity financing, construction of improvements, and project marketing. Efficient operations, building enhancements, the retention of tenants, and ultimately the capitalization of enhanced revenue upon disposition are also part of the value chain. To date, the modest economic benefits associated with distributed generation have been difficult for third-party owners and operators to realize and appreciate. Mitigating capital, technology, and operational risks is key. As a result, most deployment of distributed resources in commercial properties has been by end-users with a need for electrical and thermal output—*i.e.*, commercial and industrial customers.

Real estate developers and owners are both risk managers and opportunists, so they need a structure that reduces or eliminates capital outlays and protects owners from

[69] RMI had long urged the Bonneville Power Administration to hedge drought risks by using aluminum futures with physical delivery—an indirect form of bulk electricity storage. In a drought, the inventory could be sold, the workers furloughed, the power resold, and everyone's profits increased. Perhaps the 2001 experience will now lead BPA to take this suggestion more seriously.

technology risk. For distributed generation systems, a third-party development model is familiar to the real estate owner, and it places third-party capital and expertise between the real estate owner and any possible shortcoming with respect to new and often costly generating technologies. Of course, federal, state, and local governments could facilitate and accelerate the shift to a distributed combined-heat-and-power system, integrated with end-use efficiency, within a dispersed, diversified, and efficient competitive marketplace, thus multiplying benefits for property owners and communities.

Because the raw savings and revenues derived from distributed generation are modest when compared to other forms of building revenues, operational costs, and capital expenses, it is vital to aggregate the financial benefits and consider their impact on all these financial areas.

3.4.6.1 Challenges

Numerous challenges and barriers to entry face those who seek to deploy distributed resources, including the following:

Entitlements. The entitlements required for both installation and operation are often difficult to acquire:

- air—national and local standards are enforced by local air quality districts that issue permits and oversee air emission standards and testing

- building—local governments remain in tight control of engineering and construction standards, which include structural engineering, fire life safety, and electrical and mechanical design and execution

- utility interconnection (§§ 2.3.2.10 and 3.3.3.1.1)

- water and discharge (*e.g.*, cooling water)—where needed

Utility barriers to entry. As an incumbent monopoly, a utility is well positioned to protect its market through aggressive regulatory and legislative activities, in addition to exercising substantial and oftentimes irrational economic powers and imposing burdensome interconnection requirements (§ 3.3.3.1.1). The most viable and robust distributed energy systems are generally grid-interconnected, but this requires a complex series of life safety measures, and utility review and acceptance of the system sequencing and protocols.

Choice of technology and manufacturer. Unbiased information is rare on the many diverse distributed energy alternatives, and especially on how to combine, configure, and integrate them with existing HVAC and power infrastructure. Test data are spotty in quality and quantity.

Capital intensity. With few financing alternatives, those who wish to own or host distributed energy resources often face the expensive choice of self-funding a purchase with equity, which places all capital and operating risk on the host. Because the industry is maturing, equipment financing and third-party debt alternatives remain poorly organized and relatively expensive, though some manufacturers are developing financing options.

Building integration. Not only are there complexities in choosing the right technology, but combining technologies and integrating them successfully into the host (building) mechanical and utility infrastructure is critical and difficult. Scaling systems to mitigate risk and optimize operations requires accurate data, an acute understanding of building systems and operations, complex analytical capabilities, and considerable time and effort. The physical act of placing systems in buildings is often a "shoe-horning" exercise, requiring cooperation from property owners, an appreciation and knowledge of the built environment, its limitations (*e.g.*, statutory, structural, etc.), and occupants. This also requires knowledge of the financial obligations and agreements of building owners, including their debt and equity instruments and their development and lease documents. Such basic services as power, water and sewer, physical connection to the building HVAC plant or system(s), parasitic load, pump sizing/speed/location, metering and monitoring capabilities, and proper controls integration and sequencing are merely a short list of critical factors that affect the building's integrated distributed energy system.

Retention of revenue/savings. Historically, energy efficiency did not harm or reward building owners, but was simply capitalized by the owner, and the resulting savings passed through to tenants. One of the critical elements to creating a durable beachhead for distributed generation in the commercial property sector is a formula that rewards building owners, preserves the primacy of their relationship to tenants, adds to their suite of tenant services, and differentiates the property. A third-party development model, properly structured, can achieve these important goals.

Scaling. This is an issue both for individual systems and for their allocation over multiple locations, adjacent or otherwise.

Optimization of thermal applications and system operations. From ongoing maintenance to fuel procurement to reserve and replacement of equipment, the multiple skills and tasks associated with optimal operation of a power plant (of whatever size) are well outside the scope and core competence of commercial property owners. It is therefore essential to leverage off of service and product suppliers with resident financial and technical capabilities. Proper direction and an alignment of interests are also required.

Inefficient commodity purchasing. Owners are faced with a double challenge: purchase system fuel efficiently; provide for the physical purchase, scheduling and delivery; and effectively manage the building's residual loads.

Surplus sales. In select locations where surplus power can be sold back to the grid, owners might need or be able to sell power "off campus" to a broker or a power management entity. This might take the form of ancillary or balancing services, including VARs or voltage support, spinning and non-spinning reserves, frequency controls (regulation), and replacement reserves, often with "black start" capabilities to make the building's electric supply more secure. This is an opportunity, but also an umfamiliar complication.

3.4.6.2 Owner benefits

New and durable revenue source. If structured properly, hosting a distributed energy system can provide an excellent diversification of resources and source of revenue. Energy is a staple of building operations, and demand for onsite energy is ubiquitous. As mentioned above, the challenge is developing a system and structure that provides financial incentives to building owners while empowering them, if they wish, to share the financial benefits with tenants.

Peak demand/peak price load reductions. This results in effective load shaping and appears to the utility as demand-side management. It can also serve as an effective commodity management tool because the residual load of a property becomes more stable and constant, improving the property's economics and its owner's purchasing power.

Reduced grid uncertainty. Notwithstanding the general reliability of grid power, nothing protects owners from outages caused by mechanical failure, weather, or other force majeure events. Owners who augment their facilities' capacity to function through an interruption of utility services at best create a premium service for which they might be able to charge additional fees or rent or improve recruitment and retention, and at least protect tenants and property revenue from interruption.

Satisfied investor and tenant demand for energy solutions and management. Insurers, capital providers, and tenants are all focused on the cost of power failures and interrupted operations. Those protected against service interruptions will positively differentiate themselves and their property, and will often reduce the cost of operations associated with these key constituents.

Positive environmental statement. The marquée value of reducing a property's environmental footprint is easily translated into an economic windfall. Marketing is improved, properties are more eligible for grants and rebates, in some cases entitlements are more easily acquired, and tenant and community relations are improved.

Enhancement of HVAC infrastructure and capacity. As mentioned below, no-cost cogeneration or trigeneration infrastructure adds to a property's ambient conditions and value. Elimination or postponement of major capital costs for new HVAC capacity and/or replacement and enhancement (*e.g.,* associated with CFC displacement) is another major plus.

3.4.6.3 Tenant benefits

Standby services. These supplement UPS and support mission critical or general operations during a grid failure.

Comfort. Additional heating and cooling capacity augments distressed building electric infrastructure and HVAC systems. Ambient conditions are improved by new, more efficient systems.

Lower operating expenses. Load shaping supports lower commodity costs for residual load. New diversified plant with higher efficiency helps control or reduce common-area maintenance charges.

Reduced capital expenses. Reduced building investment in new equipment and capacity reduces passthrough expense of capital outlays (permitted under most lease structures over the useful life of the equipment).

Increased power quality. Though sometimes difficult to measure, power at the point of consumption eliminates grid power's losses, sags, and surges. Many distributed generation systems effectively act as a capacitor, or at the very least "average up" the quality of building power.

Environmental solution. The value to companies and employees who affiliate with environmentally responsible behavior is real. Corporate image and employee satisfaction are enhanced when the working environment is more efficient and designed to improve the community. This is in turn helps with staff recruitment, retention, motivation, and ultimately performance.

3.4.6.4 Systematic and societal benefits

Notwithstanding the challenges faced by property owners, the private capitalization and deployment of distributed generation resources has a positive local, regional, national and global impact:

- *Local* because onsite energy combined-heat-and-power systems diversify and multiply energy resources with more efficient and fungible technologies.

- *Regional* because onsite CHP clearly reduces demand on the electric transmission distribution system, thereby deferring and in some cases eliminating required maintenance and system improvements. Centralized generation and its associated distribution systems tend to be divisive and costly.

- *National* because the doubling and tripling of energy efficiency through on-site CHP reduces reliance on imported and expensive fuels. Additionally, and in addition to systematic diversity, small disperse plants are less likely to be targets of terrorist attack, thereby increasing overall security for those infrastructures and the activities and services that rely on those infrastructures.

- *Global* because the environmental benefit of clean CHP is so large.

3.5 Why Distributed Generation Matters to Every Citizen

Distributed generation is a disruptive technology, with an enormous (and enormously complex) range of hidden economic benefits. So what?

Why does distributed technology matter to the individual customer? After all, the electricity bill is a small part of the individual's monthly budget, far less than other essential services such as gasoline or telecommunications. Electricity crises are very rare events, and, even when they do happen, they are resolved within a matter of months: the "electricity crisis of 2001" was over by 2002. Power market prices are low, and the issue is off the media radar screen. Besides, decisions about distributed power are ultimately determined by utilities, their regulators and business customers, not individuals. So what is all the fuss about?

Distributed generation matters to each citizen for four simple reasons:

- First, there is a lot of money at stake. If distributed generators were to capture 6% of the power market, American customers would save $15 billion in energy costs alone, potentially twice that in avoided distribution costs—that's $180–360 per family *each year*.[70] And that doesn't even count the distributed benefits identified in Part 2!

- Second, the digital economy needs reliable power. Your family's Internet service, financial services, medical services, and more all depend on distributed power to serve you whenever required, without interruption. Distributed power's importance to our society is as pervasive as computing and telecommunications themselves.

- Third, distributed power makes the energy system more democratic. As a customer, you can increasingly choose what sources of electricity you want, whom you get it from, and whether to make it yourself. This real empowerment of the individual, household, firm, and community will ultimately lead to the adoption of cleaner, more reliable, and more accountable technologies.

- Fourth, distributed power makes the world safer and fairer. It is less inviting and rewarding to attack than vulnerable centralized systems are. It helps to make the energy system so resilient that major failures, accident or deliberate, become impossible by design. And it increases the availability and affordability of the kinds of local power systems that will bring the extraordinary benefits of electricity to the two billion people who don't yet have any, and the three billion who've never even made a telephone call.

And, as we noted at the beginning of this book, it's not just about distributed *generation*. Distributed *resources* also include the cheapest, fastest, most benign options of all—those on the demand side, which are such powerful and natural partners with dispersed, diverse, and especially renewable power supplies. While this book has focused largely on making electricity, how efficiently it is used is even more important and more valuable. As the economist Lord Keynes remarked, "If a thing is not worth doing, it is not worth doing well." The most perfect way to produce electricity is just a needless expense if the electricity is then wasted. The ultimate challenge, and opportunity, isn't merely to displace centralized with right-sized electric generators; it's first to *use* electricity in a way that saves money and improves our lives.

[70] In the analysis by Justin Colledge *et al.*, "Power by the Minute," (137) McKinsey consultants recognized that by shifting 5–8 percent of energy consumption to off-peak hours and cutting an extra 4–7 percent of peak demand altogether, the savings would amount to $15 billion per year. Since this analysis only looked at energy costs, not avoided distribution capital, the savings are probably twice as high.

As an individual customer, you can make a difference through the purchase decisions you make and how you express your political views. This book has made the case for the vast hidden value of distributed resources to the energy system. It is now up to you to take action to see that those benefits are recognized, monetized, and captured in the economic marketplace and throughout our society—and are made equitably available around the world.

Parenthetical reference numbers appear in blue throughout the text.

Long dashes listed after a publication in the Reference List signify repeated citations of that publication with different page numbers.

References 1–59

1 Abdukarim, A. J., and N. J. D. Lucas. 1977. "Economies of Scale in Electricity Generation in the United Kingdom." *Energy Research* 1: 223–231.

2 Adams, J. 1977. *The Appraisal of Road Schemes: Half a Baby Is Murder.* London: Geography Department, University College.

3 Aitken, D. 2002. Personal communication (11 July).

4 Alberta Energy Utilities Board. 2001 (May). www.eub.gov.ab.ca/bbs/new/newsrel/2001/nr2001-09.htm. Thanks to J. Edworthy for this reference.

5 Alderfer, R. B., T. J. Starrs, M. Monika Eldridge, PE. 2000. "Making Connections: Case Studies of Interconnection Barriers and Their Impact on Distributed Power Projects." SR-200-28053. Golden, CO: NREL (National Renewable Energy Laboratory), July.

6 Aldrich, R., EPRI. 1992. "The Distributed Utility: Strategic Implications in an IRP Setting." Distributed Utility—Is This the Future? EPRI, PG&E, and NREL conference, Berkeley, CA, 2–3 December.

7 Allison Gas Turbine Division of General Motors. 1993. *Research and Development of Proton-Exchange Membrane (PEM) Fuel Cell System for Transportation Applications: Initial Conceptual Design Report to Office of Transportation Technologies, U. S. Department of Energy.* EDR 16194 (30 November), p. C-4.

8 Almeida, A. *et al.,* "Source Reliability in a Combined Wind-Solar-Hydro System." *IEEE Transactions on Power Apparatus and Systems* PAS-102(6): 1515–1520 (June). Cited in Y. Wan and B. K. Parsons, *Factors Relevant to Utility Integration of Intermittent Renewable Technologies* (Golden, CO: NREL, 1993), p. 67.

9 Anderson, P. M. 1984. "The Effect of Photovoltaic Power Generation on Utility Operation." SAND84-7000 (February). Cited in Y. Wan and B. K. Parsons, *Factors Relevant to Utility Integration of Intermittent Renewable Technologies* (Golden, CO: NREL, 1993), p. 39.

10 Angel, J. 2001. "Emerging Technology: Energy consumption and the New Economy." *Network Magazine* (January 5). See also www.cool-companies.org/energy/.

11 An-Jen, S., J. Thorp, and R. Thomas. 1985. "An AC/DC/AC Interface Control Strategy to Improve Wind Energy Economics." *IEEE Transactions on Power Apparatus and Systems* PAS-104, no. 12 (December).

12 Anson, D. 1977. Availability Patterns in Fossil-Fired Steam Power Plants. FP-583-SR. Palo Alto, CA: EPRI (Electric Power Research Institute).

13 Aristotle. 1952. *Aristotle: II.* Great Books of the Western World. Edited by R. M. Hutchins and M. J. Adler. Verse translated by A. B. Lovins. Chicago: University of Chicago Press, chap. 7, p. 343, verses 26–29.

14 Arthur D. Little, Inc. 1999. "Distributed Generation: Policy Framework for Regulators." An Arthur D. Little White Paper. Boston, MA: Arthur D. Little, Inc.

15 ———, pp. 8–9.

16 Arthur D. Little, Inc. 2000. "Reliability and Distributed Resources." An Arthur D. Little White Paper. Boston, MA: Arthur D. Lttle, Inc.

17 Aschenbrenner, P.A. 1995. "Solar Electric Output vs. Average Residential Use." In D. Osborn, "Implementation of Utility PV: A Tutorial." Carbondale, CO: Solar Energy International (March), p. 3, Part II.

18 Aspen Institute for Humanistic Studies. 1979. "Summary Report: Decentralized Electricity and Cogeneration Options." Second Annual Energy R&D Priorities Workshop, Aspen, CO, 13–17 July. R. W. Fri and J. D. Sawhill, Co-chairmen.

19 Audin, L. 1996. "New Metal Halide Track Lights Cut Display Lighting Wattage." TU-96-2. Boulder, CO: E SOURCE (January).

20 Audin, L., D. Houghton PE, M. Shepard, and W. Hawthorne. 1994. *Lighting: Technology Atlas.* Boulder, CO: E SOURCE, p. 138.

21 Austin Energy. 2002. "AMD becomes largest subscriber of green power from Austin Energy." Austin Energy Press Release (February 26). www.austinenergy.com/press/amd.html.

22 AWEA (American Wind Energy Association). 2000. "Wind Farming: A New Cash Crop For Iowa." www.iowawind.org/factsheets/agriculture.pdf

23 AWEA. 1998. "FAQ: How Reliable are Wind Turbines?" www.awea.org/faq/reliab.html.

24 AWEA. 1998. "Island Province Study Finds Substantial Credit For Wind." www.awea.org/faq/cap.html

25 Awerbuch, S. 1993. "The Surprising Role of Risk in Utility Integrated Resource Planning." *Electricity Journal* (April).

26 Awerbuch, S. 1995. "Market-Based IRP: It's Easy!" *Electricity Journal* (April).

27 Awerbuch, S. 1995. "New Economic Cost Perspectives for Solar Technologies." In *Advances in Solar Energy, An Annual Review of Research and Development.* Edited by K. W. Boer. Boulder, CO: American Solar Energy Society (October).

28 Awerbuch, S. 1996. "Know Thy Costs: Improving Resource Selection Under Restructuring: Risk, Accounting and the Value of New Technologies." *Energy* (September): 3–7.

29 Awerbuch, S. 1996. *How to Value Renewable Energy: A Handbook and Action-Oriented Matrix for State Energy Officials* (draft edition 1.0). Interstate Renewable Energy Council (March).

30 ———, pp. 16–17, 77–85.

31 ———, p. 36.

32 ———, p. 23 n 4.

33 ———, p. 28.

34 ———, pp. 25–26.

35 ———, p. 37.

36 ———, pp. 13–14 and 65–71.

37 ———, pp. 87–88.

38 ———, pp. 66–67, n. 21.

39 ———, p. 47

40 Awerbuch, S. 1997. "The Virtual Utility: Some Introductory Thoughts on Accounting, Technological Learning and the Valuation of Radical Innovation," In *The Virtual Utility: Accounting, Technology & Competitive Aspects of the Emerging Industry.* Topics in Regulatory Economics and Policy Series, 26. Edited by S. Awerbuch and A. Preston. Boston: Kluwer Academic.

41 Awerbuch, S. 1997. Personal communication (11 August).

42 Awerbuch, S. 1997. Personal communication (13 August).

43 Awerbuch, S. 2001. "Investing in Distributed Generation: Technology is Not the Problem!" In *Decentralized Energy Alternatives.* Edited by E. Bietry and J. Donaldson, with J. Gururaja, J. Hurt and V. Mubayi. New York: Columbia University Press.

44 Awerbuch, S. 2002. *Estimating Electricity Costs and Prices: New Insights on the Effects of Market Risk and Taxes.* In production for June 2002. Paris: International Energy Agency.

45 Awerbuch, S., and A. Preston. 1993. "We Do Not Have the Measurement Concepts Necessary to Correctly Implement IRP: A Synthesis and Research Agenda." Draft paper. Billerica, MA: Mobil Solar Energy Corporation (August). Portions of this paper appear in the following publications: S. Awerbuch *et al.,* "The Virtual Utility: Some Introductory Thoughts on Accounting, Technological Learning and the Valuation of Radical Innovation," in S. Awerbuch and A. Preston, eds., *The Virtual Utility: Accounting, Technology & Competitive Aspects of the Emerging Industry* (Boston: Kluwer Academic, 1997); S. Awerbuch, "Editor's Introduction," *Energy Policy, Special Issue, "Valuing the Benefits of Renewables"* 24, no. 2 (February 1996); and S. Awerbuch *et al.,* "Capital Budgeting, Technological Innovation and the Emerging Competitive Environment of the Electric Power Industry," *Energy Policy* 24, no. 2 (February 1996).

46 ———, p. 16.

47 ———, p. 30, n. 20.

48 ———, p. 9.

49 Baard. 2002. "Choo-Choo Trains on Energy Crunch." *Wired* (3 July). www.wired.com/news/technology/0,1282,53591,00.html.

50 Baldwin, J. 1997. Personal communication (August).

51 Baldwin, S. F. 1986. "New Opportunities in Electric Motor Technology." *IEEE Technology & Society* 5:11.

52 Bancroft B., M. Shepard, A. B. Lovins, and R. C. Bishop. 1991. "The State of the Art: Water Heating." Snowmass, CO: RMI (Rocky Mountain Institute), September.

53 Barenson, A. 2002. "Power Giants Have Trouble Raising Cash for Plants," *N.Y. Times,* 19 May.

54 Barnes, P. R., J. W. Dyke, F. M. Van Tesche, and H. W. Zaininger. 1994. *The Integration of Renewable Energy Sources into Electric Power Distribution Systems. Volume I: National Assessement.* 6775. Oak Ridge, TN: ORNL (Oak Ridge National Laboratory), June.

55 ———, p. 30.

56 ———, p. 27.

57 ———, p. xv.

58 Barnett, D. L., and W. D. Browning. 1998. *A Primer on Sustainable Building.* RMI Pub. D95-2. Snowmass, CO: RMI.

59 Baughman, M. L., and D. J. Bottaro. 1976. "Electric Power Transmission and Distribution Systems: Costs and Their Allocation." *IEEE Transactions On Power Apparatus And Systems* 3:782–790.

References 60–141

60   Bentley, J., and B. Mitchell. 1995. "Fuel Cell Technology for Transportation Applications: Status and Prospects (Executive Summary)." Arthur D. Little (30 June).

61   Bergey WindPower Co. "Bergey Turbines Really are 'Tornado Tuff.'" www.bergey.com/Tornado.htm.

62   Berning, J., G. Booras, V. Longo, and C. Smyser. 1995. *Technical Assessment Guide (TAG[TM]) Volume 5: Distributed Resources.* TR-105124. Palo Alto, CA: EPRI (May).

63   ——, section 3.

64   ——, p. 3–29.

65   ——, section 3.3.

66   Bernstein, M. 2002. "Policy Issues in Energy, Security, Climate, and the Environment." Montreux Conference on Sustainable Mobility, Aspen, CO, 21 March.

67   Beyer, H. G., J. Luther, and R. Steinberger-Willms. 1989. "Power Fluctuations from Geographically Diverse, Grid Coupled Wind Energy Conversion Systems." European Wind Energy Conference, Glasgow, Scotland, 10–13 July.

68   Biewald, B. 1997. "Competition and Clean Air: the Operating Economics of Electricity Generation" *The Electricity Journal* 10(1), January.

69   Bird, L. and B. Swezey, NREL. 2002. "Estimates of Renewable Energy Developed to Serve Green Power Markets." Green Power Network website (January). www.eren.doe.gov/greenpower/new_gp_cap.shtml.

70   Birnbaum, L., J. M. Del Aguila, D. German Orive, and P. Lekander. 2002. "Why Electric Markets Go Haywire." *McKinsey Quarterly* 1.

71   ——, p. 21.

72   Bland, F. P. 1986. "Problems of Price and Transportation: Two Proposals to Encourage Competition from Alternative Energy Resources." *Harvard Law Review* 10:345, 386–387.

73   Boardman, R. W., R. Tatton, and D. H. Curtice. 1981. *Impact of Dispersed Solar and Wind Systems on Electric Distribution Planning and Operation.* Sub-7662/1. Oak Ridge, TN: ORNL (February).

74   Bower, W., Sandia National Laboratory. 2002. Personal communication (29 April).

75   Boyd, R., and R. Thompson. 1980. "The Effect of Demand Uncertainty on the Relative Economics of Electrical Generation Technologies with Differing Lead Times." *Energy Systems and Policy* 4 (1–2):99–124.

76   Brown, L. R. 2002. "World Wind Generating Capacity Jumps 31 Percent in 2001." *Earth Policy Institute Bulletin* (8 January). www.earth-policy.org/Updates/Update5.htm.

77   Brown, R. 2001. "Distribution Generation for High Value, High Reliability Service." Electric Utility Consultants, Inc. Distributed Generation Conference, Denver, CO, 8–9 August.

78   Bupp, I. C. 1981. *The Failed Promise of Nuclear Power: The Story of Light Water.* New York: Basic Books.

79   Bupp, I. C., and J. C. Derian. 1975. *Light Water: How the Nuclear Dream Dissolved.* New York: Basic Books.

80   Bureau of Reclamation. 1981. Wind-Hydroelectric Energy Project. Wyoming, Feasibility Report (June). Cited in Y. Wan and B. K. Parsons, *Factors Relevant to Utility Integration of Intermittent Renewable Technologies* (Golden, CO: NREL, 1993), p. 67.

81   Burke, A. 2000. "Ultracapacitors: Why, How and Where is the Technology?" *Journal of Power Sources* 91:37–50 (October).

82   Bushnell, J. 2000. "Dissecting California's Electricity Industry Restructuring Process." Presented to the National Association of Business Economists, Federal Reserve Bank of San Francisco, October.

83   *BusinessWeek.* 1980: "The Utilities are Building Small" (17 March), p. 148.

84   *BusinessWeek.* 1996. "Attention Pentagon Shoppers" (27 May), p. 128.

85   Bzura, J. J. 1995. "Photovoltaic Research and Demonstration Activities at New England Electric." *IEEE Transactions on Energy Conversion* 10: 169–174 (March).

86   Carroll, C. 2001. "PG&E and Santa Cruz Couple in Power Struggle." *San Jose Mercury News,* 30 July.

87   California Energy Commission, and W. J. Keese. 2000. *Supplemental Recommendation Regarding Distributed Generation Interconnection Rules.* Docket no. 99-dist-gen (2); CPUC docket no. R. 99-10-025 (October).

88   California Energy Commission. 2000. *Guidebook for Combined Heat and Power* (October).

89   Calwell, C. 1996. "Halogen Torchieres: Cold Facts and Hot Ceilings." TU-96-10. Boulder, CO: E SOURCE (September).

90   Cantley, M.F. 1979. "Questions of Scale." In *Options '79 #3.* Laxenburg, Austria: International Institute for Applied Systems Analysis, pp. 4–5.

91   Capstone Turbine Corporation. 2002. "Applications: Power Quality and Reliability." www.capstoneturbine.com/applications/power.asp.

92   Carbon Mitigation Initiative. www.princeton.edu/~cmi/.

93   Carruthers, J. 1996. Personal communication (9 October).

94   Casten, T., Chairman and CEO of Private Power LLC amd formerly of Trigen. 2002. Personal communication (June).

95   Caterpillar, Inc., Peoria, Illinois. 2001. "Caterpillar's Engine Division Offers Solutions to Critical Energy Situations." PR Newswire (2 February).

96   Cavallo, A. J. 1995. "Transforming Intermittent Wind Energy to a Baseload Power Supply Economically." *Journal of Solar Energy Engineering* 117: 137–143 (May).

97   Cavallo, A. J. 1996. "Transforming Intermittent Wind Energy to a Baseload Power Supply Economically." In *Power-Gen '96,* 4–6 December. Orlando, FL: Penwell.

98   Cavanagh, R. 1997. Personal communication (27 July).

99   Cavanagh, R. 2000. Re: The Western Electricity and Natural Gas Crisis. Memorandum from Mr. Cavanagh, NRDC Energy Program Director, on Energy Advocates List email (4 January).

100   Cazalet, E. G., C. Clark, and T. Keelin. 1978. *Costs and Benefits Of Over/Under Capacity in Electric Power System Planning.* Palo Alto, CA: EPRI (October).

101   Center for Resource Solutions. www.resource-solutions.org.

102   Chapel, S. W., L. R. Coles, J. Iannucci, and R. L. Pupp. 1993. "Distributed Utility Valuation Project Monograph." EPRI, NREL, and PG&E (August), p. 1.

103   ——, p. 106.

104   ——, p. ix.

105   ——, p. 109.

106   ——, p. 115.

107   ——, p. 118.

108   ——, p. 140.

109   ——, p. 2.

110   ——, p. 27.

111   ——, p. v.

112   ——, p. xiii.

113   ——, p. xviii

114   ——, pp. 114–115.

115   Chapman, C., and S. Ward. 1996. "Valuing the Flexibility of Alternative Sources of Power Generation." *Energy Policy* 24, no. 2 (February).

116   ——, p. 35.

117   ——, references omitted from cited passage.

118   Chauham, S. K., Technical Manager, Joe Wheeler Electric. 1997. Personal communication (2 January).

119   Chernick, P. L. 1988. "Quantifying the Economic Benefits of Risk Reduction: Solar Energy Supply Versus Fossil Fuels." American Solar Energy Society Annual Solar Conference, Cambridge, MA, 20–24 June.

120   ——, pp. 553–557.

121   Chernick, P. L. 1994. Testimony of Paul L. Chernick on behalf of the Massachusetts Office of Attorney General, 26 August. Exh. AG-PLC-3, p. 3.

122   ——, p. 22.

123   ——, p. 24.

124   Chinery, G. T., and J. M. Wood. 1985. "Estimating the Value of Photovoltaics to Electric Power Systems." 18th IEEE Photovoltaic Specialists Conference, Las Vegas, NV, 21–25 October.

125   Chinery, G. T., J. M. Wood, and A. L. Larson. 1987. "The Case for Photovoltaics in the Tennessee Valley." *Proceedings of the 19th IEEE Photovoltaic Specialists Conference,* 4–8 May, New Orleans, LA. Cited in Y. Wan and B. K. Parsons, *Factors Relevant to Utility Integration of Intermittent Renewable Technologies* (Golden, CO: NREL, 1993), p. 62.

126   Chowdhury, B. H., and S. Rahman. 1988. "Is Central Station Photovoltaic Power Dispatchable?" 88 WM 233-9. January 31–February 5. Cited in Y. Wan and B. K. Parsons, *Factors Relevant to Utility Integration of Intermittent Renewable Technologies* (Golden, CO: NREL, 1993), p. 42.

127   Christensen, C. M. 1997. *The Innovator's Dilemma: When New Technologies Cause Great Firms to Fail.* Cambridge, MA: Harvard Business School Press.

128   Clark, D. 2002. "Specific Companies are Targeted as Attacks on Computers Increase." *Wall Street Journal,* 28 January, p. B3.

129   Clemmensen, J. M. 1989. "A Systems Approach to Power Quality." Power Quality 1989 Conference. Cited in N. Lenssen, *Distributed Load Control: How Smart Appliances Could Improve Transmission Grid Operations* (Boulder, CO: E SOURCE, 1997). ref. 7.

130   Clemmer, S. 1995. "Fueling Wisconsin's Economy with Renewable Energy." Wisconsin Department of Administration, Energy Bureau. *Proceedings of the American Solar Energy Society's Solar '95 Conference,* Minneapolis, MN, 15–20 July. Updates a longer 1994 state report by S. Clemmer and D. Wicher, "The Economic Impacts of Renewable Energy Use in Wisconsin."

131   Clemmer, S., S. Olsen, and M. Arny. 1998. "The Economic and Greenhouse Gas Emission Impacts of Electric Energy Efficiency Investments: A Wisconsin Case Study." Wisconsin Department of Natural Resources. Prepared for the Energy Fitness Program of the U. S. Department of Energy and ORNL (February).

132   Cler, G. L. 1996. "The ONSI PC25 C Fuel Cell Power Plant." PP-96-2. Boulder, CO: E SOURCE (26 March).

133   Cler, G. L. 1999. "Packaging Distributed Energy Solutions." DE-9. Boulder, CO: E SOURCE (December).

134   Cler, G. L. and M. Shepard. 1996. "Distributed Generation: Good Things Are Coming in Small Packages." TU-96-12. Boulder, CO: E SOURCE (November).

135   Coelingh, J. P., B.G.C. van der Ree, and A.J.M. van Wijk. 1989. "The Hourly Variability in Energy Production of 1000 MW Wind Power in the Netherlands." In *Proceedings of the European Wind Energy Conference,* 10–13 July, Glasgow, Scotland. Cited in Y. Wan and B. K. Parsons, *Factors Relevant to Utility Integration of Intermittent Renewable Technologies* (Golden, CO: NREL, 1993), p. 46.

136   ——, p. 63.

137   Colledge, J., J. Hicks, J. B. Robb, and D. Wagle. 2002. "Power by the Minute." *The McKinsey Quarterly,* No. 1.

138   Collingridge, D. 1992. *The Management of Scale: Big Organizations, Big Technologies, Big Mistakes.* London: Unwin Hyman.

139   ——, p. 79.

140   Comtois, W. H. 1977. "Economy of Scale in Power Plants." *Power Engineering* (August): 51–53.

141   Continental Power Exchange. 1996. Press Release (31 October).

References 142–234

142 Cowart, R. 2001. "Efficient Reliability: The Critical Role of Demand Side Resources in Competitive Markets." Montpelier, VT: Regulatory Assistance Project for The National Association of Regulatory Utility Comissioners (June). www.rapmaine.org/rely.html.

143 Cyganski, D., J.A. Orr, A. K. Chakravorti, A. E. Emanuel, E. M. Gulachenski, C. E. Root, R. C. Bellemare. 1989. "Current and Voltage Harmonic Measurements and Modeling at the Gardner Photovoltaic Project." *IEEE Transactions on Power Delivery* 4(1): 800–809 (January). Cited in D. Shugar, *et al.*, *Benefits of Distributed Generation in PG&E's Transmission and Distribution System: A Case Study of Photovoltaics Serving Kerman Substation* (PG&E, 1992), p. B9.

144 Dagle, J., D.W. Winiarski, and M.K. Donnelly. 1997. "End-Use Load Control for Power System Dynamic Stability Enhancement." PNNL-11488. Richland, WA: PNNL (Pacific Northwest National Laboratory), February.

145 Danish Wind Industry Association. "Income from Wind Turbines." www.windpower.org/tour/econ/income.htm.

146 Danish Wind Industry Association. "Wind Energy News from Denmark." www.windpower.org/news/index.htm, downloaded 3 May 2002.

147 Danish Wind Industry Association. 2002. "21 Frequently Asked Questions About Wind Energy." www.windpower.org/faqs.htm#anchor295666

148 Darnell.com, Inc. 2001. "Uninterruptable Power Supplies: Global Market Forecasts, Energy Technologies and Competitive Environment." 3rd edn. www.darnell.com/services/01-ups_backup.stm.

149 Datta, K., and D. Gabaldon. 2001. "Risky Business: The Business Customer's Perspective on U.S. Electricity Deregulation." San Francisco, CA: Booz, Allen & Hamilton.

150 Deb, R. 2000. "Rethinking Asset Values in a Competitive Environment." *Public Utilities Fortnightly* (1 February).

151 Demeter, C. 1992. "Economic Impacts of a Photovoltaic Module Manufacturing Facility." Washington, DC: U. S. Department of Energy (7 May). Cited in H. Wenger, *et al. Photovoltaic Economics and Markets: The Sacramento Municipal Utility District as a Case Study* (SMUD, California Energy Commission, and USDOE PV Compact Program via NCSC, September, 1996).

152 Diesendorf, M. 1981. "As Long As the Wind Shall Blow." *Soft Energy Notes* 4(1):16–17 (February).

153 Donnelly, M. K., J. E. Dagle, and D. J. Trunowski. 1995. *Impacts of the Distributed Utility on Transmission System Stablility.* Richland, WA: PNL (Pacific Northwest Laboratory), January.

154 ———, pp. 101–106.

155 DSIRE (Database of State Incentive for Renewable Energy). 2002. "States with a Renewables Portfolio Standard" (April). www.ies.ncsu.edu/dsire/library/docs/RPS_map.pdf.

156 DSIRE. www.ies.ncsu.edu/dsire.

157 E SOURCE. 1999. *Providing Power Reliability and Power Quality Services with Distributed Generation.* EB-99-3. Boulder, CO: E SOURCE (March).

158 Eastwood, C. D. 1993. "Survey of Cost-Effective Photovoltaic Applications at US Electric Utilities." TR-102648. Palo Alto, CA: EPRI (August).

159 Edworthy, J. 1999/2002. Personal communications (May).

160 EEI (Edison Electric Institute) Economic Regulation and Competition Committee. 2000. *Unbundling Distribution Rates: A Technical Guide.* Washington DC: EEI (September). www.eei.org; www.rap.org.

161 EEI. 1997. *Statistical Yearbook of the Electric Utility Industry 1995.* Washington, DC: EEI, p. 14, table 8.

162 ———, p. 87.

163 ———, table 9.

164 EEI. 2000. *Statistical Yearbook of the Electric Utility Industry 1999.* Washington, DC: EEI.

165 ———, p 84.

166 EEI. 2001. *Statistical Yearbook of the Electric Utility Industry 2000.* Washington, DC: EEI, p. 76.

167 EEI. 2002. *Statistical Yearbook of the Electric Utility Industry 2001.* Washington, DC: EEI, p. 69.

168 EFI Inc. 2001. *Guidebook of Funds and Incentives for Distributed Energy Resources, Vol. 2.* www.efinc.com.

169 EIA (Energy Information Administration). 1994. "Form 860 data for 1994." File name: f86094.exe. ftp.eia.doe.gov/pub/electricity/.

170 EIA. 1995. "Household Energy Consumption and Expenditures 1993." File Grouping: Conel (July). www.eia.doe.gov.

171 EIA. 1995. Electric Power Annual 1994. Washington, DC: EIA, p. 65.

172 ———, p. 65n.

173 EIA. 1995. *Electric Industry Power Annual 1994, Vol II.* Washington, DC: EIA, pp. 65, 75, 76. www.eia.doe.gov.

174 ———, p.157.

175 EIA. 1995. "Performance Issues for a Changing Electric Power Industry." Washington, DC: EIA (January), pp. 2–13.

176 ———, fig. 8, p. 16.

177 EIA. 1996. *Annual Energy Review 1995.* Washington, DC: EIA (July), p. 229, diagram 5, converted at 3,413 BTU/kWh. www.eia.doe.gov.

178 ———, p. 231, table 8.1.

179 ———, p. 256, n. 1.

180 EIA. 1996. *Financial Statistics of Major U.S. Investor-Owned Electric Utilities 1995.* DOE/EIA-0437(95)/1. Washington, DC: EIA (December).

181 ———, table 28.

182 EIA. 1996. *Inventory of Power Plants in the United States as of January 1, 1996.* Washington, DC: EIA (December).

183 EIA. 1997. *Annual Energy Review 1996.* Washington, DC: EIA (July).

184 ———, p. 225.

185 ———, p. 277.

186 ———, table 3.1, p. 81.

187 EIA. 1997. *Renewable Energy Annual 1996.* Washington, DC: EIA (April), p. 72.

188 EIA. 2000. "EIA Electricity Profiles (State Information)." Washington, DC: EIA. www.eia.doe.gov/cneaf/electricity/st_profiles/maine/me.html#t1; www.eia.doe.gov/cneaf/electricity/st_profiles/new_hampshire/nh.html; www.eia.doe.gov/cneaf/electricity/st_profiles/california/ca.html.

189 EIA. 2000. *Annual Energy Review 1999.* Washington, DC: EIA (July). http://tonto.eia.doe.gov/FTPROOT/multifu-el/038499.pdf.

190 ———, p. 15.

191 ———, p. xxix.

192 EIA. 2000. *Electric Power Annual 1999, Vol. II.* DOE/EIA-0348(99)/2. Washington, DC: EIA (October). www.eia.doe.gov/cneaf/electricity/epav2/epav2.pdf.

193 ———, p. 10.

194 ———, p. 11.

195 ———, p. 9.

196 ———, pp. 5, 10.

197 ———, p. 12.

198 EIA. 2000. *Electric Power Annual 1999,* Washington DC: EIA.

199 EIA. 2001. *Annual Energy Outlook 2002.* Washington, DC: EIA (December). www.eia.doe.gov/oiaf/aeo/.

200 EIA. 2001. *Annual Energy Review 2000.* Washington, DC: EIA (August), p. 219. www.eia.doe.gov/emeu/aer/.

201 ———, p. 219.

202 ———, p. 247.

203 ———, p. 248 n. 1.

204 ———, p. xxix, fig. 48.

205 ———, p. 271.

206 EIA. 2001. *Electric Power Annual 2000, Vol. 1.* Washington, DC: EIA (August), p. 5. www.eia.doe.gov/cneaf/electricity/epav1/.

207 ———, p. 14.

208 EIA. 2002. *Electric Power Monthly.* DOE/EIA-0226. Washington, DC: EIA (February). ww.eia.doe.gov/cneaf/electricity/epm/epm_sum.html.

209 EIA. 2002. *Monthly Energy Review.* Washington, DC: EIA (February), p 97. www.eia.doe.gov/emeu/mer/.

210 EIA. 2002. *Monthly Energy Review.* Washington, DC: EIA (March), p. 97. www.eia.doe.gov/emeu/mer/

211 ———, p. 114.

212 El-Gasseir, M. M. 1992. *Molten Carbonate Fuel Cells as Distributed-Generation Resources.* Case Studies for the Los Angeles Department of Water and Power. TR-100686. Palo Alto, CA: EPRI (May).

213 ———, p. 4-8.

214 ———, p. 7-14.

215 ———, p. 7-17.

216 ———, p. 8-2.

217 ———, pp. 7-16–7-17.

218 ———, pp. 7-28 & 8-5; cf. p. 4-10.

219 ———, pp. 7-9–7-27.

220 El-Gasseir, M.M., Rumla, Inc. 1997. Personal communication (31 July).

221 *Energy Daily.* 1997. Washington International Energy Group study (5 February).

222 EPRI (Electric Power Research Institute). 1990. *Strategic Assessment of Storage Plants, Economic Studies.* GS-6646. Palo Alto, CA: EPRI (January). Cited in R. L. Ottinger *et al., Environmental Costs of Electricity* (New York: Oceana, 1991), p. 26, n. 43.

223 EPRI. 2000. "Customer Needs for Electric Power Reliability and Power Quality." *Journal of Contingency Planning and Management.* Palo Alto, CA: EPRI. www.epri.com/OrderableitemDesc.asp?product_id=1000428

224 Epstein, N. 1993. "Pioneers of Power." *Rural Electrification* 52(2):20–24 (November).

225 ERCOT. 2002. Procedures, Guides, and Criteria. www.ercot.com

226 ERCOT.com Newsroom. 2002. www.ercot.com

227 Fancher, R. B., *et al.*, Decision Focus, Inc. 1986. *Dynamic Operating Benefits of Energy Storage.* EPRI AP-4875. Palo Alto, CA: EPRI (October).

228 Faruqui, A., and K. Seiden. 2001. "Tomorrow's Electric Distribution Companies." *Business Economics* (1 January), pp. 54–62.

229 Feder, B. J. 2000, "Digital Economy's Demand for Steady Power Strains Utilities." *New York Times,* 3 July.

230 Federal Power Commission. 1971. *1970 National Power Survey. Vol. 4.* USGPO (U.S. Government Printing Office).

231 Federal Reserve Economic Database. 1997. "SP500 Historic Values" (August). www.stls.frb.org/fred/data/business/trsp500.

232 Felder, Frank A. 1996 "Integrating Financial Thinking with Strategic Planning to Achieve Competitive Success." *Electricity Journal* (May): 62–67 .

233 FERC (Federal Energy Regulatory Commission). Docket Nos. EL00-95, EL00-98, RT01-85, EL01-68.

234 Fialka, J. J. 2000. "Electric Power Grids' Reliability Erodes." *Wall Street Journal,* 13 January.

References 235–300

235 Fickett, A. C. W. Gellings, and A. B. Lovins. 1990. "Efficient Use of Electricity." RMI Pub. E90-19. *Scientific American* (September), pp. 64–74.

236 Firestone, D. 2002. "Board Votes to Restart Nuclear Plant in Alabama," *New York Times,* 17 May.

237 Fisher, J. C. 1979. "Optimum Size of Subcritical Fossil Fueled Electric Generating Units." In *Size and Productive Efficiency: The Wider Implications.* Laxenburg, Austria: International Institute for Applied Systems Analysis (June).

238 Fisher, J. C. 1979. "The Optimal Size of Subcritical Fossil Fueled Electric Generating Units." In *Scale In Production Systems.* Edited by J. A. Buzacott, M. F. Cantley, V. N. Glagolev, and R. Tomlinson. IIASA Proceedings Series, vol. 15. Tarrytown, NY/Oxford: Pergamon Press, p. 61–71, chap. 4.

239 Fisher, J. C. 1985. Letter to A. B. Lovins (15 September).

240 Flaim, T., and S. M. Hock. 1984. *Wind Energy Systems for Electric Utilities: A Synthesis of Value Studies.* SERI/TR-211-2318. Golden, CO: Solar Energy Research Institute (May). Cited in Y. Wan and B. K. Parsons, *Factors Relevant to Utility Integration of Intermittent Renewable Technologies* (Golden, CO: NREL, 1993), p. 63.

241 Flanigan, T. 1992. "Results Center Profile #9." Washington, DC: Renewable Energy Policy Project and the Center for Renewable Energy and Sustainable Technology. http://sol.crest.org/efficiency/irt/9.htm.

242 Flavin, C., and N. Lenssen. 1994. "Powering the Future: Blueprint for a Sustainable Electricity Industry." World Watch Paper 119. Washington, DC: Worldwatch Institute (June).

243 Ford, A. 1978. *Expanding Generating Capacity for an Uncertain Future: The Advantage of Small Power Plants.* Los Alamos National Laboratory (December).

244 Ford, A. 1979. Testimony of A. Ford to New Mexico Public Service Commission, case #1454, Santa Fe, NM, 30 March.

245 Ford, A. 1985. "The Financial Advantages of Shorter Lead Time Generating Technologies and the R&D Cost Goals of the Southern California Edison Company." Proprietary study prepared for Southern California Edison (May). Cited in W. R. Meade, D. F. Teitelbaum, "A Guide to Renewable Energy and Least Cost Planning" (Interstate Solar Coordination Council [ISCC], September 1989), p. 11, ex. 8.

246 Ford, A., and A. Polyzou. 1981. *Simulating the Planning Advantages of Short Lead Time Generating Technologies Under Irregular Demand Growth.* Los Alamos National Laboratory (3 July).

247 Ford, A., and A. Youngblood. 1982. "Simulating the Planning Advantages of Shorter Lead Time Generating Technologies." *Energy Systems and Policy* 6(4).

248 Ford, A., and A. Youngblood. 1983. "Simulating the Spiral of Impossibility in the Electric Utility Industry." *Energy Policy* (March).

249 Ford, A., and I. W. Yabroff. 1978. *Defending Against Uncertainty in the Electric Utility Industry.* Los Alamos National Laboratory (December).

250 Ford, A., and T. Flaim. 1979. *An Economic and Environmental Analysis of Large and Small Electric Power Stations in the Rocky Mountain West.* Los Alamos National Laboratory (October).

251 ———, ch. 10.

252 ———, p. 28.

253 ———, p. 37 fn

254 ———, pp. 9–11.

255 ———, pp. 30–31.

256 ———, pp. 31–32.

257 ———, pp. 32–33.

258 Ford, A., F. Roach, and M. D. Williams. 1984. *A New Look at Small Power Plants: An Update on the Environmental Issues of Interest to Italy.* Los Alamos National Laboratory (10 May), p.12.

259 Foster, D. 1996. "More Blackouts Predicted." Associated Press, 14 August.

260 Fryer, L. 1996. "The New Hampshire Retail Wheeling Pilot Program Learning Experience, Chaos, or Price War?" SM-96-9. Boulder, CO: E SOURCE (November), p. 5.

261 Fuchs, E. F. 1987. *Investigations on the Impact of Voltage and Current Harmonics on End-Use Devices and Their Protection.* DOE-RA-50150-23. DOE.

262 Fuldner, A. H. 1997. *"Upgrading Transmission Capacity for Wholesale Electric Power Trade."* Washington, DC: EIA (9 April). www.eia.doe.gov/cneaf/pubs_html/feat_trans_capacity/w_sale.html.

263 ———, p. 4-3.

264 ———, p. 4-7.

265 ———, table FE2.

266 Galloway C. D., and L. K. Kirchmayer. 1958. "Comment." *Transactions of the American Institute of Electrical Engineers* 39:1142–1144, fig. 4-3. Cited in A. Ford and T. Flaim, *An Economic and Environmental Analysis of Large and Small Electric Power Stations in the Rocky Mountain West* (Los Alamos National Laboratory, 1979), p 29.

267 Garman, D. 2001. Address to UPEx '01: The Photovoltaic Experience Conference, Sacramento, CA, 2 October.

268 George, K. L. 1996. *CELECT: Intelligent and Invisible Load Control.* PP-96-3. Boulder, CO: E SOURCE (September).

269 Gibe, P. 1992. "How Much Energy Does it Take to Build a Wind System in Relation to the Energy it Produces?" *Wind Energy Weekly* 521 (9 November). www.awea.org/faq/bal.html.

270 Gillette, S. 2001. "The Utility as Standby Power." *Capstone Turbine Cogeneration and On-site Power Production.* 2(6):42–47 (November/December). www.capstoneturbine.com/applications/cogeneration.asp.

271 Ginsburg, J. 2001."Reinventing the Power Grid: Companies Are Warming Up to Decentralized Production." *BusinessWeek* (26 February), p. 106.

272 Gipe, P. 2001. "Noise from Small Wind Turbines: An Unaddressed Issue." www.chelseagreen.com/Wind/articles/noiseswt.htm

273 Glynn, P. W., and A. S. Manne. 1988. "On the Valuation of Payoffs from a Geometric Random Walk of Oil Prices." *Pacific and Asian Journal of Energy* 2(1):47-48.

274 Goldberg, M. 2000. "Federal Energy Subsidies: Not All Technologies Are Created Equal." Renewable Energy Policy Project (July). www.repp.org/articles/static/1/1010437456_1008081187.html.

275 Government of Alberta. 2002. "How Transmission is Paid For." *New Power Generation in Alberta. Section 1: Opportunities for New Generation.* www.energy.gov.ab.ca/com/Electricity/Key+Publications/Key+Publications/New+Generation+Brochure+-+Section+1.htm. Thanks to J. Edworthy for this reference.

276 Grainge, J. J. and T. P. Mauldin. 1992. "The Distributed Utility: Aspects of Systems Integration." Distributed Utility—Is This the Future? EPRI, PG&E, and NREL conference, Berkeley, CA, 2–3 December.

277 Green-e, Renewable Electricity Certification Program. 2001. "Green-e Standard for Electricity Products." www.green-e.org/ipp/standard_for_marketers.html.

278 Grove, A. 1999. *Only the Paranoid Survive: How to Exploit the Crisis Points that Challenge Every Company and Career.* New York: Bantam Books.

279 Grubb, M. J. 1987. "The Integration and Analysis of Intermittent Source on Electricity Supply Systems." Ph.D. thesis, Kings College, University of Cambridge. Cited in Y. Wan and B. K. Parsons, *Factors Relevant to Utility Integration of Intermittent Renewable Technologies* (Golden, CO: NREL, 1993), p. 47.

280 ———, p. 48.

281 Haase, P. 1996. "Breakthrough in Stability Assessment." *EPRI Journal* (July).

282 Hadley S., L. J. Hill, and R. D. Perlack. 1993. "Report on the Study of the Tax and Rate Treatment of Renewable Energy Projects." 6772. Oak Ridge, TN: ORNL (December).

283 Halberg, N. 1990. "Wind Energy Research Activities of the Dutch Electric Generation Board." European Community Wind Energy Conference, Madrid, Spain, 10–14 September.

284 Halberg, N. 1990. "Wind Energy Research Activities of the Dutch Electric Generation Board." Paper presented at the EC Wind Conference (September). Cited in Y. Wan and B. K. Parsons, *Factors Relevant to Utility Integration of Intermittent Renewable Technologies* (Golden, CO: NREL, 1993), p. 63.

285 Hall, S., R. Gass, and S. Smith. 1992. *Avilability Assessment of Energy Research Corporation's 2-MW Carbonate Fuel Cell Demonstration Power Plant.* TR-101107. Palo Alto, CA: EPRI (September).

286 Hamrin, J., and N. A. Rader. 1994. "Affected with the Public Interest: Electric Utility Restructuring in an Era of Competition." Washington, DC: National Association of Regulatory Utility Commissioners.

287 Hawawini, G., and D. B. Keim. 1997. "Better Ways to Calculate Beta." *Financial Times,* 27 May, pp. 13–14.

288 Hawken, P., A. B. Lovins, and L. H. Lovins. 1999. *Natural Capitalism: Creating the Next Industrial Revolution.* Boston, New York, and London: Little, Brown and Company.

289 Hayes, W.C. 1978. "What Happened?" *Electrical World* 190(7):3 (1 October).

290 Hayward, D., and M. Schmidt. 1999. *Valuing an Electric Utility: Theory and Application.* Vienna, VA: Public Utilities Reports, Inc.

291 Heede, R. H. 1985. "A Preliminary Assessment of Federal Energy Subsidies in FY1984." Testimony to the U.S. House Subcommittee on Energy Conservation and Power. RMI Pub. CS85-7. Snowmass, CO: RMI.

292 Heede, R. H., and A. B. Lovins. 1985. "Hiding the True Costs of Energy Sources." *Wall Street Journal,* 17 September, p. 28.

293 Heffner, G., C. K. Woo, B. Horii, and D. Lloyd-Zannetti. 1998. "Variations in Time- and Area-Specific Marginal Capacity Costs of Electricity Distribution." *IEEE Transactions on Power Systems* 13: 560–565.

294 Henrichsen, E. N., and J. Nolan. 1981. *Dynamics of Single and Multi-Unit Wind Energy Conversion Plants Supplying Electric Utility Systems.* DOE/ET/20466-78/1 (August).

295 Herrera, J. I., J. S. Lawer, T. W. Reddoch, and R. L. Sullivan. 1986. Status Report on Utility Interconnection Issues for Wind Power Generation. DOE/NASA/4105-3 (June). Cited in Y. Wan and B.K. Parsons, *Factors Relevant to Utility Integration of Intermittent Renewable Technologies* (Golden, CO: NREL, 1993), p. 30.

296 Hilson, D. W., Sadanandan, N. D. *et al.* 1983. "Impact Assessment of Wind Generation on the Operation of A Power System," *IEEE Transactions on Power Apparatus and Systems* PAS-102, no. 9 (September), pp. 2905–2911. Cited in Y. Wan and B. K. Parsons, *Factors Relevant to Utility Integration of Intermittent Renewable Technologies* (Golden, CO: NREL, 1993), p. 30.

297 Hirsh, R. F. 1989. *Technology and Transformation in the American Electric Utility Industry.* Cambridge, MA: Cambridge University Press.

298 ———, preface.

299 Hirsh, R. F. 1999. *Power Loss: The Origins of Deregulation and Restructuring in the American Electric Utility System.* Cambridge, MA: The MIT Press.

300 Hirst, E. 2001. "Price-Responsive Retail Demand." *Public Utilities Fortnightly* (March).

References 301–369

301   Hoag, J. and K. Terasawa. 1981. "Some Advantages of Systems With Small Module Size." JPL 900-990, vol. 8, review draft, Economics Group. Pasadena, CA: CalTech Jet Propulsion Laboratory (June).

302   Hochanadel, J. R., and D. W. Aitken. 1996. "Utility Restructuring: Boon or Bane for Renewables?" *Solar Today* (November), pp. 23–25.

303   Hodge, G., and M. Shepard. 1997. "The Distributed Utility." E SOURCE Strategic Issues Paper IX. Boulder, CO: E SOURCE (June).

304   ———, p. 22.

305   Hoff, T. E. 1987. "The Value of Photovoltaics: A Utility Perspective." 19th IEEE Photovoltaic Specialists Conference, New Orleans, LA, 4–8 May, pp. 1145–1149.

306   Hoff, T. E. 1997. "Integrating Renewable Energy Technologies in the Electric Supply Industry: A Risk Management Approach." NREL/SR-520-23089. Golden, CO: NREL (March).

307   ———, pp. 3–4.

308   ———, p 5n.

309   ———, p. 10.

310   ———, p. 20.

311   ———, p. 42.

312   ———, p. 25, n. 17

313   ———, pp. 27–29.

314   ———, pp. 33–34.

315   Hoff, T. E. 1997. "Using Distributed Resources to Manage Risks Caused by Demand Uncertainty." PEG (May).

316   Hoff, T. E. 2000. "The Benefits of Distributed Resources to Local Governments: An Introduction." Draft report to NREL (12 September), pp.11–12. www.clean-power.com./research/distributedgeneration/DGandLocalGoverments.pdf.

317   Hoff, T. E., and C. Herig. 1997. "Managing Risk Using Renewable Energy Technologies." In *The Virtual Utility: Accounting, Technology and Competitive Aspects of the Emerging Industry.* Edited by S. Awerbuch and A. Preston. Boston: Kluwer Academic. www.clean-power.com./research/riskmanagement/mrur.pdf.

318   Hoff, T. E., and C. Herig. 2000. "The Market for Photovoltaics in New Homes Using Micro-Grids." Golden, CO: NREL (27 January). www.clean-power.com/research/microgrids/NewHomeMarketReport.pdf.

319   Hoff, T. E., and J. J. Iannucci. 1993. "Siting PV Plants: A Value Based Approach." 20th IEEE Photovoltaic Specialists Conference, Las Vegas, NV, 26–30 September, pp. 1056–1061. Cited in Y. Wan and B. K. Parsons, *Factors Relevant to Utility Integration of Intermittent Renewable Technologies* (Golden, CO: NREL, 1993), p. 64.

320   Hoff, T. E., and M. Cheney. 1999 "The Electric Co-op Market: Replacing Rural Lines

with PV." UPEx '99, Tucson, AZ, 5 October. www.clean-power.com/research/microgrids/ReplacingRuralLinesPV-UPEX99.pdf.

321   Hoff, T. E., and M. Cheney. 2000. "An Historic Opportunity for Photovoltaics and Other Distributed Resources in Rural Electric Cooperatives." www.clean-power.com.

322   Hoff, T. E., and M. Cheney. 2000. "The Potential Market for Photovoltaics and Other Distributed Resources in Rural Electric Cooperatives," *Energy Journal* 21(3):113–127. www.clean-power.com/research/microgrids/HistoricOpportunity.pdf.

323   Hoff, T. E., C. Herig, and R. W. Shaw, Jr. 1997. "Distributed Generation and Microgrids." 18th USAEE IAEE, September. www.clean-power.com/research/microgrids/MicroGrids.pdf.

324   Hoff, T. E., H. J. Wenger, and B. K. Farmer. 1996. "Distributed Generation: An Alternative to Electric Utility Investments in System Capacity." *Energy Policy* 24(2).

325   ———, pp.137–147.

326   Hoff, T. E., H. J. Wenger, C. Herig, and R. W. Shaw, Jr. 1997. "Distributed Generation and Micro-Grids." 18th USAEE/IAEE Conference, San Francisco, CA, September. www.clean-power.com/research/microgrids/MicroGrids.pdf

327   Hoff, T. E., H. J. Wenger, C. Herig, and R. W. Shaw, Jr. 1998. "A Micro-Grid with PV, Fuel Cells, and Energy Efficiency." Speech to ASES Solar '98, Albuquerque, NM, 16 June. www.clean-power.com/research/microgrids/MicroGrids2.pdf.

328   Hoffman, S. 1996. "Enhancing Power Grid Reliability." *EPRI Journal* 21 (November).

329   ———, pp. 6–15.

330   Hoffman, S., and B. Banerjee. 1997. "The Written-Pole Revolution." *EPRI Journal* 22(3): 27–34 (May–June).

331   ———, p. 32

332   Hogan, W. 2000. "Regional Transmission Organizations: Millennium Order on Designing Institution." Cambridge, MA: Center for Business and Government, JFK School of Government, Harvard University (May).

333   Hohmeyer, O., R. L. Ottinger, K. Rennings, eds. 1997. *Social Costs and Sustainability · Valuation and Implementation in the Energy and Transport Sector · Proceedings of an International Conference, Held at Ladenburg, Germany, May 27–30, 1995.* Heidelberg, Germany: Springer-Verlag.

334   Hollands, K. G. T., and J. F. Orgill. 1977. "Potential for Solar Heating in Canada." 77-01/4107-1, 2. Waterloo, Ontario: University of Waterloo Research Institute.

335   Holt, H. R. 1988. Personal communication (18 March).

336   Holy Cross Electric Association. 2000. Interview (September).

337   Houshmand, M. 2001. "Distributed Peaking: A New DG Concept." Electric Utility

Consultants, Inc. Distributed Generation Conference, Denver, CO, 9 August.

338   Howard, G. 2002. *Report on Calpine.* Crédit Lyonnais Securities (22 January).

339   Howe, B. 1993. *Distribution Transformers: A Growing Energy Savings Opportunity.* Boulder, CO: E SOURCE (December).

340   Howe, B. 1994. *Communications Protocols for Home Automation: From the Power Line to the Television.* TM-94-5. Boulder, CO: E SOURCE (August).

341   Howe, B. 1995. " Protecting Facilities From Power Outages." TU-95-2. Proprietary Technical Report. Boulder, CO : E SOURCE (April).

342   Howe, B. 1995. "Selecting Dry-Type Transformers: Getting the Most Energy Efficiency for the Dollar." TU-95-6. Boulder, CO: E SOURCE (August).

343   Howe, B. 1998. "Corporate Energy Managers Express Their Views in Third Annual E SOURCE Survey." Boulder, CO: E SOURCE (May).

344   Howell, D. G. President, Howell-Mayhew Engineering. Edmonton, Alberta, ghowell@ualberta.net, 780-484-0476. 2002. Personal communication (29 and 31 May). www.powerlight.com.

345   Howell, D. G., and G. M. Rekken. 1984. "Analysis of the Non-Economic Benefits of Low-Energy Passive Solar Housing—A Case Example," *Proceedings of the 10th Annual Conference of the Solar Energy Society of Canada Inc.* Alberta, Canada.

346   Huethner, D. 1973. "Shifts in Long Run Average Costs Curves: Theoretical and Managerial Implications." *Omega* 1(4).

347   Humm, O., and P. Toggweiler. 1993/96. *Photovoltaik und Architektur / Photovoltaics in Architecture: The Integration of Photovoltaic Cells in Building Envelopes.* Basel, Switzweland: Birkhäuser Verlag (Architectural).

348   ICF Consulting. 2002. "Economic Assessment of RTO Policy." Report to FERC (26 February). www.ferc.gov/electric/rto/mrkt-strct-comments/rtostudy_final_0226.pdf.

349   Idaho Power. 1993. *Questions and Answers About Solar Energy Systems From Idaho Power/ Consumer Information Pamphlet.* ESD-231. Idaho Power (August).

350   IEA (International Energy Agency). 2001. *Electricity Information 2001 with 2000 Data.* German system capacity and load taken from document; apparent Belgium tabular error corrected. Paris: IEA.

351   IEEE (Institute of Electrical and Electronics Engineers, Inc.) Distributed Resources and Electric Power Systems Interconnection (P1547). http://grouper.ieee.org/groups/scc21/1547/.

352   IEEE. 2000. *Recommended Practice for Utility Interface of Photovoltaic Systems.* IEEE Std 929-2000. IEEE.

353   Interstate Renewable Energy Council. 2002. "Status of Interconnection Rules for Renewables and Distributed Generation." Latham, NY: IREC. www.irecusa.org/connect/state-by-state.pdf

354   Jaffe, S. 2001. "Street Wise: Distributed Energy for Investing." *BusinessWeek* (2 July).

355   Jenkins A. F., R. A. Chapman, and H. E. Reilly. 1996. "Tax Barriers to Four Renewable Electric Generation Technologies." California Energy Commission (June). www.energy.ca.gov/development/tax_neutrality_study/.

356   Jenkins, N., R. Allan, P. Crossley, D. Kirschen, and G. Strbac. 2000. *Embedded Generation.* Power and Energy Series 31. London: The Institution of Electrical Engineers.

357   ———, p. 7.

358   ———, p. 13.

359   ———, p. 2.

360   ———, p. 4.

361   ———, p. 5.

362   Jensen, D. 1994. "Studies of the Environmental Costs of Electricity." OTA-BP-ETI-134 Office of Technology Assessment (September).

363   Jewell, W. T. 1986. "The Effects of Moving Cloud Shadows on Electric Utilities with Dispersed Solar Photovoltaic Generation." Ph.D. Thesis, Oklahoma State University, December. Cited in Y. Wan and B. K. Parsons, *Factors Relevant to Utility Integration of Intermittent Renewable Technologies* (Golden, CO: NREL, 1993), pp. 40–42.

364   Jewell, W. T., and T. D. Unruh. 1990. "Limits on Cloud-induced Fluctuation in Photovoltaic Generation." IEEE Transactions on Energy Conversion. 5(1):8–14 (March). Cited in Y. Wan and B. K. Parsons, *Factors Relevant to Utility Integration of Intermittent Renewable Technologies* (Golden, CO: NREL, 1993), p. 41.

365   Joskow, P. L. 2001. Statement before the Committee of Government Affairs, United States Senate.

366   Joskow, P. L., and E. Kahn. 2001. "A Quantitative Analysis of Pricing Behavior in California's Wholesale Electricity Market During the Summer of 2000." Washington, DC: AEI-Brookings Joint Center for Regulatory Studies (15 January). www.aei.brookings.org/publications/working/working_01_01.pdf.

367   Junger, P. 1976. "A Recipe for Bad Water: Welfare Economics and Nuisance Law Mixed Well." *Case Western Reserve Law Review* 27: 3–335 (Fall).

368   Jurewitz, J. 2000. "Fear and Loathing on the California Grid." Presentation to Harvard Electricity Power Group, 8 December. www.ksg.harvard.edu/hepg/papers.htm.

369   Kahn, E. 1977. Testimony Before the Board of Public Utility Commissioners, State of New Jersey. Construction Hearings Docket No. 762-194. Cited in A. Ford and T. Flaim, *An Economic and Environmental Analysis of Large and Small Electric Power Stations in the Rocky Mountain West* (Los Alamos National Laboratory, 1979), p. 3.

References 370–440

370 Kahn, E. 1978. *Reliability Planning in Distributed Electric Energy Systems*. LBL-7877. Berkely, CA: LBL (Lawrence Berkeley Laboratory).

371 ———, pp. 333ff.

372 ———, p. 33.

373 Kahn, E. 1979. *Project Lead Times and Demand Uncertainty: Implications for Financial Risk of Electric Utilities*. Berkeley, CA: LBL/University of California (8 March).

374 ———, p. 319ff.

375 Kahn, E., and S. Schutz. 1978. *Utility Investment in On-Site Solar: Risk and Return Analysis for Capitalization and Financing*. LBL-7876. Berkeley, CA: LBL.

376 Kammen, D. M., and R. Margolis. 1999. "Evidence of Under-Investment in Energy R&D Policy in the United States and Impact of Federal Policy." *Energy Policy* 27(10): 575–584.

377 Kaslow, T. W., and Pindyck, R. S. 1994. "Valuing Flexibility in Utility Planning." *The Electricity Journal* (March): 60–65.

378 Kelly, K. 1994. *Out of Control: The Rise of Neo-Biological Civilization*. New York, NY: Addison-Wesley.

379 ———, chap. 24.

380 ———, p. 470.

381 Kitamura, A. , H. Matsuda, F. Yamamoto, and T. Masuoka. 2000. "Islanding Phenomenon of Grid Connected PV Systems." In *28th PV Specialists' Conference Proceedings*. Anchorage, AK, September 15–22. IEEE, pp. 1591–4.

382 Knapp, K. E., and T. L. Jester. 2000. "An Empirical Perspective on the Energy Payback Time for Photovoltaic Modules." Solar 2000, Madison, WI, 16–21 June. www.pv.bnl.gov/keystone.htm.

383 Komanoff, C. 1977. "Comparative Economics of Nuclear and Fossil Fueld Generating Facilities." Testimony Before the New York State Public Service Commission, Case #26974. New York: Komanoff Energy Associates (28 October). Cited in A. Ford and T. Flaim, An Economic and Environmental Analysis of Large and Small Electric Power Stations in the Rocky Mountain West (Los Alamos National Laboratory, 1979), p. 3.

384 Komanoff, C. 1981. "Power Plant Cost Escalation: Nuclear and Coal Capital Costs, Regulations and Economics." New York, NY: Komanoff Energy Associates.

385 Koomey, J. G. *et al.* "Information Technology and Resource Use." Webpage at http://enduse.lbl.gov/Projects/InfoTech.html.

386 Koomey, J. G. *et al.* 2002. "Sorry, Wrong Number." *Annual Review of Energy and the Environment, Volume 27*, in press; also LBNL-50499. Berkely, CA: LBNL.

387 Koplow, D. 1996. Personal Communication (20 September).

388 Krepchin, I., and B. Howe. 1999. Storage Technologies for Ride-Through Capability. PQ-3. Boulder, CO: E SOURCE (May).

389 Krohn, S. 2000. "Seasonal Variation in Wind Energy." Danish Wind Industry Association website. www.windpower.org/tour/grid/season.htm

390 Krohn, S. 2002. "Birds and Wind Turbines." Danish Wind Industry Association website. www.windpower.dk/tour/env/birds.htm.

391 Ku, W. S., and N. E. Nour. 1983. "Economic Evaluation of Photovoltaic Generation Application in a Large Electric Utility System." *IEEE Transactions on Power Apparatus and Systems* PAS-102(8): 2811–2816 (August).

392 LADWP (Los Angeles Department of Water and Power). 2002."Tops in Nation—LADWP Green Power Program, Most Customers in U.S." LADWP Press Release, DWP News (March 13). www6.dwp.ci.la.ca.us/whatnew/dwp-news/031302.htm.

393 LBNL (Lawrence Berkeley National Laboratory). Heat Island Group webpage. http://eetd.lbl.gov/HeatIsland.

394 Leay, B. 1997. Personal communication (7 July).

395 Lee, S. T., and Z. A. Yamayee. 1981. "Load Following and Spinning Reserve Penalties for Intermittent Generation." *IEEE Transactions on Power Apparatus and Systems* PAS-100(3):1203–1211 (March). Cited in Y. Wan and B. K. Parsons, *Factors Relevant to Utility Integration of Intermittent Renewable Technologies* (Golden, CO: NREL, 1993), p. 38.

396 Leiter, D. 2000. "Distributed Energy Resources." Prepared by the U.S. Department of Energy for Fuel Cell Summit IV (May 10).

397 Lenssen, N. 1995. "Local Integrated Resource Planning: A New Tool for a Competitive Era." Boulder, CO: E SOURCE (November).

398 Lenssen, N. 1996. "Local Integrated Resource Planning: A New Tool for a Competitive Era." *The Electricity Journal* 9(6):26–36 (November).

399 Lenssen, N. 1997. *Distributed Load Control: How Smart Appliances Could Improve Transmission Grid Operations*. SM-97-3. Boulder, CO: E SOURCE (March).

400 Lind, R. C., K. J. Arrow, G. R. Corey, P. Dasgupta, A. K. Sen, and T. Stauffer, J. E. Stiglitz, J. A. Stockfisch, and R. Wilson. 1982. *Discounting for Time and Risk in Energy Policy*. Washington, DC: Resources for the Future, pp. 68, 102, 448.

401 ———, p. 63.

402 ———, p.251.

403 ———, p. 205.

404 ———, pp. 14–15.

405 Linden, H. R., Retired President of the Gas Research Institute, Chicago. 1997. Personal communication (7 July).

406 Lineweber, D. and S. McNulty. 2001. "The Cost of Power Disturbances to Industrial and Digital Economy Companies" (June).Palo Alto, CA: EPRI. http://ceids.epri.com/ceids/Docs/outage_study.pdf.

407 Lineweber, D. and S. McNulty. 2001. "The Cost of Power Disturbances to Industrial and Digital Economy Companies." *Primen* (July 29).

408 Lomax Jr., F. D., B. D. James, and R. P. Mooradian. 1997. "PEM Fuel Cell Cost Minimization Using 'Design for Manufacture and Assembly' Techniques." National Hydrogen Association's 8th Annual U.S. Hydrogen Meeting, 11–13 March, Alexandria, Virginia.

409 Lotspeich, C., and Y. Santo. 2002. "Liquid Electricity: Flow Batteries Expand Large-Scale Energy Storage Market." Proprietary report DR-18. Boulder, CO: E SOURCE (June).

410 Lovins, A. B. 1976. "Energy Strategy: The Road Not Taken?" *Foreign Affairs* 55(1) (October): 66ff.

411 Lovins, A. B. 1977. "Cost-Risk-Benefit Assessments in Energy Policy." *George Washington Law Review* 45 (August): 911–943.

412 Lovins, A. B. 1977. "Energy Strategy: The Road Not Taken?" *Friends of the Earth's Not Man Apart* 6(20):2–15 (November). Edited by Stephanie Mills.

413 Lovins, A. B. 1977. "Technology is the Answer! (But What Was the Question?)" In *Proceedings of the Conference on Future Strategies for Energy Development*, Oak Ridge Associated Universities, 20–21 October 1976.

414 Lovins, A. B. 1977. *Soft Energy Paths: Toward a Durable Peace*. New York, NY: Harper and Row.

415 ———, p. 86.

416 Lovins, A. B. 1978. *Annual Review of Energy and the Environment*.Palo Alto, CA: Annual Reviews.

417 Lovins, A. B. 1981. "Electric Utility Investments: *Excelsior* or Confetti?" *Journal of Business Administration* 12(2): 91–114 (Spring).

418 Lovins, A. B. 1982. "How To Keep Electric Utilities Solvent." *Energy Journal*.

419 Lovins, A. B. 1983. "The Fragility of Domestic Energy." *Atlantic Monthly* (November). RMI Pub. S83-8. www.rmi.org/images/other/S-FragileEnergy.pdf.

420 Lovins, A. B. 1985. "The Electric Industry." *Science* 229: 914–915 (6 September).

421 Lovins, A. B. 1986. "Should Utilities Promote Energy Conservation?" RMI Pub. U86-7. *Electric Potential* (March–April).

422 Lovins, A. B. 1986. "The Origins of the Nuclear Power Fiasco." RMI Pub. E86-29. *Energy Policy Studies* 3:7–34.

423 Lovins, A. B. 1988. "If Customers Save Electricity, Must Rates Rise?" RMI Pub. U88-16. Snowmass, CO: RMI.

424 Lovins, A. B. 1989. "The Great Demand-Side Bidding Debate Rages On." RMI Pub. U89-15. Snowmass, CO: RMI.

425 Lovins, A. B. 1989. "The State of the Art: Drivepower." Snowmass, CO: RMI/Competitek (April).

426 ———, p. xi–xiv.

427 Lovins, A. B. 1993. "Spotlight on Direct Access: Perspectives on DR Planning Under Competition." *DR Connection, EPRI* (November): 3.

428 Lovins, A. B. 1994. "Negawatts: Is There Life After the CPUC Order?" Keynote address to the National Association of Regulatory Utility Commissioners, Kalispell, MT, 16 May. RMI Pub. U94-17.

429 Lovins, A. B. 1995. "The Super-Efficient Passive Building Frontier." RMI Pub. E95-28. *ASHRAE Journal* 37(6):79–81 (June).

430 Lovins, A. B. 1995. Comments on FERC Mega-NOPR of 29 March 1995.RMI Pub. U95-37.

431 Lovins, A. B. 1995. Comments on FERC's Mega-NOPR. RMI Pub. U95-36 (July).

432 Lovins, A. B. 1995. Letter to the FERC (24 July).

433 Lovins, A. B. 1996. "Negawatts: Twelve Transitions, Eight Improvements, and One Distraction." *Energy Policy* 24(4):331–343 (April). RMI Pub. U96-11. www.rmi.org/images/other/E-Negawatts12-8-1.pdf.

434 Lovins, A. B. 1996. "Hypercars: The Next Industrial Revolution." RMI Pub. T96-9. Snowmass, CO: RMI. www.rmi.org/images/other/HC-NextIndRev.pdf.

435 Lovins, A. B. 1998. *Negawatts for Fabs: Advanced Energy Productivity for Fun and Profit*.RMI Pub. E98-3. Snowmass, CO: RMI. www.rmi.org/sitepages/pid171.asp.

436 Lovins, A. B. 1999. "Profiting from a Nuclear-Free Third Millenium." *Power Economics* (November). www.rmi.org/images/other/E-ProfitNukeFree.pdf.

437 Lovins, A. B. 2000. "Hypercars: Uncompromised Vehicles, Disruptive Technologies, and the Rapid Transition to Hydrogen." Redefining the Global Automotive Industry: Technologies and Fuels for the Future, 16 June, Washington, DC RMI Pub. T00-26. www.rmi.org/images/other/HC-UncompVehDisTech.pdf

438 Lovins, A. B. 2001. "California Dreaming." *American Spectator* (May).

439 Lovins, A. B. 2001. "California Electricity: Facts, Myths, and National Lessons." Address to Worldwatch, Aspen, CO, 22 July. RMI Pub. U01-2. www.rmi.org/images/other/E-WorldwatchPPT.pdf. Slightly updated from Commonwealth Club of San Francisco address, 11 July. www.rmi.org/images/other/E-CwealthClub.pdf.

440 Lovins, A. B., and B. D. Williams. 1999. "A Strategy for the Hydrogen Transition." National Hydrogen Association's 10th Annual U. S. Hydrogen Meeting, Vienna, VA, 7–9 April. RMI Pub. T99-7. www.rmi.org/images/other/HC-StrategyHCTrans.pdf.

References 441–523

441 Lovins, A. B., and D. Yoon. 1993. "Renewables in Integrated Energy Systems." Keynote address to the ANZ Solar Energy Society Solar '93 Conference, Perth, W. Australia (by PictureTel). Snowmass, CO: RMI.

442 Lovins, A. B., and L. H. Lovins. 1982. *Brittle Power.* Andover, MA: Brick House. Reposted with related readings at www.rmi.org/sitepages/pid533.php.

443 ———, appendices 2–3.

444 ———, appendix 2.

445 ———, chap. 13.

446 ———, chap. 14 and app. 1.

447 ———, p. 215ff.

448 ———, p. 218.

449 ———, p. 219.

450 ———, p. 220.

451 ———, p. 221.

452 ———, p. 222.

453 ———, p. 231.

454 ———, p. 233.

455 ———, p. 269.

456 ———, p. 332.

457 ———, p. 353.

458 ———, p. 359.

459 ———, p. 69.

460 ———, pp. 221–224.

461 ———, pp. 223–234.

462 ———, pp. 275–277.

463 ———, pp. 339–340.

464 ———, pp. 350–351.

465 ———, pp. 360–361.

466 ———, pp. 372–375.

467 Lovins, A. B., and L. H. Lovins. 1984. "Reducing Vulnerability: The Energy Jugular." In *Nuclear Arms Ethics, Strategy, Politics.* Edited by R. J. Woolsey. San Francisco: ICS Press. RMI Pub. S84-23. www.rmi.org/images/other/S-EnergyJugular.pdf.

468 Lovins, A. B., and L. H. Lovins. 1997. *Climate: Making Sense and Making Money.* Snowmass, CO: RMI (13 November). www.rmi.org/images/other/C-ClimateMSMM.pdf.

469 Lovins, A. B., and L. H. Lovins. 2002. "Energy Forever." *American Prospect* 13(3):30–34 (11 February). RMI Pub. E02-1a. www.rmi.org/images/other/E-EnergyForever.pdf.

470 Lovins, A. B., and L. H. Lovins. 2002. "Mobilizing Energy Solutions." *American Prospect* 13(2):18–21 (28 January). RMI Pub. E02-1. www.rmi.org/sitepages/pid171.php.

471 Lovins, A. B., and R. Sardinsky, R. 1988. "The State of the Art: Lighting." Snowmass, CO: RMI/Competitek (March).

472 Lovins, A. B., L. H. Lovins, and E. von Weizsäcker. 1997. *Factor Four: Doubling Wealth, Halving Resources.* London, UK: Earthscan Publications Limited, pp. 158–161.

473 ———, p. 165–176.

474 Lovins, A. B., M. M. Brylawski, D. C. Cramer, and T. C. Moore. 1996. *Hypercars: Materials, Manufacturing, and Policy Implications.* Snowmass, CO: RMI (March).

475 Ma, F. S. and Curtice, D. H. 1982. "Distribution Planning and Operations with Intermittent Power Production." *IEEE Transactions on Power Apparatus and Systems* PAS-101(8):2931–2940 (August). Cited in Y. Wan and B. K. Parsons, *Factors Relevant to Utility Integration of Intermittent Renewable Technologies* (Golden, CO: NREL, 1993), p. 70.

476 MacArthur, D. 2002. "Success of Britain's NETA Raises Concerns about Ontario Power Market." In *Canadian Natural Gas Market Report* 18(3):3ff (3 April).

477 MacFarlane, T. W. B., Ontario Hydro. 1997. Personal communication (11 July).

478 Mainzer, E. 2000. Telephone interview with Mr. Mainzer, Manager, West Power Structuring, Enron North America, 121 SW Salmon, 3WTC-0306, Portland OR 97204 (13 October).

479 Marchetti, C. 1975. "Geoengineering and the Energy Island." In *Second Status Report on the IIASA Project on Energy Systems.* RR-76-1. Edited by W. Hafale. Laxenburg, Austria: International Institute for Applied Systems Analysis.

480 Mariyappan, J., J. Gregerson, J. Kreider, and P. Curtis. 2001. "Enabling Distributed Generation: Real Time Control and Communications." DE-14. Boulder, CO: E SOURCE (March).

481 Maycock, P. 2001. Personal communication (9 April).

482 McArthur, D. 2002. "Success of Britain's NETA Raises Concerns about Ontario Power Market." In *Canadian Natural Gas Market Report* 18(3): 3ff (3 April).

483 McKinsey & Co., Inc., T. Kopland, T. Koller, and J. Murrin. 2000. *Valuation: Measuring and Managing the Value of Companies.* Third Edition. New York, NY: John Wiley & Sons.

484 Messing, M., H. P. Friesema, and D. Morrell. 1979. *Centralized Power: The Politics of Scale in Electricity Generation.* Cambridge, MA: Oelgeschager, Gunn & Hain, pp. 204–206.

485 ———, p. 206.

486 Metz, W. D. 1978. "Energy Storage and Solar Power: An Exaggerated Problem." *Science* 200:1471–1473.

487 Meyer, N. 2000. "Renewable Energy in Liberalised Energy Markets." EuroSun Conference, Copenhagen, Denmark, June.

488 Meyer, R. 1996. "Distributed Generation: What's the Role of Natural Gas." EPRI Second Annual Distributed Resources Conference, Vancouver, Canada, 6 November. TR-107585. www.epri.com.

489 Meyers, E., and M. Hu. 2001. "Clean Distributed Generation: Policy Options to Promote Clean Air and Reliability." *Electricity Journal* (January/February): 92.

490 ———, p. 94.

491 Ministry of Commerce of New Zealand. 1998. *Auckland Power Supply Failure 1998: The Report of the Minsiterial Inquiry into the Auckland Power Supply Failure* (July). www.med.govt.nz/inquiry/final_report.

492 Moore, T. C. 1996. "Ultralight Hybrid Vehicles: Principles & Design." RMI Pub. T96-10. Snowmass, CO: RMI.

493 Mooz, W. E. 1978. Cost Analysis of Light Water Reactor Power Plants. R-2304-DOE. Santa Monica: The RAND Corporation.

494 Morris, P. 1996. "Optimal Strategies for Distribution Investment Planning." EPRI Second Annual Distributed Resources Conference, Vancouver, Canada, 6 November. TR-107585. www.epri.com.

495 Morris, P. 1996. Address to EPRI Second Annual Distributed Resources Conference, Vancouver, Canada, 6 November. TR-107585. www.epri.com.

496 Morris, P., A. Cohn, W. Wong, R. Wood, 1994. *Value of Flexibility & Modularity of Distributed Generation.* EPRI Investment Strategies Project (October).

497 Moskovitz, D., C. Harrington, R. Cowart, W. Shirley, and F. Weston. 2000. "Profits and Progress through Distributed Resources." Gardiner, ME: Regulatory Assistance Program for the National Association of Regulatory Utility Comissioners, pp. 16–18, 20–22. www.rapmaine.org/P&pdr.htm.

498 Moskovitz, D., C. Harrington, W. Shirley, R. Cowart, R. Sedano, F. Weston. 2001. "Distributed Resource Distribution Credit Pilot Programs: Revealing the Value to Consumers and Vendors." Montpelier, VT: Regulatory Assistance Project (September). www.rapmaine.org/DPDeaveragedCredits.pdf.

499 Mott, L. H. 2001. Northern Power Systems, Waitsfield, VT, Personal communication (30 October).

500 Mungenast, J., *Power Quality* magazine Editor. 1997. Personal communication (July).

501 Mussavian, M. 1997. "An APT Alternative to Assessing Risk in Mastering Finance." *Financial Times,* 27 May, pp. 10–11.

502 Nash, H. 1979. "The Energy Controversy: Soft Path Questions and Answers by Amory Lovins and His Critics." San Francisco: Friends of the Earth.

503 National Energy Marketers Association. 2001. *National Guidelines for Implementing Distributed Generation and Related Services.* Washington, DC: National Energy Marketers Association (June). www.energymarketers.com/Documents/NEM_DG_Guidelines_final_7-12-01.PDF.

504 National Energy Policy Development Group. 2001. *National Energy Strategy: Reliable, Affordable, and Environmentally Sound Energy for America's Future.* Washington, DC: National Energy Policy Development Group (May). www.whitehouse.gov/energy/.

505 National Energy Policy Initiative. 2002. *National Energy Policy Initiative Endorsements.* Snowmass, CO: RMI and Consensus Building Institute. www.nepinitiative.org/endorsements.html.

506 National Rural Electric Cooperative Association. www.nreca.org/coops/elecoop3.html.

507 Neal, J. 1996. "Transportable 200 KW Fuel Cell for Rural Dispersed Generation." EPRI Second Annual Distributed Resources Conference, Vancouver, Canada, 6 November. TR-107585. www.epri.com.

508 Neal, J. 1996. Address to the EPRI Second Annual Distributed Resources Conference, Vancouver, Canada, 6 November. TR-107585. www.epri.com

509 NEPOOL Congestion Report. 2002. www.nepool.com.

510 NEPOOL Market Rules & Procedures. 2002. www.nepool.com

511 NERC (North American Electric Reliability Council). 1995. *Generating Availability Report 1990–1994.* Princeton, NJ: NERC (June). www.nerc.com/~filez/gar.html.

512 NERC. *Electric Power Supply and Demand 1984–1993.* Princeton, NJ.: Princeton, NJ.: NERC. www.nerc.com.

513 Netherlands Agency of Energy and the Environment. 1995. *Building With Photovoltaics.* Amsterdam: Netherlands Agency of Energy and the Environment (October).

514 New Jersey Department of Energy. 1981. *Energy Management Workbook for New Jersey's Industries: IV. Electricity and other Utilities.*

515 Newcomb, J., and W. Byrne. 1995. "Real-time Pricing and Electric Utility Industry Restructuring: Is the Future 'Out of Control?'" E SOURCE Strategic Issues Paper. Boulder, CO: E source (April).

516 ———, pp. 5–6.

517 ———, p 6.

518 Nishizawa-sensei, President of Sendai University. 1989. Personal communication.

519 NRDC. 1996. "Risky Business: Hidden Environmental Liabilities Of Power Plant Ownership." Natural Resources Defence Council. www.nrdc.org/nrdcpro/rbr/rbtinx.html

520 NYISO Congestion Contracts. 2002. www.nyisco.com.

521 NYISO Day-Ahead, Real-Time and Ancillary. 2002. wws.nyiso.com

522 Oldt, T. 1978. "How Well Do the Utilities Forecast Electricity Demand?" *Energy Daily* 6(209): 3–4.

523 Orans, R. C. K. Woo, J. N. Swisher. 1992. "Targeting DSM for Transmission and Distribution Benefits: A Case Study of PG&E's Delta District." EPRI and PG&E (May).

References 524–589

524 Osborn, D. E. 1995. "Implementation of Utility PV: A Tutorial." Carbondale, CO: Solar Energy International (March).

525 ———, bullet list adapted from Part IV, p. 37.

526 ———, end of Part III.

527 Osborn, D. E., SMUD. 1997. Personal communication (22 July).

528 Osborn, D. E., SMUD. 1997. Personal communication (July).

529 Osborn, D. E., SMUD. 2000. Email interview (29 December).

530 Osborn, D. E., SMUD. 2001. Personal communication (13 December).

531 Osborn, D. E., SMUD. 2001. Personal communication (13 December). Citing *Builder OnLine*, October 2001.

532 OTA (U.S. Congress, Office of Technology Assessment). 1985. New Electric Power Technologies: Problems and Prospects for the 1990s. OTA-E-246. OTA (July), p. 177. A fuller account is in Y. Wan and B.K. Parsons. 1993. *Factors Relevant to Utility Integration of Intermittent Renewable Technologies.* NREL/TP-4634953. Golden, CO: NREL (August). www.nrel.gov/docs/legosti/old/4953.pdf.

533 ———, p. 32.

534 OTA. 1978. *Application of Solar Technology to Today's Energy Needs—Vol. I.* OTA (June).

535 OTA. 1978. *Application of Solar Technology to Today's Energy Needs—Vol. II.* OTA (September).

536 OTA. 1978. *Application of Solar Technology to Today's Energy Needs.* OTA.

537 OTA. 1985. *New Electric Power Technologies: Problems and Prospects for the 1990s.* OTA-E-246. OTA (July), p. 173. www.wws.princeton.edu/~ota/ns20/year_f.html.

538 ———, p. 172.

539 Ottinger, R. L., D. R. Wooley, N. A. Robinson, D. R. Hodas, and S. E. Babb. 1991. *Environmental Costs of Electricity.* New York: Oceana.

540 Ouwens, K. D. 1992. Personal communication.

541 Ouwens, K. D. 1993. "Cheap Electricity with Autonomous Solar Cell Systems." *Energy Policy* (November): 1085–1092.

542 Parrella, E.A. Merrill Lynch Capital Markets Report 8326478 for Mirant, dated 22 January 2002 and Report 8313540, dated 24 December 2001.

543 Pasqualetti, M.J., P. Gipe, R.W. Righter, eds. 2002. *Wind Power in View.* San Diego, CA: Academic Press.

544 *PC Magazine.* 2002. "Getting to the Source of Downtime." iBiz section (29 January), pp. 1–7.

545 Peerenboom, J.P., R.E. Fisher, S.M. Rinaldi, and T.K. Kelly. 2002. "Studying the Chain Reaction." Electric Perspectives. Washington DC: EEI (January/February). www.eei.org/editorial/Jan_02/0102CHAIN.html.

546 Perez R., R. Seals, R. Stewart. 1994. *"Solar Resource—Utility Load-Matching Assessment."* Golden, CO: NREL (March).

547 Perez, R., W. Berkheiser III, R. Stewart, M. Kapner, and G. Stillman. 1989. "Photovoltaic Load Matching Potential for Metropolitan Utilities and Large Commercial Users in the Northeastern United States." 9th European Photovoltaic Conference, Greiburg, Federal Republic of Germany. pp. 903–907. Cited in Y. Wan and B. K. Parsons, *Factors Relevant to Utility Integration of Intermittent Renewable Technologies* (Golden, CO: NREL, 1993), p. 64.

548 PG&E Management Committee. 1989. PG&E Management Committee meeting minutes (30 October). Cited in D. Shugar, *et al., Benefits of Distributed Generation in PG&E's Transmission and Distribution System: A Case Study of Photovoltaics Serving Kerman Substation.* (PG&E, 1992), p. B9.

549 Piller, C. 2001. "Power Grid Vulnerable to Hackers." *Los Angeles Times,* 13 August. www.latimes.com/la-000065693aug13.story.

550 PJM Capacity Credit Markets. 2002. www.pjm.com

551 PJM Transmission Adequacy Assessment Report. 2002. www.pjm.com/trans_exp_plan/downloads/assessment.pdf

552 Pope III, C. A., R. T. Burnett, M. J. Thun, E. E. Calle, D. Krewski, K. Ito, and G. D. Thurston. 2002. "Lung Cancer, Cardiopulmonary Mortality, and Long-term Exposure to Fine Particulate Air Pollution." *JAMA* 287 (6 March): 1132–1141.

553 Pratt, R. G., Z. T. Taylor, L. A. Klevgard, and A. G. Wood. 1994. "Potential for Feeder Equipment Upgrade Deferrals in a Distributed Utility." *American Council for an Energy Efficient Economy* (July): 2229–2240.

554 Pratt, R. G., Z. T. Taylor, L. A. Klevgard, and A. G. Wood. 1994. *Using Distributed Utility Technologies to Defer Equipement Upgrades.* Richland, WA: PNL (7 December).

555 Preliminary Report of Working Group 04. 1999. "Distributed Generation." CIRED (The International Conference on Electricity Distribution Networks), Nice, France, June. Reproduced from p. 5 of N. Jenkins, *et al. Embedded Generation* (London:IEEE, 2000).

556 President's Committee of Advisors on Science & Technology. 1997. "Federal Energy Research and Development for the Challenges of the Twenty-First Century." The White House (September).

557 Procaccia, H. 1975. "Probabilité de Défaillance des Circuits de Refroidissement Normaux et des Circuits de Refroidissement de Secours des Centrales Nucléaires." IAEA-SM-195/3. In *Reliability of Nuclear Power Plants.* Vienna, Austria: International Atomic Energy Agency, pp. 351–372, 392.

558 Public Utilities Commission of California 9th circ. v. FERC. 2001. No. 01-71051 (29 June).

559 Public Utility Commission of Texas. 2000. Substantive Rule, Chapter 25, Electric, section 25.211. www.puc.state.tx.us/rules/subrules/electric/25.211/25.211ei.cfm.

560 Puntel, W. R., and V. J. Longo. 1996. "Distributed Resources: Reliability Implications." EPRI Second Annual Distributed Resources Conference, Vancouver, Canada, 7 November. TR-107585. www.epri.com.

561 Ranade, S. J., N. R. Prasad, and S. R. Omick. 1989. *Islanding in Dispersed, Utility-Interactive Photovoltaic Systems.* SAND88-7042. Sandia National Laboratories (May). Cited in Y. Wan and B. K. Parsons, *Factors Relevant to Utility Integration of Intermittent Renewable Technologies* (Golden, CO: NREL, 1993), ref . 24, p. 26 .

562 Rau, N. S. and Y. Wan. 1994. "Optimum Location of Resources in Distributed Planning." *IEEE Transactions on Power Systems* 9(4): 2014–2020 (November). www.ieee.org/organizations/pubs/transactions/tps2.htm.

563 Reading, M. 1995. "Momentary Outage Costs for PG&E Customers >500kW." PG&E (July). Cited in N. Lenssen, *Distributed Load Control: How Smart Appliances Could Improve Transmission Grid Operations* (Boulder, CO: E SOURCE, 1997).

564 Reason, J. 1996. "Solid State Transfer: Speed Alone Won't Solve Power-Quality Problems." *Electrical World* 120(8): 20 (August). Cited in N. Lenssen, *Distributed Load Control: How Smart Appliances Could Improve Transmission Grid Operations* (Boulder, CO: E SOURCE, 1997).

565 RMI. 1991. "Visitors' Guide." RMI Pub. H-1. Snowmass, CO: RMI.

566 RMI. 2002. "RMI Helping the Cutting Edge of 'Turbo-machinery'." *RMI Solutions Newsletter* (Spring), p. 15. www.rmi.org/images/other/NLRMIspring02.pdf.

567 Robertson, C., and J. Romm. 2002. "Data Centers, Power, and Pollution Prevention." www.cool-companies.org.

568 Romm, J. 1999. "The Internet Economy and Global Warming." www.cool-companies.org/energy/.

569 Romm, J. 2001. Personal communication (19 July).

570 Romm, J. 2001. Personal communication (20 July).

571 Romm, J., and W. D. Browning. 1994. *Greening the Building and the Bottom Line: Increasing Productivity Through Energy-Efficient Design.* RMI Pub. D94-27. Snowmass, CO: RMI. www.rmi.org/images/other/GDS-GBBL.pdf

572 Rosenthal, A. L. 1992. *Photovoltaic System Performance Assessment for 1990.* Palo Alto, CA: EPRI (November).

573 Rosenzweig, R., M. Varilek, J. Janssen, B. Feldman, R. Kuppalli, and Natsource, LLC. 2002. *The Emerging International Greenhouse Gas Market.* Arlington, VA: The Pew Center on Global Climate Change (July). www.pewclimate.org/projects/trading.cfm.

574 Ross, S. A., R. W. Westerfield, and J. F. Jaffe. 1993. *Corporate Finance.* Third Edition. Boston, MA: IRWIN.

575 Rowland, L. R. 1996. BACnet Opens Doors for Controls: But Watch Your Step! TU-96-6. Boulder, CO: E SOURCE (May).

576 Rueger, G. M., and G. Manzoni. 1991. "Utility Planning and Operational Implications of Photovoltaic Power Systems." San Ramon, CA: PG&E.

577 Ryle, M. 1977. "The Economics of Alternative Energy Sources." *Nature* 267: 111–117.

578 Safir, A., President, RECON Research Corp. 1997. Personal communication (10 August).

579 *San Jose Business Journal.* 2001. "Chipmaker to Use Solar Power," 14 November.

580 Sant, R., AES Corporation. 1997. Personal communication (19 February and 7 July).

581 Schlueter, R. A., and G. L. Park, et al. 1984. "Simulation and Assessment of Wind Array Variations Based on Simultaneous Wind Speed Measurements." *IEEE Transactions on Power Apparatus and Systems* PAS-103(5):1008–1016 (May). Cited in Y. Wan and B. K. Parsons, *Factors Relevant to Utility Integration of Intermittent Renewable Technologies* (Golden, CO: NREL, 1993), p. 45.

582 Schlueter, R. A., and G. L. Park, T. W. Reddoch, P. R. Barnes, and J. S. Lawler. 1985. "A Modified Unit Commitment and Generation Control for Utilities with Large Wind Generation Penetrations." *IEEE Transactions on Power Apparatus and Systems* PAS-104(7):1630–1636 (July). Cited in Y. Wan and B. K. Parsons, *Factors Relevant to Utility Integration of Intermittent Renewable Technologies* (Golden, CO: NREL, 1993), p. 45.

583 Schoengold, D. 1995. *Application of the Distributed Utility Concept to the Boston Edison Company / Creating Additional Value for the Customer.* Middleton, WI: MSB Energy Associates.

584 Schwartz, F. H. 2000. Interview with Mr. Schwartz, Marketing Consultant, AFS Trinity Power Corporation, 6724D Preston Avenue, Livermore CA 94550 (10 December).

585 Seitz, N. E. 1990. *Capital Budgeting and Long-term Financing Decisions.* Orlando, FL: Dryden Press

586 ———, pp. 257–258.

587 Shah, A., P. Torres, R. Tscharner, N. Wyrsch, and H. Keppner. 1999. "Photovoltaic Technology: The Case for Thin-Film Solar Cells." *Science* 285:692–698 (30 July).

588 Sharkey, W. W., and M. Hill. 1977. *Efficient Production When Demand Is Uncertain.* Murray Hill, NJ: Bell Laboratories.

589 Sharpe, W. F. 2001. Personal communication (19 March).

References 590–698

590   Shell International Limited, Global Business Environment. 2001. "Exploring the Future: Energy Needs, Choices and Possibilities—Scenarios to 2050." London: Shell Centre. www. shell.com.

591   Shugar, D. S. 1996. EPRI Second Annual Distributed Resources Conference, Vancouver, Canada, 6 November. TR-107585. www.epri.com.

592   Shugar, D. S., H. J. Wenger, and G. J. Ball. 1993. "Photovoltaic Grid Support: A New Screening Methodology." *Solar Today* (September), pp. 21–24.

593   ———, p. 23.

594   ———, p. 24.

595   Shugar, D., R. Orans, A. Jones, M. El-Gassier, and A. Suchard. 1992. "Benefits of Distributed Generation in PG&E's Transmission and Distribution System: A Case Study of Photovoltaics Serving Kerman Substation." PG&E (November).

596   ———, p. 2.

597   ———, p. 2-10, 4-20n.

598   ———, p. 2-15.

599   ———, p. 2-3.

600   ———, p. 2-4.

601   ———, p. 2-5.

602   ———, p. 2-6.

603   ———, p. 2-7n

604   ———, p. 2-8.

605   ———, p. 2-9.

606   ———, p. 3.

607   ———, p. 3-10.

608   ———, p. 3-2.

609   ———, p. 3-20.

610   ———, p. 3-4.

611   ———, p. 3-5.

612   ———, p. 4-12.

613   ———, p. 4-12.

614   ———, p. 4-13.

615   ———, p. 4-14.

616   ———, p. 4-16.

617   ———, p. 4-18.

618   ———, p. 4-20.

619   ———, p. 4-21.

620   ———, p. 4-22.

621   ———, p. 4-23.

622   ———, p. 4-3.

623   ———, p. 4-4n.

624   ———, p. 4-5.

625   ———, p. 4-6.

626   ———, p. 4-7.

627   ———, p. 5-12.

628   ———, p. 5-2

629   ———, p. 5-6.

630   ———, p. 5-7.

631   ———, p. A7.

632   ———, p. B10.

633   ———, p. B11.

634   ———, p. B12.

635   ———, p. B8.

636   ———, p. B9.

637   ———, p. 2-10.

638   ———, pp. 2-11–2-15.

639   ———, pp. 2-6–2-7.

640   ———, pp. 4-17–4-18.

641   ———, pp. 4-22–4-33.

642   ———, pp. 4-24–4-25.

643   ———, pp. 4-26–4-27.

644   ———, pp. 5-7–5-8.

645   ———, pp. 5-7–6-9.

646   ———, pp. A4–A5.

647   ———, pp. B11–12.

648   Shula, W. 1991. *Photovoltaic Generation Effects on Distribution Feeder. Volume 2: Analysis Methods and Results.* EL-6754. Palo Alto, CA: EPRI (September).

649   ———, p. 1-6.

650   ———, p. 1-2.

651   ———, p. 1-7.

652   ———, p. 1-8.

653   ———, pp. 172–173.

654   ———, section 4.

655   Sick, F. and T. Erge, eds. 1997. *Photovoltaics in Buildings: A Design Handbook for Architects and Engineers.* London: James & James. www.jxj.com/catofpub/NEWphotovoltaics_buildings.html.

656   Sillin, J. O. 1995. "Judging Past Policies: Where Are We in the Electricity Debate?" *Electricity Journal* 8 (October).

657   Sinden, F. W. 1960. "The Replacement and Expansion of Durable Equipment." *Journal of the Society for Industrial and Applied Mathematics* 8(3):466–480 (September).

658   Smith, R. 2002. "Power Industry Cuts Plans for New Plants, Posting Risks for Post-Recessionary Period." *Wall Street Journal,* 4 January, p. A3.

659   Smock, R. W. 1996. "New Generating Capacity is Needed." *Power Engineering* (November): 3. Pennwell.

660   SMUD (Sacramento Municipal Utility District). 2001. "SMUD Solar Program: 2001 Year in Review" (February). www.smud.org/pv/2001Review.pdf.

661   SMUD, Sacramento, California. 1997. Press Release on Solar Programs (17 July). www.smud.org.

662   Solar Energy Research Institute. 1990. *The Potential of Renewable Energy: Interlaboratory Whitepaper.* Golden, CO: SERI (March).

663   Sørensen, B. 1976. "Dependability of Wind Energy Generators with Short-Term Energy Storage." *Science* 194, pp. 935–937.

664   Sørensen, B. 1979. *Renewable Energy.* New York, NY: Academic Press.

665   Starrs, T. J. 2001. "Summary of State 'Net Metering' Programs" (26 June). Seattle, WA: Kelso Starrs & Associates. www.newrules.org/electricity/netmeteringstarrs.pdf; www.irecusa.org.

666   Starrs, T. J. 2001. Personal communication (March).

667   Stevens, J. 1988. *The Interconnection Issues of Utility-Intertied Photovoltaic System.* SAND87-3146. Sandia National Laboratories (November). Cited in Y. Wan and B. K. Parsons, *Factors Relevant to Utility Integration of Intermittent Renewable Technologies* (Golden, CO: NREL, 1993), p. 18, ref. 5. www.nrel.gov/docs/legosti/old/4953.pdf.

668   Stevens, J. 1992. *EPRI Coordinated Project Lead to Cost-Effective Applications of PV for T&D Sectionalizing Switches* IN-100488. Palo Alto, CA: EPRI (May). Cited in D. E. Osborn, "Implementation of Utility PV: A Tutorial" (Carbondale, CO: Solar Energy International, 1995).

669   Stevens, J. 1994. "Distributed Photovoltaic Benefits; Four Case Studies." Address to EPRI Conference: Research Results and Utility Experience—Distributed Utility Valuation, Baltimore, MD, 15–16 March.

670   Stipp, D. 1992. "Power Glitches Become Critical as World Computerizes." *Wall Street Journal,* 18 May.

671   Stover, J.G. 1978. *Incorporating Uncertainty in Energy Supply Models.* EA-703. Palo Alto, CA: EPRI (February).

672   SurePower Corporation. www.hi-availability.com

673   Sutherland, R. J., A. Ford, S. V. Jackson, C. A. Mangeng, R. W. Hardie, and R. E. Malenfant. 1985. The Future Market for Electric Generating Capacity: Technical Documentation. Los Alamos, NM: Los Alamos National Laboratory (March).

674   ———, pp. 147–151.

675   ———, pp. 145–146.

676   ———, pp. 48–49.

677   ———, pp. 77–185.

678   ———, p. 146.

679   Swisher, J. N., and R. Orans. 1996. "A New Utility DSM Strategy Using Intensive Campaigns Based on Area-Specific Costs." *Utilities Policy* 5:185–197.

680   Swisher, J. N., G. Jannuzzi, and R. Redlinger. 1998. *Tools and Methods for Integrated Resource Planning: Improving Energy Efficiency and Protecting the Environment.* Roskilde, Denmark: UNEP Collaborating Centre on Energy and Environment. www.uccee.org/IRPManual/index.htm.

681   Swisher, J., and R. Orans. 1995. "A New Utility DSM Strategy Using Intensive Campaigns Based on Area Specific Costs." Paper submitted for ECEEE Summer Study, 11 January.

682   Systems Control, Inc. 1980. "Decentralized Energy Technology Integration Assessment Study: Second Principal Report." SCI 5278. Washington, DC: USDOE Office of Policy & Evaluation (December).

683   TECC Group. 1995. "U.S. Electric IOU Research, Development, and Demonstration Expense Trends 1990–94." Littleton, CO: TECC Group (August).

684   Teisberg, E. O. 1994. "An Option Value Analysis of Investment Choices by a Regulated Firm." *Management Science* (April): 24.

685   Thomas, R. J. 1986. "An Integration Methodology for Large Wind-Energy Conversion Systems, Final Report." DOE/RA/50664-1. Ithaca, NY: Cornell University (June). Cited in Y. Wan and B. K. Parsons, *Factors Relevant to Utility Integration of Intermittent Renewable Technologies* (Golden, CO: NREL, 1993), p. 46.

686   Thomas, R., ed. 2001. *Photovoltaics and Architecture.* UK: Spon Press.

687   Thomas, S., Directed Technologies, Inc. 1997. Personal communication.

688   Thresher, R. W. 1996. "Wind as a Distributed Resource." EPRI Second Annual Distributed Resources Conference, Vancouver, Canada, 6 November. TR-107585. www.epri.com.

689   Treat, J. E. 1991. *Energy Futures: Trading Opportunities for the 1990´s.* Tulsa, OK: Pennwell Books.

690   Treat, J. E. 2002. *Energy Futures: Past and Present.* Tulsa, OK: Pennwell Books.

691   Turner, J. A. 1999. "A Realizable Renewable Energy Future." *Science* 285:687–689 (30 July).

692   U.S. Department of Commerce, Economics and Statistics Administration, Bureau of the Census. 1992. *1992 Economic Census.* EC92-PR-2. Wahington, DC: U.S. Government Printing Office. www.census.gov/econ/www/img/92ec.pdf.

693   USDOE (U.S. Department of Energy). "Institute of Electrical and Electronics Engineers Standards." Washington, DC: USDOE. www.eren.doe.gov/distributedpower/issue_ieee.html.

694   USDOE. 1998. "Distributed Power Barriers." Washington, DC: Distributed Energy Resources. www.eren.doe.gov/distributedpower/whatis_barriers.html.

695   V.A.C.S. Art. 1446c-0, Sec. V.A.C.S. Art. 1446c-0, Sec. 39.101(b)(3) (Added by 1999 Amendments: SB 7, § 39).

696   Vishwanath, P. 1995. *Smart Residential Appliances: Will the Information Superhighway Dead End into a Dumb House?* TU-95-4. Boulder, CO: E SOURCE (June).

697   Wallace, D. B. 1985. "PV Power Conditioner Harmonics." *Proceedings of the Joint ASME-ASES Solar Energy Conference,* Knoxville, TN, March. Cited in Y. Wan and B. K. Parsons, *Factors Relevant to Utility Integration of Intermittent Renewable Technologies* (Golden, CO: NREL, 1993), p. 17.

698   Wallach, J., P. Chernick, and J. Plunkett. 1993. *From Here to Efficiency: Securing Demand-Management Resources, Vol. 5.* Harrisburg, PA: Pennsylvania Energy Office, pp. 121–126.

References 699–782

699 Wan, Y., and B. K. Parsons. 1993. *Factors Relevant to Utility Integration of Intermittent Renewable Technologies.* TP-4634953. Golden, CO: NREL (August), p. 3. www.nrel.gov/docs/legosti/old/4953.pdf.

700 ———, p. 24.

701 ———, p. 62.

702 ———, pp. 35–36.

703 ———, p. 22.

704 ———, p. 17.

705 ———, p. 18.

706 ———, p. 20.

707 ———, p. 21.

708 ———, p. 23.

709 ———, p. 24, emphasis added.

710 ———, p. 25.

711 ———, p. 30.

712 ———, p. 37.

713 ———, p. 40.

714 ———, p. 41.

715 ———, p. 47.

716 ———, p. 48.

717 ———, p. 49.

718 ———, p. 50.

719 ———, p. 50.

720 ———, p. 64.

721 ———, p. 66.

722 ———, p. 67.

723 ———, p. v.

724 ———, pp. 20–21.

725 ———, pp. 30–31.

726 ———, pp. 33–35.

727 ———, pp. 38–50.

728 ———, pp. 44–45.

729 Wan, Y., and S. Adelman. 1995. "Distributed Utility Technology Cost, Performance, and Environmental Characteristics." NREL/TP-463-7844. Golden, CO: NREL (June). www.nrel.gov.

730 *Washington Post National Weekly Edition.* 1997. "Changing the Electric Power Industry." Advertising Feature, 17 February.

731 Wayne, G., Powerlight Corp., Berkeley, CA. 2002. Personal communication (31 May). www.powerlight.com.

732 Wayne, M. 1983. "New Capacity in Smaller Packages." *EPRI Journal* (May): 7–13.

733 Weinberg, C. J. *Cost-Effective Photovoltaic Applications.* Presentation for the Regulatory Assistance Project, Gardiner, ME.

734 Weinberg, C. J., J. Iannucci, and M. M. Reading. 1991. "The Distributed Utility: Technology, Customer, and Public Policy Changes Shaping the Electrical Utility of Tomorrow." *Energy Systems and Policy* 15 (4):307–322.

735 Wenger, H. J., T. E. Hoff, and B. K. Farmer. 1994. "Measuring The Value of Distributed Photovoltaic Generation: Final Results of the Kerman Grid-Support Project." First World Conference on Photovoltaic Energy Conversion, Waikoloa, HI, December.

736 ———, p. 3.

737 Wenger, H. J., T. E. Hoff, and J. Pepper. 1996. *Photovoltaic Economics and Markets: The Sacramento Municipal Utility District as a Case Study.* SMUD, California Energy Commission, and USDOE PV Compact Program via NCSC (September). www.energy.ca.gov/development/solar/SMUD.PDF.

738 ———, p. 10-1ff.

739 ———, p. 3-2.

740 ———, p. 8-27.

741 ———, p. 8-42.

742 ———, p. v.

743 ———, pp. 2-7–2-8.

744 ———, pp. 8–26.

745 ———, pp. 8–35.

746 Wenger, H., T. Hoff, and R. Perez. 1992. "Photovoltaics as a Demand-Side Management Option: Benefits of a Utility-Customer Partnership." World Energy Engineering Congress (October), p. iv.

747 Weston, F., C. Harrington, D. Moskovitz, W. Shirley, and R. Cowart. 2000. "Charging for Distribution Utility Services: Issues in Rate Design." Montpelier, VT: Regulatory Assistance Project (December). www.rapmaine.org/RateDesign.pdf.

748 ———, p. 25.

749 Weston, F., C. Harrington, D. Moskovitz, W. Shirley, R. Cowart, R. Sedano. 2001. "Accommodating Distributed Resources in Wholesale Power Markets." Montpelier, VT: Regulatory Assistance Project (September). www.rapmaine.org/DPWholesaleMarkets.pdf.

750 ———, p. 13.

751 ———, p. 24.

752 ———, p. 6.

753 ———, pp. 11–12.

754 ———, pp. 17–19.

755 Weston, F., N. Siedman, and C. James. 2001. "Model Regulations for the Output of Specified Air Emissions From Smaller-Scale Electrical Generation Resources." Montpelier, VT: Regulatory Assistance Project (November). www.rapmaine.org/DREmissionsRuleNovDraft.PDF.

756 Whitlock, J. *et al.* 2000. "Distributed Generation Creeps into the Electricity and Distribution World." *Standard & Poor's Creditweek* (27 November).

757 Wiegner, E. 1977. "Tax Incentives and Utility Cash Flow." Atomic Industrial Forum Conference on Nuclear Financial Considerations, Seattle, WA, 24–27 July.

758 Williams, B. D., T. C. Moore, A. B. Lovins. 1997. "Speeding the Transition: Designing A Fuel-Cell Hypercar." National Hydrogen Association's 8th Annual U.S. Hydrogen Meeting, Alexandria, VA, 11–13 March. RMI Pub. T97-9. www.rmi.org/images/other/HC-SpeedTrans.pdf.

759 Williams, R. H. 1996. "Fuel Decarbonization for Fuel Cell Applications and Sequestration of the Separated CO2." PU/CEES 295. Princeton, NJ: CEES, Princeton University (January).

760 Williams. 2002. www.williams.com.

761 Willis, H. L., and W. G. Scott. 2000. *Distributed Power Generation.* New York: Marcel Dekker.

762 ———, p. 15.

763 ———, p. 17.

764 ———, p. 373.

765 ———, p. 541.

766 ———, p. 64.

767 ———, p. 9.

768 ———, pp. 30–31.

769 Wilson, A., J. Uncapher, L. McManigal, and M. Cureton. 1998. *Green Development: Integrating Ecology and Real Estate.* New York: John Wiley & Sons.

770 *Wind Energy Weekly* 20(955), 27 July 2001.

771 Wiser, R., M. Bolinger, E. Holt, B. Swezey. 2001. *Forecasting the Growth of Green Power Markets in the United States.* NREL/TP-620-30101/LBNL-48611. Golden, CO: NREL. (October). www.eren.doe.gov/greenpower/pdf/30101.pdf

772 Woo, C. K., R. Orans, B. Horii, R. Pupp, and G. Heffner. 1994. "Area- and Time-Specific Marginal Capacity Costs of Electricity Distribution." *Energy: The International Journal* 19(12):1213–1218.

773 www.clean-power.com/research.htm. See under *Micro-Grids.*

774 www.eia.doe.gov/cneaf/electricity/chg_str/regmap.html.

775 www.hypercar.com.

776 www.SolarAccess.com. 2001. "Small Renewable Generators Get Low-Cost Connection" (9 September).

777 www.stern.nyu.edu/%7Eadamodar/pc/datasets/histretSP.xls, downloaded 8 July 2001.

778 www.stirlingtech.com.

779 www.utcfuelcells.com/index.shtml. See under Commercial Power Systems (Reliability Statistics).

780 Yaeger, K. E. 2001. "Electricity Technology Development for a Sustainable World." World Energy Council 18th Congress, Buenos Aires, 21–25 October.

781 Zaininger, H. W. 1987. *Benefits of Battery Storage as Spinning Reserve: Quantitative Analysis.* EPRI AP-5327. Palo Alto, CA: EPRI (July). Cited in Y. Wan and B. K. Parsons, *Factors Relevant to Utility Integration of Intermittent Renewable Technologies* (Golden, CO: NREL, 1993), p. 66.

782 Zaininger, H. W. 1992. "Distributed Renewables Project." Distributed Utility—Is This the Future? EPRI, PG&E, and NREL conference, Berkeley, CA, 2–3 December.

About the authors

E. Kyle Datta

CEO, New Energy Partners

Kyle Datta is a former Vice President of Booz, Allen & Hamilton where he was the Managing Partner of the firm's Asia Energy Practice and later led the U.S. Utilities practice. Mr. Datta holds BS, MES, and MPPM degrees from Yale University. He is currently CEO of New Energy Partners, an energy consulting and renewable development firm located in Hawai'i.

Thomas Feiler

Tom Feiler is a former Managing Director and Principal of Rocky Mountain Institute's Energy and Resources team. He is an authority on industry structure, resource planning, and competitive strategy development for the electric power and natural gas sectors. Mr. Feiler has been an international consultant, speaker, expert witness, and author on the electric power and natural gas industries, addressing such issues as competitive markets, strategic planning, industry trends and environmental policy analysis, market development, risk analysis, integrated resource planning, and demand-side management. He is a University of Denver graduate *summa cum laude* in Political Science and in Philosophy, and holds an MA from the Fletcher School of Law and Diplomacy.

André Lehmann

Associate, Batinergie

André Lehmann, an electrical engineer holding an MSEE from the Federal Polytechnic of Lausanne, is a former Research Associate at Rocky Mountain Institute. After leaving RMI, he completed a PhD in atmospheric physics at the Federal Institute of Technology in Zürich. Dr. Lehmann consults on and implements energy efficiency improvements for new homes in the Lake Geneva area of Switzerland.

Amory B. Lovins

CEO, Rocky Mountain Institute

Physicist Amory Lovins is cofounder and CEO of Rocky Mountain Institute and Chairman of its fourth spinoff, Hypercar, Inc. (www.hypercar.com). Published in 27 previous books and hundreds of papers, his work has been recognized by the "Alternative Nobel," Onassis, Nissan, Shingo, and Mitchell Prizes, a MacArthur Fellowship, the Happold Medal, eight honorary doctorates, and the Heinz, Lindbergh, World Technology, and "Hero for the Planet" Awards. He has advised the energy industries (including scores of utilities), major energy customers, and governments worldwide for three decades, and has been analyzing distributed benefits since the mid-1970s. His previous book, with Paul G. Hawken and L. Hunter Lovins, is *Natural Capitalism: Creating the Next Industrial Revolution* (www.natcap.org).

Karl R. Rábago

Sustainability Alliances Leader, Cargill Dow LLC

Karl R. Rábago is a former Deputy Assistant Secretary for the U.S. Department of Energy, a former public utility commissioner for the State of Texas, and a former Managing Director and Principal of the Energy and Resources team at Rocky Mountain Institute. He is a Board member of the Center for Resource Solutions and Chairman of the Green Power Board. His JD is from the University of Texas, and he holds LLM degrees from Pace University School of Law (Environmental Law) and the U.S. Army Judge Advocate General's School (Military Law).

Joel N. Swisher

Leader and Principal, Energy and Resources Services, Rocky Mountain Institute

Joel Swisher is a registered Professional Engineer and holds a PhD in Energy and Environmental Engineering from Stanford University. With 25 years' experience in research and consulting on many aspects of clean energy technology, Dr. Swisher is an internationally recognized expert in the analysis, design, and evaluation of utility energy efficiency, distributed generation, and emission reduction programs and in the development and finance of carbon offset projects. His hundred-odd professional publications include RMI's 2002 monograph *Cleaner Energy, Greener Profits: Fuel Cells as Cost-Effective Distributed Energy Resources*.

Ken Wicker

Senior Research Associate for the E SOURCE Distributed Energy Service

Ken Wicker provides analysis, conducts research, and reports on current emerging distributed energy technologies. He is the primary author of "High Temperature Fuel Cells for Stationary Power: A Long, Hot Road to the Promised Product." He is a former Rocky Mountain Institute researcher and has been a consultant to small renewable energy companies in Bhutan and Sri Lanka. Mr. Wicker holds an MA in environmental and energy policy from the University of Delaware and a certificate in energy management and design from Sonoma State University.

About the publisher

Rocky Mountain Institute (www.rmi.org), founded by Hunter and Amory Lovins in 1982, is an independent, entrepreneurial, nonpartisan, nonprofit applied research center.

Its ~50 staff foster the efficient and restorative use of natural and human capital to help make the world secure, prosperous, and life-sustaining. The Institute's ~$6-million annual budget comes roughly half from programmatic enterprise earnings, chiefly private-sector consultancy, and half from grants and donations.

RMI is known worldwide for its work in advanced resource productivity, business innovations related to natural capitalism, and highly original transdisciplinary syntheses at the nexus of energy, resources, environment, development, and security.

207 BENEFITS OF DISTRIBUTED RESOURCES (Continued from front endpapers.) **95** Distributed substitutes for traditional spinning reserve capacity can reduce its operating hours—hence the mechanical wear, thermal stress, corrosion, and other gradual processes that shorten the life of expensive, slow-to-build, and hard-to-repair central generating equipment. **96** When distributed resources provide "virtual spinning reserve," they can reduce cycling, turn-on/shutdown, and low-load "idling" operation of central generating units, thereby increasing their lifetime. **97** Such life extension generally incurs a lower risk than supply expansion, and hence merits a more favorable risk-adjusted discount rate, further increasing its economic advantage. **98** Distributed resources can help reduce the reliability and capacity problems to which an aging or overstressed grid is liable. **99** Distributed resources offer greater business opportunities for profiting from hot spots and price spikes, because time and location-specific costs are typically more variable within the distribution system than in bulk generation. **100** Strategically, distributed resources make it possible to position and dispatch generating and demand-side resources optimally so as to maximize the entire range of distributed benefits. **101** Distributed resources (always on the demand side and often on the supply side) can largely or wholly avoid every category of grid costs on the margin by being already at or near the customer and hence requiring no further delivery. **102** Distributed resources have a shorter haul length from the more localized (less remote) source to the load, hence less electric resistance in the grid. **103** Distributed resources reduce required net inflow from the grid, reducing grid current and hence grid losses. **104** Distributed resources cause effective increases in conductor cross-section per unit of current (thereby decreasing resistance) if an unchanged conductor is carrying less current. **105** Distributed resources result in less conductor and transformer heating, hence less resistance. **106** Distributed resources' ability to decrease grid losses is increased because they are close to customers, maximizing the sequential compounding of the different losses that they avoid. **107** Distributed photovoltaics particularly reduce grid loss load because their output is greatest at peak hours (in a summer-peaking system), disproportionately reducing the heating of grid equipment. **108** Such onpeak generation also reduces losses precisely when the reductions are most valuable. **109** Since grid losses avoided by distributed resources are worth the product of the number times the value of each avoided kWh of losses, their value can multiply rapidly when using area- and time-specific costs. **110** Distributed resources can reduce reactive power consumption by shortening the electron haul length through lines and by not going through as many transformers—both major sources of inductive reactance. **111** Distributed resources can reduce current flows through inductive grid elements by meeting nearby loads directly rather than by bringing current through lines and transformers. **112** Some end-use-efficiency resources can provide reactive power as a free byproduct of their more efficient design. **113** Distributed generators that feed the grid through appropriately designed DC-to-AC inverters can provide the desired real-time mixture of real and reactive power to maximize value. **114** Reduced reactive current improves distribution voltage stability, thus improving end-use device reliability and lifetime, and enhancing customer satisfaction, at lower cost than for voltage-regulating equipment and its operation. **115** Reduced reactive current reduces conductor and transformer heating, improving grid components' lifetime.

116 Reduced reactive current, by cooling grid components, also makes them less likely to fail, improving the quality of customer service. **117** Reduced reactive current, by cooling grid components, also reduces conductor and transformer resistivity, thereby reducing real-power losses, hence reducing heating, hence further improving component lifetime and reliability. **118** Reduced reactive current increases available grid and generating capacity, adding to the capacity displacement achieved by distributed resources' supply of real current. **119** Distributed resources, by reducing line current, can help avoid voltage drop and associated costs by reducing the need for installing equipment to provide equivalent voltage support or step-up. **120** Distributed resources that operate in the daytime, when sunlight heats conductors or transformers, help to avoid costly increases in circuit voltage, reconductoring (replacing a conductor with one of higher ampacity), adding extra circuits, or, if available, transferring load to other circuits with spare ampacity. **121** Substation-sited photovoltaics can shade transformers, thereby improving their efficiency, capacity, lifetime, and reliability. **122** Distributed resources most readily replace distribution transformers at the smaller transformer sizes that have higher unit costs. **123** Distributed resources defer or avoid adding grid capacity. **124** Distributed resources, by reducing the current on transmission and distribution lines, free up grid capacity to provide service to other customers. **125** Distributed resources help "decongest" the grid so that existing but encumbered capacity can be freed up for other economic transactions. **126** Distributed resources avoid the siting problems that can occur when building new transmission lines. **127** These siting problems tend to be correlated with the presence of people, but people tend to correlate with both loads and opportunities for distributed resources. **128** Distributed resources' unloading, hence cooling, of grid components can disproportionately increase their operating life because most of the life-shortening effects are caused by the highest temperatures, which occur only during a small number of hours.

129 More reliable operation of distribution equipment can also decrease periodic maintenance costs and outage costs. **130** Distributed resources' reactive current, by improving voltage stability, can reduce tapchanger operation on transformers, increasing their lifetime. **131** Since distributed resources are nearer to the load, they increase reliability by reducing the length the power must travel and the number of components it must traverse. **132** Carefully sited distributed resources can substantially increase the distribution system operator's flexibility in rerouting power to isolate and bypass distribution faults and to maintain service to more customers during repairs. **133** That increased delivery flexibility reduces both the number of interrupted customers and the duration of their outage.

134 Distributed generators can be designed to operate properly when islanded, giving local distribution systems and customers the ability to ride out major or widespread outages. **135** Distributed resources require less equipment and fewer procedures to repair and maintain the generators. **136** Stand-alone distributed resources not connected to the grid avoid the cost (and potential ugliness) of extending and connecting a line to a customer's site. **137** Distributed resources can improve utility system reliability by powering vital protective functions of the grid even if its own power supply fails. **138** The modularity of many distributed resources enables them to scale down advantageously to small loads that would be uneconomic to serve with grid power because its fixed connection costs could not be amortized from electricity revenues. **139** Many distributed resources, notably photovoltaics, have costs that scale far more closely to their loads than do the costs of distribution systems.

140 Distributed generators provide electric energy that would otherwise have to be generated by a centralized plant, backed up by its spinning reserve, and delivered through grid losses to the same location. **141** Distributed resources available on peak can reduce the need for the costlier to-keep-warm centralized units. **142** Distributed resources very slightly reduce spinning reserves' operational cost. **143** Distributed resources can reduce power stations' startup cycles, thus improving their efficiency, lifetime, and reliability. **144** Inverter-driven distributed resources can provide extremely fast ramping to follow sudden increases or decreases in load, improving system stability and component lifetimes. **145** By combining fast ramping with flexible location, often in the distribution system, distributed resources may provide special benefits in correcting transients locally before they propagate upstream to affect more widespread transmission and generating resources.

146 Distributed resources allow for net metering, which in general is economically beneficial to the distribution utility (albeit at the expense of the incumbent generator). **147** Distributed resources may reduce utilities' avoided marginal cost and hence enable them to pay lower buyback prices to Qualifying Facilities. **148** Distributed resources' ability to provide power of the desired level of quality and reliability to particular customers—rather than just a homogeneous commodity via the grid—permits providers to match their offers with customers' diverse needs and to be paid for that close fit.

149 Distributed resources can avoid harmonic distortion in the locations where it is both more prevalent (*e.g.*, at the end of long rural feeders) and more costly to correct.

150 Certain distributed resources can actively cancel harmonic distortion in real time, at or near the customer level.